Image Fusion

Image Fusion: Algorithms and Applications

Edited by

Tania Stathaki

Amsterdam • Boston • Heidelberg • London • New York
Oxford • Paris • San Diego • San Francisco • Singapore
Sydney • Tokyo

Academic Press is an imprint of Elsevier

Academic Press is an imprint of Elsevier
84 Theobald's Road, London WC1X 8RR, UK
Radarweg 29, PO Box 211, 1000 AE Amsterdam, The Netherlands
30 Corporate Drive, Suite 400, Burlington, MA 01803, USA
525 B Street, Suite 1900, San Diego, CA 92101-4495, USA

First edition 2008

Library of Congress Cataloging-in-Publication Data
A catalog record for this book is available from the Library of Congress

ISBN: 978-0-12-372529-5

British Library Cataloguing-in-Publication Data
A catalogue record for this book is available from the British Library

For information on all Academic Press publications
visit our web site at www.books.elsevier.com

Printed and bound in Great Britain
08 09 10   10 9 8 7 6 5 4 3 2 1

Contents

Preface

The need for Image Fusion in current image processing systems is increasing mainly due to the increased number and variety of image acquisition techniques. Image Fusion is defined as the process of combining substantial information from several sensors using mathematical techniques in order to create a single composite image that will be more comprehensive and thus, more useful for a human operator or other computer vision tasks. Current technology in imaging sensors offers a wide variety of information that can be extracted from an observed scene. Images which have been acquired using different sensor modalities exhibit diverse characteristics, such as type of degradation, salient features, texture properties etc. Representative examples of available sensors are radar, sonar and other acoustic sensors, infrared and thermal imaging cameras, seismic, magnetic, lidar and other types of sensors. Multi-sensor information is jointly combined to provide an enhanced representation in many cases of experimental sciences. The automated procedure of conveying all the meaningful information from the input sensors to a final composite image is the goal of a fusion system, which appears to be an essential pre-processing stage for a number of applications, such as aerial and satellite imaging, medical imaging, robot vision and vehicle or robot guidance.

This book provides a representative collection of the recent advances in research and development in the field of Image Fusion and thereby promotes the synergism among the numerous disciplines that are contributing to its growth. It demonstrates both spatial domain and transform domain fusion methods such as, Bayesian methods, statistical approaches, ICA and wavelet domain techniques and others. It also includes valuable material on image mosaics, remote sensing applications and performance evaluation. Chapters emphasise one or more of the three facets: theory (algorithms), design (architectures) and applications. They deal with fundamental theoretical analyses of image fusion methods as well as their application to real world problems and furthermore, they raise various issues associated with their implementation.

The contributing authors are all established leaders in Image Fusion and they have provided a representative and thorough overview of the available algorithms and applications of this active and fast developing scientific and technological field.

Primary audience of this book will be academic and industrial researchers and system developers involved in various Image Fusion applications and project managers requiring an overview of on-going research in the field. Finally, much of the material would also be of value for Masters and Ph.D. level students who attend related courses or do research in Image Fusion and related fields.

List of contributors

Bruno Aiazzi Institute of Applied Physics 'Nello Carrara' of the National Research Council, Florence, Italy.

Stefano Baronti Institute of Applied Physics 'Nello Carrara' of the National Research Council, Florence, Italy.

Jürgen Beyerer Fraunhofer-Institut für Informations- und Datenverarbeitung IITB and Universität Karlsruhe (TH), Institut für Technische Informatik, Lehrstuhl für Interaktive Echtzeitsysteme, Karlsruhe, Germany.

M. Cacciola University Mediterranea of Reggio Calabria, Faculty of Engineering, Dimet, Italy.

Luca Capobianco Department of Information Engineering, University of Siena, Siena, Italy.

Jan G.P.W. Clevers Wageningen University, Centre for Geo-Information, Wageningen, The Netherlands.

Tim Cootes Imaging Science and Biomedical Engineering, University of Manchester, Oxford Road, Manchester, UK.

Andrea Garzelli Department of Information Engineering, University of Siena, Siena, Italy.

Ioana Gheţa Universität Karlsruhe (TH), Institut für Technische Informatik, Lehrstuhl für Interaktive Echtzeitsysteme, Karlsruhe, Germany.

Stamatia Giannarou Communications and Signal Processing Group, Imperial College London, London, UK.

Michael Heizmann Fraunhofer-Institut für Informations- und Datenverarbeitung IITB, Karlsruhe, Germany.

Jing Jin Department of Control Science and Engineering, Harbin Institute of Technology, P.R. China.

Antonis Katartzis Communication and Signal Processing Group, Department of Electrical and Electronic Engineering, Imperial College, London, UK.

Roger L. King Department of Electrical and Computer Engineering, Mississippi State University, USA.

Shutao Li College of Electrical and Information Engineering, Hunan University, Changsha, China.

David Looney Imperial College London, UK.

Danilo P. Mandic Imperial College London, UK.

Nikolaos Mitianoudis Communications and Signal Processing Group, Imperial College London, London, UK.

F.C. Morabito University Mediterranea of Reggio Calabria, Faculty of Engineering, Dimet, Italy.

Shree K. Nayar Department of Computer Science, Columbia University, New York, USA.

Filippo Nencini Department of Information Engineering, University of Siena, Siena, Italy.

Xavier Otazu Computer Vision Center, Universitat Autònoma de Barcelona, Cerdanyola del Vallès, Barcelona, Spain.

Laurent Oudre Imperial College London, UK.

Maria Petrou Communication and Signal Processing Group, Department of Electrical and Electronic Engineering, Imperial College, London, UK.

Vladimir Petrović Imaging Science and Biomedical Engineering, University of Manchester, Oxford Road, Manchester, UK.

Pushkar Pradham Department of Electrical and Computer Engineering, Mississippi State University, USA.

Jennifer Sander Universität Karlsruhe (TH), Institut für Technische Informatik, Lehrstuhl für Interaktive Echtzeitsysteme, Karlsruhe, Germany.

Yoav Y. Schechner Department of Electrical Engineering, Technion – Israel Institute of Technology, Haifa, Israel.

Massimo Selva Institute of Applied Physics 'Nello Carrara' of the National Research Council, Florence, Italy.

Yi Shen Department of Control Science and Engineering, Harbin Institute of Technology, P.R. China.

G. Simone University Mediterranea of Reggio Calabria, Faculty of Engineering, Dimet, Italy.

Tania Stathaki Communications and Signal Processing Group, Imperial College London, London, UK.

Qiang Wang Department of Control Science and Engineering, Harbin Institute of Technology, P.R. China.

Costas Xydeas Head of DSP research, Department of Communication Systems, Infolab21, Lancaster University, Lancaster, UK.

Bin Yang College of Electrical and Information Engineering, Hunan University, Changsha, China.

Nicolas H. Younan Department of Electrical and Computer Engineering, Mississippi State University, USA.

Raul Zurita-Milla Wageningen University, Centre for Geo-Information, Wageningen, The Netherlands.

Current trends in super-resolution image reconstruction

Antonis Katartzis and Maria Petrou

Communication and Signal Processing Group, Department of Electrical and Electronic Engineering, Imperial College, London, UK

Super-resolution (SR) reconstruction is a branch of image fusion for bandwidth extrapolation beyond the limits of traditional electronic imaging systems. This chapter describes the main principles of SR reconstruction, and provides an overview of the most representative methodologies in the domain. We analyse the advantages and limitations of each set of techniques, present a promising new approach based on Normalised Convolution and robust Bayesian estimation, and perform quantitative and qualitative comparisons using real video sequences.

1.1 Introduction

Super-resolution (SR) is a fusion process for reconstructing a high resolution (HR) image from several low resolution (LR) images covering the same region in the world. It extends classical single frame image reconstruction/restoration methods by simultaneously utilising information from multiple observed images to achieve resolutions higher than that of the original data. These observations can be LR images captured simultaneously or at different times by a single or multiple imaging devices. This methodology, also known as *multiframe super-resolution reconstruction*, registers the observed images to a common high resolution reference frame in order to formulate the problem of fusion as one of constrained image reconstruction with missing data.

The general strategy that characterises super-resolution comprises three major processing steps [1]:

1. *LR image acquisition*: Acquisition of a sequence of LR images from the same scene with non-integer (in terms of inter-pixel distances) geometric displacements between any two of the images.

2. *Image registration/motion compensation*: Estimation of the sub-pixel geometric transformation of each source image with respect to the reference HR desirable grid.
3. *HR image reconstruction*: Solution of the problem of reconstructing a HR image from
   the available data supplied by the source images.

The theoretical basis for super-resolution was laid by Papoulis [2], with the *Generalised
Sampling Theorem*. It was shown that a continuous band-limited signal $z(\mathbf{x})$ may be reconstructed from samples of convolutions of $z(\mathbf{x})$ with different filters, assuming these
filters satisfy certain conditions. For example, if these filters kill some high frequencies,
then there is no unique solution [3]. This is one of the factors that make SR an *ill-posed*
problem. The solution in general does not fulfil Hadamard's classical requirements of
existence, uniqueness and stability: solutions may not exist for all data, they may not be
unique (which raises the practically relevant question of *identifiability*, i.e. the question
of whether the data contain enough information to determine the desired quantity), and
they may be unstable with respect to data perturbations. The last aspect is very important,
since in real-world measurements the presence of noise is inherent. As a consequence,
the reconstruction must rely on *natural constraints*, that is, general *a priori* assumptions
about the physical world, in order to derive an unambiguous output. However, as demonstrated in [4], the quality of reconstruction of the HR image has an upper limit defined by
the degree of degradation of the involved LR frames.

This chapter provides a description of the main principles of super-resolution, together
with an overview of the most representative methodologies in the domain. We analyse
the advantages and limitations of each set of techniques and present a promising new approach based on Normalised Convolution and robust Bayesian estimation. The chapter is
organised as follows. Section 1.2 presents a general formulation of the SR problem from
the point of view of image acquisition. A general overview of the existing SR methods
is presented in Section 1.3, whereas the new alternative is described in Section 1.4. Both
qualitative and quantitative results for real video sequences, alongside comparisons with
a series of SR methodologies are presented in Section 1.5. Finally, general conclusions
are drawn in Section 1.6.

1.2 Modelling the imaging process

The solution of the SR problem requires the modelling of the relationship between the
sought HR frame and the available LR images. The latter can be considered as geometrically distorted and degraded versions of the ideal HR frame we wish to construct. The
first relation that has to be modelled is that of the geometric transformation between the
LR grids and the HR grid we select to express the HR image. The next relationship is
established by modelling the degradation process that takes place during image acquisition. As the unknown HR frame is supposed to be captured by a fictitious perfect camera
with a desired high resolution, we may assume that the observed LR frames are the result of degradations this ideal image suffered due to imperfect imaging conditions. These
may involve blurring, downsampling or the presence of noise. The following sections describe in detail the most commonly used models that relate the ideal HR image with the
available LR frames.

1.2.1 Geometric transformation models

A highly accurate registration of the LR images in a reference HR grid is essential to the success of any SR algorithm. The accuracy of this process determines, to a high extent, the efficiency of the overall reconstruction. Image registration is a widely used procedure in the field of image analysis [5,6]. Although the in-depth investigation of the several methodologies is beyond the scope of this chapter, some of their main principles are listed below.

The objective is the identification of a local or global geometric transformation $\mathcal{T}$, which maps the coordinates $\mathbf{x} = [x, y]^T$ of a current frame to a new coordinate system of the reference frame, with coordinates $\mathbf{x}' = [x', y']^T$:

$$\mathbf{x}' = \mathcal{T}(\mathbf{x})$$

Transformation $\mathcal{T}$ is determined through the minimisation of a similarity criterion between the two images. This similarity measure can be based on:

1. *normalised cross correlation* [6];
2. *mean square intensity difference* (optical flow constraint) [7];
3. *mutual information*[1] [9].

Depending on the form of the transformation function $\mathcal{T}$, we may classify the registration methodologies as parametric (global) and non-parametric (local).

1.2.1.1 Global parametric approaches
In this case, $\mathcal{T}$ has an analytical form, which is explicitly identified with a set of unknown parameters [10]. Two representative parametric distortion models often used in SR reconstruction are the following.

- *Affine transform*
  The six-parameter affine transform

$$\mathcal{T}(\mathbf{x}) = A\mathbf{x} + \mathbf{b}, \quad A \in \mathbb{R}^{2 \times 2}, \ \mathbf{b} \in \mathbb{R}^2 \tag{1.1}$$

  includes rotation, translation and scaling as special cases. This mapping preserves straight lines and straight line parallelism. It may be used for multiview registration assuming the distance of the camera from the scene is large in comparison with the size of the scanned area, the camera is perfect (a pin-hole camera), the scene is flat, and the geometric transformation between the two frames is purely global, with no local extra components.

---

[1]Registration via mutual information is generally used in SR applications related to multi-modal, rather than multi-frame image fusion (e.g. [8]).

- *Perspective transform*
  If the condition on the infinite distance of the camera from the scene is not satisfied, the eight-parameter perspective transform should be used:

$$T(\mathbf{x}) = \frac{A\mathbf{x} + \mathbf{b}}{\mathbf{c}^{\mathsf{T}}\mathbf{x} + 1} - \mathbf{x}, \quad A \in \mathbb{R}^{2 \times 2}, \ \mathbf{b}, \mathbf{c} \in \mathbb{R}^2 \qquad (1.2)$$

This model describes exactly the deformation of a flat scene photographed by a pin-hole camera the optical axis of which is not perpendicular to the scene. It can map a general quadrangle onto a square while preserving straight lines. The perspective model can accommodate more general quadrilateral deformations and includes the affine models as special cases.

1.2.1.2 Local non-parametric approaches

These methods do not assume any global, parametric model for the image deformation. Instead, they try to identify directly the *motion vector* for each individual pixel. This is carried out via the minimisation of a functional that assesses the overall discrepancy between the two images, using one of the three similarity measures described above (e.g. [11,12]). Such non-parametric approaches, although being computationally intensive, can characterise a wide range of geometric distortions, including *non-rigid* or *elastic* deformations. Shen et al. [13] proposed a method for super-resolving scenes with multiple independently moving objects, in which motion estimation is followed by a motion segmentation process.

1.2.2 Image degradation models

1.2.2.1 Blurring

This source of degradation includes three main types.

1. *Camera blurring.* This accounts for two sources of degradation: (a) imperfect imaging optics, and (b) limitations in resolution capabilities of the sensor, as specified by its Modulation Transfer Function (MTF). These two factors determine an overall *point spread function* (PSF) for the imaging system.

   A well known artifact of the first class is the *out-of-focus* blurring [14]. This type of blurring is primarily due to effects of the camera aperture that result in a point source being imaged as a blob. As accurate knowledge of all factors that create such an effect is generally unknown (e.g. focal length, camera's aperture size and shape, etc.), several uniform models have been adopted to approximate the resulting PSF, as follows.[2]
   - *Uniform out-of-focus blurring.* This models the simple defocussing found in a variety of imaging systems as a uniform intensity distribution within a circular disk:

---

[2]For simplicity, the blurring kernels are represented in the continuous domain.

$$h(x, y) = \begin{cases} \frac{1}{\pi r^2}, & \text{if } \sqrt{x^2 + y^2} \leqslant r, \\ 0, & \text{otherwise} \end{cases} \tag{1.3}$$

- *Uniform 2-D blurring.* This is a more severe form of degradation that approximates an out-of-focus blurring, and is used in many research simulations:

$$h(x, y) = \begin{cases} \frac{1}{R^2}, & \text{if } -\frac{R}{2} < x, y < \frac{R}{2}, \\ 0, & \text{otherwise} \end{cases} \tag{1.4}$$

2. *Atmospheric scattering.* This is mainly evident in the case of remotely sensed imagery and can be modelled by a Gaussian PSF:

$$h(x, y) = \mathcal{K} e^{-\frac{x^2 + y^2}{2\sigma^2}} \tag{1.5}$$

where $\mathcal{K}$ is a normalising constant ensuring that the PSF integrates to 1 and σ^2 is the variance that determines the severity of the blurring.
3. *Motion blurring.* This effect is the result of slow camera shutter speed relative to rapid camera motion. In general, it represents the 1-D uniform local averaging of neighbouring pixels. An example of horizontal motion can be expressed as:

$$h(x) - \begin{cases} \frac{1}{R}, & \text{if } -\frac{R}{2} < x < \frac{R}{2}, \\ 0, & \text{otherwise} \end{cases} \tag{1.6}$$

A SR restoration technique that considers this type of degradation is described in [15].

Image blurring may be modelled by a convolution with a lowpass kernel that comprises all three degradation processes and may be approximated by a matrix of the following form: $\mathbf{H} = \mathbf{H}_{cam} \mathbf{H}_{atm} \mathbf{H}_{motion}$. Matrix $\mathbf{H}$ may represent either Linear, Shift-Invariant (LSI) blurring or Linear, Shift-Varying (LSV) blurring.

Some of the ways of estimating the overall PSF include the use of camera manufacturer information (which generally is hard to obtain) or the analysis of the degradation for a picture of a known object [16,17] (e.g. a white dot on a black background). The methodology of super-resolving from LR data without any information about the degradation process is called *blind super-resolution*. It belongs to the general group of *blind deconvolution* techniques, where the problem is the restoration of an original image from a degraded observation, without any information about the blurring. There has been extensive work on blind deconvolution. A good survey on the topic can be found in the survey paper of Kundur and Hatzinakos [18]. Existing blind deconvolution methods can be categorised into two main classes:

1. methods that separate blurring identification as a disjoint procedure from reconstruction;
2. methods that combine blurring identification and reconstruction in one procedure.

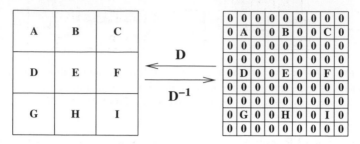

Figure 1.1 *Effect of upsampling matrix D^{-1} on a 3×3 image and downsampling matrix D on the corresponding 9×9 upsampled image.*

Blind deconvolution methods may also be generalised to handle multiple observations. Multi-frame blind deconvolution is better at suppressing noise and edge artifacts and preventing PSF estimates from converging to the trivial delta function. Using multiple LR frames, Shekarforoush and Chellappa [19] proposed estimating the optical transfer function by finding spikes in the magnitude of the cross power spectrum of consecutive frames. Other approaches involve the *expectation maximisation* (EM) algorithm in order to find a maximum likelihood estimate of the parameters under consideration [20,21]. Another popular method is the one proposed in [22], which is based on the concept of *generalised cross-validation* (GCV). This method considers a parameterised estimate of the blurring based on the minimisation of a restoration residual over the image and blurring parameter space. In the case of SR reconstruction, a similar approach was proposed in [23].

1.2.2.2 Spatial sampling

Subsampling is the main difference between the models related to super-resolution and classical image restoration/reconstruction. In the SR framework, each pixel of the LR image may be considered as the result of the averaging of a block of HR pixels. This models the spatial integration of light intensity over a surface region performed by CCD image acquisition sensors [24]. An alternative to the averaging form of D is to consider a downsampling matrix D by performing a homogeneous sampling as shown in Figure 1.1 [25].

1.2.2.3 Additive noise

In super-resolution, as in similar image processing tasks, it is usually assumed that the noise is additive and normally distributed with zero-mean. The assumption of normal distribution of the noise is not accurate in most of the cases, as most of the noise in the imaging process is non-Gaussian (quantisation, camera noise, etc.), but modelling it in a more realistic way would end in a very large and complex optimisation problem which is usually hard to solve. Some methods use signal-dependent noise, something that leads to non-linear approaches to image restoration [26].

1.2.3 Observation model – Mathematical formulation

An observation model for the SR restoration process may be created using the aforementioned principles of image formation. We assume that we have in our disposal a

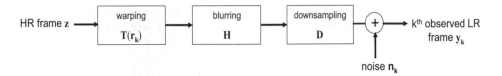

Figure 1.2 *Block diagram of the degradation process relating the HR frame with a LR one.*

set of K overlapping LR frames/images of size $M_1 \times M_2$. Using a lexicographic ordering, each LR image may be expressed as $\mathbf{y}_k = [y_{k,1}, y_{k,2}, \ldots, y_{k,M}]^T$, $\forall k = 1, 2, \ldots, K$, where $M = M_1 M_2$. Hereafter, the set of all LR samples (all available observations) will be denoted by $\mathbf{y} \equiv [\mathbf{y}_1^T, \ldots, \mathbf{y}_K^T]^T$.

Considering a given LR image as the reference frame, our objective is the estimation of a HR version of this frame given the whole LR sequence. We assume that the targeted HR frame is of size $LM_1 \times LM_2$, where L is the upsampling factor in both directions. Following similar notations, the HR frame may be written as a lexicographically ordered vector $\mathbf{z} = [z_1, z_2, \ldots, z_N]^T$, where $N = L^2 M_1 M_2$.

The observed LR frames are assumed to have been produced by a degradation process that involves geometric warping, blurring, and uniform downsampling performed on the HR image $\mathbf{z}$ (see Figure 1.2). Moreover, each LR frame is typically corrupted by additive noise which is uncorrelated between the different LR frames. Thus, the kth LR frame may be written as

$$\mathbf{y}_k = \mathbf{DHT}(\mathbf{r}_k)\mathbf{z} + \mathbf{n}_k = \mathbf{W}(\mathbf{r}_k)\mathbf{z} + \mathbf{n}_k, \quad \forall k = 1, 2, \ldots, K \qquad (1.7)$$

where $\mathbf{W}(\mathbf{r}_k) \equiv \mathbf{DHT}(\mathbf{r}_k)$ is an $M \times N$ matrix that represents the contribution of the HR pixels of $\mathbf{z}$ to the LR pixels of $\mathbf{y}_k$, via motion, blurring and downsampling. In particular, the $N \times N$ warping matrix $\mathbf{T}(\mathbf{r}_k)$ represents the 2-D geometrical transformation that maps the HR image to each full resolution LR frame and depends on a parameter vector $\mathbf{r}_k$. $\mathbf{H}$ is an $N \times N$ matrix that represents the effects of LSI or LSV blurring and $\mathbf{D}$ is an $M \times N$ downsampling matrix, which models the effect of creating aliased LR frames from the warped and blurred HR image. Finally, $\mathbf{n}_k$ denotes the lexicographically ordered M-dimensional noise field. In a more condensed version, (1.7) may be expressed as:

$$\mathbf{y} = \mathbf{Wz} + \mathbf{n} \qquad (1.8)$$

1.3 State-of-the-art SR methods

1.3.1 Frequency domain methods

Frequency-domain SR methods typically rely on familiar Fourier transform properties, specifically the shifting and sampling theorems, for the removal of alias. Tsai and Huang in their seminal paper [27] were the first to introduce the concept of multi-frame SR reconstruction. They proposed a frequency domain observation model, by disregarding the effects of blurring and observation noise during image acquisition. The observed images were modelled as under-sampled images of a static, unknown scene with continuous spatial variables, which was subject to global translational motion. Ideal impulse sampling was assumed, but with a sampling rate below the Nyquist rate. The shift and aliasing properties were used to formulate a system of equations which related the aliased Discrete Fourier Transform (DFT) coefficients of the observed images to samples of the Continuous Fourier Transform (CFT) of the unknown scene from which the observations were derived.

Starting from the original continuous scene $z(x, y)$, global translational motion yields K shifted images:

$$z^{(k)}(x, y) = f(x + \Delta_{x^{(k)}}, y + \Delta_{y^{(k)}})$$

with $k \in \{1, 2, \ldots, K\}$. The CFT of the scene is given by $\mathcal{Z}(u, v)$ and that of the translations by $\mathcal{Z}^{(k)}(u, v)$. The shifted images are impulse sampled to yield K observed LR images:

$$y^{(k)}[m, n] = z(mT_x + \Delta_{x^{(k)}}, nT_y + \Delta_{y^{(k)}})$$

where T_x and T_y denote the sampling periods in the x and y dimensions, respectively. The K corresponding 2-D DFTs are denoted by $\mathcal{Y}^{(k)}[u, v]$. The CFT of the scene and the DFT of the shifted and sampled images are related via aliasing by

$$\mathcal{Y}^{(k)}[u, v] = \frac{1}{T_x T_y} \sum_{p_1 = -\infty}^{\infty} \sum_{p_2 = -\infty}^{\infty} \mathcal{Z}^{(k)} \left(\frac{u}{M_1 T_x} + \frac{p_1}{T_x}, \frac{v}{M_2 T_y} + \frac{p_2}{T_y} \right) \qquad (1.9)$$

Due to the shifting property of the Fourier transform, spatial shifting appears as phase shifting:

$$\mathcal{Z}^{(k)}(u, v) = e^{i2\pi[\Delta_{x^{(k)}} u + \Delta_{y^{(k)}} v]} \mathcal{Z}(u, v) \qquad (1.10)$$

If $z(x, y)$ is band limited, there exists $\Omega_x, \Omega_y \in \mathbb{N}$ such that $\mathcal{Z}(u, v) \to 0$ for $|u| \geqslant \Omega_x / T_x$ and $|v| \geqslant \Omega_y / T_y$ and the infinite summations in (1.9) are reduced to finite sums. Using the shifting property of (1.10) the relationship in (1.9) obtains the following matrix form:

$$\mathbf{Y} = \Phi \mathbf{Z} \qquad (1.11)$$

where $\mathbf{Y}$ is a $K \times 1$ vector, the kth element of which contains the DFT coefficients $\mathcal{Y}^{(k)}[u, v]$ of the observed frame $y^{(k)}[m, n]$. Φ is a matrix that relates the DFTs of the observed frames to the samples of the unknown CFT of $z(x, y)$ contained in vector $\mathbf{Z}$. Therefore, the method may be summarised as follows: SR reconstruction is reduced to finding the DFTs of the K observed images, determining Φ, solving the system of equations in (1.11) for $\mathbf{Z}$ (based on the least-squares approach) and then using the inverse DFT to obtain the reconstructed image.

The frequency-based model described in [27], although establishing the foundations for future research in the field of super-resolution restoration, has several important limitations [28]. The assumption of ideal sampling is unrealistic since real world imaging systems are characterised by sampling sensors which perform spatial as well as temporal integration (during the *aperture time*). Additionally, no consideration is given to the effects of the optical system on the recorded image data, while the issue of noise degradation is also ignored. Finally, the global translational motion model that describes the geometric transformations between the several frames is very restrictive.

Several extensions of the basic frequency-based model have been proposed in the literature, but all of them are limited to the case of translational motion models. Kim et al. [29] extend the model of [27] to consider observation noise as well as the effects of spatial blurring, resulting in a weighted Recursive Least-Squares (RLS) solution of the linear system of equations in (1.11). Kim and Su [30] addressed the issue of the ill-posedness of the restoration inverse problem and proposed replacing the RLS solution for Equation (1.11) with a method based on Tikhonov regularisation [31]. Finally, a method that accommodates non-global translational motion models was presented in [32].

1.3.2 Projection Onto Convex Sets (POCS)

The method of projection onto convex sets (POCS) is a powerful, iterative technique, with its prominent feature being the ease with which prior knowledge about the solution may be incorporated into the reconstruction process. In set theoretic estimation each piece of information is represented by a *property set* in a solution space Ξ and the intersection of these sets represents the feasible class of solutions, the so-called *feasibility set* [33]. Given a set of C prior constraints Ψ_c, $\forall c = 1, 2, \ldots, C$, the resultant property sets $S_c \subset \Xi$ are

$$S_c = \{\mathbf{z} \in \Xi \mid \mathbf{z} \text{ satisfies } \Psi_c\}, \quad \forall c = 1, 2, \ldots, C \tag{1.12}$$

whereas the feasibility set S is given by

$$S = \bigcap_{c=1,2,\ldots,C} S_c \tag{1.13}$$

A feasible solution in S may be reached via the principle of successive *projections onto convex sets* (POCS), the theory of which was originally proposed in [34] and first applied to the domain of image processing in [35] and [36]. For any $\mathbf{z} \in \Xi$, the projection $P_c \mathbf{z}$ of $\mathbf{z}$ onto each set S_c is the element in S_c closest to $\mathbf{z}$. For closed and convex sets S_c, the sequence $(\mathbf{z}_p)_{p \geqslant 0}$ of the successive projections

$$\mathbf{z}_{p+1} = P_C P_{C-1} \ldots P_1 \mathbf{z}_p, \quad \forall p = 0, 1, \ldots \tag{1.14}$$

converges weakly to a point in S [37]. In a more general form, (1.14) is expressed as

$$\mathbf{z}_{p+1} = Q_C Q_{C-1} \ldots Q_1 \mathbf{z}_p, \quad \forall p = 0, 1, \ldots \tag{1.15}$$

with $Q_c \triangleq I + \lambda_c (P_c - I)$ and I the identity operator. The λ_c's, $c = 1, \ldots, C$, are relaxation parameters that control the rate of convergence. According to [34] this is guaranteed for $0 < \lambda_c < 2$.

In the SR framework, Tekalp et al. [38] proposed a POCS technique based on the observation model in (1.8), assuming also a global translational motion between the LR frames. Let us consider the residual vector

$$\varrho = \mathbf{y} - \mathbf{Wz} \tag{1.16}$$

which actually follows the characteristics of the observation noise $\mathbf{n}$. Using the assumption that each component of ϱ should be below a confidence level δ, which may be set according to the noise statistics [39], KM data consistency constraints C_i and corresponding convex property sets S_{C_i} may be generated (KM being the total number of pixels in all LR available frames):

$$S_{C_i} = \left\{ \mathbf{z} \in \Xi \mid |\varrho_i| \leqslant \delta \right\}, \quad \forall i = \{1, \ldots, KM\} \tag{1.17}$$

The projection of $\mathbf{z}$ onto S_{C_i} is given by

$$P_i \mathbf{z} = \begin{cases} z_j + \frac{\varrho_i - \delta}{\sum_k w^2(i,k)} w(i,j), & \text{if } \varrho_i > \delta, \\ z_j, & \text{if } -\delta < \varrho_i < \delta, \\ z_j + \frac{\varrho_i + \delta}{\sum_k w^2(i,k)} w(i,j), & \text{if } \varrho_i < -\delta \end{cases} \tag{1.18}$$

These projection operators are applied in turn for all LR pixels and the sequence

$$\mathbf{z}^{(n+1)} = P_{KM} P_{KM-1} \ldots P_2 P_1 \mathbf{z}^{(n)} \tag{1.19}$$

converges to the desired HR image $\mathbf{z}$ for an initial estimate $\mathbf{z}^{(0)}$. Note that the reached solution is in general non-unique and depends on the initial guess. In this case, additional constraints may be imposed from prior knowledge to favour a particular HR image. These constraints may include:

- band limiting constraints

$$S_{C_b} = \left\{ \mathbf{z} \in \Xi \mid \mathcal{Z}(u, v) = 0, \; |u|, |v| \geqslant \Omega \right\} \tag{1.20}$$

- amplitude constraints

$$S_{C_a} = \{ \mathbf{z} \in \Xi \mid \alpha \leqslant z_i \leqslant \beta, \; \alpha < \beta \} \tag{1.21}$$

- energy constraint

$$S_{C_e} = \left\{ \mathbf{z} \in \Xi \mid \|\mathbf{z}\|^2 \leqslant E \right\} \tag{1.22}$$

where E is the maximum permissible energy of the HR image.

A more elaborate POCS technique, which can deal with motion blurring distortions due to non-zero aperture time of image acquisition, can be found in [15].

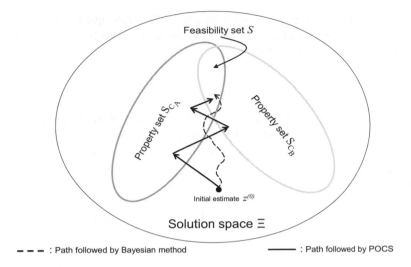

Figure 1.3 *Schematic representation of the convergence of a Bayesian MAP technique compared with a POCS method (using an adapted relaxation parameter λ_c that varies at each iteration). POCS methods seek the solution in the intersection of the two sets that represent the solutions that satisfy two different constraints. This is done by following a path in the solution space Ξ that alternatively jumps from one possible solution to another that satisfies one or the other constraint. On the other hand, Bayesian approaches follow a path in Ξ that is designed to reach the solution without satisfying the constraints exactly. In cases where the constraint subsets do not intersect POCS methods are expected to fail, while Bayesian methods will yield a solution that is the best compromise between the two (or more) incompatible constraints.*

1.3.3 Bayesian/variational methods

Contrary to POCS-based methodologies, where the objective is the convergence to a set of feasible solutions, methods based on the Bayesian/variational framework belong to the family of point estimation techniques, which are based upon solving an optimisation problem that provides a single solution via the minimisation/maximisation of a criterion function (see Figure 1.3). Given that the registration parameters are known, this function corresponds to the posterior probability $P(\mathbf{z}|\mathbf{y}, \mathbf{r})$, and the optimal solution is given by

$$\hat{\mathbf{z}} = \arg\max_{\mathbf{z}} P(\mathbf{z}|\mathbf{y}, \mathbf{r}) = \arg\max_{\mathbf{z}} P(\mathbf{y}|\mathbf{z}, \mathbf{r}) P(\mathbf{z}) \tag{1.23}$$

The Maximum a Posteriori Probability (MAP) criterion in (1.23) is equivalent to the variational problem of minimising a posterior energy function $U(\mathbf{z}|\mathbf{y}, \mathbf{r})$:

$$\hat{\mathbf{z}} = \arg\min_{\mathbf{z}} U(\mathbf{z}|\mathbf{y}, \mathbf{r}) = \arg\min_{\mathbf{z}} \left\{ -\log P(\mathbf{y}|\mathbf{z}, \mathbf{r}) - \log P(\mathbf{z}) \right\} \tag{1.24}$$

The data likelihood $P(\mathbf{y}|\mathbf{z}, \mathbf{r})$ corresponds to the data fidelity term of the energy function and depends on the statistical behaviour of the noise. Assuming that the elements of the noise field are independent and identically distributed (iid) samples, the data likelihood may be expressed as

$$P(\mathbf{y}|\mathbf{z}, \mathbf{r}) \propto e^{-1/(2\sigma_n^2) \sum_k \|\mathbf{y}_k - \mathbf{W}(\mathbf{r}_k)\mathbf{z}\|^p} \tag{1.25}$$

where $\| \cdot \|^p$ represents the L_p norm. Under the assumption of Gaussian noise, $p = 2$ and the data consistency energy term obtains a quadratic form [13,21,24,40]. In [25] the authors used instead a more robust expression of the data likelihood by employing an L_1 norm in (1.25) and subsequently assuming that the noise follows a Laplacian distribution.

On the other hand, the prior distribution $P(\mathbf{z})$ incorporates contextual constraints on the elements of the unknown HR image $\mathbf{z}$ and describes our a priori knowledge about the solution. A simple prior model has the following form:

$$P(\mathbf{z}) = \frac{1}{Z} e^{-\mathbf{z}^T Q \mathbf{z}} \tag{1.26}$$

where Q is a symmetric, positive-definite matrix [41]. Another way to think about (1.26) is as a multivariate Gaussian distribution over $\mathbf{z}$, in which Q is the inverse of the covariance matrix. More common priors are based on smoothness constraints that penalise high intensity fluctuations between adjacent pixels. These may be obtained by considering $\mathbf{z}$ to be a Markov random field (MRF), characterised by a Gibbs distribution of the following form:

$$P(\mathbf{z}) \propto \exp\left\{ -\sum_{s=1}^{N} f\left(\sum_{t=1}^{N} Q_{st}(z_t) \right) \right\} \tag{1.27}$$

where Q represents an $N \times N$ high pass operator (e.g. gradient or Laplacian). Hardie et al. [40] used a quadratic form of $f(x)$, with Q being the Laplacian operator (thin-plate model). Quadratic models, although behaving well in the removal of noise, generally produce over-smoothed versions of the solution. On the other hand, discontinuity adaptive priors may be introduced by choosing $f(x)$ to be a robust error norm, which considers predominant discontinuities in the signal as outliers. In the SR framework, popular forms of robust functions are the Huber [24], Total Variation (L_1 norm) [25] or Lorentzian [42].

The strategies followed for minimising the energy function in (1.24) depend on the form of the function itself. In the case of convex functions (e.g. quadratic energy functionals), a unique global minimum may be identified, using, for example, gradient descent [40], conjugate gradient methods [23] or the robust minimisation approach of [3]. However, this is not the case when the energy is non-convex, with several local minima (e.g. use of robust error norms in (1.27)). A very accurate estimation of the global minimum may be obtained via stochastic relaxation methods, such as simulated annealing. However, such methods are painfully slow, especially in optimisation problems with continuous sets of labels (like image intensity). In Section 1.4 we present an attractive method for approximating the global minimum solution, based on successive convex approximations of the energy function.

Concluding this section, it is worthwhile mentioning an interesting technique proposed in [43], which is based on a hybrid Bayesian/POCS reconstruction principle. In particular, the authors proposed an efficient numerical scheme for the minimisation of quadratic posterior energy functions, using additional non-quadratic constraints in the form of convex sets.

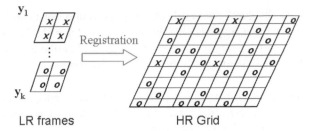

Figure 1.4 *Registration of LR samples in the HR grid.* x*: samples of the reference LR frame;* o*: samples of neighbouring LR frames.*

1.3.4 Interpolation-based approaches

This section refers to methodologies that tackle the SR problem from the perspective of image interpolation. Following registration of the LR frames, the available samples are distributed irregularly on a reference HR grid $\mathcal{R} \subset \mathbb{Z}^2$ (see Figure 1.4). This results in an irregularly sampled image $z_{\mathcal{D}}(\mathbf{x})$, where $\mathbf{x} = [x, y]^T$ denotes a position vector in $\mathcal{R}$. The SR problem may then be formulated as a reconstruction problem from irregularly sampled data. The non-uniform sampling problem arises in many scientific fields, such as geophysics, astronomy, oceanography, medical imaging and speech processing. An interesting overview of reconstruction techniques from non-uniform samples via several ways of interpolation may be found in [44]. Among these one can distinguish the trigonometric polynomials approach of [45], the local polynomial fitting method of Normalised Convolution (NC) [46] or techniques based on Delaunay triangulation. Interesting extensions of such approaches employ also information related to the degradation processes of Section 1.2.2, via the use of post-processing steps that involve deblurring and/or denoising (see [47] and [48]).

In this chapter we focus our attention on the NC method, an interesting extension of which is presented in Section 1.4.2. NC is a technique for local signal reconstruction, using an additional *certainty* map that describes our confidence in the data that constitute the unknown signal. Given an irregularly sampled image $z_{\mathcal{D}}(\mathbf{x})$, its reconstruction $\hat{z}(\mathbf{x})$ is obtained via projections onto a set of basis functions, using local weighted least-squares in square neighbourhoods of v pixels. The most common basis functions are polynomials: $\{\mathbf{1}, \mathbf{X}, \mathbf{Y}, \mathbf{X}^2, \mathbf{Y}^2, \mathbf{XY}, \ldots\}$, where $\mathbf{1} = [1, 1, \ldots, 1]^T$ (v entries), $\mathbf{X} = [x_1, x_2, \ldots, x_v]^T$, $\mathbf{X}^2 = [x_1^2, x_2^2, \ldots, x_v^2]^T$, and so on. These polynomials are constructed from local coordinates of v input samples. Given a set of m polynomials, within a local neighbourhood centred at $\mathbf{x}_0 = [x_0, y_0]^T$, the intensity value at position $\mathbf{x} = [x_0 + x', y_0 + y']^T$ is approximated by a polynomial expansion:

$$\hat{z}(\mathbf{x}, \mathbf{x}_0) = u_0(\mathbf{x}_0) + u_1(\mathbf{x}_0)x' + u_2(\mathbf{x}_0)y' + u_3(\mathbf{x}_0)x'^2$$
$$+ u_4(\mathbf{x}_0)x'y' + u_5(\mathbf{x}_0)y'^2 + \cdots \tag{1.28}$$

where $[x', y']^T$ are the local coordinates with respect to the centre $\mathbf{x}_0$ of the given neighbourhood. $\mathbf{u}(\mathbf{x}_0) \equiv [u_0(\mathbf{x}_0), u_1(\mathbf{x}_0), \ldots, u_m(\mathbf{x}_0)]^T$ are the projection coefficients onto the set of m polynomial basis functions at $\mathbf{x}_0$.

The identification of the coefficients $\mathbf{u}$ is performed using a weighted least-squares approach. The objective is the minimisation of the following approximation error

$$\varepsilon(\mathbf{x}_0) = \sum_{\mathbf{x}} \left(z_{\mathcal{D}}(\mathbf{x}) - \hat{z}(\mathbf{x}) \right)^2 c(\mathbf{x}) \alpha(\mathbf{x} - \mathbf{x}_0) \tag{1.29}$$

where $0 \leqslant c(\mathbf{x}) \leqslant 1$ is the *signal certainty* that specifies the reliability of the signal data at each point $\mathbf{x}$. Common practise suggests that missing data in the irregularly sampled image have a certainty equal to zero, while the observed samples have a certainty equal to one. On the other hand, $\alpha(\mathbf{x} - \mathbf{x}_0)$ is the so-called *applicability function* that localises the polynomial fit. A commonly used form of this function is an isotropic Gaussian kernel, the size of which depends on the scale of the analysis. Both the applicability function and the signal certainty control the impact of a particular sample to the local polynomial fit.

The least-squares solution for the polynomial coefficients $\mathbf{u}$ is then given by

$$\mathbf{u} = (\mathcal{B}^{\mathrm{T}} \mathcal{W} \mathcal{B})^{-1} \mathcal{B}^{\mathrm{T}} \mathcal{W} \mathbf{z}_{\mathcal{D}_{\nu}} \tag{1.30}$$

where $\mathbf{z}_{\mathcal{D}_{\nu}}$ is a $\nu \times 1$ vector representing the sampled image at the given neighbourhood, $\mathcal{B} = [\mathbf{b}_1 \ \mathbf{b}_2 \ \dots \ \mathbf{b}_m]$ is an $\nu \times m$ matrix of the m basis functions sampled at local coordinates of ν input samples, and $\mathcal{W} = \mathrm{diag}(\mathbf{c}) \cdot \mathrm{diag}(\boldsymbol{\alpha})$ is an $\nu \times \nu$ diagonal matrix constructed from an element-by-element product of the signal certainty $\mathbf{c}$ and the sampled applicability $\boldsymbol{\alpha}$ (each of them represented by a $\nu \times 1$ vector). Having identified the coefficients $\mathbf{u}$, the image may be reconstructed locally using the approximation in (1.28).

An interesting property of this formulation is that for zero-order polynomials, NC may be implemented very efficiently using simple convolution operations. In this case, the least-squares solution in (1.30) gives an approximated image equal to

$$\hat{z}(\mathbf{x}) = \frac{\alpha(\mathbf{x}) * (c(\mathbf{x}) z(\mathbf{x}))}{\alpha(\mathbf{x}) * c(\mathbf{x})} \tag{1.31}$$

with $*$ denoting the convolution operator. However, despite its simplicity, zero-order NC with a constant basis function is not capable of modelling image features such as edges or ridges. In such cases higher-order polynomial basis functions are required.

1.4 A new robust alternative for SR reconstruction

Some of the limitations of the existing SR techniques refer to the fact that they use simple image prior models, they formulate the solution using a simplistic assumption of translational motion between the frames and they do not account for possible errors during registration. In this section, we present a new approach that circumvents, to some degree, some of the above limitations and may be used in realistic scenarios with more complex geometric distortions (e.g. affine distortions) [49]. The SR reconstruction is formulated under the Bayesian framework as a joint registration/reconstruction problem, using a discontinuity adaptive robust kernel that characterises the image's prior distribution. In addition, the initialisation of the optimisation is performed using an adapted Normalised Convolution (NC) technique that incorporates the uncertainty due to misregistration.

1.4.1 Sub-pixel registration

For each frame k we consider the case of a general affine transformation parameterised by a 2×2 matrix A_k and a translation vector $\mathbf{S}_k$:

$$A_k = \begin{bmatrix} a_1^k & a_2^k \\ a_3^k & a_4^k \end{bmatrix}, \qquad \mathbf{S}_k = \begin{bmatrix} s_1^k \\ s_2^k \end{bmatrix} \tag{1.32}$$

Using a vector notation, the entire set of affine parameters is represented by $\mathbf{r} \equiv [\mathbf{r}_1^T, \mathbf{r}_2^T, \ldots, \mathbf{r}_K^T]^T$, with $\mathbf{r}_k \equiv [a_1^k, a_2^k, a_3^k, a_4^k, s_1^k, s_2^k]^T$. For the identification of a first estimate $\hat{\mathbf{r}}^0$ of the affine parameters we employed the multiresolution approach of [50]. This is a subpixel registration method using a pyramidal representation of the image based on splines and its objective is the minimisation of the mean square intensity between frames. Although we focus our attention on affine geometric transformations, the proposed SR technique is generic and can deal with any sub-pixel motions.

1.4.2 Joint Bayesian registration/reconstruction

The low quality of the available LR frames imposes an upper limit on the accuracy of every registration scheme [51]. This requires a sequential updating of the SR estimates and the registration parameters. Several Bayesian SR approaches work in this direction [13,21,40,52]. The objective is to form a maximum a posteriori (MAP) estimate of both $\mathbf{z}$ and the registration parameters $\mathbf{r}$, given the observations $\mathbf{y}$. In particular, the estimates of $\hat{\mathbf{z}}$ and $\hat{\mathbf{r}}$ are given by

$$(\hat{\mathbf{z}}, \hat{\mathbf{r}}) = \arg \max_{\mathbf{z}, \mathbf{r}} P(\mathbf{z}, \mathbf{r} | \mathbf{y}) = \arg \max_{\mathbf{z}, \mathbf{r}} P(\mathbf{y} | \mathbf{z}, \mathbf{r}) P(\mathbf{z}) \tag{1.33}$$

where we assume a uniform distribution for $P(\mathbf{r})$. This is equivalent to the minimisation of a posterior energy function $U(\mathbf{z}, \mathbf{r} | \mathbf{y})$:

$$(\hat{\mathbf{z}}, \hat{\mathbf{r}}) = \arg \min_{\mathbf{z}, \mathbf{r}} U(\mathbf{z}, \mathbf{r} | \mathbf{y}) = \arg \min_{\mathbf{z}, \mathbf{r}} \{ -\log P(\mathbf{y} | \mathbf{z}, \mathbf{r}) - \log P(\mathbf{z}) \} \tag{1.34}$$

In our scheme, the minimisation of $U(\mathbf{z}, \mathbf{r} | \mathbf{y})$ is performed iteratively using a deterministic relaxation method, which is described in detail in Section 1.4.2.3. At each iteration n, the registration parameters $\mathbf{r}$ are sequentially refined each time a new estimate of the unknown HR image is obtained:

$$\hat{\mathbf{r}}^{n+1} = \arg \min_{\mathbf{r}} U(\hat{\mathbf{z}}^n, \mathbf{r} | \mathbf{y}), \qquad \hat{\mathbf{z}}^{n+1} = \arg \min_{\mathbf{z}} U(\mathbf{z}, \hat{\mathbf{r}}^{n+1} | \mathbf{y}) \tag{1.35}$$

Note that $U(\mathbf{z}, \mathbf{r} | \mathbf{y})$ is not readily differentiable with respect to $\mathbf{r}$ for several motion models. In our case, the update of $\mathbf{r}$ is performed using the method of [50] and the current estimate $\hat{\mathbf{z}}^n$.

1.4.2.1 Data likelihood

The observation model in (1.7) may be rewritten as:

$$y_{k,s} = \sum_{t=1}^{N} W_{st}(\mathbf{r}_k) z_t + n_{k,s}, \quad \forall k = 1, 2, \ldots, K, \; \forall s = 1, 2, \ldots, M \quad (1.36)$$

Considering that the elements of the noise field are iid Gaussian samples with variance σ_n^2, the data likelihood term may be expressed as

$$P(\mathbf{y}|\mathbf{z}, \mathbf{r}) = \prod_{k=1}^{K} \prod_{s=1}^{M} P(y_{k,s}|\mathbf{z}, \mathbf{r})$$

$$= \prod_{k=1}^{K} \prod_{s=1}^{M} \frac{1}{\sqrt{2\pi}\sigma_n} \exp\left\{ -\frac{1}{2\sigma_n^2} \left(y_{k,s} - \sum_{t=1}^{N} W_{st}(\mathbf{r}_k) z_t \right)^2 \right\} \quad (1.37)$$

1.4.2.2 Prior distribution

Being an ill-posed problem, the estimation of $\mathbf{z}$ requires the introduction of prior constraints that restrict the solution space and introduce prior knowledge to the reconstruction. In particular, we employ a discontinuity adaptive smoothness constraint, which is frequently used in the domain of signal reconstruction.

In our method, we consider $\mathbf{z}$ as being a Markov Random Field (MRF), characterised by a Gibbs distribution of the following form:

$$P(\mathbf{z}) = \exp\left\{ -\lambda \sum_{s=1}^{N} \rho \left(\sum_{t=1}^{N} \mathcal{Q}_{st}(z_t) \right) \right\} \quad (1.38)$$

where λ is a constant and $\mathcal{Q}$ represents an $N \times N$ Laplacian operator. In the MRF-Gibbs framework, $\rho(x)$ corresponds to a clique potential function, which penalises high intensity fluctuations between adjacent pixels. From the robust statistics point of view, $\rho(x)$ is a robust error norm, which considers predominant discontinuities in the signal as outliers. Similar discontinuity adaptive constraints may be introduced via the notion of line-processes [53] or weak membrane or plate models [54]. In our method, as error norm we adopted the Lorentzian function [55] (see Figure 1.5):

$$\rho(x, \tau) = \log\left(1 + \frac{x^2}{2\tau^2} \right) \quad (1.39)$$

which depends on a single scale parameter τ. In robust statistics, its derivative

$$\psi(x, \tau) = \frac{2x}{2\tau^2 + x^2} \quad (1.40)$$

is called the *influence function* and characterises the effect that a particular discontinuity has on the solution.

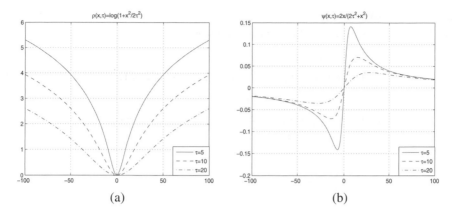

Figure 1.5 *Lorentzian robust norm function $\rho(x)$ and its derivative for several values of τ.*

1.4.2.3 Minimisation of the posterior energy

According to (1.37) and (1.38), the posterior energy is expressed as

$$U(\mathbf{z}, \mathbf{r}|\mathbf{y}) = \frac{1}{2\sigma_n^2} \sum_{k=1}^{K} \sum_{s=1}^{M} \left(y_{k,s} - \sum_{t=1}^{N} W_{st}(\mathbf{r}_k)z_t \right)^2 + \lambda \sum_{s=1}^{N} \rho \left(\sum_{t=1}^{N} \mathcal{Q}_{st}(z_t), \tau \right) \quad (1.41)$$

with

$$\frac{\partial U(\mathbf{z}, \mathbf{r}|\mathbf{y})}{\partial z_\ell} = \frac{1}{\sigma_n^2} \sum_{k=1}^{K} \sum_{s=1}^{M} \left(\sum_{t=1}^{N} W_{st}(\mathbf{r}_k)z_t - y_{k,s} \right) W_{s\ell}(\mathbf{r}_k)$$

$$+ \lambda \sum_{s=1}^{N} \psi \left(\sum_{t=1}^{N} \mathcal{Q}_{st}(z_t), \tau \right) \mathcal{Q}_{s\ell} \quad (1.42)$$

It is evident that the resultant energy function is non-convex, with several local minima. We opted for a deterministic *continuation* method, able to minimise non-convex functions via the construction of convex approximations. In particular, we used the *Graduated Non-Convexity* (GNC) approach of [54]. The general principle of the GNC method is depicted in Figure 1.6. Initially, we start with a convex approximation of the energy function, denoted by $U^0(\mathbf{z}, \mathbf{r}|\mathbf{y})$, which theoretically contains only one minimum. The minimum is obtained using a gradient descent algorithm. The same approach is performed with a series of successive approximations $U^p(\mathbf{z}, \mathbf{r}|\mathbf{y})$, with $p = \{1, 2, \ldots\}$, until the desired form of $U(\mathbf{z}, \mathbf{r}|\mathbf{y})$ is reached.

The employed gradient descent approach is the *Simultaneous Over-Relaxation* (SOR) [54]. The scale parameter τ of the error norm is successively decreased, starting from a high value that assures the convexity of $U(\mathbf{z}, \mathbf{r}|\mathbf{y})$ (see Figure 1.6). The SOR algorithm is an iterative approach, where the value of each site ℓ at iteration $n + 1$ is updated as follows:

$$z_\ell^{n+1} = z_\ell^n - \omega \frac{1}{T(z_\ell)} \frac{\partial U(\mathbf{z}, \mathbf{r}|\mathbf{y})}{\partial z_\ell} \quad (1.43)$$

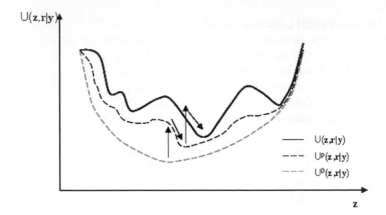

Figure 1.6 *Schematic representation of the Graduated Non-Convexity method. The minimum identified in the first convex approximation of the energy function $U^0(\mathbf{z}, \mathbf{r}|\mathbf{y})$ is used as the starting point for the minimisation of the second approximation $U^1(\mathbf{z}, \mathbf{r}|\mathbf{y})$. This process is repeated until a good starting point is identified for minimising the original non-convex energy function $U(\mathbf{z}, \mathbf{r}|\mathbf{y})$.*

where $0 < \omega < 2$ is an over-relaxation parameter[3] and $T(z_\ell)$ is an upper bound of the second derivative of $U(\mathbf{z}, \mathbf{r}|\mathbf{y})$:

$$T(z_\ell) \geqslant \frac{\partial^2 U(\mathbf{z}, \mathbf{r}|\mathbf{y})}{\partial z_\ell^2} \tag{1.44}$$

Using (1.41), we have

$$\frac{\partial^2 U(\mathbf{z}, \mathbf{r}|\mathbf{y})}{\partial z_\ell^2} = \frac{1}{\sigma_n^2} \sum_{k=1}^{K} \sum_{s=1}^{M} W_{s\ell}(\mathbf{r}_k)^2 + \lambda \sum_{s=1}^{N} \frac{\partial \psi \left(\sum_{t=1}^{N} \mathcal{Q}_{st}(z_t), \tau \right)}{\partial z_s} \mathcal{Q}_{s\ell}^2 \tag{1.45}$$

Knowing that

$$\max_{x} \frac{\partial \psi(x, \tau)}{\partial x} = \frac{1}{\tau^2}$$

we may use

$$T(z_s) = \frac{1}{\sigma_n^2} \sum_{k=1}^{K} \sum_{s=1}^{M} W_{s\ell}(\mathbf{r}_k)^2 + \frac{\lambda}{\tau^2} \sum_{s=1}^{N} \mathcal{Q}_{s\ell}^2 \tag{1.46}$$

Finally, the iterative updating of $\mathbf{z}$ is terminated if the following condition is satisfied:

$$\frac{\|\mathbf{z}^n - \mathbf{z}^{n-1}\|^2}{\|\mathbf{z}^{n-1}\|^2} < 10^{-5} \tag{1.47}$$

---

[3]For $\omega = 1$, SOR coincides with the Gauss–Seidel method.

1.4.2.4 Obtaining a first good approximation $\hat{z}^0$

The deterministic optimisation method presented in the previous section requires a good first approximation $\hat{z}^0$ of the reference HR frame. This fact is generally neglected by the existing SR methodologies, which mainly resort to simple interpolation techniques using only the LR reference frame. A fast and efficient way of obtaining $\hat{z}^0$ is the method of Normalised Convolution (see Section 1.3.4), which additionally offers the possibility of modelling the uncertainty regarding possible registration errors.

Common practise suggests that missing data in the irregularly sampled image have a certainty equal to zero, while the observed samples have a certainty equal to one. We propose to use an alternative approach, which accounts for errors related to sub-optimal registration. In particular, we use a non-binary set of certainties, where samples of the reference frame get a certainty value of one, whereas samples from neighbouring frames (see Figure 1.4) get a positive value equal to $\epsilon < 1$, which reflects the accuracy of the registration method. On the other hand, the applicability function $\alpha(\mathbf{x})$ in (1.29) corresponds to an isotropic Gaussian kernel, the size of which equals the support of the considered PSF. For the sake of speed, we use zero-order polynomials and the reconstruction result is given by (1.31).

1.5 Comparative evaluations

We conducted a series of experiments using two realistic video sequences, which are parts of the well-known CITY MPEG sequence, which is acquired using an airborne sensor with random jitter motion and a frame-rate equal to 60 Hz. Each of the tested sequences comprises 17 overlapping frames. Considering the set of HR frames as ground truth, we produce a sequence of LR frames (downsampling factor $L = 3$) using the degradation process of Section 1.2.3 that simulates the frame acquisition with a low quality camera. We considered a linear shift invariant PSF represented by a Gaussian kernel with a standard deviation equal to 1. Each frame was contaminated with Gaussian noise, corresponding to a SNR equal to -12.47 dB.[4] Figure 1.7 shows both the HR (ground truth) and LR reference frames of the two sequences.

The performance of our SR method is quantitatively assessed using the root mean square error (RMSE) and the structural similarity index (SSIM) [56] between the reconstructed image and the corresponding ground truth. The SSIM measure takes advantage of known characteristics of the human visual system and incorporates luminance, contrast and structural information to assess image quality (as opposed to RMSE that is based only on luminance similarities). Good reconstruction results in low values for the RMSE and high values for the SSIM.

We compared our method with the POCS approach of [38], the non-robust MAP approach of [40], together with the robust SR methods of [3] and [25]. For the latter method the

---

[4]We use the definition SNR $= 10\log_{10}[\text{var}(\text{image})/(M_1 M_2 \sigma_n^2)]$, with M_1 and M_2 being the image dimensions, and σ_n^2 the noise variance.

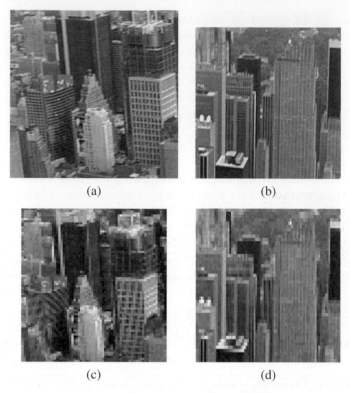

Figure 1.7   *(a, b) Original HR frames (ground truth) of the two considered sequences; (c, d) simulated LR frames.*

Table 1.1   *Quantitative results using NC as initialisation (RMSE/SSIM).*

	Sequence 1	Sequence 2
Tekalp et al. [38]	5.72/0.893	5.09/0.889
Hardie et al. [40]	4.89/0.915	4.08/0.924
Zomet et al. [3]	4.67/0.915	4.44/0.914
Farsiu et al. [25]	4.47/0.928	4.54/0.915
Proposed method	4.10/0.935	3.61/0.939

L_2 norm is used to describe the data consistency term. The parameters for these methods were tuned for a best quality measure (RMSE and SSIM) between the reconstructed and the original image. In order to have a common ground for comparisons, for all techniques the registration parameters were automatically updated during optimisation. Table 1.1 provides a quantitative assessment of the obtained results using NC as initialisation. It is evident that our approach gives the best results, something that can be also visually verified by looking at Figures 1.8 and 1.9. Finally, in Table 1.2 we show how the quality of reconstruction for all methods deteriorates if the initialisation is based on a linear interpolation instead of the proposed NC approach. Figure 1.10 shows the initial reconstruction using both alternatives.

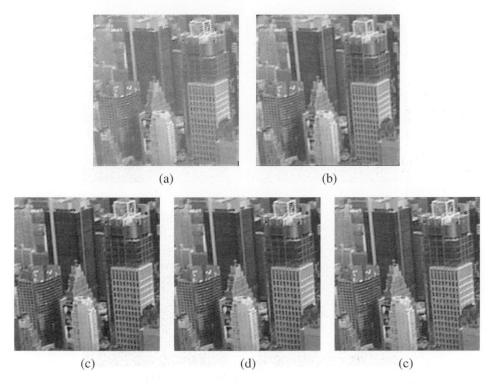

Figure 1.8 *Sequence 1 reconstruction results using NC as initialisation: (a) Tekalp et al. [38]; (b) Hardie et al. [40]; (c) Zomet et al. [3]; (d) Farsiu et al. [25]; (e) proposed method.*

Table 1.2 *Quantitative results using linear interpolation as initialisation (RMSE/SSIM).*

	Sequence 1	Sequence 2
Tekalp et al. [38]	6.59/0.855	6.04/0.832
Hardie et al. [40]	4.83/0.917	4.12/0.923
Zomet et al. [3]	5.45/0.894	4.77/0.897
Farsiu et al. [25]	4.56/0.927	4.72/0.911
Proposed method	4.17/0.932	3.73/0.938

1.6 Conclusions

SR reconstruction is a special branch of image fusion capable of bandwidth extrapolation beyond the limits of traditional electronic imaging systems. This chapter provides an overview of the existing techniques in the domain, analyses their characteristics, highlights their limitations and provides an insight into the steps required to increase the accuracy of reconstruction. We show that the latter may be obtained using a stochastic framework for joint reconstruction and sub-pixel registration, with edge-adaptive prior constraints and an efficient scheme for initialising the optimisation process involving misregistration uncertainties. Experimental results on realistic sequences and comparisons with several representative SR techniques justify these conclusions. An interesting topic for further investigation is to explicitly model the registration errors for their incorporation in our stochastic scheme and the automatic identification of the NC certainties.

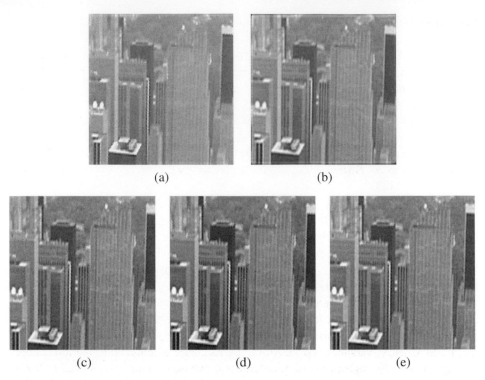

Figure 1.9 *Sequence 2 reconstruction results using NC as initialisation: (a) Tekalp et al. [38]; (b) Hardie et al. [40]; (c) Zomet et al. [3]; (d) Farsiu et al. [25]; (e) proposed method.*

Figure 1.10 *Sequence 1: (a) Reconstruction using non-binary NC with $\epsilon = 0.3$ (RMSE $= 6.64$/SSIM $= 0.86$). (b) Reconstruction using linear interpolation of the reference LR frame (RMSE $= 10.26$/SSIM $= 0.73$).*

Acknowledgements

This work has been carried out with the support of the UK MOD Data and Information Fusion Defence Technology Centre under project DTC Applied Multidimensional Fusion (AMDF).

References

[1] S. Park, M. Park and M. Kang, 'Super-resolution image reconstruction: A technical overview', *IEEE Signal Processing Magazine*, Vol. 20, No. 3, 2003, pp. 21–36.

[2] A. Papoulis, 'Generalized sampling expansion', *IEEE Transactions on Circuits and Systems*, Vol. 24, No. 11, 1977, pp. 652–654.

[3] A. Zomet and S. Peleg, *Super-Resolution from Multiple Images Having Arbitrary Mutual Motion,* Vol. 1, Kluwer Academic, 2001, pp. 195–209.

[4] S. Baker and T. Kanade, 'Limits on super-resolution and how to break them', *IEEE Transactions on Pattern Analysis and Machine Intelligence*, Vol. 24, No. 9, 2002, pp. 1167–1183.

[5] B. Zitova and J. Flusser, 'Image registration methods: A survey', *Image and Vision Computing*, Vol. 21, 2003, pp. 977–1000.

[6] M. Petrou, 'Image registration: An overview', *Advances in Imaging and Electron Physics*, Vol. 130, 2004, pp. 243–291.

[7] B. Horn and B. Schunck, 'Determining optical flow', *Artificial Intelligence*, Vol. 17, No. 7, 1981, pp. 185–203.

[8] M. Eismann and R. Hardie, 'Hyperspectral resolution enhancement using high-resolution multispectral imagery with arbitrary response functions', *IEEE Transactions on Geoscience and Remote Sensing*, Vol. 43, No. 3, 2005, pp. 455–465.

[9] II.-M. Chen, P. Varshney and M. Arora, 'Performance of mutual information similarity measure for registration of multitemporal remote sensing images', *IEEE Transactions on Geoscience and Remote Sensing*, Vol. 41, No. 11, 2003, pp. 2445–2454.

[10] C. Stiller and J. Konrad, 'Estimating motion in image sequences', *IEEE Signal Processing Magazine*, Vol. 16, No. 4, 1999, pp. 70–91.

[11] I. Pratikakis, C. Barillot, P. Hellier and E. Memin, 'Robust multiscale deformable registration of 3d ultrasound images', *International Journal of Image and Graphics*, Vol. 3, No. 4, 2003, pp. 547–565.

[12] R. Fransens, C. Strecha and L.V. Gool, 'Multimodal and multiband image registration using mutual information', in *ESA-EUSC 2004: Theory and Applications of Knowledge Driven Image Information Mining, with Focus on Earth Observation*, Madrid, Spain, 2004.

[13] H. Shen, L. Zhang, B. Huang and P. Li, 'A MAP approach for joint motion estimation, segmentation, and super resolution', *IEEE Transactions on Image Processing*, Vol. 16, No. 2, 2007, pp. 479–490.

[14] M. Banham and A. Katsaggelos, 'Digital image restoration', *IEEE Signal Processing Magazine*, Vol. 14, No. 2, 1997, pp. 24–41.

[15] A. Patti, M. Sezan and A. Tekalp, 'Superresolution video reconstruction with arbitrary sampling lattices and nonzero aperture time', *IEEE Transactions on Image Processing*, Vol. 6, No. 8, 1997, pp. 1064–1076.

[16] M. Irani and S. Peleg, 'Improving resolution by image registration', *CVGIP: Graphical Models and Image Processing*, Vol. 53, 1991, pp. 231–239.

[17] M. Petrou and P. Bosdogianni, *Image Processing: The Fundamentals*, Willey, New York, 1999.

[18] D. Kundur and D. Hatzinakos, 'Blind image deconvolution', *IEEE Signal Processing Magazine*, Vol. 13, No. 3, 1996, pp. 43–64.

[19] H. Shekarforoush and R. Chellappa, 'Data-driven multichannel superresolution with application to video sequences', *Journal of the Optical Society of America*, Vol. 16, No. 3, 1990, pp. 481–492.

[20] R. Lagendijk, J. Biemond and D. Boekee, 'Identification and restoration of noisy blurred images using the expectation-maximization algorithm', *IEEE Transactions on Acoustics, Speech, and Signal Processing*, Vol. 38, No. 7, 1990, pp. 1180–1191.

[21] N. Woods, N. Galatsanos and A. Katsaggelos, 'Stochastic methods for joint registration, restoration, and interpolation of multiple undersampled images', *IEEE Transactions on Image Processing*, Vol. 15, No. 1, 2006, pp. 201–213.

[22] S. Reeves and R. Mersereau, 'Blur identification by the method of generalized cross-validation', *IEEE Transactions on Image Processing*, Vol. 1, No. 3, 1992, pp. 301–311.

[23] N. Nguyen, P. Milanfar and G. Golub, 'A computationally efficient image superresolution algorithm', *IEEE Transactions on Image Processing*, Vol. 10, No. 4, 2001, pp. 573–583.

[24] R. Schultz and R. Stevenson, 'Extraction of high-resolution frames from video sequences', *IEEE Transactions on Image Processing*, Vol. 5, No. 6, 1996, pp. 996–1011.

[25] S. Farsiu, M. Robinson, M. Elad and P. Milanfar, 'Fast and robust multiframe super resolution', *IEEE Transactions on Image Processing*, Vol. 13, No. 10, 2006, pp. 1327–1344.

[26] D. Kuan, A. Sawchuk, T. Strand and P. Chavel, 'Adaptive noise smoothing filter for images with signal dependent noise', *IEEE Transactions on Pattern Analysis and Machine Intelligence*, Vol. 7, 1985, pp. 653–665.

[27] R. Tsai and T. Huang, *Multiframe Image Restoration and Registration,* Vol. 1, JAI Press, 1984, pp. 317–339.

[28] S. Borman, 'Topics in multiframe superresolution restoration', Ph.D. thesis, University of Notre Dame, IN, 2004.

[29] S. Kim, N. Bose and H. Valenzuela, 'Recursive reconstruction of high-resolution image from noisy undersampled frames', *IEEE Transactions on Acoustics Speech, and Signal Processing*, Vol. 38, 1990, pp. 1013–1027.

[30] S. Kim and W. Su, 'Recursive high-resolution reconstruction of blurred multiframe images', *IEEE Transactions on Image Processing*, Vol. 2, 1993, pp. 534–539.

[31] A. Tikhonov and V. Arsenin, *Solutions of Ill-Posed Problems*, Winston & Sons, Washington, DC, 1977.

[32] W. Su and S. Kim, 'High-resolution restoration of dynamic image sequences', *International Journal of Imaging Systems and Technology*, Vol. 5, No. 4, 1994, pp. 330–339.

[33] P. Combettes and M. Civanlar, 'The foundations of set theoretic estimation', in *Proc. IEEE Int. Conf. on Acoustics, Speech and Signal Processing (ICASSP'91)*, Vol. 4, 1991, pp. 2921–2924.

[34] L. Gubin, B. Polyak and E. Raik, 'The method of projections for finding the common point if convex sets', *USSR Computational Mathematics and Mathematical Physics*, Vol. 7, No. 6, 1967, pp. 1–24.

[35] D. Youla and H. Webb, 'Image restoration by the method of convex projections: Part 1, Theory', *IEEE Transactions on Image Processing*, Vol. MI-1, No. 2, 1982, pp. 81–94.

[36] M. Sezan and H. Stark, 'Image restoration by the method of convex projections: Part 2, Applications and numerical analysis', *IEEE Transactions on Med. Imaging*, Vol. 1, 1982, pp. 95–101.

[37] L. Bregman, 'The method of successive projection for finding a common point of convex sets', *Soviet Mathematics – Doklady*, Vol. 6, No. 3, 1965, pp. 688–692.

[38] A. Tekalp, M. Ozkan and M. Sezan, 'High-resolution image reconstruction from lower-resolution image sequences and space-varying image restoration', in *IEEE Int. Conf. on Acoustics, Speech, and Signal Processing*, San Francisco, CA, 1992, pp. 169–172.

[39] H. Trussell and M. Civanlar, 'The feasible solution in signal restoration', *IEEE Transactions on Acoustics, Speech, and Signal Processing*, Vol. 32, No. 2, 1984, pp. 201–212.

[40] R. Hardie, K. Barnard and E. Armstrong, 'Joint MAP registration and high-resolution image estimation using a sequence of undersampled images', *IEEE Transactions on Image Processing*, Vol. 6, No. 12, 1997, pp. 1621–1633.

[41] D. Capel and A. Zisserman, 'Computer vision applied to super resolution', *IEEE Signal Processing Magazine*, Vol. 20, No. 3, 2003, pp. 75–86.

[42] A. Lettington, M. Rollason, S. Tzimopoulou and E. Boukouvala, 'Image restoration using a two-dimensional lorentzian probability model', *Journal of Modern Optics*, Vol. 47, No. 5, 2000, pp. 931–938.

[43] M. Elad and A. Feuer, 'Restoration of a single superresolution image from several blurred, noisy, and undersampled measured images', *IEEE Transactions on Image Processing*, Vol. 6, No. 12, 1997, pp. 1646–1658.

[44] R. Piroddi and M. Petrou, 'Analysis of irregularly sampled data: A review', *Advances in Imaging and Electron Physics*, Vol. 132, 2004, pp. 109–165.

[45] T. Strohmer, 'Computationally attractive reconstruction of bandlimited images from irregular samples', *IEEE Transactions on Image Processing*, Vol. 4, No. 6, 1997, pp. 540–548.

[46] H. Knutsson and C. Westin, 'Normalized and differential convolution', in *IEEE Computer Society Conference on Computer Vision and Pattern Recognition (CVPR'93)*, New York, USA, 1993, pp. 515–523.

[47] S. Lertrattanapanich and N. Bose, 'High resolution image formation from low resolution frames using Delaunay triangulation', *IEEE Transactions on Image Processing*, Vol. 11, No. 12, 2002, pp. 1427–1441.

[48] T. Pham, L. van Vliet and K. Schutte, 'Robust fusion of irregularly sampled data using adaptive normalized convolution', *EURASIP Journal on Applied Signal Processing*, Vol. 2006, 2006, pp. 1–12.

[49] A. Katartzis and M. Petrou, 'Robust Bayesian estimation and normalized convolution for super-resolution image reconstruction', in *IEEE CVPR Workshop on Image Registration and Fusion*, 2007.

[50] P. Thevenaz, U. Ruttimann and M. Unser, 'A pyramid approach to subpixel registration based on intensity', *IEEE Transactions on Image Processing*, Vol. 7, No. 1, 1998, pp. 27–41.

[51] D. Robinson and P. Milanfar, 'Fundamental performance limits in image registration', *IEEE Transactions on Image Processing*, Vol. 13, No. 9, 2004, pp. 1185–1199.

[52] C. Segall, A. Katsaggelos, R. Molina and J. Mateos, 'Bayesian resolution enhancement of compressed video', *IEEE Transactions on Image Processing*, Vol. 13, No. 7, 2004, pp. 898–910.

[53] S. Geman and D. Geman, 'Stochastic relaxation, Gibbs distributions, and Bayesian restoration of images', *IEEE Transactions on Pattern Analysis and Machine Intelligence*, Vol. 6, 1984, pp. 721–741.

[54] A. Blake and A. Zisserman, *Visual Reconstruction*, MIT Press, Cambridge, MA, 1987.

[55] M. Black and P. Anandan, 'The robust estimation of multiple motions: Parametric and piecewise-smooth flow fields', *Computer Vision and Image Understanding*, Vol. 63, No. 1, 1996, pp. 75–104.

[56] Z. Wang, A. Bovik, H. Sheikh and E. Simoncelli, 'Image quality assessment: From error visibility to structural similarity', *IEEE Transactions on Image Processing*, Vol. 13, No. 4, 2004, pp. 600–612.

2

Image fusion through multiresolution oversampled decompositions

Bruno Aiazzi, Stefano Baronti and Massimo Selva

Institute of Applied Physics 'Nello Carrara' of the National Research Council, Florence, Italy

2.1 Introduction

Image fusion has been receiving increasing attention in the research community with the aim of investigating general formal solutions to a wide spectrum of applications. In the remote sensing field, the increasing availability of spaceborne imaging sensors, operating in a variety of ground scales and spectral bands, undoubtedly provides strong motivations. Because of the trade-off imposed by the physical constraint between spatial and spectral resolutions, spatial enhancement of poor-resolution multispectral (MS) data is desirable. In a different perspective, spectral enhancement of data collected with adequate ground resolution but poor spectral selection (as a limit case, a single panchromatic Pan image) can be obtained.

A great number of data fusion algorithms have been proposed in the literature starting from the second half of the eighties [1–3] and applied to imaging sensors that have been progressively launched.

A typology of simple and fast well established algorithms is known as component substitution (CS). When exactly three multispectral (MS) bands are concerned, the most straightforward CS fusion approach is to resort to an intensity–hue–saturation (IHS) transformation [4]. The I component is then *substituted* by the Pan image before the inverse IHS transform is applied. This procedure is nothing other than an *injection*, i.e. an addition, of the difference between the sharp Pan image and the smooth intensity I into the resampled MS bands [5]. Since the Pan image, histogram-matched to the I component, does not generally have the same local radiometry as the latter, when the fusion result is displayed in colour composition a large spectral distortion (colour changes) may be noticed.

If more than three bands are available, a viable solution is to define a generalised IHS (GIHS) transform by including the response of the near-infrared (NIR) band into the intensity component [6].

An alternative to IHS-based techniques is Principal Component Analysis (PCA). Analogously to the IHS scheme, the Pan image is *substituted* to the first principal component (PC1). Histogram-matching of Pan to PC1 is mandatory before substitution, because the mean and variance of PC1 are generally far greater than those of Pan. It is well established that PCA performs better than IHS [7], and in particular, that the spectral distortion in the fused bands is usually less noticeable, even if it cannot be completely avoided. Generally speaking, if the spectral responses of the MS bands are not perfectly overlapped with the bandwidth of Pan, as it happens with the most advanced very-high-resolution imagers, Ikonos and QuickBird, IHS- and PCA-based methods may yield poor results in terms of spectral fidelity [8].

Another well known CS technique reported in the literature is Gram–Schmidt (GS) spectral sharpening invented by Laben and Brover in 1998 and patented by Eastman Kodak [9]. The GS method is widely used since it has been implemented in the ENVI package. It has two operational modes. In one case (*mode 1*) the low pass approximation is computed as the average of the MS bands. In the other case (*mode 2*) the approximation is preliminarily obtained by low-pass filtering and decimating the Pan image.

Although the spectral quality of CS fusion results may be sufficient for most applications and users, methods based on injecting *zero-mean* high-pass spatial details, taken from the Pan image without resorting to any transformation, have been extensively studied to definitely overcome the inconvenience of spectral distortion.

In fact, since the pioneering *high-pass filtering* (HPF) technique [7], fusion methods based on injecting high-frequency components into resampled versions of the MS data have demonstrated a superior spectral fidelity [10]. HPF basically consists of an addition of spatial details, taken from a high-resolution Pan observation, into a bicubically resampled version of the low resolution MS image. Such details are obtained by taking the difference between the Pan image and its lowpass version achieved through a simple local pixel averaging, i.e. a box filtering.

Later improvements have been obtained with the introduction of multiresolution analysis (MRA), by employing several decomposition schemes, specially based on the *discrete wavelet transform* (DWT) [11,12], uniform rational filter banks (borrowed from audio coding) [13], and Laplacian pyramids (LP) [14,15]. Although never explicitly addressed by most of the literature, the rationale of highpass detail injection as a problem of spatial frequency spectrum substitution from a signal to another, was formally developed in a multiresolution framework as an application of filter banks theory [16].

The DWT has been extensively employed for remote sensing data fusion [17–19]. According to the basic DWT fusion scheme [20], couples of sub-bands of corresponding frequency content are merged together. The fused image is synthesised by taking the inverse transform. Fusion schemes based on the 'à trous' wavelet algorithm and Laplacian pyramids were successively proposed [21–23]. Actually, unlike the DWT which is *critically subsampled*, the 'à trous' wavelet and the LP are *oversampled*. Omitting the decimation step allows an image to be decomposed into nearly disjointed bandpass channels in the spatial frequency domain, without losing the spatial connectivity (*translation invariance property*) of its highpass details, e.g. edges and textures. This property is fundamental because, for critically subsampled schemes, *spatial* distortions, typically *ringing* or *aliasing* effects may be present in the fused products and originate shifts or blur of contours and textures.

As a simple outcome of multirate signal processing theory [24], the LP can be easily *generalised* (GLP) to deal with scales whose ratios are whatsoever integer or even fractional numbers [25–27].

A further goal of an advanced fusion method is to increase spectral information, by *unmixing* the coarse MS pixels through the sharp Pan image. This task requires the definition of a model establishing how the missing highpass information to be injected is extracted from the Pan image. It may be accomplished either in the domain of *approximations* between each of the resampled MS bands and a lowpass version of the Pan image having the same spatial frequency content as the MS bands, or in that of medium frequency *details*, in both cases by measuring local matching [28]. High frequency details are not available for MS bands, and must be inferred through the model, starting from those of Pan.

Quantitative results of data merge are provided thanks to the availability of reference originals obtained either by simulating the target sensor by means of high resolution data from an airborne platform, or by degrading all available data to a coarser resolution and carrying out the merge from such data. In most cases, only the latter strategy is feasible; the underlying idea, however, is that if an algorithm is optimised to yield best results at coarser scales, i.e. on spatially degraded data, it should still be optimal when the data are considered at finer scales, as it happens in practice. This assumption may be reasonable in general, but unfortunately is questionable for very high resolution data, especially when a highly detailed urban environment is concerned. The reason of this behaviour lies in the characteristics of the *modulation transfer function* (MTF) of the imaging systems [29]. Any interscale injection model should take into account that the MTF of real systems is generally bell-shaped and its magnitude value at the cutoff Nyquist frequency is far lower than 0.5, to prevent aliasing. Furthermore, the MTFs of the MS sensors may be significantly different from one another in terms of the decay rate, and especially are different from that of the Pan sensor. Hence, models empirically optimised at a coarser scale on data degraded by means of digital filters that are close to be ideal, may yield little enhancement when are utilised at a finer scale. If a model of MTF is assumed, results of fusion schemes based on non-critically subsampled MRA are improved, since in such schemes the reduction filter can be modelled on the basis of the MTF [29].

The aim of this chapter is to present and discuss the key points of image fusion based on MRA. In Section 2.2, the basic principles of MRA are reported by focusing on *non-*

critically subsampled decompositions that guarantee the best results for image fusion. In Section 2.3, the relevance of the MTF of the acquisition system is addressed while modelling of the injected spatial detail is discussed in Section 2.4. Quality issues are considered in Section 2.5 while qualitative and quantitative results are reported in Section 2.6.

2.2 Multiresolution analysis

The theoretical fundamentals of multiresolution analysis will be briefly reviewed with specific reference to the dyadic case, i.e. an analysis in which the scales vary as powers of two. Thus, the outcome frequency bands exhibit octave structure, that is, their extent doubles with increasing frequency. Although this constraint may be relaxed to allow more general analyses [30], such an issue will not be addressed here, for the sake of clarity and conciseness. The goal is to demonstrate that multiresolution analysis is a unifying framework in which novel and existing image fusion schemes can be easily accommodated.

2.2.1 Fundamental principles

Let $L^2(\mathbb{R})$ denote the Hilbert space of real square summable functions, with a scalar product $\langle f, g \rangle = \int f(x)g(x)\, dx$. Multiresolution analysis with J levels of a continuous signal f having finite energy is a projection of f onto a basis $\{\phi_{J,k}, \{\psi_{j,k}\}_{j \leqslant J}\}_{k \in \mathbb{Z}}$ [31]. Basis functions $\phi_{j,k}(x) = \sqrt{2^{-j}}\phi(2^{-j}x - k)$ result from translations and dilations of a same function $\phi(x)$ called the *scaling* function, verifying $\int \phi(x)\, dx = 1$. The family $\{\phi_{j,k}\}_{k \in \mathbb{Z}}$ spans a subspace $V_j \subset L^2(\mathbb{R})$. The projection of f onto V_j gives an *approximation* $\{a_{j,k} = \langle f, \phi_{j,k} \rangle\}_{k \in \mathbb{Z}}$ of f at the scale 2^j.

Analogously, basis functions $\psi_{j,k}(x) = \sqrt{2^{-j}}\psi(2^{-j}x - k)$ are the result of dilations and translations of the same function $\psi(x)$ called the *wavelet* function, which fulfils $\int \psi(x)\, dx = 0$. The family $\{\psi_{j,k}\}_{k \in \mathbb{Z}}$ spans a subspace $W_j \subset L^2(\mathbb{R})$. The projection of f onto W_j yields the wavelet coefficients of f, $\{w_{j,k} = \langle f, \psi_{j,k} \rangle\}_{k \in \mathbb{Z}}$, representing the *details* between two successive approximations: the data to be added to V_{j+1} to obtain V_j. Hence, W_{j+1} is the complement of V_{j+1} in V_j, that is, if the symbol $\oplus$ denotes the union operator:

$$V_j = V_{j+1} \oplus W_{j+1} \tag{2.1}$$

The subspaces V_j realise the multiresolution analysis. They present the following properties [32]:

- $V_{j+1} \subset V_j, \forall j \in \mathbb{Z}$;
- $f(x) \in V_{j+1} \Leftrightarrow f(2x) \in V_j$;
- $f(x) \in V_j \Leftrightarrow f(2^j x - k) \in V_0, \forall k \in \mathbb{Z}$;
- $\bigcup_{-\infty}^{+\infty} V_j$ is dense in $L^2(\mathbb{R})$ and $\bigcap_{-\infty}^{+\infty} V_j = 0$;
- $\exists \phi \in V_0$ such that $\{\sqrt{2^{-j}}\phi(2^{-j}x - k)\}_{k \in \mathbb{Z}}$ is a basis of V_j;
- a wavelet function $\psi(x)$ exists such that $\{\psi_{j,k}(x) = \sqrt{2^{-j}}\psi(2^{-j}x - k)\}_{k \in \mathbb{Z}}$ is a basis for W_j.

Eventually, multiresolution analysis with J levels yields the following decomposition of $L^2(\mathbb{R})$:

$$L^2(\mathbb{R}) = \left(\bigoplus_{j \leqslant J} W_j \right) \oplus V_J \tag{2.2}$$

All functions $f \in L^2(\mathbb{R})$ can be decomposed as follows:

$$f(x) = \sum_k a_{J,k} \tilde{\phi}_{J,k}(x) + \sum_{j \leqslant J} \sum_k w_{j,k} \tilde{\psi}_{j,k}(x) \tag{2.3}$$

The functions $\tilde{\phi}_{J,k}(x)$ and $\{\tilde{\psi}_{j,k}(x)\}_{k \in \mathbb{Z}}$ are generated from translations and dilations of dual functions, $\tilde{\phi}(x)$ and $\tilde{\psi}(x)$, that are to be defined in order to ensure a perfect reconstruction.

The connection between filter banks and wavelets stems from dilation equations allowing us to pass from a finer scale to a coarser one [31]:

$$\phi(x) = \sqrt{2} \sum_i h_i \phi(2x - i), \qquad \psi(x) = \sqrt{2} \sum_i g_i \phi(2x - i) \tag{2.4}$$

with $h_i = \langle \phi, \phi_{-1,i} \rangle$ and $g_i = \langle \psi, \phi_{-1,i} \rangle$.

Normalisation of the scaling function implies $\sum_i h_i = \sqrt{2}$. Analogously, $\int \psi(x)\,\mathrm{d}x = 0$ implies $\sum_i g_i = 0$. Multiresolution analysis of a signal f can be performed with a filter bank composed of a lowpass analysis filter $\{h_i\}$ and a highpass analysis filter $\{g_i\}$:

$$a_{j+1,k} = \langle f, \psi_{j+1,k} \rangle - \sum_i h_{i-2k} a_{j,i},$$

$$w_{j+1,k} = \langle f, \psi_{j+1,k} \rangle = \sum_i g_{i-2k} a_{j,i} \tag{2.5}$$

As a result, successive coarser approximations of f at scale 2^j are provided by lowpass filtering, with a downsampling operation applied on each filter output. This type of analysis consists in a series of *decimations*. Wavelet coefficients at scale 2^j are obtained by highpass filtering an approximation of f at the scale 2^{j-1}, followed by a downsampling.

The signal reconstruction is directly derived from (2.1):

$$a_{j,k} = \langle f, \phi_{j,k} \rangle = \sum_i \tilde{h}_{k-2i} a_{j+1,i} + \sum_i \tilde{g}_{k-2i} w_{j+1,i} \tag{2.6}$$

where the coefficients $\{\tilde{h}_i\}$ and $\{\tilde{g}_i\}$ define the synthesis filters.

If wavelet analysis is applied to a discrete sequence, the original signal samples, $\{f_n = f(nX)\}$, with $X = 1$, are regarded as the coefficients of the projection of a continuous function $f(x)$ onto V_0. The coefficients relative to the lower resolution subspace and to

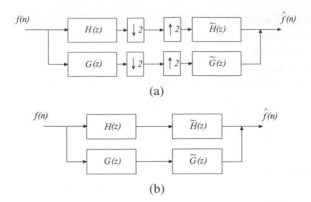

Figure 2.1 *Dyadic wavelet decomposition (analysis) and reconstruction (synthesis): (a) decimated; (b) un-decimated.*

its orthogonal complement can be obtained through the subsampling of the discrete convolution of f_n with the coefficients of the impulse response of the two digital filters $\{h_i\}$ and $\{g_i\}$, lowpass and highpass, respectively [32]. The two output sequences represent a smoothed version of $\{f_n\}$, or *approximation*, and the rapid changes occurring within the signal, or *detail*.

To achieve reconstruction of the original signal, the coefficients of the approximation and detail signals are upsampled and filtered by the dual filter of $\{h_i\}$ and $\{g_i\}$, i.e. *synthesis filters* $\{\tilde{h}_i\}$ and $\{\tilde{g}_i\}$, which are still lowpass and highpass filters, respectively. The scheme of a decimated wavelet coefficient decomposition and reconstruction is depicted in Figure 2.1(a), in which $\{f_n\}$ is a discrete 1-D sequence and $\{\hat{f}_n\}$ the sequence reconstructed after the analysis/synthesis stages. As it can be seen, the wavelet representation is closely related to a sub-band decomposition scheme [24].

2.2.1.1 Orthogonal wavelets

The functions $\psi(x)$ and $\phi(x)$ can be constructed in such a way to realise an orthogonal decomposition of the signal; then W_{j+1} is the orthogonal complement of V_{j+1} in V_j. These filters cannot be chosen independently of each other if *perfect reconstruction* (PR) is desired. The synthesis bank must be composed by filters having an impulse response that is a reversed version of that of the analysis ones [32,24], that is, $\tilde{h}_n = h_{-n}$ and $\tilde{g}_n = g_{-n}$. Quadrature mirror filters (QMF) satisfy all these constraints [33,31] with $g_n = (-1)^{n-1}h_{-n}$; hence, $G(\omega) = H(\omega + \pi)$. Thus, the *power-complementary* (PC) property, stated in the frequency domain as $|H(\omega)|^2 + |G(\omega)|^2 = 1$, which allows cancellation of aliasing created by downsampling in the dyadic analysis/synthesis scheme shown in Figure 2.1(a), becomes $|H(\omega)|^2 + |H(\omega + \pi)|^2 = 1$. Despite the mathematical elegance of the decomposition, constraints imposed on QMF do not allow the design of filters with impulse response symmetric around the zeroth coefficient, i.e. with null phase, since the number of coefficients is necessarily *even*. Furthermore, the bandwidth value is fixed to be exactly one half (in the dyadic case) of the available one.

2.2.1.2 Bi-orthogonal wavelets

If the orthogonality constraint is relaxed, symmetric (zero-phase) filters can be designed, which are suitable for image processing. Furthermore, the filters of the bank are no longer constrained to have the same size and may be chosen independently of each other. In order to obtain PR, two conditions must be met on the conjugate filters of the filter bank [34]:

$$H(\omega)\tilde{H}^*(\omega) + G(\omega)\tilde{G}^*(\omega) = 1,$$

$$H(\omega + \pi/2)\tilde{H}^*(\omega) + G(\omega + \pi/2)\tilde{G}^*(\omega) = 0 \qquad (2.7)$$

The former implies a correct data restoration from one scale to another, the latter represents the compensation of recovery effects introduced by downsampling, i.e. the *aliasing* compensation. Synthesis filters are derived from the analysis filters with the aid of the following relations:

$$\tilde{h}_n = (-1)^{n+1}g_{-n}, \qquad \tilde{g}_n = (-1)^{n+1}h_{-n} \qquad (2.8)$$

2.2.2 Undecimated discrete wavelet transform

The multiresolution analysis described above does not preserve the translation invariance. In other words, a translation of the original signal does not necessarily imply a translation of the corresponding wavelet coefficients. This property is essential in image processing. On the contrary, wavelet coefficients generated by an image discontinuity could disappear arbitrarily. This *non-stationarity* in the representation is a direct consequence of the downsampling operation following each filtering stage. In order to preserve the translation invariance property, some authors have introduced the concept of *stationary* wavelet transforms [35]. The downsampling operation is suppressed as shown in Figure 2.1(b) but filters are *upsampled* by 2^j, i.e. dilated by inserting $2^j - 1$ zeroes between any couple of consecutive coefficients:

$$h_k^{[j]} = h_k \uparrow 2^j = \begin{cases} h_{k/2^j}, & k = 2^j m, \text{ if } m \in \mathbb{Z}, \\ 0, & \text{otherwise}, \end{cases}$$

$$g_k^{[j]} = g_k \uparrow 2^j = \begin{cases} g_{k/2^j}, & k = 2^j m, \text{ if } m \in \mathbb{Z}, \\ 0, & \text{otherwise} \end{cases} \qquad (2.9)$$

The frequency response of (2.9) will be $H(2^j\omega)$ and $G(2^j\omega)$, respectively [24].

2.2.3 Multi-level decomposition of wavelet transforms

The decimated and undecimated wavelet decompositions may be recursively applied, i.e. the lowpass output of the wavelet transform may be further decomposed into two sequences. This process creates a set of *levels of wavelet decomposition* that represent the signal viewed at different scales. If the decomposition of the lowpass signal is repeated J times, $J + 1$ sequences are obtained: one sequence represents the *approximation* of the original signal containing a fraction $(1/2^J)$ of the original spectrum around zero; the other

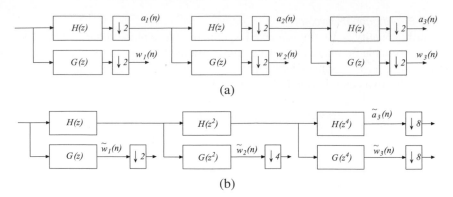

(a)

(b)

Figure 2.2   *(a) Three-level scheme ($J = 3$) for decimated wavelet decomposition; (b) equivalent scheme with undecimated wavelet sub-bands (denoted with a tilde).*

J sequences are the detail information that allow to reconstruct the original signal. The scheme of the decomposition of a signal into three levels ($J = 3$) is shown in Figure 2.2(a) in the case of *decimated*, or *critically subsampled*, wavelet. In the decimated domain, $a_{j,n}$ and $w_{j,n}$ denote the *approximation*, i.e. lowpass, and the *detail*, i.e. highpass or bandpass, sequences at the output of the jth stage, respectively. An equivalent representation is given in Figure 2.2(b), obtained from that of Figure 2.2(a) by shifting the downsamplers towards the output and by using upsampled filters [24]. The coefficients before the last downsamplers will be denoted with $\tilde{a}_{j,n}$ and $\tilde{w}_{j,n}$ and this representation will be referred to as an *undecimated*, or *oversampled*, discrete wavelet transform (UDWT).

Notice that the coefficients $a_{j,n}$ ($w_{j,n}$) can be obtained by downsampling $\tilde{a}_{j,n}$ ($\tilde{w}_{j,n}$) by a factor 2^j. PR is achieved in both cases. In the undecimated domain, lowpass and highpass coefficients are obtained by filtering the original signal. In fact, from Figure 2.2(b) it can be noticed that at the jth decomposition level, the sequences $\tilde{a}_{j,n}$ and $\tilde{w}_{j,n}$ can be obtained by filtering the original signal f_n with a bank of *equivalent filters* given by

$$H_j^{eq}(\omega) = \prod_{m=0}^{j-1} H(2^m \omega), \qquad G_j^{eq}(\omega) = \left[\prod_{m=0}^{j-2} H(2^m \omega) \right] \cdot G(2^{j-1} \omega) \qquad (2.10)$$

The frequency responses of the equivalent analysis filters are shown in Figure 2.3. As it appears, apart from the lowpass filter (leftmost), all the other filters are bandpass with bandwidths roughly halved as j increases by one. The prototype filters h and g are Daubechies-4 [33] with $L = 8$ coefficients.

A sequence $\{\hat{f}_n \equiv f_n\}$ can be reconstructed from the wavelet sub-bands, either decimated (see Figure 2.1(a)), or not (Figure 2.1(b)), by using the synthesis filters $\{\tilde{h}_i\}$ and $\{\tilde{g}_i\}$.

2.2.4 Translation-invariant wavelet decomposition of a 2-D image

Image multiresolution analysis was introduced by Mallat [32] in the decimated case. However, the 1-D filter bank used for the stationary wavelet decomposition can still be

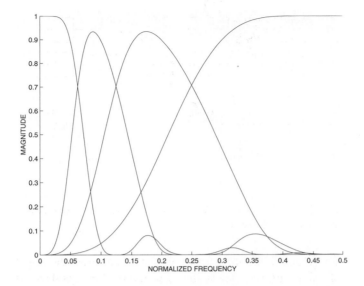

Figure 2.3　*Frequency responses of the equivalent analysis filters of an undecimated wavelet decomposition, for J = 3.*

applied in the 2-D case. Image rows and columns are then filtered separately. Filtering relationships to obtain the level $j + 1$ from the jth level are the following, in which (m, n) stands for pixel position:

$$A_{j+1}(m, n) = \sum_k \sum_l h_k^{[j]} h_l^{[j]} A_j(m + k, n + l),$$

$$W_{j+1}^{LH}(m, n) = \sum_k \sum_l g_k^{[j]} h_l^{[j]} A_j(m + k, n + l),$$

$$W_{j+1}^{HL}(m, n) = \sum_k \sum_l h_k^{[j]} g_l^{[j]} A_j(m + k, n + l),$$

$$W_{j+1}^{HH}(m, n) = \sum_k \sum_l g_k^{[j]} g_l^{[j]} A_j(m + k, n + l) \tag{2.11}$$

where A_j is the approximation of the original image at the scale 2^j, giving the low-frequency content in the sub-band $[0, \pi/2^j]$. Image details are contained in three high-frequency zero-mean 2-D signals W_j^{LH}, W_j^{HL}, and W_j^{HH}, corresponding to horizontal, vertical, and diagonal detail orientations, respectively. Wavelet coefficients of the jth level give high-frequency information in the sub-band $[\pi/2^j, \pi/2^{j-1}]$. For each decomposition level, in the undecimated case images preserve their original size since down-sampling operations after each filter have been suppressed. Thus, such a decomposition is highly redundant. In a J-level decomposition, a number of coefficients $3J + 1$ times greater than the number of pixels is generated. Figure 2.4 shows examples of DWT and UDWT applied on Landsat TM band 5 image of the Elba Isle in Tuscany (Italy).

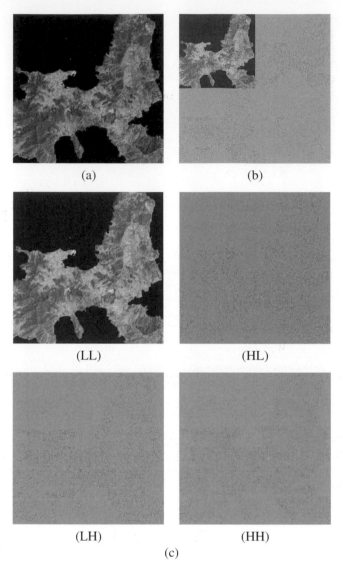

Figure 2.4   *Landsat TM band #5 of the Elba island: (a) original; (b) DWT and (c) UDWT, both with $J = 1$.*

2.2.5 'À trous' wavelet decomposition of an image

The 'à trous' wavelet (ATW) [36] is a non-orthogonal multiresolution decomposition defined by a filter bank $\{h_i\}$ and $\{g_i = \delta_i - h_i\}$, with the Kronecker operator δ_i denoting an allpass filter. Such filters are not QMF; thus, the filter bank does not allow PR if the output is decimated. In the absence of decimation, the lowpass filter is upsampled by 2^j, as in (2.9), before processing the jth level; hence the name 'à trous' which means 'with holes.' In two dimensions, the filter bank becomes $\{h_k h_l\}$ and $\{\delta_k \delta_l - h_k h_l\}$, which means that the 2-D detail signal is given by the pixel difference between two successive approximations, which have all the same scale 2^0, i.e. 1.

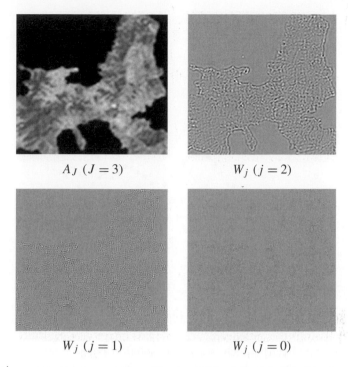

$A_J \ (J=3)$ $W_j \ (j=2)$

$W_j \ (j=1)$ $W_j \ (j=0)$

Figure 2.5 '*À trous' wavelet decomposition of Landsat TM band #5 of the Elba island portrayed in Figure 2.4(a).*

The prototype lowpass filter is usually zero-phase symmetric. For a J-level decomposition, the 'à trous' wavelet accommodates a number of coefficients $J + 1$ times greater than the number of pixels. Due to the absence of the decimation step, the synthesis is simply obtained by summing the detail levels to the approximations:

$$A_0(m,n) = A_J(m,n) + \sum_{j=0}^{J-1} W_j(m,n) \qquad (2.12)$$

in which $A_J(m,n)$ and $W_j(m,n)$, $j = 0, \ldots, J - 1$, are obtained through 2-D separable linear convolution between the original image $A_0(m,n)$ with the equivalent lowpass and highpass filters, respectively.

Figure 2.5 displays an example for the image reported in Figure 2.4(a) for $J = 3$. Details are coarse at the upper level and progressively become thinner when J decreases.

Incidentally, HPF [7] uses a frequency decomposition identical to an ATW with $J = 1$, apart from the analysis filter which, however, is odd-sized with constant coefficients and thus zero-phase. The frequency responses of box filters of different sizes, plotted in Figure 2.6, show that a smooth transition band is accompanied by a large ripple outside the passband.

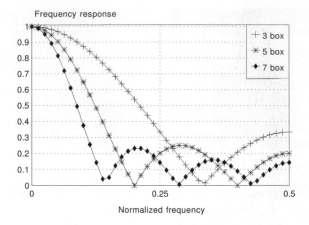

Figure 2.6 *Frequency responses of zero-phase box filters of HPF fusion: 7-box was specifically used by Chavez for fusion of Landsat-TM and SPOT-P (1:3).*

2.2.6 Laplacian pyramid

The Laplacian pyramid, originally proposed by Burt and Adelson [37] before multiresolution wavelet analysis was introduced, is a bandpass image decomposition derived from the Gaussian pyramid (GP) which is a multiresolution image representation obtained through a recursive *reduction* (lowpass filtering and decimation) of the image data set.

In the present multiresolution framework, a modified version of the LP, known as *enhanced* LP (ELP) [38] can be regarded as an ATW in which the image is recursively lowpass filtered and downsampled to generate a lowpass sub-band, which is re-expanded and subtracted pixel by pixel from the original image to yield the 2-D detail signal having zero-mean. The output of a separable 2-D filter is downsampled along rows and columns to yield the next level of approximation. Again, the detail is given as the difference between the original image and an *expanded* version of the lowpass approximation. Unlike the baseband approximation, the 2-D detail signal cannot be decimated if PR is desired.

The attribute *enhanced* depends on the zero-phase expansion filter (denoted as e_p) being forced to cut off at exactly one half of the bandwidth, and chosen independently of the reduction filter (denoted as r_p), which may be *half-band* as well or not. The ELP outperforms the former Burt's LP [37] when image compression is concerned [39], thanks to its layers being almost completely uncorrelated with one another.

Figure 2.7 shows the GP and ELP applied on the Landsat TM image of Figure 2.4(a). Notice the *lowpass* octave structure of GP layers, as well as the *bandpass* octave structure of ELP layers. An octave ELP is oversampled by a factor $4/3$ at most (when the baseband is one pixel wide). The data overhead is kept moderate thanks to decimation of the lowpass component.

In the case of a scale ratio $p = 2$, i.e. frequency octave decomposition, *polynomial* kernels with 3 (linear), 7 (cubic), 11 (fifth-order), 15 (seventh-order), 19 (ninth-order), and 23 (eleventh-order) coefficients have been assessed [25]. The term polynomial stems from

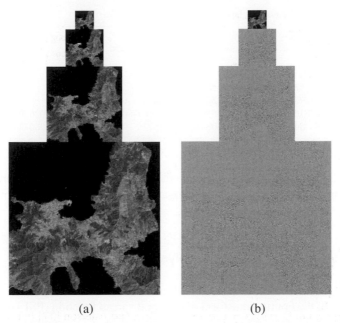

Figure 2.7 *Examples of GP (a) and ELP (b) applied to the Landsat TM image of Figure 2.4(a).*

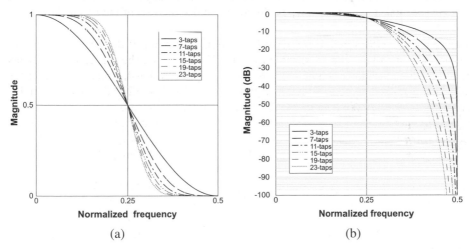

(a)   (b)

Figure 2.8 *Frequency responses of pyramid-generating filters: (a) linear scale; (b) logarithmic scale. Reduction filters are also used for expansion, with dc gain doubled.*

interpolation and denotes fitting an nth-order polynomial to the non-zero samples. The 7-tap kernel is widespread to yield a *bicubic* interpolation. It is noteworthy that *half-band* filters have the even order coefficients, except the zeroth one, all identically null [24]. The frequency responses of all the filters are plotted in Figure 2.8. Frequency is normalised to the sampling frequency f_S which is known to be twice the bandwidth available to the discrete signal. The above kernels are defined by the coefficients reported in Table 2.1. The filter design stems from a trade off between selectivity (sharp frequency cut off) and

Table 2.1   *Non-zero coefficients of polynomial 1-D kernels $r_2(i) = \frac{1}{2}e_2(i)$.*

Ord.	Taps	0	±1	±3	±5	±7	±9	±11
1	3	0.5	0.25	–	–	–	–	–
3	7	0.5	0.28125	−0.03125	–	–	–	–
5	11	0.5	0.29296875	−0.048828125	0.005859375	–	–	–
7	15	0.5	0.299072265	−0.059814453	0.011962890	−0.001220703	–	–
9	19	0.5	0.302810668	−0.067291259	0.017303466	−0.003089904	0.000267028	–
11	23	0.5	0.305334091	−0.072698593	0.021809577	−0.005192756	0.000807762	−0.000060081

computational cost (number of non-zero coefficients). In particular, the absence of ripple, which can be appreciated in the plots with logarithmic scale, is one of the most favourable characteristics.

2.3 MTF-tailored multiresolution analysis

Figure 2.9(a) shows the theoretical MTF of an imaging system. The MTF is defined as the magnitude of the Fourier transform of the point spread function (PSF) of the imaging system. In principle, two spectral replicas originated by 2-D sampling of the radiance signal with sampling frequency along- and across-track equal to the Nyquist rate, should cross each other at the Nyquist frequency (half of Nyquist rate) with magnitude values equal to 0.5. However, the scarce selectivity of the response prevents from using a Nyquist frequency with magnitude equal to 0.5. As a trade-off between maximum spatial resolution and minimum aliasing of the sampled signal, the Nyquist frequency is usually chosen such that the corresponding magnitude value is around 0.2. This situation is depicted in Figure 2.9(b) portraying the true MTF of an MS channel, which depends on several system parameters as the optical assembly, the platform motion and such external condition as atmospheric effects and viewing angle. A different situation occurs for the MTF of the Pan channel, whose extent is mainly dictated by diffraction limits (at least for instruments with resolution around 1 m). In that case the cutoff magnitude may also be lower (e.g. 0.1) and the appearance of the acquired images rather blurred. However, whereas the enhancing Pan image that is commonly available has been already processed for MTF restoration, the MS bands cannot be preprocessed analogously because of SNR constraints. In fact, restoration implies a kind of inverse filtering, that has the effect of increasing the noisiness of the data.

Eventually, the problem may be stated in the following terms: an MS band resampled at the finer scale of the Pan image lacks high spatial frequency components, that may be inferred from the Pan image via a suitable interscale injection model. If the highpass filter used to extract such frequency components from the Pan image is taken such as to approximate the complement of the MTF of the MS band to be enhanced, then the high frequency components, that are present in the MS band but have been damped by its MTF, can be restored. Otherwise, if spatial details are extracted from the Pan image by using a filter having normalised frequency cutoff at exactly the scale ratio between Pan and MS (e.g. 1/4 for 1 m Pan and 4 m MS), such frequency components will not be injected. This occurs with critically sub-sampled wavelet decompositions, whose filters are constrained

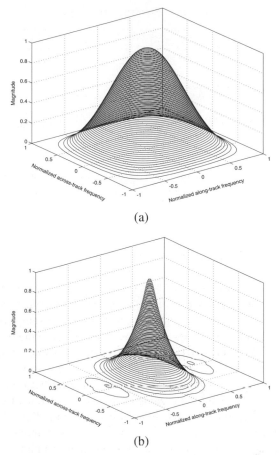

Figure 2.9 *(a) Ideal (isotropic) MTF with magnitude equal to 0.5 at cutoff Nyquist frequency; (b) typical (anisotropic) MTF with magnitude 0.2 at cutoff Nyquist frequency (across-track). All frequency scales are normalised to sampling frequency, or Nyquist rate (twice the Nyquist frequency).*

to cutoff at exactly an integer fraction (usually a power of two) of the Nyquist frequency of Pan data, corresponding to the scale ratio between Pan and MS. An attractive characteristic of the redundant pyramid and wavelet decompositions firstly proposed by the authors [40] and successively by Garzelli and Nencini [41] is that the lowpass reduction filter used to analyse the Pan image may be easily designed such that it matches the MTF of the band in which the extracted details will be injected. Figure 2.10(a) shows examples for three values of magnitude cutoff. The resulting benefit is that the restoration of spatial frequency content of the MS bands is provided by the multiresolution analysis of Pan through the injection model.

2.4 Context-driven multiresolution data fusion

The work by Núñez et al. [21] can be considered a forerunner of this rationale, even if no considerations on MTF were made in that paper. Figure 2.10(b) shows that the Gaussian-

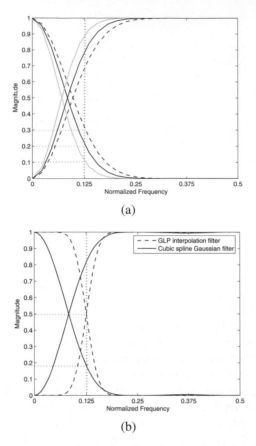

(a)

(b)

Figure 2.10 *1-D frequency responses of equivalent filters for separable 2-D 1:4 multiresolution analysis (lowpass and highpass filters generate approximation and detail, respectively). (a) Sample MTF-adjusted Gaussian-shaped filters with magnitudes 0.1, 0.2 and 0.3 at cutoff Nyquist frequency; (b) ELP-generating filters (dashed) and cubic spline filter for ATWT generation (solid).*

like frequency response of the cubic-spline wavelet filter used to generate the 'à trous' wavelet transform (ATWT) matches the shape of the MTF of a typical V-NIR band, with cutoff magnitude value equal to 0.185. The complementary highpass filter, yielding the detail level to be injected for 1:4 fusion, retains a greater amount of spatial frequency components than an ideal filter, such the one used by the standard ELP and ATWT [40, 41], thereby resulting in a greater spatial enhancement.

From the multiresolution analysis reviewed in Section 2.2, it appears that DWT, UDWT, ATW, and (E)LP may be regarded as particular cases of multiresolution analysis. The critical subsampling property, featured by DWT only, though essential for data compression, is not required for other applications, e.g. for data fusion. Actually, redundant decomposition structures show superior performances for image fusion [40,42] by avoiding aliasing phenomena and strongly reducing ringing effects. As a consequence, fusion schemes based on non-critically subsampled decompositions should be preferred.

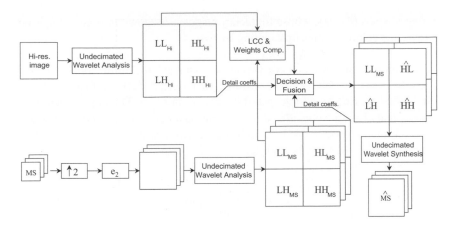

Figure 2.11 *Flowchart of undecimated wavelet-based image fusion procedure for a 1:2 scale ratio.*

Data fusion based on multiresolution analysis requires the definition of a model establishing how the missing highpass information to be injected into the MS bands is extracted from the P band [21]. Such a model can be global over the whole image or depend on context, either spectral [43–45], or spatial [28,29,46,47]. The goal is to make the fused bands the most similar to what the narrow-band MS sensor would image if it had the same resolution as the broadband one sensing the P band. The higher-frequency coefficients taken from the higher-resolution image are selected based on statistical congruence and weighted by a space-varying factor to achieve gain equalisation of otherwise different sensors. This is accomplished by measuring the degree of matching between each of the expanded MS bands and a lowpass version of the P band having the same spatial scale, i.e. the smaller one. The matching function can be thresholded to establish whether detail injection should occur or, more in general, to weight injection details. A gain factor mapping the highpass coefficients from the P image into the resampled MS band is locally given by the ratio of standard deviations of the target (one MS band) to the source (P image). The approach is similar to the one used by Hill et al. [46].

In the following of this section, three MRA schemes, respectively based on UDWT, ELP, and ATWT, will be described and briefly discussed.

2.4.1 Undecimated wavelet-based data fusion scheme

Figure 2.11 outlines a procedure based on UDWT, suitable for fusion of MS and P image data whose scale ratio is two [48]. For ratios greater than two, but still powers of two, the UDWT is achieved from an octave wavelet transform by omitting all decimators and upsampling the filter bank, as shown in Figure 2.2(b).

With reference to Figure 2.11, both the higher-resolution P image, and the lower-resolution MS image data set are decomposed by the one-level UDWT. The MS images have been previously interpolated by two along rows and columns, in order to process MS images having the same spatial scale as the P image. To this purpose, the 23-tap pyramid generating lowpass filter e_2, whose frequency response is shown in Figure 2.8, is applied along rows and columns, after upsampling by two.

Two sets of undecimated wavelet coefficients are obtained, including *approximation* (LL) and *detail* (HL, LH, and HH) signals of the original data. A decision fusion strategy usually establishes the rule by which spatial details are injected. The approximation coefficients LL_{Hi} and LL_{MS} are considered for computing *local correlation coefficients* (LCC) over a square sliding window [48]. An LCC map is computed between the approximation LL_{MS} of each of the MS bands, and that of the P image, LL_{Hi}, both at the scale of the latter, i.e. of the fused image.

When injection of the smaller scale details takes place, the higher-frequency coefficients from the high-resolution image are weighted to equalise contributions coming from different sensors according to the context model that has been chosen. The ratio of the local standard deviation of the low-resolution image to the corresponding value of the high-resolution image is used as the local weight. To avoid the insertion of unlikely details, injection is accomplished only if the context-based criterion is locally fulfilled. For each pixel (m, n) of the approximation of a MS band expanded at the scale of the P image, if $LCC(m, n)$ is greater than a threshold θ in the range $[-1, 1]$, the three detail coefficients at location (m, n) of the P image are multiplied by the local weight and replace the corresponding coefficients in the UDWT of that MS band. The threshold value may be selected, e.g. by minimising the standard deviation between the fused image (obtained from a P image and a degraded version of an MS data set) and the original MS data. Finally, undecimated wavelet synthesis is performed for each band of the MS image by applying the synthesis filter bank.

The above context model can also be extended to DWT (decimated case). The LCC is now calculated between the *approximation* of P and the MS band itself, both having the same scale (twice that of the undecimated case for one level of dyadic analysis).

2.4.2 Pyramid-based data fusion scheme

Since its first appearance in the literature [49], LP-based fusion has been progressively upgraded and generalised (GLP) to different application cases [26]. The block diagram reported in Figure 2.12 describes the multirate data fusion algorithm for the more general case of two image data sets, whose scale ratio is p/q, p, q being integer numbers with $p > q$. The two image data sets are preliminarily registered on each other. Let $A^{(P)}$ be the data set constituted by a single image having smaller scale (i.e. finer resolution) and size $Mp/q \times Np/q$, and $A^{(l)}$, $l = 1, \ldots, L$, the data set made up of L multi-spectral observations having scale larger by a factor p/q (i.e. coarser resolution), and thus size $M \times N$. The goal is to get a set of L multi-spectral images, each having the same spatial resolution as $A^{(P)}$. The upgrade of $A^{(l)}$ to the resolution of $A^{(P)}$ is the zero-mean GLP of $A^{(P)}$, computed for $J = 0$ (J denotes the resolution level, $J = 0$ being the full resolution level). The images of the set $A^{(l)}$ have to be interpolated by p and then reduced by q to match the finer resolution. Then, the highpass component from $A^{(P)}$ is added to the

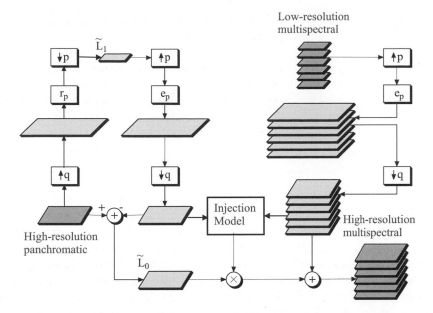

Figure 2.12 *Flowchart of GLP data fusion for an MS image and a P one, whose integer scale ratio is $p/q > 1$. r_p is the p-reduction lowpass filter with frequency cutoff at $1/p$ of bandwidth and unity DC gain. e_p is the p-expansion lowpass filter with cutoff at $1/p$ and DC gain equal to p. $\downarrow p$ (q) denotes down-sampling by p (q) while $\uparrow p$ (q) denotes up-sampling by p (q).*

expanded $A^{(l)}$, $l = 1, \ldots, L$, which constitute the lowpass component, in order to yield a spatially enhanced set of multispectral observations, $\hat{A}^{(l)}$, $l = 1, \ldots, L$.

It is noteworthy that in the case of 1:p fusion the interpolation by a factor q is no more needed and this scheme is referred as ELP. For all the sensors currently used, ELP is sufficient. Hereinafter, ELP acronym will be used.

In the particular case of $q = 1$ and $p = 4$, the scheme yields 1:4 fusion, which is the most interesting case because of the availability of such satellite data as those of Ikonos and QuickBird for which the ratio between P and MS is 1:4. In addition, as outlined in Section 2.3, the reduction filter r_p can be designed to match the system MTF, thus improving the quality of fused images. For computational convenience, $J = 2$ and $p = 2$ are preferable, since 1:4 filters are much longer than 1:2 filters of same characteristics. Furthermore, less data are to be processed at the second level, thanks to decimation after the first level.

As in the case of the previous section, a check on the congruence of the injection (*injection model*) is present in the fusion scheme with the aim of preventing the introduction of 'ghost' details in some of the MS bands and of equalising the values of P details to MS images before merging.

Figure 2.13 shows the flowchart of general case of MS + Pan fusion with 1:p scale ratio, based on the ELP [40]. In this context, emphasis will be given to the detail injection

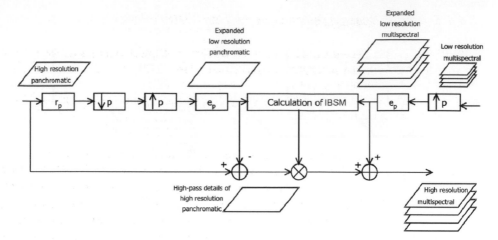

Figure 2.13   *Flowchart of multiresolution image fusion based on ELP and suitable for merging multispectral and panchromatic image data, whose integer scale ratio is $p > 1$. $\downarrow p$ denotes down-sampling by p; $\uparrow p$ denotes up-sampling by p; r_p is the MTF-matching p-reduction lowpass filter with cutoff at $1/p$ of the spectrum extent; e_p is ideal p-expansion lowpass filter with $1/p$ cutoff.*

model, indicated as interband structure model (IBSM), for which two improvements of the solutions introduced by Garzelli and Nencini [41] and by Aiazzi et al. [40] will be described.

2.4.3 'À trous' wavelet data fusion scheme

The ATW fusion scheme is obtained by means of ATWT decomposition reported in Section 2.2.5 and by synthesising the fused image according to (2.12). An injection model is usually inserted in the fusion chain in order to adaptively select the details to be added to the expanded PAN image. In the case of a single level decomposition ($J = 1$), the ATW fusion scheme is the same of Figure 2.13, in which upsampling and downsampling have been eliminated.

As a matter of fact, ELP and ATW decompositions guarantee the same quality in the fused images once the interpolation kernel e_p of ELP is correctly designed. ATW is computationally advantageous when only one level of decomposition is requested because decimation and interpolation are avoided. In case of multiple decomposition level, ELP is computationally more efficient and conceptually simpler, since downsampling guarantees a lower number of pixels to be processed with a shorter kernel that does not change along the decomposition levels. Actually, for ATW equivalent filters have to be considered, whose length increases with decomposition level.

Also in the case of ATW, r_p can take into account the MTF of the acquisition system to improve the quality of results. This is immediate when only one decomposition level is involved.

2.4.4 Enhanced Spectral Distortion Minimising (ESDM) model

Given two spectral vectors, $\mathbf{v}$ and $\hat{\mathbf{v}}$, both having L components, in which $\mathbf{v} = \{v_1, v_2, \ldots, v_L\}$ is the original spectral pixel vector $v_l = A^{(l)}(m, n)$ while $\hat{\mathbf{v}} = \{\hat{v}_1, \hat{v}_2, \ldots, \hat{v}_L\}$ is the distorted vector obtained by applying fusion to the coarser resolution MS data, i.e. $\hat{v}_l = \hat{A}^{(l)}(m, n)$, the spectral angle mapper (SAM) denotes the absolute value of the spectral angle between the two vectors:

$$\text{SAM}(\mathbf{v}, \hat{\mathbf{v}}) \triangleq \arccos\left(\frac{\langle \mathbf{v}, \hat{\mathbf{v}} \rangle}{\|\mathbf{v}\|_2 \cdot \|\hat{\mathbf{v}}\|_2}\right) \tag{2.13}$$

The *Enhanced Spectral Distortion Minimising* (ESDM) model is defined through its injection gain, which is both spatially and spectrally varying; in a vector notation, the injection gain is defined as $\vec{\alpha}_S(m, n) = \{\alpha_S^{(l)}(m, n), \ l = 1, \ldots, L\}$. Let $\vec{\tilde{A}}(m, n) \triangleq \{\tilde{A}^{(l)}(m, n), \ l = 1, \ldots, L\}$ denote the pixel vector of the expanded MS image; let also $\vec{W}_0^{\text{MS}}(m, n) \triangleq \vec{\alpha}_S(m, n) \cdot W_0^{(P)}(m, n)$ denote the MS detail vector to be injected, $W_0^{(P)}$ being the image of spatial details extracted from the Pan image. In order to minimise the SAM distortion between resampled MS bands and fused products, the injected detail vector at pixel position (m, n) must be parallel to the resampled MS vector, i.e. to $\vec{\tilde{A}}(m, n)$. At the same time, each component $\alpha_S^{(l)}(m, n)$ should be designed in such a way as to minimise the radiometric distortion when the detail component $W_0^{(l)}(m, n)$ is injected into $\tilde{A}^{(l)}(m, n)$ to generate the fused image $\vec{\hat{A}}$. Starting from the vector merge expression

$$\vec{\hat{A}}(m, n) = \vec{\tilde{A}}(m, n) + \vec{\alpha}_S(m, n) \cdot W_0^{(P)}(m, n) \tag{2.14}$$

let us define the lth components of $\vec{\alpha}_S(m, n)$ as

$$\alpha_S^{(l)}(m, n) = \beta(m, n) \cdot \frac{\tilde{A}^{(l)}(m, n)}{\tilde{A}_2^{(P)}(m, n)}, \quad l = 1, \ldots, L \tag{2.15}$$

in which $\tilde{A}_2^{(P)}(m, n)$ denotes the approximation of the Pan image produced by the equivalent lowpass filter while the coefficient $\beta(m, n)$, which does not depend on the band index l, is defined in such a way the length of the fused vector is statistically close to that of the (unavailable) true high-resolution vector. A suitable definition of $\beta(m, n)$ can be taken as the ratio between average local standard deviations of resampled MS bands and local standard deviation of lowpass approximation of Pan:

$$\beta(m, n) = \sqrt{\frac{(1/L) \sum_{l=1}^{L} \text{var}[\tilde{A}^{(l)}](m, n)}{\text{var}[\tilde{A}_2^{(P)}](m, n)}} \tag{2.16}$$

In the original SDM model [50], the correcting factor β, aimed at mitigating the unlikely overenhancement, mainly noticed on vegetated areas, was not considered.

From (2.14) and (2.15) it stems that

$$\vec{A} \times \vec{\tilde{A}} = \vec{A} \times [\vec{A} + \vec{W}_0^{MS}] = \vec{A} \times [\vec{A} + \vec{\alpha_S} \cdot W_0^{(P)}]$$

$$= \vec{A} \times \vec{A} \cdot \left[1 + \beta \cdot \frac{W_0^{(P)}}{\tilde{A}_2^{(P)}} \right] = 0 \qquad (2.17)$$

in which $\times$ stands for the vector product and indexes (m, n) have been omitted. Equation (2.17) states that the spectral angle (SAM) is unchanged when a vector pixel in the expanded MS image, $\vec{\tilde{A}}(m, n)$, is enhanced to yield the fused product, $\vec{\tilde{A}}(m, n)$, because the upgrade vector $\vec{W}_0^{MS}(m, n)$ is always parallel to $\vec{\tilde{A}}(m, n)$.

2.4.5 Enhanced Context-Based (ECB) model

The *Enhanced Context-Based* (ECB) injection model rules the insertion of the details extracted from the Pan image into the target *l*th MS band. Although substantially similar to the context-based decision (CBD) model introduced by Aiazzi et al. [40], it lacks the threshold decision on injection and introduces a normalisation factor given by the global correlation coefficient defined below. The ECB model may be stated as

$$W_0^{(l)}(i, j) = \alpha_C^{(l)}(m, n) \cdot W_0^{(P)}(m, n), \quad l = 1, \ldots, L \qquad (2.18)$$

The space-varying model $\alpha_C^{(l)}(m, n)$ is calculated between the *l*th MS band resampled to the scale of the Pan image, $\tilde{A}^{(l)}$, and the approximation of the Pan image at the resolution of the MS bands, $\tilde{A}_2^{(P)}$, as

$$\alpha_C^{(l)}(m, n) = \min \left\{ \frac{\rho_{l,P}(m, n)}{\bar{\rho}_{l,P}} \cdot \frac{\sigma_{\tilde{A}^{(l)}}(m, n)}{\sigma_{\tilde{A}_2^{(P)}}(m, n)}, c \right\} \qquad (2.19)$$

in which $\rho_{l,P}(m, n)$ is the *local* linear correlation coefficient between $\tilde{A}^{(l)}$ and $\tilde{A}_2^{(P)}$, calculated on a square sliding window of size $N \times N$ centred on pixel (m, n); $\bar{\rho}_{l,P}$ is the *global* one and is constant throughout.

The ECB model is uniquely defined by the window size N depending on the spatial resolutions and scale ratio of the images to be merged, as well as on the landscape characteristics (typically, $7 \leqslant N \leqslant 11$). A clipping constant c was introduced to avoid numerical instabilities ($2 \leqslant c \leqslant 3$).

2.5 Quality

2.5.1 Quality assessment of fusion products

Quality assessment of Pan-sharpened MS images is a hard task [7,10]. Even when spatially degraded MS images are processed for Pan-sharpening, and therefore reference MS images are available for comparisons, assessment of *fidelity* to the reference usually requires computation of a number of different indexes, the most significant of which are recalled in Section 2.5.2 and utilised for the reported experiments.

A general paradigm usually accepted in the research community for quality assessment of fused images was first proposed by Wald et al. [10] and re-discussed in [28]. Such a paradigm is found on three properties the fused data have to cope with as much as possible.

By using the notation of the previous sections, the first property requires that *any fused image $\hat{A}$, once degraded to its original resolution, should be as identical as possible to the original image A*. To achieve this, the fused image $\hat{A}$ is spatially degraded to the same scale of A, thus obtaining an image $\hat{A}^*$. $\hat{A}^*$ has to be very close to A.

The second property states that *any image $\hat{A}$ fused by means of a high resolution (HR) image should be as identical as possible to the ideal image A_I that the corresponding sensor, if existent, would observe at the resolution of the HR image.*

The third property considers the multispectral properties of the whole set of fused images: *the multispectral vector of images $\vec{\hat{A}}$ fused by means of a high resolution (HR) image should be as identical as possible to the multispectral vector of ideal images $\vec{A}_I$ that the corresponding sensor, if existent, would observe at the spatial resolution of the HR image.*

The second and third properties cannot usually be verified directly, since $\vec{A}_I$ is not commonly available. Thus, a change of scale is usually performed to check these properties. The multispectral image $\vec{A}^*$ and the panchromatic image P* are created from the original sets of images $\vec{A}$ and P. P is degraded to the resolution of multispectral image (i.e. an Ikonos panchromatic image at 4 m) and $\vec{A}$ to a lower resolution depending on the scale ratio for which fusion is assessed (i.e. an Ikonos multispectral image at 8 m or 16 m, as considered in Section 2.6). The fusion method is applied to these two sets of images, resulting into a set of fused images at the resolution of the original MS image. The MS image serves now as reference and the second and third properties can be tested. The quality observed for the fused products is assumed to be close to the quality that would be observed for the fused products at the full scale. This point has been largely discussed by Wald et al. [10].

In testing the first property, an important point is the way the fused image $\hat{A}$ is degraded to $\hat{A}^*$, since the results depend on the filtering operator used. Wald et al. [10] showed that relative discrepancies between the results are on the order of a very few per cent. In conclusion, there is an influence of the filtering operator upon the results, but it can be kept very small provided the filtering operation is performed in a suitable way.

In any case, all the MRA methods discussed in Section 2.6 are by essence built to satisfy this first property and this issue, although important in general, is not critical for the analysed schemes.

2.5.2 Quality indices

Once a quality assessment protocol is chosen, quality indices and distortion measurements can be defined.

In this chapter mean bias ($\Delta\mu$), root mean square error (RMSE), spectral angle mapper (SAM), and *relative dimensionless global error in synthesis* (ERGAS) have been utilised as distortion measurements. Cross-correlation coefficient (CC) and a quaternion-based coefficient (Q4) have been adopted as quality indices.

$\Delta\mu$, RMSE, CC refer to single MS bands and are suitable to test the first and second properties of Wald's protocol. ERGAS, Q4, and SAM are cumulative global indices and are suitable to assess the third property.

Given two images A and B with means $\mu(A)$ and $\mu(B)$ given by E[A] and E[B], respectively, where E($\cdot$) denotes the expectation operator, the mean bias between A and B is defined as

$$\Delta\mu \triangleq \mu(A) - \mu(B) \qquad (2.20)$$

In order to guarantee that images A and B are similar, $\Delta\mu$ has to assume values close to zero.

The correlation coefficient between A and B is defined as

$$CC \triangleq \frac{\sigma^2_{A,B}}{\sigma_A \sigma_B} \qquad (2.21)$$

where $\sigma^2_{A,B}$ is the covariance between A and B, given by E[$(A - \mu(A))(B - \mu(B))$], and σ_A is the standard deviation of A given by $\sqrt{E[(A - \mu(A))^2]}$. In the same way $\sqrt{E[(B - \mu(B))^2]}$ represents the standard deviation of B. CC can assume values in the range $[-1, 1]$. When the images are similar, CC is close to 1 and is exactly 1, when A is equal to B.

The RMSE between A and B is defined as

$$RMSE \triangleq \sqrt{E[(A - B)^2]} \qquad (2.22)$$

When A and B are similar, RMSE is close to zero and is zero if and only if $A = B$.

SAM is suitable to characterise multispectral images. SAM(A, B) is defined according to (2.13) as E[SAM(a, b)], where a and b denote the generic pixel vector element of multispectral image A and B, respectively. SAM is usually expressed in degrees and is close to zero, when images A and B are similar.

ERGAS was proposed by Wald et al. [10] as an error index that offers a global indication of the quality of a fused product, and is given by

$$\text{ERGAS} \triangleq 100 \frac{d_h}{d_l} \sqrt{\frac{1}{L} \sum_{l=1}^{L} \left(\frac{\text{RMSE}(l)}{\mu(l)} \right)^2} \qquad (2.23)$$

where d_h/d_l is the ratio between pixel sizes of Pan and MS, e.g. 1/4 for Ikonos and QuickBird data, $\mu(l)$ is the mean (average) of the lth band, and L is the number of bands. Low values of ERGAS indicates similarity between multispectral data.

Another quality index, namely Q4, suitable for images having four spectral bands, was recently proposed by one of the authors for quality assessment of Pan-sharpened MS imagery [51]. For MS images with four spectral bands, let a, b, c, and d denote the radiance values of a given image pixel in the four bands, typically acquired in the B, G, R, and NIR wavelengths. Q4 is made up of different factors accounting for correlation, mean bias, and contrast variation of each spectral band, as well as of spectral angle. Since the modulus of the hypercomplex correlation coefficient (CC) measures the alignment of spectral vectors, its low value may detect when radiometric distortion is accompanied by spectral distortion. Thus, both radiometric and spectral distortions may be encapsulated in a unique parameter. Let

$$\mathbf{z}_A = a_A + \mathbf{i}b_A + \mathbf{j}c_A + \mathbf{k}d_A, \qquad \mathbf{z}_B = a_B + \mathbf{i}b_B + \mathbf{j}c_B + \mathbf{k}d_B \qquad (2.24)$$

denote the 4-band reference MS image and the fusion product, respectively, both expressed as quaternions or hypercomplex numbers. The Q4 index is defined as

$$Q4 \triangleq \frac{4|\sigma_{z_A z_B}| \cdot |\overline{\mathbf{z}_A}| \cdot |\overline{\mathbf{z}_B}|}{(\sigma_{z_A}^2 + \sigma_{z_B}^2)(|\overline{\mathbf{z}_A}|^2 + |\overline{\mathbf{z}_B}|^2)} \qquad (2.25)$$

Equation (2.25) may be written as product of three terms:

$$Q4 = \frac{|\sigma_{z_A z_B}|}{\sigma_{z_A} \cdot \sigma_{z_B}} \cdot \frac{2\sigma_{z_A} \cdot \sigma_{z_B}}{\sigma_{z_A}^2 + \sigma_{z_B}^2} \cdot \frac{2|\overline{\mathbf{z}_A}| \cdot |\overline{\mathbf{z}_B}|}{|\overline{\mathbf{z}_A}|^2 + |\overline{\mathbf{z}_B}|^2} \qquad (2.26)$$

the first of which is the modulus of the hypercomplex CC between $\mathbf{z}_A$ and $\mathbf{z}_B$ and is sensitive both to loss of correlation and to spectral distortion between the two MS data sets. The second and third terms, respectively, measure contrast changes and mean bias on all bands simultaneously. Ensemble expectations are calculated as averages on $N \times N$ blocks. Hence, Q4 will depend on N as well. Eventually, $Q4$ is averaged over the whole image to yield the *global* score index. Alternatively, the minimum attained by $Q4$ over the whole image may represent a measure of *local* quality. $Q4$ assumes values in the range [0, 1] and is equal to 1, when A and B are equal.

2.6 Experimental results

2.6.1 Data set and compared methods

The reported fusion procedures have been assessed on two very high-resolution image data sets, collected by QuickBird and Ikonos satellites. The former displays the urban and suburban areas of Pavia in Italy and was acquired on June 23, 2002; the latter was taken on May 15, 2000 and shows the city of Toulouse in France.

The four MS bands of QuickBird span the visible and near infrared (NIR) wavelengths and are spectrally disjoint: blue (B1 = 450–520 nm), green (B2 = 520–600 nm), red (B3 = 630–690 nm), and NIR (B4 = 760–900 nm). The Pan band approximately covers the whole interval (Pan = 500–900 nm). The four MS bands of Ikonos span the visible and NIR (VNIR) wavelengths and are non-overlapped, with the exception of B1 and B2: B1 = 440–530 nm, B2 = 520–600 nm, B3 = 630–700 nm, and B4 = 760–850 nm. The bandwidth of Pan covers the interval 450–950 nm.

Both data sets are geometrically and radiometrically calibrated. They are available as geocoded products, resampled to uniform ground resolutions of 2.8 m (MS)–0.7 m (Pan) and 4 m (MS)–1 m (Pan), for QuickBird and Ikonos, respectively. All data are acquired on a dynamic range of 11 bits and packed in 16-bit words. Square regions of about 2 km^2 (QuickBird) and 4 km^2 (Ikonos) were analysed. The original Pan images are of size 2048×2048, while the original MS images are of size 512×512.

To allow quantitative distortion measures to be achieved, the Pan and MS bands are preliminarily lowpass filtered and decimated by 4, to yield 2.8 m Pan–11.2 m MS, and 4 m Pan–16 m MS, for QuickBird and Ikonos, respectively. Such spatially degraded data are used to re-synthesise the four spectral bands at 2.8 m and 4 m, respectively. Thus, the true 2.8 m/4 m 512×512 MS data are available for objective distortion measurements. To highlight the trend in performance varying with scale ratio, also 2:1 fusion simulations are carried out on both data sets, after the original MS data have been spatially degraded by 2; the Pan images are still degraded by 4, i.e. QuickBird: 5.6 m MS, 2.8 m Pan; Ikonos: 8 m MS, 4 m Pan.

Mean values, standard deviations, and inter-band CCs are reported in Tables 2.2, 2.3, 2.4, and 2.5, for the QuickBird and Ikonos data sets, respectively. Notice that the mean radiance significantly changes between the two data sets, being two to three times larger for QuickBird. Instead, standard deviations are comparable, except for that of the NIR band, which is almost double for QuickBird. Also, the NIR band (B4) of QuickBird is almost uncorrelated with the visible spectral bands, unlike that of Ikonos.

True-colour composites (bands 3-2-1) of 11.2 m and 2.8 m MS data are reported in Figures 2.14(a) and 2.14(h), and in Figures 2.16(a) and 2.16(h), for QuickBird and Ikonos, respectively. Local misalignments between MS and degraded Pan, that are likely to affect fusion results, are introduced by cartographic resampling and are hardly noticeable.

Table 2.2 *Means (μ) and standard deviations (σ) of the test QuickBird image (radiance units): original 2.8 m MS and 0.7 m Pan, and spatially degraded by 4 (11.2 m Ms and 2.8 m Pan).*

Original	B1	B2	B3	B4	Pan
μ_O	356.40	488.59	319.66	463.73	415.79
σ_O	15.13	34.15	48.96	90.28	57.79
Degraded	B1	B2	B3	B4	Pan
μ_D	356.10	488.17	319.31	463.30	415.79
σ_D	12.82	28.30	40.99	72.64	52.67

Table 2.3 *Means (μ) and standard deviations (σ) of the test Ikonos image (radiance units): original 4 m MS and 1 m Pan, and spatially degraded by 4 (16 m MS and 4 m Pan).*

Original	B1	B2	B3	B4	Pan
μ_O	194.92	187.42	146.81	138.60	151.04
σ_O	21.17	37.50	46.23	52.34	51.67
Degraded	B1	B2	B3	B4	Pan
μ_D	194.92	187.43	146.82	138.62	151.07
σ_D	19.41	35.59	43.09	44.63	46.34

Table 2.4 *QuickBird: CCs between original 2.8 m MS and Pan images degraded to the same pixel size.*

	B1	B2	B3	B4	Pan
B1	1.0	0.950	0.823	0.060	0.505
B2	0.950	1.0	0.911	0.182	0.625
B3	0.823	0.911	1.0	0.192	0.650
B4	0.060	0.182	0.192	1.0	0.798
Pan	0.505	0.625	0.650	0.798	1.0

Table 2.5 *Ikonos: CCs between original 4 m MS and Pan images degraded to the same pixel size.*

	B1	B2	B3	B4	Pan
B1	1.0	0.959	0.903	0.614	0.784
B2	0.959	1.0	0.965	0.662	0.827
B3	0.903	0.965	1.0	0.701	0.856
B4	0.614	0.662	0.701	1.0	0.854
Pan	0.784	0.827	0.856	0.854	1.0

A thorough performance comparison is reported for a set of state-of-the-art MRA image fusion methods, which are listed in the following. The verification of the first property of Wald's protocol has not been explicitly reported, since it is directly verified by all the considered MRA methods because of their operative modality.

As expected from the previous sections, ELP and ATWT decompositions featuring a reduction filter designed to match the MTF of the imaging devices produced the best results. Results of ELP- and ATWT-based decompositions were practically the same and

Table 2.6 *QuickBird: mean bias* $(\Delta\mu)$ *between 2.8 m originals and fused MS bands obtained from 11.2 m MS through 4:1 fusion with 2.8 m Pan and from 5.6 m MS through 2:1 fusion with 2.8 m Pan.*

4:1	EXP	AWL	SDM	RWM	CBD	GS	HPF
B1	0.002	−0.002	−0.234	−0.080	−0.027	0.002	−0.005
B2	0.007	−0.002	−0.310	−0.127	−0.058	0.007	−0.001
B3	0.011	−0.007	−0.134	−0.118	−0.117	0.011	−0.004
B4	0.027	−0.025	−0.097	−0.024	−0.204	0.028	−0.020
Avg.	0.012	−0.009	−0.194	−0.087	−0.101	0.012	−0.007
2:1	EXP	AWL	SDM	RWM	CBD	GS	HPF
B1	0.000	−0.002	−0.197	−0.028	−0.000	0.000	−0.005
B2	0.001	−0.000	−0.245	−0.046	−0.002	0.001	−0.003
B3	0.001	−0.000	−0.122	−0.089	0.006	0.001	−0.003
B4	0.007	0.005	−0.103	−0.013	0.001	0.007	0.002
Avg.	0.002	−0.001	−0.167	−0.044	−0.001	0.002	−0.002

very little differences existed between their fused products. The methods that have been compared are:

- Multiresolution IHS by Núñez et al. [21] with additive model (AWL), based on ATWT;
- ELP-based method with spectral distortion minimising (SDM) model [50];
- ATWT-based method with Ranchin–Wald–Mangolini (RWM) model [28];
- ELP-based method with context-based decision (CBD) model [40];
- Gram–Schmidt spectral sharpening (GS) method [9], as implemented in [52], 'mode 2';
- High-Pass Filtering (HPF) [7]: 3×3 and 5×5 box filters for 2:1 and 4:1 fusion.

GS spectral sharpening in '*mode 2*' has been selected because of its availability in the ENVI package and because its injection strategy is similar to MRA-based methods. Actually, details are obtained as the difference between the original Pan image and a low resolution Pan provided by the user and expanded by the algorithm to the full resolution. Spatial details are weighted by an injection model that is derived by Gram–Schmidt decomposition.

Also the case in which the MS data are simply resampled (through the 23-tap filter) and no details are added is presented (under the label EXP) to discuss the behaviour of the different fusion methods.

2.6.2 Performance comparison on QuickBird data

Mean bias $(\Delta\mu)$ and Root Mean Squared Error (RMSE) between each pair of fused and original MS bands are reported in Tables 2.6 and 2.7, respectively. Mean biases are practically zero for all entries, as expected, since all methods actually perform injection of zero-mean edges and textures into the resampled MS data. What immediately stands out by looking at the values in Table 2.7 is that, for 2:1 fusion, only few methods are capable of providing average RMSE values lower than those obtained by plain resampling of

Table 2.7 *QuickBird: RMSE between original 2.8 m MS bands and fusion products obtained from 11.2 m MS through 4:1 fusion with 2.8 m Pan and from 5.6 m MS through 2:1 fusion with 2.8 m Pan. Avg. = $\sqrt{Avg.\ MSE}$.*

4:1	EXP	AWL	SDM	RWM	CBD	GS	HPF
B1	7.31	15.85	25.51	9.78	7.45	6.35	31.97
B2	17.51	15.74	31.03	19.98	16.48	14.05	27.57
B3	25.12	20.74	20.06	29.66	23.70	19.55	26.24
B4	50.83	44.85	32.76	40.12	36.26	36.66	34.12
Avg.	29.89	27.11	27.79	27.31	23.47	22.16	30.15
2:1	EXP	AWL	SDM	RWM	CBD	GS	HPF
B1	4.57	13.54	14.95	4.92	4.71	4.24	22.89
B2	10.47	12.81	19.08	10.33	10.34	9.26	21.05
B3	13.93	14.44	13.39	13.65	14.21	12.47	20.47
B4	29.07	28.86	22.80	25.01	25.18	24.83	24.13
Avg.	17.10	18.63	17.94	15.35	15.53	14.80	22.18

Table 2.8 *QuickBird: CC between 2.8 m originals and fused MS bands obtained from 11.2 m MS through 4:1 fusion with 2.8 m Pan and from 5.6 m MS through 2:1 fusion with 2.8 m Pan.*

4:1	EXP	AWL	SDM	RWM	CBD	GS	HPF
B1	0.860	0.706	0.598	0.794	0.860	0.899	0.588
B2	0.843	0.876	0.771	0.829	0.870	0.907	0.812
B3	0.852	0.904	0.912	0.837	0.878	0.919	0.833
B4	0.819	0.879	0.929	0.898	0.912	0.921	0.922
Avg.	0.843	0.841	0.803	0.840	0.880	0.911	0.789
2:1	EXP	AWL	SDM	RWM	CBD	GS	HPF
B1	0.947	0.761	0.713	0.939	0.944	0.955	0.655
B2	0.946	0.923	0.866	0.948	0.958	0.959	0.874
B3	0.957	0.953	0.961	0.958	0.955	0.966	0.931
B4	0.944	0.950	0.966	0.959	0.958	0.960	0.964
Avg.	0.948	0.897	0.877	0.951	0.954	0.960	0.856

MS bands (EXP). In this sense, the best algorithms are the Gram–Schmidt (GS) spectral sharpening and ELP featuring the CBD model (CBD). The multiresolution IHS (AWL) and the RWM model, both exploiting ATWT, are comparable with plain resampling, even if slightly superior. The ELP-SDM algorithm, devised to optimise the spectral error with respect to the resampled MS bands, and especially HPF, yields larger average errors, mostly concentrated in the blue band. The main reason for that is the absence of a space-varying model (IBSM) varying with bands: all details are injected, thereby leading to over-enhancement. We also can see that RMSE is directly related to the square root of variance of the original band and inversely related to the correlation with Pan. Two things can be noticed: RMSE is approximately reduced by a factor equal to $\sqrt{2}$ when the scale ratio is halved, i.e. when passing from 4:1 to 2:1 fusion; for the resampled MS bands (EXP), it holds that $\sigma_O^2(B_l) \approx \sigma_D^2(B_l) + \text{RMSE}^2(B_l)$ (see Table 2.2).

Table 2.9 *QuickBird: average cumulative quality/distortion indexes between original 2.8 m MS bands and fusion products obtained from 11.2 m MS through 4:1 fusion with 2.8 m Pan and from 5.6 m MS through 2:1 fusion with 2.8 m Pan.*

4:1	EXP	AWL	SDM	RWM	CBD	GS	HPF
Q4	0.750	0.827	0.864	0.865	0.881	0.880	0.857
SAM (deg.)	2.14°	2.59°	2.14°	2.07°	1.85°	1.80°	2.33°
ERGAS	1.760	1.611	1.676	1.694	1.430	1.316	1.904
2:1	EXP	AWL	SDM	RWM	CBD	GS	HPF
Q4	0.930	0.924	0.932	0.948	0.947	0.949	0.914
SAM (deg.)	1.20°	1.32°	1.20°	1.15°	1.14°	1.13°	1.35°
ERGAS	2.005	2.237	2.152	1.828	1.858	1.745	2.815

Table 2.8 reports CCs between fused products and reference original MS bands. From a general view, the blue wavelengths band (B1) is the most difficult to synthesise, while the NIR band (B4) is the easiest. The simple explanation is that the enhancing Pan is weakly correlated with B1 and strongly with B4 (see Table 2.4). For 4:1 fusion, GS and CBD yield average CCs greater than that of the resampled bands (EXP). For 2:1 fusion, also RWM outperforms EXP.

Practically, all schemes can adequately enhance B4, but only GS and CBD provide acceptable enhancement of B1 for 4:1 fusion (also RWM for 2:1). The best average scores in terms of CCs between true and fused products follow similar trends as for RMSE, GS now being the best for 4:1, and HPF being the poorest. Obviously, the correlation loss of the resampled data is proportional to the amount of detail possessed by each band.

Although CC measurements between fused MS and reference originals may be valid detectors of fusion artifacts and especially of possible misalignments, the parameters in Table 2.9 measuring the global distortion of pixel vectors, either radiometric (ERGAS) or spectral (SAM), and both radiometric and spectral (Q4), will give a more comprehensive measure of quality.

GS attains global scores better than those of the other methods. The SAM attained by CBD and RWM is lower than the one of SDM (identical to that of resampled MS data). This means that the two space-varying injection models are capable of *unmixing* the coarse MS pixels using the Pan data, even if to a small extent. The ranking of methods confirms that HPF is spectrally better than AWL (lower SAM), but radiometrically poorer (higher ERGAS). The novel index Q4 [51] trades off both types of distortion and yields a unique quality index, according to which HPF is slightly better than AWL.

The case of 2:1 fusion points out the favourable performance of RWM, which is comparable with CBD, while it was slightly poorer for the 4:1 case (not surprisingly, because it was originally conceived for 2:1 fusion). The inter-scale comparison highlights that Q4, representing the fidelity to the reference original, and SAM, measuring the average spectral distortion, are perfectly in trend. Instead, ERGAS yields higher values for the 2:1 fusion case. The factor d_h/d_l in (2.23) was presumably introduced to obtain invariance with the scale ratio, such that the value of ERGAS may be thresholded to decide whether fusion products are satisfactory or not. However, given the trend of RMSE with the scale ratio, shown in Table 2.7, the factor d_h/d_l should be taken inside the square root to yield values of ERGAS almost independent of the scale ratio, for a given test image.

Only visual results of 4:1 fusion, carried out on 1:4 degraded images, are presented in this chapter. Actually, in the 2:1 case the extent of enhancement is hardly perceivable. Figure 2.14 is constituted of eight 64×64 tiles representing the 11.2 m MS expanded to 2.8 m scale (a), and its spatially enhanced fusion products achieved by means of SDM, RWM, CBD, HPF, GS, and AWL (b–g). The original 2.8 m MS is also included for visual reference in Figure 2.14(h). The icons are shown in R-G-B true-colour composites. True-colour visualisation has been deliberately chosen, because Pan-sharpening of MS bands falling partly outside the bandwidth of Pan, as in the case of the blue band B1, is particularly critical [8]. The displayed area is one fourth of the 512×512 area in which performance scores have been calculated.

All fused images are more similar to the reference original (Figure 2.14(h)) than to the expanded MS (Figure 2.14(a)). HPF (Figure 2.14(e)) shows marked spectral and geometric distortions, as well as over-enhancement. Since HPF implicitly relies on undecimated MRA, the same as the other schemes, its poor performance is mainly due to the *boxcar* filter having little frequency selection, as well as to the absence of an IBSM. The RWM model (Figure 2.14(c)) and the CBD model (Figure 2.14(d)) try to recover spectral signatures by unmixing pixels. However, artifacts originated by statistical instabilities of the model give a grainy appearance to the former. The most visible effect of the SDM model is that the colour hues of the MS image are exactly transferred to the fused image in Figure 2.14(b). AWL is spatially rich, but spectrally poorer than SDM, notwithstanding the low spatial frequency components of the Pan image do not affect the fusion, unlike conventional IHS approaches [5,6]. GS yields a result apparently different from the other methods, yet similar to the reference.

Eventually, Figure 2.15 shows the results of CBD, GS, and SDM on 0.7 m fusion products, as it happens in practice. Figure 2.15 displays the resampled 2.8 m MS bands and the enhanced bands at 0.7 m. A visual inspection highlights that all the spectral signatures of the original MS data are carefully incorporated in the sharpened bands. SDM is geometrically rich and detailed, but over-enhanced. CBD and GS products are visually superior and substantially similar to each other. A basic consideration is that aliasing impairments are already present in the original MS data (see the colour stripes along the main road in the upper left part). Unfortunately, they are carefully preserved by all fusion methods. This drawback motivates the decision of using another test set of images taken from a different sensor, but again of an urban landscape.

2.6.3 Performance comparison on Ikonos data

The experiments carried out on the Ikonos data set are aimed at validating the examined MRA algorithms and at revealing differences between the two satellite sensors that might contribute to further refinement of the fusion methods and the performance indexes utilised in this work. For sake of conciseness, band-to-band parameters have been

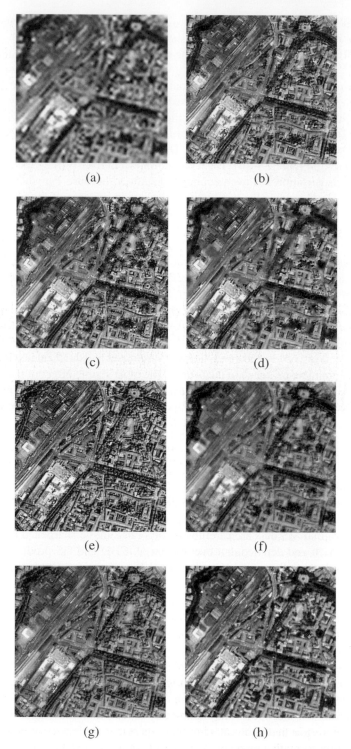

Figure 2.14   *Detail of fused QuickBird MS image (256 × 256). (a) Resampled 11.2 m MS; (b) SDM fusion; (c) RWM fusion; (d) CBD fusion; (e) HPF fusion; (f) GS fusion; (g) AWL fusion; (h) true 2.8 m MS.*

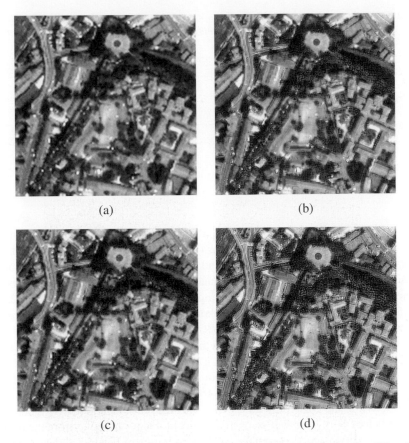

(a)

(b)

(c)

(d)

Figure 2.15 *Examples of full-scale spatial enhancement of fusion algorithms displayed as 512 × 512 true colour compositions at 0.7 m pixels spacing for the QuickBird image. (a) Original MS bands (2.8 m) re-sampled to the scale of Pan image (0.7 m); (b) ELP-CBD fusion product; (c) Gram–Schmidt fusion product; (d) ELP-SDM fusion product.*

calculated but are not reported here. Mean biases are negligible and perfectly in trend with those calculated on the QuickBird data set. Also RMSEs are slightly larger on an average, but similar. The same changes hold for CCs, which are slightly lower, but still comparable.

The global indexes Q4, SAM, and ERGAS are reported in Table 2.10 for 4:1 and 2:1 fusion carried out on spatially degraded data. The values of Q4 are slightly lower than those measured on the fused QuickBird images, as expected, since band-to-band scores are less favourable as well. The average SAM is almost doubled in value, thereby reveal-ing that the extremely detailed urban area imaged with a slightly lower resolution exhibits a considerably higher spectral information, that can be recovered only in small part by the fusion process. Indeed, the aliasing artifacts appearing in the QuickBird MS image, and totally missing in the Ikonos MS image, suggest that the former is more spatially but less spectrally informative than the latter. This idea is sustained by the scores achieved in the case of plain resampling (EXP). The numerical values of ERGAS are somewhat in

Table 2.10 *Ikonos: average cumulative quality/distortion indexes between original 4 m MS bands and fusion products obtained from 16 m MS through 4:1 fusion with 4 m Pan and from 8 m MS through 2:1 fusion with 4 m Pan.*

4:1	EXP	AWL	SDM	RWM	CBD	GS	HPF
Q4	0.608	0.846	0.867	0.886	0.880	0.839	0.824
SAM (deg.)	4.85°	4.82°	4.85°	3.52°	3.60°	3.99°	4.46°
ERGAS	5.936	4.214	4.273	3.691	3.740	4.066	4.568
2:1	EXP	AWL	SDM	RWM	CBD	GS	HPF
Q4	0.875	0.914	0.907	0.928	0.927	0.926	0.912
SAM (deg.)	3.07°	3.33°	3.07°	2.69°	2.68°	2.66°	2.85°
ERGAS	7.460	6.426	6.505	5.894	5.974	5.795	6.989

trend with Q4 and SAM values for all the algorithms, also across scales. Q4 and SAM values approach their *best* value when the fusion scale ratio changes from 4:1 to 2:1. Conversely, ERGAS shows an opposite trend due to the d_h/d_l normalisation factor. If the normalisation factor is inserted into the square root of (2.23), the ERGAS values would be almost scale-independent. However, compared with those reported in Table 2.10, the ERGAS values are abnormally higher for Ikonos than for QuickBird. The obvious explanation is that ERGAS is proportional to the RMSE value across bands normalised to band means. Given the mean radiances of Ikonos that are two to three times lower than those of QuickBird, as shown in Tables 2.2 and 2.3, it is not surprising that in practice the interval of ERGAS values is strictly sensor-dependent. No conclusions about which values are suitable to guarantee the quality of fusion products can be inferred from one MS scanner to another. While the change in values with the scale ratio can be easily modelled and hence predicted, there is no way to overcome the problem stemming from the difference in mean value across sensors, given the intrinsic definition of ERGAS. Conversely, the Q4 index is steady between the two data sets. The values of Q4 reveal the good performance of RWM, which is comparable with CBD method on the Ikonos data set.

The numerical values in Table 2.10 are confirmed by the visual analysis of the fusion results. Figure 2.16 is constituted of eight 64×64 tiles representing the 16 m MS expanded to 4 m scale (a), and its spatially enhanced fusion products achieved by means of the same methods considered for QuickBird (b–g). The original 4 m MS is displayed for reference in Figure 2.16(h). Despite the lower performance scores, the quality of Ikonos fusion products seems better than that of the QuickBird data. The image produced by RWM lacks the artifacts appearing in the QuickBird data set (perhaps statistical instabilities are due to aliasing). Hence, RWM is now comparable with CBD. The visual (and numerical) performance of GS is somewhat poorer than that achieved on QuickBird.

Eventually, CBD is compared with GS and with SDM on full resolution fusion products (1 m). Figure 2.17 displays the resampled 4 m MS bands and the spatially enhanced bands at 1 m. A visual inspection reveals that the quality of Pan-sharpened Ikonos products is better than that of the corresponding QuickBird products. No spectral distortion with respect to the resampled original MS data can be noticed. SDM is still over-textured. The CBD and GS products are visually superior and similar to each other, notwithstanding GS was poorer than CBD in the simulation on degraded data.

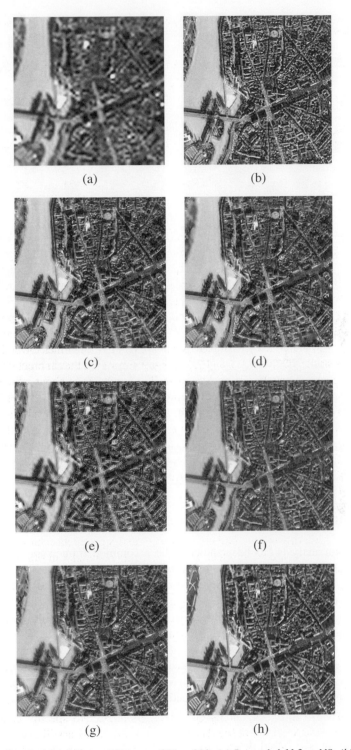

(a) (b)

(c) (d)

(e) (f)

(g) (h)

Figure 2.16 *Detail of fused Ikonos MS image (256 × 256). (a) Resampled 11.2 m MS; (b) SDM fusion; (c) RWM fusion; (d) CBD fusion; (e) HPF fusion; (f) GS fusion; (g) AWL fusion; (h) true 2.8 m MS.*

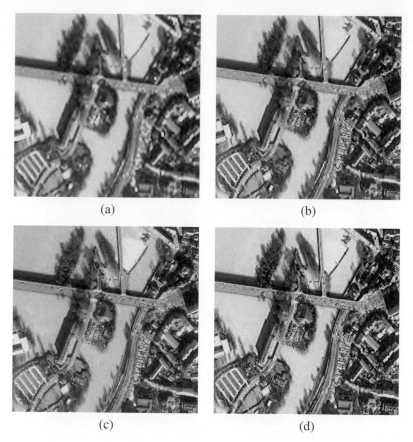

Figure 2.17 *Ikonos: full-scale spatial fusion; images are displayed as 512 × 512 true colour compositions at 1 m pixels spacing. (a) Original MS bands (4 m) resampled to 1 m; (b) CBD fusion; (c) Gram–Schmidt fusion; (d) SDM fusion.*

2.7 Concluding remarks

Multiresolution analysis has been shown to be a common framework for fusion algorithms. The absence of detail decimation featured by non-critically subsampled schemes is the key to avoid artifacts and spatial distortions in general. Actually, the translation-invariance property inherited by the missing decimation step is invaluable in practical cases concerning different sensors, since possible misregistrations of the data may be emphasised if the transformation achieving the multiresolution analysis is not shift-invariant.

Non-critically subsampled decompositions such as ELP and ATW can adopt a reduction filter matched onto the system MTF of each MS band. Spatial frequency content of fused images is thus enhanced by restoring the spatial frequencies that had been damped by the MTF of the MS sensor.

Thanks to the model-based detail selection and the space-varying sensor equalisation, multiresolution methods achieve impressive scores. As a peculiarity common to most of multiresolution schemes, spectral distortions, regarded as changes in colour hues of the composite image, never occur in any of the fused images. Ringing artifacts are completely missing for ELP and ATW. The most notable benefit of the context-based injection strategy is that spectral signatures of small size (2 or 3 pixels) may be restored [53], even though they appear to be heavily smeared in the expanded XS image.

As a conclusion, ELP and ATW can be considered the most flexible and efficient MRA decomposition schemes for image fusion, since they are translation invariant and can easily take into account the MTF of the acquisition system and an effective modelling of spatial detail injection. They can also fit any (reasonable) scaling requirements and are practically equivalent in performance. Besides the lower computational effort, ELP and ATW do not require advanced signal processing expertises to set up the filter bank when the scale ratios are not powers of two, or even non-integer.

Acknowledgements

The authors wish to thank Dr. Luciano Alparone, Dr. Andrea Garzelli, and Dr. Filippo Nencini for their valuable contribution and stimulating discussions.

References

[1] P.S. Chavez Jr., 'Digital merging of Landsat TM and digitised NIIAP data for 1:24,000 scale image mapping', *Photogrammetric Engineering and Remote Sensing*, Vol. 52, No. 10, 1986, pp. 1637–1646.

[2] R. Welch and M. Ehlers, 'Merging multiresolution SPOT HRV and Landsat TM data', *Photogrammetric Engineering and Remote Sensing*, Vol. 53, No. 3, 1987, pp. 301–303.

[3] J.C. Price, 'Combining panchromatic and multispectral imagery from dual resolution satellite instruments', *Remote Sensing of Environment*, Vol. 21, No. 2, 1987, pp. 119–128.

[4] W. Carper, T. Lillesand and R. Kiefer, 'The use of Intensity–Hue–Saturation transformations for merging SPOT panchromatic and multispectral image data', *Photogrammetric Engineering and Remote Sensing*, Vol. 56, No. 4, 1990, pp. 459–467.

[5] T.-M. Tu, S.-C. Su, H.-C. Shyu and P.S. Huang, 'A new look at IHS-like image fusion methods', *Information Fusion*, Vol. 2, No. 3, 2001, pp. 177–186.

[6] T.-M. Tu, P.S. Huang, C.-L. Hung and C.-P. Chang, 'A fast Intensity–Hue–Saturation fusion technique with spectral adjustment for IKONOS imagery', *IEEE Geoscience and Remote Sensing Letters*, Vol. 1, No. 4, 2004, pp. 309–312.

[7] P.S. Chavez Jr., S.C. Sides and J.A. Anderson, 'Comparison of three different methods to merge multiresolution and multispectral data: Landsat TM and SPOT panchromatic', *Photogrammetric Engineering and Remote Sensing*, Vol. 57, No. 3, 1991, pp. 295–303.

[8] Y. Zhang, 'Understanding image fusion', *Photogrammetric Engineering and Remote Sensing*, Vol. 70, No. 6, 2004, pp. 657–661.

[9] C.A. Laben and B.V. Brower, 'Process for enhancing the spatial resolution of multispectral imagery using pan-sharpening', US Patent #6,011,875, Eastman Kodak Company, 2000.

[10] L. Wald, T. Ranchin and M. Mangolini, 'Fusion of satellite images of different spatial resolutions: Assessing the quality of resulting images', *Photogrammetric Engineering and Remote Sensing*, Vol. 63, No. 6, 1997, pp. 691–699.

[11] B. Garguet-Duport, J. Girel, J.-M. Chassery and G. Pautou, 'The use of multiresolution analysis and wavelet transform for merging SPOT panchromatic and multispectral image data', *Photogrammetric Engineering and Remote Sensing*, Vol. 62, No. 9, 1996, pp. 1057–1066.

[12] D.A. Yocky, 'Multiresolution wavelet decomposition image merger of Landsat Thematic Mapper and SPOT panchromatic data', *Photogrammetric Engineering and Remote Sensing*, Vol. 62, No. 9, 1996, pp. 1067–1074.

[13] B. Aiazzi, L. Alparone, F. Argenti, S. Baronti and I. Pippi, 'Multisensor image fusion by frequency spectrum substitution: Subband and multirate approaches for a 3:5 scale ratio case', in *Proc. IEEE Int. Geoscience and Remote Sensing Symposium*, 2000, pp. 2629–2631.

[14] T.A. Wilson, S.K. Rogers and M. Kabrisky, 'Perceptual-based image fusion for hyperspectral data', *IEEE Transactions on Geoscience and Remote Sensing*, Vol. 35, No. 4, 1997, pp. 1007–1017.

[15] L. Alparone, V. Cappellini, L. Mortelli, B. Aiazzi, S. Baronti and R. Carlà, 'A pyramid-based approach to multisensor image data fusion with preservation of spectral signatures', in P. Gudmandsen (ed.), *Future Trends in Remote Sensing*, Balkema, Rotterdam, 1998, pp. 418–426.

[16] F. Argenti and L. Alparone, 'Filterbanks design for multisensor data fusion', *IEEE Signal Processing Letters*, Vol. 7, No. 5, 2000, pp. 100–103.

[17] J. Zhou, D.L. Civco and J.A. Silander, 'A wavelet transform method to merge Landsat TM and SPOT panchromatic data', *International Journal of Remote Sensing*, Vol. 19, No. 4, 1998, pp. 743–757.

[18] T. Ranchin and L. Wald, 'Fusion of high spatial and spectral resolution images: The ARSIS concept and its implementation', *Photogrammetric Engineering and Remote Sensing*, Vol. 66, No. 1, 2000, pp. 49–61.

[19] P. Scheunders and S.D. Backer, 'Fusion and merging of multispectral images with use of multiscale fundamental forms', *Journal of the Optical Society of America A*, Vol. 18, No. 10, 2001, pp. 2468–2477.

[20] H. Li, B.S. Manjunath and S.K. Mitra, 'Multisensor image fusion using the wavelet transform', *Graphical Models and Image Processing*, Vol. 57, No. 3, 1995, pp. 235–245.

[21] J. Núñez, X. Otazu, O. Fors, A. Prades, V. Palà and R. Arbiol, 'Multiresolution-based image fusion with additive wavelet decomposition', *IEEE Transactions on Geoscience and Remote Sensing*, Vol. 37, No. 3, 1999, pp. 1204–1211.

[22] Y. Chibani and A. Houacine, 'Model for multispectral and panchromatic image fusion', in S.B. Serpico (ed.), *Image and Signal Processing for Remote Sensing VI*, in *Proc. SPIE, EUROPTO Series*, Vol. 4170, 2000, pp. 238–244.

[23] A. Garzelli, G. Benelli, M. Barni and C. Magini, 'Improving wavelet-based merging of panchromatic and multispectral images by contextual information', in S.B. Serpico (ed.), *Image and Signal Processing for Remote Sensing VI*, in *Proc. SPIE, EUROPTO Series*, Vol. 4170, 2000, pp. 82–91.

[24] P.P. Vaidyanathan, *Multirate Systems and Filter Banks*, Prentice Hall, Englewood Cliffs, NJ, 1992.

[25] B. Aiazzi, L. Alparone, S. Baronti, V. Cappellini, R. Carlà and L. Mortelli, 'A Laplacian pyramid with rational scale factor for multisensor image data fusion', in *Proc. Int. Conf. on Sampling Theory and Applications – SampTA 97*, 1997, pp. 55–60.

[26] B. Aiazzi, L. Alparone, A. Barducci, S. Baronti and I. Pippi, 'Multispectral fusion of multisensor image data by the generalized Laplacian pyramid', in *Proc. IEEE Int. Geoscience and Remote Sensing Symposium*, 1999, pp. 1183–1185.

[27] B. Aiazzi, L. Alparone, S. Baronti and I. Pippi, 'Fusion of 18 m MOMS-2P and 30 m Landsat TM multispectral data by the generalized Laplacian pyramid', *ISPRS International Archives of Photogrammetry and Remote Sensing*, Vol. 32, No. 7-4-3W6, 1999, pp. 116–122.

[28] T. Ranchin, B. Aiazzi, L. Alparone, S. Baronti and L. Wald, 'Image fusion – the ARSIS concept and some successful implementation schemes', *ISPRS Journal of Photogrammetry and Remote Sensing*, Vol. 58, No. 1–2, 2003, pp. 4–18.

[29] B. Aiazzi, L. Alparone, S. Baronti, A. Garzelli and M. Selva, 'MTF-tailored multiscale fusion of high-resolution MS and Pan imagery', *Photogrammetric Engineering and Remote Sensing*, Vol. 72, No. 5, 2006, pp. 591–596.

[30] P. Blanc, T. Blu, T. Ranchin, L. Wald and R. Aloisi, 'Using iterated rational filter banks within the ARSIS concept for producing 10m Landsat multispectral images', *International Journal of Remote Sensing*, Vol. 19, No. 12, 1998, pp. 2331–2343.

[31] I. Daubechies, *Ten Lectures on Wavelets*, CBMS-NSF Regional Conference Series in Applied Mathematics, Vol. 61, SIAM, Philadelphia, PA, 1992.

[32] S. Mallat, 'A theory for multiresolution signal decomposition: The wavelet representation', *IEEE Transactions on Pattern Analysis and Machine Intelligence*, Vol. PAMI-11, No. 7, 1989, pp. 674–693.

[33] I. Daubechies, 'Orthonormal bases of compactly supported wavelets', *Communications on Pure and Applied Mathematics*, Vol. 41, 1988, pp. 909–996.

[34] A. Cohen, I. Daubechies and J.C. Feauveau, 'Biorthogonal bases of compactly supported wavelets', *Communications on Pure and Applied Mathematic*, Vol. 45, 1995, pp. 485–500.

[35] G.P. Nason and B.W. Silverman, 'The stationary wavelet transform and some statistical applications', in A. Antoniadis and G. Oppenheim (eds.), *Wavelets and Statistics*, in *Lecture Notes in Statistics*, Vol. 103, Springer-Verlag, New York, 1995, pp. 281–299.

[36] M.J. Shensa, 'The discrete wavelet transform: Wedding the à trous and Mallat algorithm', *IEEE Transactions on Signal Processing*, Vol. 40, No. 10, 1992, pp. 2464–2482.

[37] P.J. Burt and E.H. Adelson, 'The Laplacian pyramid as a compact image code', *IEEE Transactions on Communications*, Vol. 31, No. 4, 1983, pp. 532–540.

[38] S. Baronti, A. Casini, F. Lotti and L. Alparone, 'Content-driven differential encoding of an enhanced image pyramid', *Signal Processing: Image Communication*, Vol. 6, No. 5, 1994, pp. 463–469.

[39] B. Aiazzi, L. Alparone, S. Baronti and F. Lotti, 'Lossless image compression by quantization feedback in a content-driven enhanced Laplacian pyramid', *IEEE Transactions on Image Processing*, Vol. 6, No. 6, 1997, pp. 831–843.

[40] B. Aiazzi, L. Alparone, S. Baronti and A. Garzelli, 'Context-driven fusion of high spatial and spectral resolution data based on oversampled multiresolution analysis', *IEEE Transactions on Geoscience and Remote Sensing*, Vol. 40, No. 10, 2002, pp. 2300–2312.

[41] A. Garzelli and F. Nencini, 'Interband structure modeling for Pan-sharpening of very high resolution multispectral images', *Information Fusion*, Vol. 6, No. 3, 2005, pp. 213–224.

[42] M. Gonzáles Audícana, J.L. Saleta, R. García Catalán and R. García, 'Fusion of multispectral and panchromatic images using improved IHS and PCA mergers based on wavelet decomposition', *IEEE Transactions on Geoscience and Remote Sensing*, Vol. 42, No. 6, 2004, pp. 1291–1299.

[43] R. Nishii, S. Kusanobu and S. Tanaka, 'Enhancement of low spatial resolution image based on high resolution bands', *IEEE Transactions on Geoscience and Remote Sensing*, Vol. 34, No. 5, 1996, pp. 1151–1158.

[44] B. Zhukov, D. Oertel, F. Lanzl and G. Reinhäckel, 'Unmixing-based multisensor multiresolution image fusion', *IEEE Transactions on Geoscience and Remote Sensing*, Vol. 37, No. 3, 1999, pp. 1212–1226.

[45] M. Gonzáles Audícana, X. Otazu, O. Fors and J.A. Alvarez-Mozos, 'A low computational-cost method to fuse IKONOS images using the spectral response function of its sensors', *IEEE Transactions on Geoscience and Remote Sensing*, Vol. 44, No. 6, 2006, pp. 1683–1691.

[46] J. Hill, C. Diemer, O. Stöver and T. Udelhoven, 'A local correlation approach for the fusion of remote sensing data with different spatial resolutions in forestry applications', *ISPRS International Archives of Photogrammetry and Remote Sensing*, Vol. 32, No. 7-4-3W6, 1999, pp. 167–174.

[47] L. Alparone, S. Baronti and A. Garzelli, 'Assessment of image fusion algorithms based on noncritically decimated pyramids and wavelets', in *Proc. IEEE Int. Geoscience and Remote Sensing Symposium*, 2001, pp. 852–854.

[48] A. Garzelli and F. Soldati, 'Context-driven image fusion of multispectral and panchromatic data based on a redundant wavelet representation', in *IEEE/ISPRS Joint Workshop on Remote Sensing and Data Fusion over Urban Areas*, 2001, pp. 122–126.

[49] B. Aiazzi, L. Alparone, S. Baronti and R. Carlà, 'A pyramid approach to fusion of Landsat TM and SPOT-PAN data to yield multispectral high-resolution images for environmental archaeology', in *Remote Sensing for Geography, Geology, Land Planning, and Cultural Heritage*, in *Proc. SPIE, EUROPTO Series*, Vol. 2960, 1996, pp. 153–162.

[50] B. Aiazzi, L. Alparone, S. Baronti, A. Garzelli and M. Selva, 'Pan-sharpening of quickbird multispectral images with spectral distortion minimization', in R. Goossens (ed.), *Remote Sensing in Transition, Proc. EARSeL*, 2004, pp. 229–235.

[51] L. Alparone, S. Baronti, A. Garzelli and F. Nencini, 'A global quality measurement of Pan-sharpened multispectral imagery', *IEEE Geoscience and Remote Sensing Letters*, Vol. 1, No. 4, 2004, pp. 313–317.

[52] ENVI®, Version 4.1 User Manual, Research System Inc., 2004.

[53] B. Aiazzi, L. Alparone, S. Baronti and I. Pippi, 'Quality assessment of decision-driven pyramid-based fusion of high resolution multispectral with panchromatic image data', in *IEEE/ISPRS Joint Workshop on Remote Sensing and Data Fusion over Urban Areas*, 2001, pp. 337–341.

3

Multisensor and multiresolution image fusion using the linear mixing model

Jan G.P.W. Clevers and Raul Zurita-Milla

Wageningen University, Centre for Geo-Information, Wageningen, The Netherlands

In satellite sensor design we observe a trade-off between sensors with a high spatial resolution having only a few spectral bands and a low revisit frequency on the one hand, and sensors with a medium to low spatial resolution having many spectral bands and a high revisit time on the other hand. Many applications require a combination of a high spatial, spectral, and temporal resolution. In this chapter image fusion of a high spatial resolution (Landsat Thematic Mapper) and a high spectral resolution (Envisat MERIS) image based on the linear mixing model is presented. This approach is also known as spatial unmixing or unmixing-based data fusion. An optimisation of the number of classes used to classify the high spatial resolution image and the size of the neighbourhood, for which the unmixing equations are solved, is presented. It is illustrated for a test area in the Netherlands. Results show the feasibility of this approach yielding fused images with the spatial resolution of the high resolution image and with the spectral information from the low spatial resolution image. The quality of the fused images is evaluated using the spectral and spatial ERGAS index. Main advantage of the presented technique based on the linear mixing model is that the fused images do not include the spectral information of the high spatial resolution image in the final result in any way.

3.1 Introduction

Terrestrial vegetation plays an important role in biochemical cycles, like the global carbon cycle. Information on the type of vegetation is important for such studies. This is thematic information referring to the biome type or the land cover type. In addition, once we have the thematic class information, we need quantitative information on vegetation properties.

This quantitative information refers to the so-called continuous fields [1]. In order to quantify the role of vegetation in biochemical cycles, variables describing the area of the green photosynthetic elements (mainly the leaves), the amount of photosynthetically active radiation (PAR) absorbed by the vegetation and the chlorophyll content of the leaves are important [2]. Key variables required for the estimation of, for instance, the primary production of vegetation are the leaf area index (LAI) and the fraction of PAR absorbed by vegetation (FAPAR) [3]. The LAI refers to the one-sided green leaf area per unit ground area. The FAPAR refers to the fraction of the incoming PAR absorbed. PAR (0.4–0.7 μm) is used as energy source in the photosynthesis process, whereby atmospheric CO_2 and H_2O from the soil are converted into carbohydrates, enabling growth of the vegetation.

The estimation of both LAI and FAPAR from spectral measurements also depends on other vegetation parameters, like leaf properties and canopy structure. Therefore, vegetation class-dependent relationships are often used. In general, to describe the relationship between spectral measurements and biophysical and chemical variables of vegetation both statistical and physical approaches have been used. As an example of statistical methods, numerous vegetation indices (VIs) have been developed for estimating variables like biomass and LAI for a range of vegetation types [4–9]. Physical-based methods often use radiative transfer models describing the interaction of radiation with the plant canopy based on physical principles. Subsequently, model inversion is used for estimating biophysical and chemical properties of the canopy [10–12]. Depending on the application, these techniques can be applied from local to global scales [13].

On the one hand, datasets derived from coarse resolution sensors provide information on land cover and vegetation properties globally. In the US, global land cover products have been derived using time series with 1 km data obtained from the National Oceanic and Atmospheric Administration's (NOAA) Advanced Very High Resolution Radiometer (AVHRR) [14,15]. In addition, a 1 km land cover map has been compiled using a time series of data from the Moderate Resolution Imaging Spectroradiometer (MODIS) on the Terra platform [16]. An example of continuous fields is the MODIS LAI product, where biome specific algorithms are used for obtaining global maps of LAI [2]. Photosynthetic activity of vegetation canopies can be shown by the fraction of APAR, which can be estimated from satellite observation [17]. Using time-series of satellite data it is shown, for instance, that the length of the growing season is increasing at the northern hemisphere, which may be caused by global warming [2]. This coarse scale imagery is limiting the use for monitoring purposes due to the finer scale at which most land cover changes take place [18]. Still these sub-kilometre scale changes are critical for monitoring changes in, e.g., sinks and sources of the carbon cycle.

On the other hand, many detailed studies at the regional and landscape scale at spatial resolutions between 10 and 30 m have been performed using the Thematic Mapper (TM) aboard the Landsat satellites [19], the multispectral imager (denoted by XS) aboard the SPOT satellites [20], or the Advanced Spaceborne Thermal Emission and Reflector Radiometer (ASTER) aboard the Terra platform [21]. The use of such data is usually not appropriate at the continental scale due to their limited spatial extent and low revisit time.

The gap between fine and coarse resolution sensors may be filled with imagery at spatial resolutions of 250 m using MODIS [22] or 300 m using the Medium Resolution Imaging Spectrometer (MERIS) [23]. MODIS and MERIS are spectrometers with a large number of spectral bands and high revisit time. However, for many applications one would like to have a spatial resolution better than 250 or 300 m as being provided by MODIS and MERIS, respectively. However, these high spatial resolution systems lack the large number of spectral bands and high revisit time.

Image fusion techniques thrive on combining the advantages of the high spatial resolution systems with those of the high temporal resolution systems (which inherently have a low to medium spatial resolution). Moreover, the latter systems at medium spatial resolution often have many spectral bands in addition to a high temporal resolution. As an example, MERIS has 15 spectral bands in the region between 400 and 1000 nm. This opens the way for deriving biophysical properties, like the leaf chlorophyll content, which cannot be derived from sensors like Landsat-TM or SPOT-XS. Clevers et al. [24] showed that the so-called red-edge index can be derived from the MERIS standard band setting. This red-edge index provides information on the leaf chlorophyll content, which cannot be derived from a combination of a near-infrared and visible broad spectral band. Concerning high spectral resolution data, this seems to be the major contribution of imaging spectrometry to applications in agriculture [25]. In this chapter we will focus on image fusion of TM and MERIS based on the linear mixing model.

3.2 Data fusion and remote sensing

The term data fusion groups all the methods that deal with the combination of data coming from different sources [26]. During recent years, data fusion has attracted a lot of attention from the remote sensing community because of the increasing need to integrate the vast amount of data being collected by Earth observation satellites. The main goal of such fusion techniques is to integrate various data sources by combining the best features of each of them. Fused images (i.e. images created by combining two or more types of data) generally offer increased interpretation capabilities and more reliable results [27].

Several data fusion methods have been described in literature. However, most of them are data type dependent [28]. For instance, most of the recent data fusion methods based on wavelet transformation [29] require that the ratio of the spatial resolutions of the images to be fused is a power of 2 [30] or they require that the images to be fused are in the same spectral domain [31]. Furthermore, most of the current data fusion methods do not properly preserve the spectral information of the input images because they are mainly concerned with the visual enhancement of the images.

In this study, we selected an unmixing-based data fusion approach. The aim of this method is to combine two images that have been acquired at a different spatial resolution to produce an image with the spatial information of the high spatial resolution image and the spectral information of the low spatial resolution image, whereby no spectral information of the high resolution image contributes to the fused image [32–34]. As a result one may fuse images acquired at different dates as well, making use of the high temporal resolution of the low resolution image. Additionally, it ensures a physical interpretation of the fused image and facilitates the retrieval of landscape properties using, for instance, radiative transfer models.

3.3 The linear mixing model

The spatial resolution of medium resolution sensors like MODIS and MERIS is such, that one pixel is mostly covering several land cover types. Thus, the value of a pixel is composed of the signals coming from the individual components. It is a so-called mixed pixel. Only large, homogeneous areas will result in pure pixels in the medium spatial resolution image.

The linear mixing model assumes that the spectrum of a mixed pixel is a linear combination of the pure spectra of the components present in that pixel weighted by their fractional coverage [35]. This linear mixing model is mathematically described in Equation (3.1):

$$p_i = \sum_{i=1}^{nc}(r_{ci} \cdot f_c) + e_i, \quad i = 1, \ldots, nb \tag{3.1}$$

p_i = reflectance (or radiance) of a mixed pixel in band i
r_{ci} = reflectance (or radiance) of endmember c in band i
f_c = fraction of endmember c
e_i = residual error in band i
nc = number of endmembers
nb = number of spectral bands

We can write this in matrix notation as follows:

$$\mathbf{P}_{(nb \times 1)} = \mathbf{R}_{(nb \times nc)} \cdot \mathbf{F}_{(nc \times 1)} + \mathbf{E}_{(nb \times 1)} \tag{3.2}$$

Despite this apparent simplicity, the linear mixing model is widely used by the remote sensing community because it offers an effective framework to analyse mixed pixels. Thus, if we have *a priori* knowledge about the components that might be present in a given scene, then we can apply linear spectral unmixing of the data to retrieve the sub-pixel proportions of these components [36,37]. The success of the model outcome relies on the quality of the *a priori* knowledge on the scene composition. In other words, the results of any linear spectral unmixing method heavily depend on a proper identification of the main components present in the scene and their pure signal. This identification might be very difficult when the image has been acquired over very heterogeneous landscapes or when we work with coarse resolution data because in these cases most of the pixels are mixed (i.e. no pure signal can be found in the scene). In addition to this, the number of components that can be unmixed is limited by the number of spectral bands of the image.

The underlying assumption of spectral unmixing using the linear mixing model is that knowledge of endmembers with known spectral profiles yields the fractions (abundances) of the endmembers within a pixel. However, in the current fusion approach we would like to know the spectral profiles of the endmembers within a pixel. This requires knowledge of the abundances of the endmembers within a pixel. If we can derive the abundances, for instance, from an image with a high spatial resolution, then we may derive the spectral profiles by solving the same Equation (3.1) or (3.2).

This application of the linear mixing model is known as spatial unmixing or unmixing-based data fusion. The aim of this kind of unmixing is to downscale the radiometric information of the low resolution image to the spatial resolution provided by the high spatial resolution image. Spatial unmixing does not require *a priori* knowledge of the main components present in the low spatial resolution scene because there is no need to identify their pure signals. In fact, these signals are the output of the spatial unmixing. Therefore, spatial unmixing can be applied to any pair of images, even if the low resolution image only has mixed pixels or a small number of spectral bands.

The unmixing-based data fusion approach consists of 4 main steps [32]:

(i) The high spatial resolution image is used to identify the different components of the scene. This is done by classifying the high spatial resolution image into *nc* unsupervised classes (the endmembers).

(ii) The proportions of each of these *nc* classes that fall within each low spatial resolution pixel are then computed.

(iii) Using these proportions and the spectral information provided by the low resolution sensor, the spectral behaviour of each *nc* class is unmixed. Here it is important to notice that in contrast to linear spectral unmixing, which is solved per pixel and for all bands at once, spatial unmixing is solved per pixel, per band, and for a given neighbourhood around the pixel that is being unmixed. This neighbourhood, represented by a square kernel of size *k* by *k*, is needed in order to get enough equations to solve the unmixing problem.

(iv) Finally, a fused image is reconstructed by joining all the low resolution pixels that have been spatially unmixed.

The third step of the unmixing-based data fusion approach can be written for a given band in a compact matrix notation as follows:

$$\mathbf{P}_{(k^2 \times 1)} = \mathbf{F}_{(k^2 \times nc)} \cdot \mathbf{R}_{(nc \times 1)} + \mathbf{E}_{(k^2 \times 1)} \qquad (3.3)$$

where $\mathbf{P}$ is a $(k^2 \times 1)$ vector that contains the values of the band i for all the low spatial resolution pixels present in the neighbourhood k, $\mathbf{F}$ is a $(k^2 \times nc)$ matrix containing the fractions of each high spatial resolution component inside each low resolution pixel present in k, $\mathbf{R}$ is the $(nc \times 1)$ unknown vector of downscaled values of the band i for each of the high spatial resolution components inside k, $\mathbf{E}$ is a $(k^2 \times 1)$ vector of residual errors.

This formulation of the unmixing-based data fusion indirectly implies that the number of classes used to classify the high resolution image (nc) and the size of the neighbourhood (k) need to be optimised. The first parameter, nc, needs to be optimised because it depends on the spectral variability present in the scene. Thus, heterogeneous scenes will require a larger nc value than homogeneous ones as they have more components.

The neighbourhood size, k, also needs to be optimised because it has great influence on the spectral quality of the fused image. On the one hand, k should be kept as small as possible so that the fused image is spectrally dynamic but, on the other hand, k should be as large as possible because it determines the number of equations that are available to solve the spatial unmixing of each pixel. In other words, Equation (3.3) is a system of k^2 equations (one equation per low resolution pixel in the neighbourhood) with up to nc unknowns (depending on the number of classes present in such a neighbourhood). This means that k^2 must be equal to or greater than the number of classes inside the neighbourhood (otherwise the system of equations is undetermined and no solution can be found). However, if we use a large k value, the output image will have low spectral variability because each system of equations results in a unique solution. For instance, if the size of the neighbourhood matches the size of the scene, then we will have one unique set of equations. This implies that a unique spectral response will be assigned to each of the nc classes identified with the high spatial resolution image independently of their position in the scene. This means that the fused image will not be spectrally dynamic and that each of the components will be represented by a kind of spectral mean response.

Minghelli-Roman et al. [33] applied this technique to a simulated MERIS image. For the high spatial resolution image (Landsat TM) an optimum number of 150 classes was obtained. Since the unmixing technique is applied to the whole medium resolution image, this results potentially in 150 different signatures. The resulting fused image will only be an approximation of what such a simulated sensor would measure, because all pixels belonging to one class have the same spectral profile. Therefore, a large number of classes are required.

However, instead of using an unsupervised classification of a high spatial resolution image like TM, we may also use another source of high spatial resolution, as long as it may serve as a source for defining the objects we are interested in. This may be topographical information where all object (e.g., field) boundaries are identified, such as a land cover data base. As an example, every couple of years an update of the Dutch land cover data base (LGN) is produced. This data base may be used instead of the high resolution image. Such a land cover data base typically will have a limited number of classes, mostly considerably less than 100. The LGN has a legend of 39 classes. Applying the above methodology of Minghelli-Roman et al. [33] would mean that all pixels belonging to one specific land cover class would get the same spectral profile, eliminating all variation within a class. Therefore, we propose a regionalised approach as an alternative procedure. So, although the approach of Minghelli-Roman et al. [33] is computationally fast, we prefer to optimise the size of the neighbourhood k so that we can account for the natural variability of the components present in the scene.

The spectral information of the high spatial resolution image or data set does not need to match that of the medium (or low) resolution image because the former spectral information is not directly used. The only requirement of the high resolution data is that it provides a mapping of all the classes of interest. Therefore, a temporal match is also not required as long as there are no significant changes in the landscape occurring.

3.4 Case study

3.4.1 Introduction

This section describes the unmixing-based data fusion of the TM and the MERIS full resolution images that were available over the study area. MERIS is one of the payload components of the European Space Agency's (ESA) environmental research satellite Envisat, launched in March 2002. MERIS is a 15 band imaging spectrometer. It is designed to acquire data at variable band width of 1.25 to 30 nm over the spectral range of 390–1040 nm [23]. Data are acquired at 300 m full resolution (FR) mode or 1200 m reduced resolution (RR) mode over land. In contrast, TM has only four spectral bands in this range of the spectrum. Moreover, the TM bands are broader than those of MERIS.

Figure 3.1 summarises the main steps of the data fusion process. First, the TM image was classified into 10, 20, 40, 60, and 80 classes using an unsupervised ISODATA classification rule. The aim of this classification was to identify, with different degrees of detail, the main components (spectral groups) of the scene. After this, a sliding window of $k \times k$ MERIS pixels was applied to each of the TM classified images to generate the corresponding class proportion matrices ($\mathbf{F}$). In this study, 14 neighbourhood sizes were tested: from $k = 5$ to $k = 53$ in steps of 4.

The sliding window used to generate the matrix of proportions $\mathbf{F}$ was also applied to all the MERIS bands to extract the k^2 elements that form the $\mathbf{P}$ vector. A constrained least-squares method was subsequently used to retrieve the per band MERIS downscaled radiances ($\mathbf{R}$). The use of a constrained method is justified because the solution should fulfil the following two conditions:

 (i) the radiance values must be positive, and
(ii) the radiance values cannot be larger than the MERIS radiance saturation value.

The next step of the unmixing-based data fusion consisted of replacing each of the TM unsupervised classes present in the central pixel of the sliding window by their corresponding spectral signal in the MERIS band that is being unmixed. By repeating this operation for all the MERIS FR pixels, for all MERIS bands and for all the possible combinations of nc and k, a series of fused images was generated.

3.4.2 Study area and data

The study area covers approximately 40 by 60 km of the central part of the Netherlands (52.19° N, 5.91° E). This area was selected considering both the heterogeneity of the landscape and the availability of cloud free high and medium spatial resolution satellite data acquired nearly simultaneously. A Landsat-5 TM image from 10 July 2003 and a MERIS full resolution level 1b image acquired 4 days later were available over this area.

The TM image was geo-referenced to the Dutch national coordinate system (RD) using a cubic convolution resampling method and a pixel size of 25 m. The MERIS full resolution

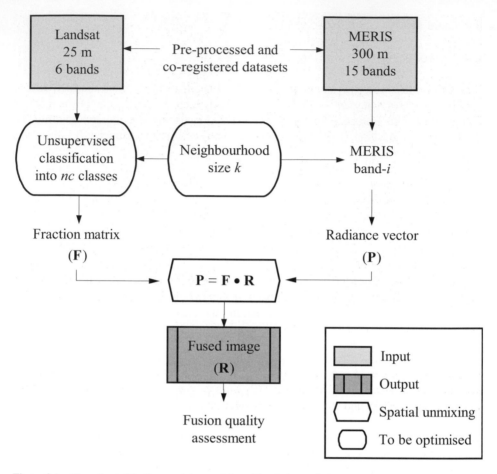

Figure 3.1 *Flow chart of the linear mixing model used for the image fusion.*

level 1b image was first transformed from digital numbers (DN) to top of atmosphere radiances (L_{TOA}) using the metadata provided with the file [38]. Then, the image was corrected for the so-called smile effect [39]. Finally, the image was re-projected into the RD coordinate system using the ground control points provided with the image file. After that, an image to image co-registration was performed using a nearest neighbour resampling method. Figure 3.2 shows the co-registered TM and MERIS FR images.

3.4.3 Quality assessment

A quantitative analysis of the quality of the fused images has to be performed in order to find out the combination of *nc* and *k* that produces the best image from a spectral point of view. In general, this kind of assessments are not straightforward because the quality of the fused images depends on many factors like the difference in spatial or spectral resolution of the input images or the type of landscape under consideration [40]. We selected the ERGAS index as a main quality indicator because it is independent of

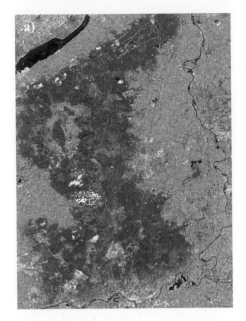

Figure 3.2 *RGB composite of the TM band 4-3-2 (a) and MERIS FR band 13-7-5 (b) over the study area.*

the units, the number of spectral bands, and the resolution of the input images [41]. The spectral and the spatial ERGAS indices [42] were used to assess the quality of the fused images at the MERIS and the TM spatial resolutions.

Bearing in mind that any fused image should be as identical as possible to the original low resolution image once degraded to that resolution, we degraded the fused images to 300 m using a mean filter. After this, we compared the degraded fused images with the original MERIS FR image. The spectral ERGAS index was used to perform such a comparison:

$$\text{ERGAS} = 100 \frac{h}{l} \sqrt{\frac{1}{N} \sum_{i=1}^{N} (\text{RMSE}_i^2 / M_i^2)} \qquad (3.4)$$

where h is the resolution of the high spatial resolution image (TM), l is the resolution of the low spatial resolution image (MERIS FR), N is the number of spectral bands involved in the fusion, RMSE_i is the root mean square error computed between the degraded fused image and the original MERIS image (for the band i), and M_i is the mean value of the band i of the reference image (MERIS FR).

The spatial ERGAS index was used to evaluate the quality of the original fused image (i.e. 15 spectral bands and 25 m pixel size). The expression of the spatial ERGAS index is basically the same as the one used to compute the spectral ERGAS (Equation (3.4)) except that (i) the RMSE_i is computed between the TM image and its spectrally corresponding band of the fused image, and (ii) the M_i is now the mean of band i of the TM image.

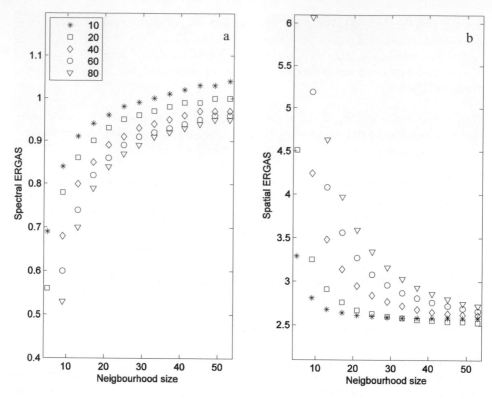

Figure 3.3   *The spectral (a) and spatial ERGAS (b) for all combinations of neighbourhood size and number of classes.*

In order to select the optimal number of classes (*nc*) and size of the neighbourhood (*k*), the average of the spectral and spatial ERGAS is used [42].

3.4.4 Results and discussion

First, the fused images were calculated for all combinations of 10, 20, 40, 60, and 80 classes with neighbourhood sizes from 5 up to 53 in steps of 4. Only the combinations of a neighbourhood of 5 with 40, 60, and 80 classes do not provide sufficient equations to solve the unmixing. The result is a series of fused images at a spatial resolution of 25 m with the MERIS spectral bands. Degrading the fused images to 300 m should yield the original MERIS image again. By comparing the degraded image with the original MERIS image one can calculate the spectral ERGAS. By comparing the fused image at 25 m with the TM image one can calculate the spatial ERGAS.

Figure 3.3 illustrates the spectral and the spatial ERGAS index values for all fused images that were generated for the different combinations of *nc* and *k*.

All fused images yielded spectral ERGAS values well below 3, which is the upper limit for a good data fusion [41]. This means that the unmixing-based data fusion succeeded in preserving the spectral information of the MERIS image. A larger neighbourhood size and a smaller number of classes both give less differentiation of the spectral signatures in the fused image, thus yielding less correspondence with the MERIS image. This results in higher values of the spectral ERGAS with increasing neighbourhood size and decreasing number of classes.

Most of the spatial ERGAS values were also below the upper limit for a good quality. However, the spatial ERGAS values of the images unmixed using a small neighbourhood exceed the empirical upper limit of 3. This might be due to two reasons:

(i) The solution of the unmixing is not stable when few equations are used (remember that k determines the number of equations). Nevertheless, it is important to notice that the use of a large neighbourhood does not automatically guarantee a stable solution because if two components have a proportional coverage in that neighbourhood, then the matrix of proportions $\mathbf{F}$ will be rank deficient. In this case, the use of regularisation techniques will also be required to find a solution [43].

(ii) The upper limit might not be applicable here because this limit was defined for the spectral ERGAS which compares two images with the same spectral configuration. In this case we compare two images that have the same spatial resolution (25 m) but different band positions and bandwidths.

Two other things can be noticed from Figure 3.3. First, the spectral and the spatial ER-GAS indices are inversely correlated: the spectral ERGAS decreases when increasing the number of classes and increases with larger neighbourhood sizes, whereas the spatial ERGAS shows the opposite behaviour. This means that there is a trade-off between the spatial and the spectral reconstruction of the fused images and that we cannot find an optimum combination of nc and k that minimises both ERGAS values. Secondly, the spectral and the spatial ERGAS indices present saturation behaviour; this means that increasing nc or k beyond the values that were tested in this study will not improve the quality of the fused images.

Figure 3.3 shows that there is a clear trade-off between the size of the neighbourhood and the number of classes. To obtain a spatial ERGAS of, e.g., 0.9, one can apply a small neighbourhood with, e.g., 10 classes, or a larger neighbourhood with a larger number of classes. Both combinations may be used for representing regional variability of the spectral profiles in the fused image. Which combination to use may depend on the high resolution data set that is available. For instance, if this data set only has a limited number of classes, a small neighbourhood should be used in order to get useful variability within the spectral profiles of the image pixels. A small neighbourhood is also feasible mathematically since a small number of classes require only a small number of equations for solving the unmixing equation.

In this study we used the average of the spectral and spatial ERGAS for selecting the optimal number of classes and neighbourhood. Several combinations yielded a similar value. We selected the combination $nc = 60$ and $k = 45$ to illustrate the fusion result. Figure 3.4 shows the fused image for the test site using this parameter combination. When comparing this figure with Figure 3.2, we may conclude that the fused image exhibits the

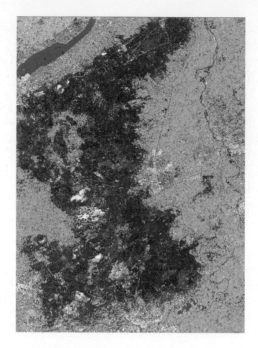

Figure 3.4 *RGB colour composite of the fused image band 13-7-5 over the study area.*

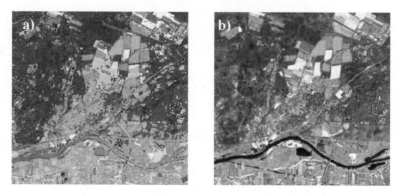

Figure 3.5 *RGB colour composite of a 25 by 25 pixels subset of the fused (a) and the Landsat TM (b) image over the study area.*

spatial detail of the Landsat-TM image, whereas it shows the spectral information of the MERIS image. In order to enable more detailed comparisons, Figure 3.5 shows a subset of the fused image (a) and the original TM image (b). Figure 3.6 shows the fused image after degradation to 300 m pixel size (a) and the original MERIS image (b).

The fused image shows individual fields like the TM image because it now has a spatial resolution of 25 m (Figure 3.5). However, it is now possible to extract spectral signatures of individual land cover types. This is illustrated in Figure 3.7, showing top-of-

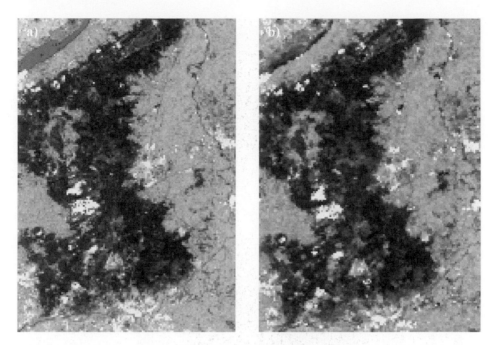

Figure 3.6 *RGB colour composite of band 13-7-5 of the fused image degraded to 300 m (a) and the MERIS image (b) over the study area.*

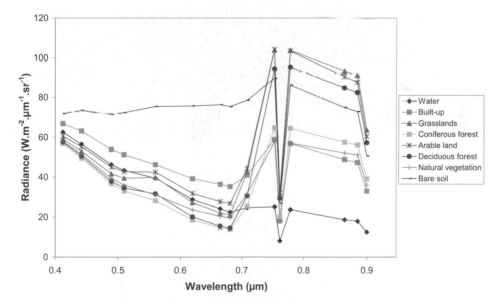

Figure 3.7 *Example of spectral signatures of various land cover types derived from the fused image.*

atmosphere radiance values. The general shape of these radiances corresponds to typical spectra of the respective classes. The first few bands show relatively high values due to atmospheric scattering in the blue. Band 11 shows very low values due to an oxygen

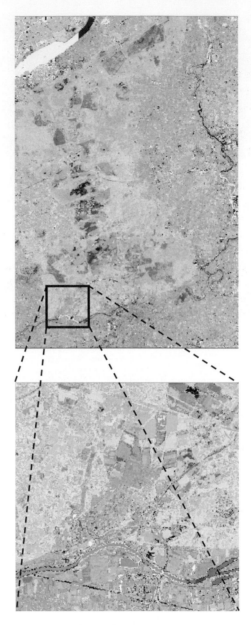

Figure 3.8   *Illustration of the red-edge index image based on the fused image over the study area (red-edge index values are linearly scaled between 710 and 730 nm). At the bottom the same subset area as in Figure 3.5 is depicted.*

absorption band. The last band shows quite low values due to a water absorption feature. These spectral bands make it possible to calculate the red-edge index as presented by Clevers et al. [24] for MERIS data. The red-edge index based on the fused image is illustrated in Figure 3.8.

3.5 Conclusions

In this chapter we have studied the applicability of the linear mixing model to fuse a Landsat TM and a MERIS full resolution level 1b image. The method, known as unmixing-based data fusion, requires the optimisation of 2 parameters: the number of classes used to classify the TM image, nc, and the size of the MERIS neighbourhood, k, used to solve the unmixing equations. Several combinations of nc and k have been tested. The spectral and the spatial ERGAS indices were used to assess the quality of the fused images and to assist with the identification of the best fused image.

The unmixing-based data fusion approach is particularly suitable for fusing MERIS FR time series with one or more TM images since the spectral information of the TM images is not included in the fused image in any way. This multitemporal data fusion exercise will be of great interest for land cover mapping and for monitoring vegetation dynamics (e.g., in terms of FAPAR, LAI, or chlorophyll content) at high spatial, spectral, and temporal resolutions. Nevertheless, it is important to notice that these fused images will only be an approximation of what the MERIS sensor would be measuring if it had a spatial resolution of 25 m. In addition, possible landscape changes between the dates of the Landsat TM acquisition and the MERIS images might reduce the quality of the fused images [34].

References

[1] M.E. Schaepman, 'Spectrodirectional remote sensing: From pixels to processes', *International Journal of Applied Earth Observation and Geoinformation*, Vol. 9, 2007, pp. 204–223.

[2] R.B. Myneni, R.R. Nemani and S.W. Running, 'Estimation of global leaf area index and absorbed par using radiative transfer models', *IEEE Transactions on Geoscience and Remote Sensing*, Vol. 35, 1997, pp. 1380–1393.

[3] A. Ruimy, B. Saugier and G. Dedieu, 'Methodology for the estimation of terrestrial net primary production from remotely sensed data', *Journal of Geophysical Research*, Vol. 99, 1994, pp. 5263–5283.

[4] N.H. Broge and E. Leblanc, 'Comparing prediction power and stability of broadband and hyperspectral vegetation indices for estimation of green leaf area index and canopy chlorophyll density', *Remote Sensing of Environment*, Vol. 76, 2001, pp. 156–172.

[5] C.S.T. Daughtry, C.L. Walthall, M.S. Kim, E. Brown De Colstoun and J.E. McMurtrey III, 'Estimating corn leaf chlorophyll concentration from leaf and canopy reflectance', *Remote Sensing of Environment*, Vol. 74, 2000, pp. 229–239.

[6] D. Haboudane, J.R. Miller, E. Pattey, P.J. Zarco-Tejada and I.B. Strachan, 'Hyperspectral vegetation indices and novel algorithms for predicting green LAI of crop canopies: Modeling and validation in the context of precision agriculture', *Remote Sensing of Environment*, Vol. 90, 2004, pp. 337–352.

[7] D. Haboudane, J.R. Miller, N. Tremblay, P.J. Zarco-Tejada and L. Dextraze, 'Integrated narrow-band vegetation indices for prediction of crop chlorophyll content for application to precision agriculture', *Remote Sensing of Environment*, Vol. 81, 2002, pp. 416–426.

[8] P.S. Thenkabail, R.B. Smith and E. De Pauw, 'Evaluation of narrowband and broadband vegetation indices for determining optimal hyperspectral wavebands for agricultural crop characterization', *Photogrammetric Engineering and Remote Sensing*, Vol. 68, 2002, pp. 607–621.

[9] M. Schlerf, C. Atzberger and J. Hill, 'Remote sensing of forest biophysical variables using HyMap imaging spectrometer data', *Remote Sensing of Environment*, Vol. 95, 2005, pp. 177–194.

[10] C. Atzberger, 'Object-based retrieval of biophysical canopy variables using artificial neural nets and radiative transfer models', *Remote Sensing of Environment*, Vol. 93, 2004, pp. 53–67.

[11] B. Combal, F. Baret, M. Weiss, A. Trubuil, D. Mace, A. Pragnere, R. Myneni, Y. Knyazikhin and L. Wang, 'Retrieval of canopy biophysical variables from bidirectional reflectance – Using prior information to solve the ill-posed inverse problem', *Remote Sensing of Environment*, Vol. 84, 2003, pp. 1–15.

[12] S. Jacquemoud, C. Bacour, H. Poilve and J.P. Frangi, 'Comparison of four radiative transfer models to simulate plant canopies reflectance: Direct and inverse mode', *Remote Sensing of Environment*, Vol. 74, 2000, pp. 471–481.

[13] W.A. Dorigo, R. Zurita-Milla, A.J.W. De Wit, J. Brazile, R. Singh and M.E. Schaepman, 'A review on reflective remote sensing and data assimilation techniques for enhanced agroecosystem modeling', *International Journal of Applied Earth Observation and Geoinformation*, Vol. 9, 2007, pp. 165–193.

[14] T.R. Loveland, B.C. Reed, J.F. Brown, D.O. Ohlen, Z. Zhu, L. Yang and J.W. Merchant, 'Development of a global land cover characteristics database and IGBP DISCover from 1 km AVHRR data', *International Journal of Remote Sensing*, Vol. 21, 2000, pp. 1303–1330.

[15] M.C. Hansen, R.S. Defries, J.R.G. Townshend and R. Sohlberg, 'Global land cover classification at 1 km spatial resolution using a classification tree approach', *International Journal of Remote Sensing*, Vol. 21, 2000, pp. 1331–1364.

[16] M.A. Friedl, D.K. McIver, J.C.F. Hodges, X.Y. Zhang, D. Muchoney, A.H. Strahler, C.E. Woodcock, S. Gopal, A. Schneider, A. Cooper, A. Baccini, F. Gao and C. Schaaf, 'Global land cover mapping from MODIS: Algorithms and early results', *Remote Sensing of Environment*, Vol. 83, 2002, pp. 287–302.

[17] G. Asrar, M. Fuchs, E.T. Kanemasu and J.L. Hatfield, 'Estimating absorbed photosynthetic radiation and leaf-area index from spectral reflectance in wheat', *Agronomy Journal*, Vol. 76, 1984, pp. 300–306.

[18] C.A. Mücher, K.T. Steinnocher, F.P. Kressler and C. Heunks, 'Land cover characterization and change detection for environmental monitoring of pan-Europe', *International Journal of Remote Sensing*, Vol. 21, 2000, pp. 1159–1181.

[19] W.B. Cohen and S.N. Goward, 'Landsat's role in ecological applications of remote sensing', *BioScience*, Vol. 54, 2004, pp. 535–545.

[20] E. Bartholome and A.S. Belward, 'GLC2000: A new approach to global land cover mapping from Earth observation data', *International Journal of Remote Sensing*, Vol. 26, 2005, pp. 1959–1977.

[21] A.R.S. Marcal, J.S. Borges, J.A. Gomes and J.F.P. Da Costa, 'Land cover update by supervised classification of segmented ASTER images', *International Journal of Remote Sensing*, Vol. 26, 2005, pp. 1347–1362.

[22] X. Zhan, R.A. Sohlberg, J.R.G. Townshend, C. Dimiceli, M.L. Carroll, J.C. Eastman, M.C. Hansen and R.S. Defries, 'Detection of land cover changes using MODIS 250 m data', *Remote Sensing of Environment*, Vol. 83, 2002, pp. 336–350.

[23] M. Rast, J.L. Bezy and S. Bruzzi, 'The ESA medium resolution imaging spectrometer MERIS – A review of the instrument and its mission', *International Journal of Remote Sensing*, Vol. 20, 1999, pp. 1681–1702.

[24] J.G.P.W. Clevers, S.M. De Jong, G.F. Epema, F.D. Van Der Meer, W.H. Bakker, A.K. Skidmore and K.H. Scholte, 'Derivation of the red edge index using the MERIS standard band setting', *International Journal of Remote Sensing*, Vol. 23, 2002, pp. 3169–3184.

[25] J.G.P.W. Clevers, 'The use of imaging spectrometry for agricultural applications', *ISPRS Journal of Photogrammetry and Remote Sensing*, Vol. 54, 1999, pp. 299–304.

[26] L. Wald, 'Some terms of reference in data fusion', *IEEE Transactions on Geoscience and Remote Sensing*, Vol. 37, 1999, pp. 1190–1193.

[27] C. Pohl and J.L. Van Genderen, 'Multisensor image fusion in remote sensing: Concepts, methods and applications', *International Journal of Remote Sensing*, Vol. 19, 1998, pp. 823–854.

[28] Y. Zhang, 'Understanding image fusion', *Photogrammetric Engineering and Remote Sensing*, Vol. 70, 2004, pp. 657–661.

[29] F.W. Acerbi-Junior, J.G.P.W. Clevers and M.E. Schaepman, 'The assessment of multi-sensor image fusion using wavelet transforms for mapping the Brazilian Savanna', *International Journal of Applied Earth Observation and Geoinformation*, Vol. 8, 2006, pp. 278–288.

[30] W. Shi, C. Zhu, Y. Tian and J. Nichol, 'Wavelet-based image fusion and quality assessment', *International Journal of Applied Earth Observation and Geoinformation*, Vol. 6, 2005, pp. 241–251.

[31] X. Otazu, M. Gonzalez-Audicana, O. Fors and J. Nunez, 'Introduction of sensor spectral response into image fusion methods. Application to wavelet-based methods', *IEEE Transactions on Geoscience and Remote Sensing*, Vol. 43, 2005, pp. 2376–2385.

[32] B. Zhukov, D. Oertel, F. Lanzl and G. Reinhackel, 'Unmixing-based multisensor multiresolution image fusion', *IEEE Transactions on Geoscience and Remote Sensing*, Vol. 37, 1999, pp. 1212–1226.

[33] A. Minghelli-Roman, M. Mangolini, M. Petit and L. Polidori, 'Spatial resolution improvement of MeRIS images by fusion with TM images', *IEEE Transactions on Geoscience and Remote Sensing*, Vol. 39, 2001, pp. 1533–1536.

[34] A. Minghelli-Roman, L. Polidori, S. Mathieu-Blanc, L. Loubersac and F. Cauneau, 'Spatial resolution improvement by merging MERIS-ETM images for coastal water monitoring', *IEEE Geoscience and Remote Sensing Letters*, Vol. 3, 2006, pp. 227–231.

[35] J.J. Settle and N.A. Drake, 'Linear mixing and the estimation of ground cover proportions', *International Journal of Remote Sensing*, Vol. 14, 1993, pp. 1159–1177.

[36] S.L. Ustin, M.O. Smith and J.B. Adams, 'Remote sensing of ecological processes: A strategy for developing and testing ecological models using spectral mixture analysis', in J.R. Ehleringer and C.B. Field (eds.), *Scaling Physicological Processes: Leaf to Globe*, Academic Press, San Diego, CA, 1993, 388 pp.

[37] J.B. Adams, D.E. Sabol, V. Kapos, R. Almeida, D.A. Roberts, M.O. Smith and A.R. Gillespie, 'Classification of multispectral images based on fractions of endmembers – Application to land-cover change in the Brazilian Amazon', *Remote Sensing of Environment*, Vol. 52, 1995, pp. 137–154.

[38] J.G.P.W. Clevers, M.E. Schaepman, C.A. Mücher, A.J.W. De Wit, R. Zurita-Milla and H.M. Bartholomeus, 'Using MERIS on Envisat for land cover mapping in the Netherlands', *International Journal of Remote Sensing*, Vol. 28, 2007, pp. 637–652.

[39] R. Zurita-Milla, J.G.P.W. Clevers, M.E. Schaepman and M. Kneubuehler, 'Effects of MERIS L1b radiometric calibration on regional land cover mapping and land products', *International Journal of Remote Sensing*, Vol. 28, 2007, pp. 653–673.

[40] C. Thomas and L. Wald, 'Assessment of the quality of fused products', in M. Oluic (ed.), *24th EARSeL Symposium, New Strategies for European Remote Sensing*, Dubrovnik, Croatia, 25–27 May 2004.

[41] L. Wald, *Data Fusion Definitions and Architectures: Fusion of Images of Different Spatial Resolutions*, Ecole des Mines Pres, 2002.

[42] M. Lillo-Saavedra, C. Gonzalo, A. Arquero and E. Martinez, 'Fusion of multispectral and panchromatic satellite sensor imagery based on tailored filtering in the Fourier domain', *International Journal of Remote Sensing*, Vol. 26, 2005, pp. 1263–1268.

[43] R. Faivre and A. Fischer, 'Predicting crop reflectances using satellite data observing mixed pixels', *Journal of Agricultural, Biological, and Environmental Statistics*, Vol. 2, 1997, pp. 87–107.

4

Image fusion schemes using ICA bases

Nikolaos Mitianoudis and Tania Stathaki

Communications and Signal Processing Group, Imperial College London, London, UK

Image fusion is commonly described as the task of enhancing the perception of a scene by combining information captured by different modality sensors. The *pyramid decomposition* and the *Dual-Tree Wavelet Transform* have been employed as analysis and synthesis tools for image fusion by the fusion community. Using various fusion rules, one can combine the important features of the input images in the transform domain to compose an enhanced image. In this study, the authors demonstrate the efficiency of a transform constructed using *Independent Component Analysis* (ICA) and *Topographic Independent Component Analysis* bases for image fusion. The bases are trained offline using images of similar context to the observed scene. The images are fused in the transform domain using novel *pixel-based* or *region-based* rules. An unsupervised adaptation ICA-based fusion scheme is also introduced. The proposed schemes feature improved performance compared to approaches based on the wavelet transform and slightly increased computational complexity.

4.1 Introduction

The need for data fusion in current image processing systems is increasing mainly due to the increase of image acquisition techniques [1]. Current technology in imaging sensors offers a wide variety of different information that can be extracted from an observed scene. This information is jointly combined to provide an enhanced representation in many cases of experimental sciences. The automated procedure of conveying all the meaningful information from the input sensors to a composite image is the topic of this article. *Fusion systems* appear to be an essential preprocessing stage for a number of applications, such as aerial and satellite imaging, medical imaging, robot vision and vehicle or robot guidance [1].

Let $I_1(x, y), I_2(x, y), \ldots, I_T(x, y)$ represent T images of size $M_1 \times M_2$ capturing the same scene. Each image has been acquired using different instrument modalities or capture techniques. Consequently, each image has different characteristics, such as degradation, thermal and visual characteristics. These images need not be perfect, otherwise fusion would not be necessary. This imperfection can appear in the form of imprecision, ambiguity or incompleteness. However, the source images should offer complementary and redundant information about the observed scene [1]. In addition, each of these images should contain information that might be useful for the composite image and is not provided by the other input images. In other words, there is no potential in fusing an image that has mainly degraded information compared to the other input images. Although the fusion system will most probably be able to reject the misleading information, it is not conceptually valid to present the system with no beneficial information, as the performance might be degraded and the computational complexity increased.

In this scenario, we usually employ multiple sensors that are placed relatively close and are observing the same scene. The images acquired by these sensors, although they should be similar, are bound to have some translational errors, i.e. miscorrespondence between several points of the observed scene. *Image registration* is the process of establishing point-by-point correspondence between a number of images, describing the same scene [2]. In the opposite case that the sensors are arbitrarily placed, all input images need to be registered. In this study, the input images $I_i(x, y)$ are assumed to have negligible registration problems, which implies that the objects in all images are geometrically aligned.

The process of combining the important features from these T images to form a single enhanced image $I_f(x, y)$ is usually referred to as *image fusion*. Fusion techniques are commonly divided into *spatial domain* and *transform domain* techniques [3]. In spatial domain techniques, the input images are fused in the spatial domain, i.e. using localised spatial features. Assuming that $g(\cdot)$ represents the 'fusion rule,' i.e. the method that combines features from the input images, the spatial domain techniques can be summarised, as follows:

$$I_f(x, y) = g\big(I_1(x, y), \ldots, I_T(x, y)\big) \tag{4.1}$$

The main motivation behind moving to a transform domain is to work in a framework, where the image's salient features are more clearly depicted than in the spatial domain. It is important to understand the underlying image structure for fusion rather than fusing image pixels independently. Most transformations used in image processing are decomposing the images into important local components, i.e. unlocking the basic image structure. Hence, the choice of the transformation is very important. Let $T\{\cdot\}$ represent a transform operator and $g(\cdot)$ the applied fusion rule. Transform-domain fusion techniques can then be outlined, as follows:

$$I_f(x, y) = T^{-1}\big\{g\big(T\{I_1(x, y)\}, \ldots, T\{I_T(x, y)\}\big)\big\} \tag{4.2}$$

The fusion operator $g(\cdot)$ describes the merging of information from the different input images. Many fusion rules have been proposed in the literature [4–6]. These rules can be categorised, as follows:

- *Pixel-based rules*: the information fusion is performed in a pixel-by-pixel basis either in the transform or spatial domain. Each pixel (x, y) of the T input images is combined with various rules to form the corresponding pixel (x, y) in the 'fused' image I_T. Several basic transform-domain schemes were proposed [4], such as:
  - *Fusion by averaging*: fuse by averaging the corresponding coefficients in each image ('mean' rule)

$$\mathcal{T}\{I_f(x, y)\} = \frac{1}{T} \sum_{i=1}^{T} \mathcal{T}\{I_i(x, y)\} \tag{4.3}$$

  - *Fusion by absolute maximum*: fuse by selecting the greatest in absolute value of the corresponding coefficients in each image ('max-abs' rule)

$$\mathcal{T}\{I_f(x, y)\} = \text{sgn}(\mathcal{T}\{I_i(x, y)\}) \max_i |\mathcal{T}\{I_i(x, y)\}| \tag{4.4}$$

  - *Fusion by denoising (hard/soft thresholding)*: perform simultaneous fusion and denoising by thresholding the transform's coefficients (sparse code shrinkage [7]).
  - *High/low fusion*, i.e. combining the 'high-frequency' parts of some images with the 'low-frequency' parts of some other images.

  The different properties of these fusion schemes will be explained later on. For a more complete review on pixel-based fusion methods, one can have always refer to Piella [5], Nikolov et al. [4] and Rockinger et al. [6].
- *Region-based fusion rules*: in order to exploit the image structure more efficiently, these schemes group image pixels to form contiguous regions, e.g. objects and impose different fusion rules to each image region. In [8], Li et al. created a binary decision map to choose between the coefficients using a majority filter, measuring activity in small patches around each pixel. In [5], Piella proposed several activity level measures, such as the absolute value, the median or the contrast to neighbours. Consequently, she proposed a region-based scheme using a local correlation measurement to performs fusion of each region. In [9], Lewis et al. produced a joint-segmentation map out of the input images. To perform fusion, they measured *priority* using *energy*, *variance*, or *entropy* of the wavelet coefficients to impose weighting on each region in the fusion process along with other heuristic rules.

In this study, the application of *Independent Component Analysis* (ICA) and *Topographic Independent Component Analysis* bases as an analysis tool for image fusion in both noisy and noiseless environments is examined. The performance of the proposed framework in image fusion is compared to traditional fusion analysis tools, such as the *wavelet transform*. Common pixel-based fusion rules are tested together with a proposed 'weighted-combination' scheme, based on the $\mathcal{L}_1$-norm. A region-based approach that segments and fuses active and non-active areas of the image is introduced. Finally, an adaptive unsupervised scheme for image fusion in the ICA domain using *sparsity* is presented.

The paper is structured, as follows. In Section 4.2, we introduce the basics of the Independent Component Analysis technique and how it can be used to generate analysis/synthesis bases for image fusion. In Section 4.3, we describe the general method for performing image fusion using ICA bases. In Section 4.4, the proposed pixel-based weighted combination scheme and a combinatory region-based scheme are introduced. In Section 4.5, we describe an unsupervised adaptive fusion scheme in the ICA framework. In Section 4.6, several issues concerning the reconstruction of the fused image from the ICA representation are discussed. In Section 4.7, the proposed transform and fusion schemes is benchmarked using common fusion testbed. Finally, in Section 4.8, we outline the advantages and disadvantages of the proposed schemes together with some suggestions about future work.

4.2 ICA and Topographic ICA bases

Assume an image $I(x, y)$ of size $M_1 \times M_2$ and a window W of size $N \times N$, centred around the pixel (x_0, y_0). An 'image patch' is defined as the product between a $N \times N$ neighbourhood centred around pixel (x_0, y_0) and the window W:

$$I_w(k, l) = W(k, l) I\left(x_0 - \lfloor N/2 \rfloor + k, \, y_0 - \lfloor N/2 \rfloor + l\right),$$

$$\forall k, l \in [0, N-1] \tag{4.5}$$

where $\lfloor \cdot \rfloor$ represents the lower integer part and N is odd. For the subsequent analysis, we will assume a rectangular window, i.e.

$$W(k, l) = 1, \quad \forall k, l \in [0, N-1] \tag{4.6}$$

4.2.1 Definition of bases

In an effort to understand the underlying structure of an image, it is common practice in image analysis to express an image as the synthesis of several *basis* images. These bases are chosen according to the image features that need to be highlighted with this analysis. A number of basis have been proposed in literature so far, such as cosine bases, complex cosine bases, Hadamard bases and wavelet bases. In this case, the bases are well-defined in order to serve some specific analysis tasks. However, one can estimate non-standard bases by training with a population of similar content images. The bases are estimated after optimising a cost function that defines the bases' desired properties.

The $N \times N$ image patch $I_w(k, l)$ can be expressed as a linear combination of a set of K basis images $b_j(k, l)$, i.e.

$$I_w(k, l) = \sum_{j=1}^{K} u_j b_j(k, l) \tag{4.7}$$

where u_j are scalar constants. The two-dimensional (2D) representation can be simplified to an one-dimensional (1D) representation, by employing *lexicographic ordering*, in order to facilitate the analysis. In other words, the image patch $I_w(k, l)$ is arranged into a vector $\underline{I}_w$, taking all elements from matrix I_w in a row-wise fashion. The vectors $\underline{I}_w$ are normalised to zero mean, to avoid the possible bias of the local greyscale levels.

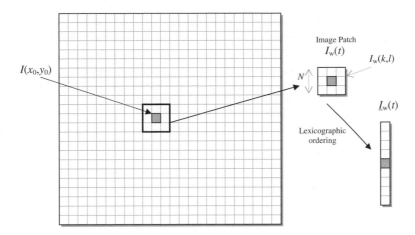

Figure 4.1 *Selecting an image patch I_w around pixel (x_0, y_0) and the lexicographic ordering.*

Assume that we have a population of patches I_w, acquired randomly from the original image $I(x, y)$. These image patches can then be expressed in lexicographic ordering, as follows:

$$\underline{I}_w(t) = \sum_{j=1}^{K} u_j(t)\underline{b}_j = [b_1 \; b_2 \; \dots \; \underline{b}_K] \begin{bmatrix} u_1(t) \\ u_2(t) \\ \dots \\ u_K(t) \end{bmatrix} \tag{4.8}$$

where t represents the tth image patch selected from the original image. The whole procedure of image patch selection and lexicographic ordering is depicted in Figure 4.1. Let $B = [\underline{b}_1 \; \underline{b}_2 \; \dots \; \underline{b}_K]$ and $\underline{u}(t) = [u_1(t) \; u_2(t) \; \dots \; u_K(t)]^T$. Then, Equation (4.8) can be simplified, as follows:

$$\underline{I}_w(t) = B\underline{u}(t) \tag{4.9}$$

$$\underline{u}(t) = B^{-1}\underline{I}_w(t) = A\underline{I}_w(t) \tag{4.10}$$

In this case, $A = B^{-1} = [\underline{a}_1 \; \underline{a}_2 \; \dots \; \underline{a}_K]^T$ represents the *analysis* kernel and B the *synthesis* kernel. This 'transformation' projects the observed signal $\underline{I}_w(t)$ on a set of basis vectors $\underline{b}_j$. The aim is to estimate a finite set of basis vectors that will be capable of capturing most of the signal's structure (energy). Essentially, we need N^2 bases for a *complete* representation of the N^2-dimensional signals $\underline{I}_w(t)$. However, with some redundancy reduction mechanisms, we can have efficient *overcomplete* representations of the original signals using $K < N^2$ bases.

The estimation of these K vectors is performed using a population of training image patches $\underline{I}_w(t)$ and a criterion (cost function), which is going to be optimised in order to select the basis vectors. In the next paragraphs, we will estimate bases from image patches using several criteria.

4.2.1.1 Principal Component Analysis (PCA) bases

One of the transform's targets might be to analyse the image patches into uncorrelated components. *Principal Component Analysis* (PCA) can identify uncorrelated vector bases [10], assuming a linear generative model, as in (4.9). In addition, PCA can be used for dimensionality reduction to identify the K most important basis vectors. This is performed by eigenvalue decomposition of the data correlation matrix $C = \mathcal{E}\{\underline{I}_{w}\underline{I}_{w}^{T}\}$, where $\mathcal{E}\{\cdot\}$ represents the expectation operator. Assume that H is a matrix containing all the eigenvectors of C and D a diagonal matrix containing the eigenvalues of C. The eigenvalue at the ith diagonal element should correspond to the eigenvector at the ith column of H. The rows of the following matrix V provide an orthonormal set of uncorrelated bases, which are called PCA bases:

$$V = D^{-0.5}H^{T} \tag{4.11}$$

The above set forms a *complete* set of bases, i.e. we have as many bases as the dimensionality of the problem (N^2). As PCA has efficient energy compaction properties, one can form a reduced (*overcomplete*) set of bases, based on the original ones. The eigenvalues can illustrate the significance of their corresponding eigenvector (basis vector). We can order the eigenvalues in the diagonal matrix D, in terms of decreasing absolute value. The eigenvector matrix H should be arranged accordingly. Then, we can select the first $K < N^2$ eigenvectors that correspond to the K most important eigenvalues and form reduced versions of $\hat{D}$ and $\hat{H}$. The reduced $K \times N^2$ PCA matrix $\hat{V}$ is calculated using (4.11) for $\hat{D}$ and $\hat{H}$. The input data can be mapped to the PCA domain via the transformation:

$$\underline{z}(t) = \hat{V}\underline{I}_{w}(t) \tag{4.12}$$

The size of the overcomplete set bases K is chosen so that the computational load of a complete representation can be reduced. However, the overcomplete set should be able to provide an almost lossless representation of the original image. Therefore, the choice of K is usually a trade-off between computational complexity and image quality.

4.2.1.2 Independent Component Analysis (ICA) bases

A stricter criterion than uncorrelatedness is to assume that the basis vectors or equivalently the transform coefficients are *statistically independent*. *Independent Component Analysis* (ICA) can identify statistically independent basis vectors in a linear generative model [11]. A number of different approaches have been proposed to analyse the generative model in (4.9), assuming statistical independence between the coefficients u_i in the transform domain. Statistical independence can be closely linked with non-Gaussianity. The *Central Limit Theorem* states that the sum of several independent random variables tends towards a Gaussian distribution. The same principal holds for any linear combination I_w of these independent random variables u_i. The Central Limit Theorem also implies that a combination of the observed signals in I_w with minimal Gaussian properties can be one of the independent signals. Therefore, statistical independence and non-Gaussianity can be interchangeable terms.

A number of different techniques can be used to estimate independent coefficients u_i. Some approaches estimate u_i by minimising the *Kullback–Leibler* (KL) divergence between the estimated coefficients u_i and *several probabilistic priors* on the coefficients. Other approaches minimise the *mutual information* conveyed by the estimated coefficients or perform approximate diagonalisation of a *cumulant tensor* of $\underline{I}_w$. Finally, some methods estimate u_i by estimating the directions of the most non-Gaussian components using *kurtosis* or *negentropy*, as non-Gaussianity measures. More details on these techniques can be found in tutorial books on ICA, such as [11,12].

In this study, we will use an approach that optimises negentropy, as a non-Gaussianity measurement to identify the independent components u_i. This is also known as FastICA and was proposed by Hyvärinen and Oja [13]. According to this technique, PCA is used as a preprocessing step to select the K most important vectors and orthonormalise the data using (4.12). Consequently, the statistical independent components can be identified using orthogonal projections $\underline{a}_i^T \underline{z}$. In order to estimate the projecting vectors $\underline{a}_i$, we have to minimise the following non-quadratic approximation of negentropy:

$$J_G(\underline{q}_i) = \left(\mathcal{E}\{G(\underline{q}_i^T \underline{z})\} - \mathcal{E}\{G(v)\} \right)^2 \tag{4.13}$$

where $\mathcal{E}\{\cdot\}$ denotes the expectation operator, v is a Gaussian variable of zero mean and unit variance and $G(\cdot)$ is practically any non-quadratic function. A couple of possible functions were proposed in [14]. In our analysis, we will use:

$$G(x) = \frac{1}{\alpha} \log \cosh(\alpha x) \tag{4.14}$$

where α is a constant that usually is bounded to $1 \leqslant \alpha \leqslant 2$. Hyvärinen and Oja produced a fixed-point method, optimising the above definition of negentropy, which is also known as the *FastICA* algorithm:

$$\underline{q}_i^+ \leftarrow \mathcal{E}\{\underline{q}_i \phi(\underline{q}_i^T \underline{z})\} - \mathcal{E}\{\phi'(\underline{q}_i^T \underline{z})\}\underline{q}_i, \quad 1 \leqslant i \leqslant K \tag{4.15}$$

$$Q \leftarrow Q(Q^T Q)^{-0.5} \tag{4.16}$$

where $\phi(x) = -\partial G(x)/\partial x$. We randomly initialise the update rule in (4.15) for each projecting vector $\underline{q}_i$. The new updates are then orthogonalised, using the symmetric orthogonalisation scheme in (4.16). These two steps are iterated, until $\underline{q}_i$ have converged.

4.2.1.3 Topographical Independent Component Analysis (TopoICA) bases

In practical applications, one can frequently observe clear violations of the independence assumption. It is possible to find couples of estimated components that they are clearly dependent on each other. This dependence structure can be very informative about the actual image structure and it would be useful to estimate it [14].

Hyvärinen et al. [14] used the residual dependency of the 'independent' components, i.e. dependencies that could not be cancelled by ICA, to define a *topographic* order between the components. Therefore, they modified the original ICA model to include a topographic order between the components, so that components that are near to each other in the topographic representation are relatively strongly dependent in the sense of higher-order correlations or mutual information. The proposed model is usually known as the *Topographic ICA* model. The topography is introduced using a neighbourhood function $h(i, k)$, which expresses the proximity between the ith and the kth component. A simple neighbourhood model can be the following:

$$h(i, k) = \begin{cases} 1, & \text{if } |i - k| \leqslant L, \\ 0, & \text{otherwise} \end{cases} \tag{4.17}$$

where L defines the width of the neighbourhood. Consequently, the estimated coefficients u_i are no longer assumed independent, but can be modelled by some generative random variables d_k, f_i that are controlled by the neighbourhood function and shaped by a non-linearity $\phi(\cdot)$ (similar to the one in the FastICA algorithm for positive numbers). The topographic source model, proposed by Hyvärinen et al. [14], is the following:

$$u_i = \phi \left(\sum_{k=1}^{K} h(i, k) d_k \right) f_i \tag{4.18}$$

Assuming a fixed-width neighbourhood $L \times L$ and a PCA preprocessing step, Hyvärinen et al. performed Maximum Likelihood estimation of the synthesis kernel B using the linear model in (4.9) and the topographic source model in (4.18), making several assumptions for the generative random variables d_k and f_i. Optimising an approximation of the derived log-likelihood, they formed the following gradient-based Topographic ICA rule:

$$\underline{q}_i^+ \leftarrow \underline{q}_i + \eta \mathcal{E}\{\underline{z}(\underline{q}_i^{\mathrm{T}} \underline{z}) r_i\}, \quad 1 \leqslant i \leqslant K \tag{4.19}$$

$$Q \leftarrow Q(Q^{\mathrm{T}} Q)^{-0.5} \tag{4.20}$$

where η defines the learning rate of the gradient optimisation scheme and

$$r_i = \sum_{k=1}^{K} h(i, k) \phi \left(\sum_{j=1}^{K} h(j, k) \left(\underline{q}_i^{\mathrm{T}} \underline{z} \right)^2 \right) \tag{4.21}$$

As previously, we randomly initialise the update rule in (4.19) for each projecting vector $\underline{q}_i$. The new updates are then orthogonalised and the whole procedure is iterated, until $\underline{a}_i$ have converged. For more details on the definition and derivation of the Topographic ICA model, one can always refer to the original work by Hyvärinen et al. [14].

Finally, after estimating the matrix Q, using the ICA or the topographic ICA algorithm, the analysis kernel is given by multiplying the original PCA bases matrix $\hat{V}$ with Q:

$$A \leftarrow Q\hat{V} \tag{4.22}$$

4.2.2 Training ICA bases

In this paragraph, we describe the training procedure of the ICA and topographic ICA bases more thoroughly. The training procedure needs to be completed only once for each data type. After we have successfully trained the desired bases for each image type, the estimated transform can be used for fusion of similar content images.

We select a set of images with similar content to the ones that will be used for image fusion. A number of $N \times N$ patches (usually around 10 000) are randomly selected from the training images. We apply lexicographic ordering to the selected images patches and normalise them to zero mean. We perform PCA on the selected patches and select the $K < N^2$ most important bases, according to the eigenvalues corresponding to the bases. It is always possible to keep the complete set of bases. Then, we iterate the ICA update rule in (4.15) or the topographical ICA rule in (4.19) for a chosen $L \times L$ neighbourhood until convergence. After each iteration, we orthogonalise the bases using the scheme in (4.16).

Some examples from trained ICA and topographic ICA bases are depicted in Figure 4.2. We randomly selected 10 000 16×16 patches from natural landscape images. Using PCA, we selected the 160 most important bases out of the 256 bases available. In Figure 4.2(a), we can see the ICA bases estimated using FastICA (4.15). In Figure 4.2(b), the set of the estimated Topographic ICA bases using the rule in (4.19) and a 3×3 neighbourhood for the topographic model are depicted. The estimated bases feature an ordering based on similarity and correlation and thus offer a more structured and meaningful representation.

4.2.3 Properties of the ICA bases

Let us explore some of the properties of the ICA and the Topographical ICA bases and the transforms they constitute. Both transforms are *invertible*, i.e. they guarantee perfect reconstruction. Using the symmetric orthogonalisation step $Q \leftarrow Q(Q^T Q)^{-0.5}$, the estimated bases remain orthogonal in the ICA domain, i.e. the transform is *orthogonal*.

We can examine the estimated example set of ICA and Topographical ICA bases in Figure 4.2. The ICA and topographical ICA basis vectors seem to be closely related to wavelets and Gabor functions, as they all represent localised edge features. However, the ICA bases have more degrees of freedom than wavelets [14]. The Discrete Wavelet transform has only two orientations and the Dual-Tree wavelet transform can give six distinct sub-bands at each level with orientation $\pm 15°$, $\pm 45°$, $\pm 75°$. In contrast, the ICA bases can get arbitrary orientations to fit the training patches. On the other hand, the ICA bases do not offer a multilevel representation as the wavelet or pyramid decomposition, but only focus on localised features.

One basic drawback of the ICA-based transformations is that they are not *shift invariant* by definition. This property is generally mentioned to be very important for image fusion in literature [4]. Piella [5] comments that the fusion result will depend on the location or orientation of objects in the input sources in the case of misregistration problems or when used for image sequence fusion. As we assume that the observed images are all registered, the lack of shift invariance should not necessarily be a problem. In addition, Hyvärinen et al. proposed to approximate shift invariance in these ICA schemes, by em-

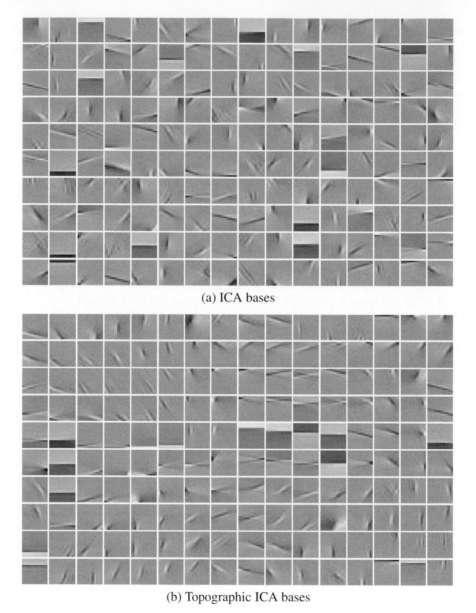

(a) ICA bases

(b) Topographic ICA bases

Figure 4.2 *Comparison between ICA and the topographical ICA bases trained on the same set of image patches. We can observe the spatial correlation of the bases, introduced by 'topography.'*

ploying a *sliding window* approach [7]. This implies that the input images are not divided into distinct patches, but instead every possible $N \times N$ patch in the image is analysed. This is similar to the *spin cycling* method, proposed by Coifman and Donoho [15]. This will also increase the computational complexity of the proposed framework. The sliding window approach is only necessary for the fusion part and not for the estimation of bases.

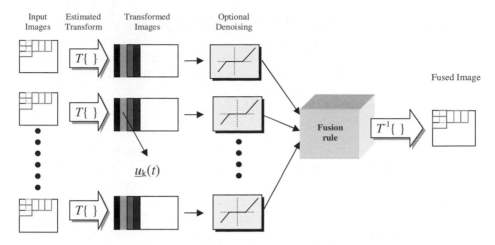

Figure 4.3 *The proposed fusion system using ICA/Topographical ICA bases.*

The basic difference between ICA and topographic ICA bases is the 'topography,' as introduced in the latter bases. The introduction of some local correlation in the ICA model enables the algorithm to uncover some connections between the independent components. In other words, topographic bases provide an ordered representation of the data, compared to the unordered representation of the ICA bases. In an image fusion framework, 'topography' can identify groups of features that can characterise certain objects in the image. One can observe the ideas comparing Figures 4.2(a) and 4.2(b). Topographic ICA seems to offer a more comprehensive representation compared to the general ICA model.

Another advantage of the ICA bases is that the estimated transform can be tailored to the application field. Several image fusion applications work with specific types of images. For example, military applications work with images of airplanes, tanks, ships, etc. Biomedical applications employ Computed Tomography (CT), Positron Emission Tomography (PET), ultra-sound scan images, etc. Consequently, one can train bases for specific application areas using ICA. These bases should be able to analyse the trained data types more efficiently than a generic transform.

4.3 Image fusion using ICA bases

In this section, we describe the whole procedure of performing image fusion using ICA or Topographical ICA bases, which is summarised in Figure 4.3 [16,17]. We assume that a ICA or Topographic ICA transform $\mathcal{T}\{\cdot\}$ is already estimated, as described in Section 4.2.2. Also, let $I_k(x, y)$ be T $M_1 \times M_2$ registered sensor images that need to be fused. From each image we isolate every possible $N \times N$ patch and using lexicographic ordering, we form the vector $\underline{I}_k(t)$. The patches' size N should be the same as the one used in the transform estimation. Therefore, each image $I_k(x, y)$ is now represented by a population of $(M_1 - N)(M_2 - N)$ vectors $\underline{I}_k(t)$, $\forall t \in [1, (M_1 - N)(M_2 - N)]$. These vectors are normalised to zero mean and the subtracted means of each vector $MN_k(t)$ are stored in order to be used in the reconstruction of the fused image. Each of these representations $\underline{I}_k(t)$ is transformed to the ICA or Topographic ICA domain representation $\underline{u}_k(t)$. Assuming that A is the estimated analysis kernel, we have:

$$\underline{u}_k(t) = \mathcal{T}\{\underline{I}_k(t)\} = A\underline{I}_k(t) \qquad (4.23)$$

Once the image representations are in the ICA domain, one can apply a 'hard' threshold on the coefficients and perform optional denoising (sparse code shrinkage), as proposed by Hyvärinen et al. [7]. The threshold can be determined by supervised estimation of the noise level in constant background areas of the image. Then, one can perform image fusion in the ICA or Topographic ICA domain in the same manner that is performed in the wavelet or dual-tree wavelet domain. The corresponding coefficients $\underline{u}_k(t)$ from each image are combined in the ICA domain to construct a new image $\underline{u}_f(t)$. The method $g(\cdot)$ that combines the coefficients in the ICA domain is called 'fusion rule':

$$\underline{u}_f(t) = g\big(\underline{u}_1(t), \dots, \underline{u}_k(t), \dots, \underline{u}_T(t)\big) \qquad (4.24)$$

Many of the proposed rules for fusion, as they were analysed in the introduction section and in literature [5,4], can be applied to this framework. The 'max-abs' and the 'mean' rules can be two very common options. However, one can use more efficient fusion rules, as will be presented in the next section. Once the composite image $\underline{u}_f(t)$ is constructed in the ICA domain, one can move back to the spatial domain, using the synthesis kernel B, and synthesise the image $I_f(x, y)$ by averaging the image patches $I_f(t)$ in the same order they were selected during the analysis step. The whole procedure can be summarised as follows:

(1) Segment all input images $I_k(x, y)$ into every possible $N \times N$ image patch and transform them to vectors $\underline{I}_k(t)$ via lexicographic ordering.
(2) Move the input vectors to the ICA/Topographic ICA domain, and get the corresponding representation $\underline{u}_k(t)$.
(3) Perform optional thresholding of $\underline{u}_k(t)$ for denoising.
(4) Fuse the corresponding coefficient using a fusion rule and form the composite representation $\underline{u}_f(t)$.
(5) Move $\underline{u}_f(t)$ to the spatial domain and reconstruct the image $I_f(x, y)$ by averaging the overlapping image patches.

4.4 Pixel-based and region-based fusion rules using ICA bases

In this section, we describe two proposed fusion rules for ICA bases. The first one is an extension of the 'max-abs' pixel-based rule, which we will refer to as the *Weight Combination* (WC) rule. The second one is a combination of the WC and the 'mean' rule in a region-based scenario.

4.4.1 A Weight Combination (WC) pixel-based method

An alternative to common fusion methods, is to use a 'weighted combination' of the transform coefficients, i.e.

$$T\{\underline{I}_f(t)\} = \sum_{k=1}^{T} w_k(t) T\{\underline{I}_k(t)\} \tag{4.25}$$

There are several parameters that can be employed in the estimation of the contribution $w_k(t)$ of each image to the 'fused' one. In [5], Piella proposed several *activity measures*. Following the general ideas proposed in [5], we propose the following scheme. As each image is processed in $N \times N$ patches, we can use the mean absolute value ($\mathcal{L}_1$-norm) of each patch (arranged in a vector) in the transform domain, as an activity indicator in each patch:

$$E_k(t) = \|\underline{u}_k(t)\|_1, \quad k = 1, \ldots, T \tag{4.26}$$

The weights $w_k(t)$ should emphasise sources that feature more intense activity, as represented by $E_k(t)$. Consequently, the weights $w_k(t)$ for each patch t can be estimated by the contribution of the kth source image $\underline{u}_k(t)$ over the total contribution of all the T source images at patch t, in terms of activity. Hence, we can choose:

$$w_k(t) = E_k(t) \Big/ \sum_{k=1}^{T} E_k(t) \tag{4.27}$$

There might be some cases, where $\sum_{k=1}^{T} E_k(t)$ is very small, denoting small edge activity or constant background in the corresponding patch. As this can cause numerical instability, the 'max-abs' or 'mean' fusion rule can be used for those patches. Equally, a small constant can be added to alleviate this instability.

4.4.2 Region-based image fusion using ICA bases

In this section, the analysis of the input images in the estimated ICA domain will be employed to perform some regional segmentation in order to fuse these regions using different rules, i.e. perform *region-based* image fusion. During, the proposed analysis methodology, we have already divided the image in small $N \times N$ patches (i.e. regions). Using the splitting/merging philosophy of region-based segmentation [18], a criterion is employed to merge the pixels corresponding to each patch in order to form contiguous areas of interest.

One could use the energy activity measurement, as introduced by (4.26), to infer the existence of edges in the corresponding frame. As the ICA bases tend to focus on the edge information, it is clear that great values for $E_k(t)$, correspond to increased activity in the frame, i.e. the existence of edges. In contrast, small values for $E_k(t)$ denote the existence of almost constant background or insignificant texture in the frame. Using

this idea, we can segment the image in two regions: (i) 'active' regions containing details and (ii) 'non-active' regions containing background information. The threshold that
will be used to characterise a region as 'active' or 'non-active' can be set heuristically
to $2\text{mean}_t\{E_k(t)\}$. Since the aim here is to create the most accurate edge-detector, we
can allow some tolerance around the real edges of the image. As a result, we form the
following segmentation map $m_k(t)$ from each input image:

$$m_k(t) = \begin{cases} 1, & \text{if } E_k(t) > 2\text{mean}_t\{E_k(t)\}, \\ 0, & \text{otherwise} \end{cases} \tag{4.28}$$

The segmentation map of each input image is combined to form a single segmentation
map, using the logical OR operator. As mentioned earlier, we are not interested in forming a very accurate edge detection map, but instead it is important to ensure that our
segmentation map contains most of the strong edge information.

$$m(t) = \text{OR}\{m_1(t), m_2(t), \dots, m_T(t)\} \tag{4.29}$$

Once the image has been segmented into 'active' and 'non-active' regions, we can fuse
these regions using different pixel-based fusion schemes. For the 'active' region, we can
use a fusion scheme that preserves the edges, i.e. the 'max-abs' scheme or the weighted
combination scheme and for the 'non-active' region, we can use a scheme that preserves
the background information, i.e. the 'mean' or 'median' scheme. Consequently, this could
form a more accurate fusion scheme that looks into the actual structure of the image itself,
rather than fuse information generically.

4.5 A general optimisation scheme for image fusion

In this section, the focus is placed on defining an unsupervised image fusion approach
based on the minimisation of a formulated cost function involving several source images.
The main aim is to achieve visual improvements over the original source images, such
that certain specific features in the original source images can be detected visually or
through various models in the fused image. Practical usage of this algorithm includes
the confirmation of a particular target in military purposes, when several different source
images are obtained from different sensors under different conditions [17].

The minimisation of a cost function involves the estimation of a set of optimal parameters
that will minimise the output value of the cost function. This concept can thus be incorporated into the process of image fusion to obtain a set of optimal coefficients that can be
used to produce a fused image of better quality than each of the original source images.

Let us assume that we are interested in the $N \times N$ patches around pixel (x_0, y_0) in the
input sensor image $I_1, \dots, I_T$. These patches are lexicographically ordered, as described
in the previous section, to form the vectors $\underline{I}_1, \dots, \underline{I}_T$. We also assume that an ICA
transform $\mathcal{T}\{\cdot\}$ has been trained, using patches of similar content images. In this case,
we will be using a complete representation, i.e. $K = N^2$, although any overcomplete
representation may also be used. The input patches in the transform domain are denoted

by $\underline{u}_i = \mathcal{T}\{\underline{I}_i\}$. The fused image $\underline{u}_f$ in the transform domain can be given by the following linear combination:

$$\underline{u}_f = w_1\underline{u}_1 + w_2\underline{u}_2 + \cdots + w_T\underline{u}_T \qquad (4.30)$$

where $w_1, \ldots, w_T$ are scalar coefficients that denote the mixing of each input sensor patch in the transform domain. We denote $\underline{w} = [w_1\ w_2\ \ldots\ w_T]^T$. All elements of vector $\underline{u}_i$ will contribute in the formation of the fused image, according to the weight w_i. Let us now define:

$$\underline{x}(n) = \left[u_1(n)\ u_2(n)\ \ldots\ u_T(n)\right]^T \quad \forall n = 1, \ldots, N^2 \qquad (4.31)$$

Hence, the fusion procedure can be equivalently described by the following product:

$$u_f(n) = \underline{w}^T\underline{x}(n) \quad \forall n = 1, \ldots, N^2 \qquad (4.32)$$

The problem of fusion can now be described as an optimisation problem of estimating $\underline{w}$, so that the fused image follows certain properties, described by the cost function. A logical assumption is that the fusion process should enhance *sparsity* in the ICA domain. In other words, the fusion should emphasise the existence of strong coefficients in the transform, whilst suppress small values. We will approach the problem of estimating $\underline{w}$, using a ML estimation approach, assuming several probabilistic priors, that describe sparsity.

The connection between *sparsity* and ICA representations has been investigated thoroughly by Olshausen [19]. The basis functions that emerge when adapted to static, whitened natural images under the assumption of statistical independence, resemble the Gabor-like spatial profiles of cortical simple-cell receptive fields. That is to say that the functions become spatially localised, oriented and bandpass. Because all of these properties emerge purely from the objective of finding sparse, independent components for natural images, the results suggest that the receptive fields of V1 neurons have been designed under the same principle. Therefore, the actual non-distorted representation of the observed scene in the ICA domain should be more sparse than the distorted or different sensor input. Consequently, an algorithm that maximises the sparsity of the fused image in the ICA domain can be justified.

4.5.1 Laplacian priors

Assuming a Laplacian model for $u_f(n)$, we can perform Maximum Likelihood (ML) estimation of $\underline{w}$. The Laplacian probability density function is given below:

$$p(u_f) \propto e^{-\alpha|u_f|} \qquad (4.33)$$

where α is a parameter that controls the width (variance) of the Laplacian. The likelihood expression for ML estimation can be given by

$$L_n = -\log p(u_f|\theta_n) \propto -\log e^{-\alpha|u_f|}$$
$$= \alpha|u_f| = \alpha\left|\underline{w}^T\underline{x}(n)\right| \qquad (4.34)$$

Maximum Likelihood estimation can be performed by maximising the cost function $J(\underline{w}) = \mathcal{E}\{L_n\}$. Hence, the optimisation problem to be solved is the following:

$$\max_{\mathbf{w}} \mathcal{E}\{\alpha|\underline{w}^{\mathrm{T}}\underline{x}|\} \tag{4.35}$$

$$\text{subject to} \quad \underline{e}^{\mathrm{T}}\underline{w} = 1 \tag{4.36}$$

$$\underline{w} > 0 \tag{4.37}$$

where $\underline{e} = [1\ 1\ \ldots\ 1]^{\mathrm{T}}$. To begin evaluate the solutions to this problem, we can firstly calculate the first derivative:

$$\frac{\partial J(\underline{w})}{\partial \underline{w}} = \frac{\partial}{\partial \underline{w}}\mathcal{E}\{\alpha|\underline{w}^{\mathrm{T}}\underline{x}|\} = \alpha\mathcal{E}\{\mathrm{sgn}(\underline{w}^{\mathrm{T}}\underline{x})\underline{x}\} \tag{4.38}$$

To solve the above optimisation problem, one has to consult methods for constraints optimisation. Using the Lagrange multipliers method for equality constraints and the Kuhn–Tucker conditions for inequality constraints is definitely going to increase the computational complexity of the algorithm. In addition, the available data points for the estimation of the expectation are limited to N^2. Therefore, we propose to solve the unconstrained optimisation problem using a gradient ascent method and impose the constraints at each stage of the adaptation. Consequently, the proposed algorithm can be summarised, as follows:

(1) Initialise $\underline{w} = \underline{e}/T$. This implies the mean fusion rule, i.e. equal importance to all input patches.
(2) Update the weight vector, as follows:

$$\underline{w}^+ \leftarrow \underline{w} + \eta\mathcal{E}\{\mathrm{sgn}(\underline{w}^{\mathrm{T}}\underline{x})\underline{x}\} \tag{4.39}$$

where η represents the learning rate.
(3) Apply the constraints, using the following update rule:

$$\underline{w}^+ \leftarrow |\underline{w}|/(\underline{e}^{\mathrm{T}}|\underline{w}|) \tag{4.40}$$

(4) Iterate steps 2, 3 until convergence.

Effectively, Equation (4.40) ensures that the weights w_i remain always positive and they sum up to one, as it is essential not to introduce any sign or scale deformation during the estimation of the fused image.

4.5.2 Verhulstian priors

The main drawback of using Laplacian priors is the use of the $\mathrm{sgn}(u)$ function in the update algorithm, that has a discontinuity at $u \to 0$ and therefore may cause numerical instability and errors during the update. Usually, this problem is alleviated by thresholding u by a small constant, so that u never gets zero values. Therefore, one can use alternate probabilistic priors that denote sparsity, such as the *generalised Laplacian* or the *Verhulstian* distribution. In the section, we will examine the use of Verhulstian priors in the ML estimation of the fused image.

The *Verhulstian* probability density function can be defined, as follows:

$$p(u) = \frac{e^{-(u-m)/s}}{s(1 + e^{-(u-m)/s})^2} \tag{4.41}$$

where m, s are parameters that control the mean and the standard deviation of the density function. In our case, we will assume zero mean and therefore $m = 0$. We can now derive the log-likelihood function for ML estimation:

$$
\begin{aligned}
L_n &= -\log \frac{e^{-u_{\mathrm{f}}/s}}{s(1 + e^{-u_{\mathrm{f}}/s})^2} \\
&= \frac{u_{\mathrm{f}}}{s} + \log s + 2\log\left(1 + e^{-u_{\mathrm{f}}/s}\right) \\
&= \frac{1}{s}\underline{w}^{\mathrm{T}}\underline{x} + \log s + 2\log\left(1 + e^{-(1/s)\underline{w}^{\mathrm{T}}\underline{x}}\right) \tag{4.42}
\end{aligned}
$$

Maximum Likelihood estimation can be performed in a similar fashion to Laplacian priors, by maximising the cost function $J(\underline{w}) = \mathcal{E}\{L_n\}$. Again, a gradient ascent algorithm is employed, as explained in the previous section with a correcting step that will constrain the solutions in the solution space, permitted by the optimisation problem. The gradient is calculated, as follows:

$$
\begin{aligned}
\frac{\partial J(\underline{w})}{\partial \underline{w}} &= \frac{\partial}{\partial \underline{w}} \mathcal{E}\left\{ \frac{1}{s}\underline{w}^{\mathrm{T}}\underline{x} + \log s + 2\log\left(1 + e^{-(1/s)\underline{w}^{\mathrm{T}}\underline{x}}\right) \right\} \\
&= \mathcal{E}\left\{ \frac{1}{s}\underline{x} - \frac{1}{s}\underline{x} \frac{2e^{-(1/s)\underline{w}^{\mathrm{T}}\underline{x}}}{1 + e^{-(1/s)\underline{w}^{\mathrm{T}}\underline{x}}} \right\} \\
&= \frac{1}{s}\mathcal{E}\left\{ \frac{1 - e^{-(1/s)\underline{w}^{\mathrm{T}}\underline{x}}}{1 + e^{-(1/s)\underline{w}^{\mathrm{T}}\underline{x}}} \underline{x} \right\} \tag{4.43}
\end{aligned}
$$

We can now perform the same algorithm as introduced for Laplacian priors, the only difference being that in Equation (4.39), we have to replace the gradient with that of Equation (4.43). Consequently, the algorithm can be outlined as follows:

(1) Initialise $\underline{w} = \underline{e}/T$. This implies the mean fusion rule, i.e. equal importance to all input patches.
(2) Update the weight vector, as follows:

$$\underline{w}^{+} \leftarrow \underline{w} + \eta \mathcal{E}\left\{ \frac{1 - e^{-(1/s)\underline{w}^{\mathrm{T}}\underline{x}}}{1 + e^{-(1/s)\underline{w}^{\mathrm{T}}\underline{x}}} \underline{x} \right\} \tag{4.44}$$

where η represents the learning rate.

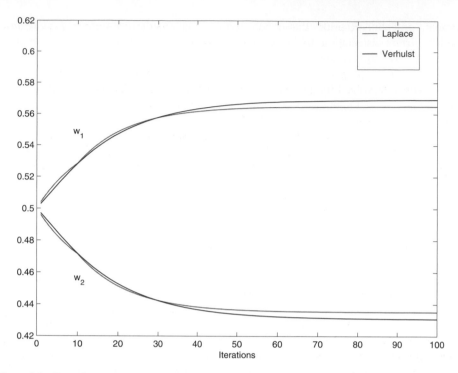

Figure 4.4   *Typical convergence of the ML-estimation fusion scheme using Laplacian and Verhulstian priors.*

(3) Apply the constraints, using the following update rule:

$$\underline{w}^{+} \leftarrow |\underline{w}| / (\underline{e}^{\mathrm{T}} |\underline{w}|) \qquad (4.45)$$

(4) Iterate steps 2, 3 until convergence.

In Figure 4.4, a typical convergence of the two ML-estimation schemes using the two proposed priors is shown. The algorithms converge smoothly after an average of 50–60 iterations.

4.6 Reconstruction of the fused image

The above algorithms have provided a number of possible methods to estimate the fused image $\underline{u}_{\mathrm{f}}(t)$ in the ICA transform domain. The next step is to estimate the spatial-domain representation of the image $I_{\mathrm{f}}(x, y)$. To reconstruct the image in the spatial domain, the process described in Section 4.2 is inverted. The vectors $\underline{u}_{\mathrm{f}}(t)$ are re-transformed to the local $N \times N$ patches $I_{\mathrm{f}}(k, l)$. The local mean of each patch is restored using the stored patches means $MN_k(t)$. The patches are consequently averaged with 1-pixel overlap to create the grid in Figure 4.1, i.e. the fused image. This averaging usually creates an artificial 'frame' around the reconstructed image, which occurs due to the reduced number of frames that are available around the image's borders. To overcome this effect, one can pad with zeros the borders of the input sensors images before the fusion stage, so that the 'framing' effect affects the zero-padded areas only.

The restoration of the patches' local means is a very important issue. Initially, all the patches were normalised to zero mean and the subtracted local intensity mean $MN_k(t)$ was stored to be used in the reconstruction of the fused image. Consequently, there exist T local intensity values for each patch of the reconstructed image, each belonging to the corresponding input sensor. In the case of performing multi-focus image fusion, it is evident that the local intensities from all input sensors will be similar, if not equal, for all corresponding patches. In this case, the local means are reconstructed by averaging the $MN_k(t)$, in terms of k. In the case of multi-modal image fusion, the problem of reconstructing the local intensities of the fused image becomes more serious, since the T input images are acquired from different modality sensors with different intensity range and values. The fused image is an artificial image, that does not exist in nature, and it is therefore difficult to find a criterion that can dictate the most efficient way of combining the input sensors intensity range. The details from all input images will be transferred to the fused image by the fusion algorithm, however, the local intensities will be selected to define the intensity profile of the fused image. In Figure 4.5, the example of a multi-modal fusion scenario is displayed: a visual sensor image is fused with an infrared sensor image. Three possible reconstructions of the fused image's means are shown: (a) the contrast (local means) is acquired from the visual sensor, (b) the contrast is acquired from the infrared image and (c) an average of the local means is used. All three reconstructions contain the same salient features, since these are dictated by the ICA fusion procedure. Each of the three reconstructions simply gives a different impression of the fused image, depending on the prevailing contrast preferences. The average of the local means seems to give a more balanced representation compared to the two extremes. The details are visible in all three reconstructions. However, an incorrect choice of local means may render some of the local details, previously visible in some of the input sensors, totally invisible in the fused image and therefore deteriorate the fusion performance. In this chapter, we will use the average of the local means, giving equal importance to all input sensors. However, there might be another optimum representation of the fused image, by perhaps emphasising means from input sensors with greater intensity range.

An additional problem can be the creation of a 'colour' fused image, as the result of the fusion process. Let us assume that one of the input sensors is a visual sensor. In most real-life situations the visual sensor will provide a colour input image or in other terms a number of channels representing the colour information provided by the sensor. The most common representation in Europe is the RGB (Red–Green–Blue) representation featuring 3 channels of the three basic colours. If the traditional fusion methodology is applied on this problem, a single channel 'fused' image will be produced featuring only intensity changes in greyscale. However, most users and operators will demand a colour rather than a greyscale representation of the 'fused' image. There are several surveillance applications, where a colour 'fused' image is expected from a visual and an infrared sensor [20]. Even in the case of a greyscale visual input sensor and other infrared, thermal sensors, the operator is more likely to prefer a synthetic colour representation of the 'fused' image, rather than a greyscale one [21]. Therefore, the problem of creating a 3-channel representation of the 'fused' image from T channels available by the input sensors can be rather demanding.

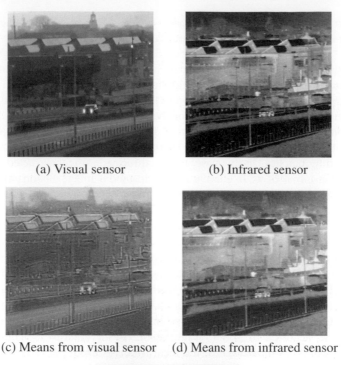

(a) Visual sensor   (b) Infrared sensor

(c) Means from visual sensor   (d) Means from infrared sensor

(e) Average means

Figure 4.5   *Effect of local means choice in the reconstruction of the fused image.*

A first thought would be to treat each of the visual colour channels independently and fuse them with the input channels from the other sensors independently to create a three channel representation of the 'fused' image. Although this technique seems rational and may produce satisfactory results in several cases, it does not utilise the dependencies between the colour channels that might be beneficial for the fusion framework [22]. Another proposed approach [20,22] was to move to another colour space, such as the YUV colour space that describes a colour image using one luminance and two chrominance channels [22] or the HSV colour space that describes a colour image using Hue, Saturation and Intensity (luminance) channels. The two chrominance channels as well as the hue–saturation channels convey colour information solely, whereas the intensity channel describes the image details more accurately. Therefore, the proposed strategy is to fuse the intensity channel with the other input sensor channels and create the intensity channel for the 'fused' image. The chrominance/hue–saturation channels can be used to provide colour information for the 'fused' image. This scheme features reduced computational complexity as one visual channel is fused instead of the original three. In addition, as all these colour transformations are linear mappings from the RGB space, one can use Principal Component Analysis to define the principal channel in terms of maximum variance. This channel is fused with the other input sensors and the resulting image is mapped back to the RGB space, using the estimated PCA matrix. The above techniques are producing satisfactory results in the case of colour out-of-focus input images, since all input images have the same chrominance channels. In the case of multi-modal or multi-exposure images, these methods may not be sufficient and then one can use more complicated colour channel combination and fusion schemes in order to achieve an enhance 'fused' image [20]. These schemes may offer enhanced performance for selected applications only but not in every possible fusion scenario.

4.7 Experiments

In this section, we test the performance of the proposed image fusion schemes based on ICA bases. It is not our intention to provide an exhaustive comparison of the many different transforms and fusion schemes that exist in literature. Instead, a comparison with fusion schemes using *wavelet packets* analysis and the *Dual-Tree (Complex) Wavelet Transform* are performed. In these examples we will test the 'fusion by absolute maximum' (max-abs), the 'fusion by averaging' (mean), the weighted combination (weighted), the region-based (regional) fusion and the adaptive (Laplacian prior) fusion rules.

We present three experiments, using both artificial and real image data sets. In the first experiment, the *Ground Truth* image $I_{gt}(x, y)$ is available, enabling us to perform explicit numerical evaluation of the fusion schemes. We assume that the input images $I_i(x, y)$ are processed by the fusion schemes to create the 'fused' image $I_f(x, y)$. To evaluate the scheme's performance, we can use the following *Signal-to-Noise Ratio* (SNR) expression to compare the ground truth image with the fused image:

$$\text{SNR}_{(\text{dB})} = 10 \log_{10} \frac{\sum_x \sum_y I_{gt}(x, y)^2}{\sum_x \sum_y (I_{gt}(x, y) - I_f(x, y))^2} \tag{4.46}$$

As traditionally employed by the fusion community, we can also use the *Image Quality Index* Q_0, as a performance measure [23]. Assume that m_I represents the mean of the image $I(x, y)$ and all images are of size $M_1 \times M_2$. As $-1 \leqslant Q_0 \leqslant 1$, the value of Q_0 that is closer to 1 indicates better fusion performance:

$$Q_0 = \frac{4\sigma_{I_{gt}I_f} m_{I_{gt}} m_{I_f}}{(m_{I_{gt}}^2 + m_{I_f}^2)(\sigma_{I_{gt}}^2 + \sigma_{I_f}^2)} \tag{4.47}$$

where

$$\sigma_I^2 = \frac{1}{M_1 M_2 - 1} \sum_{x=1}^{M_1} \sum_{y=1}^{M_2} (I(x, y) - m_I)^2 \tag{4.48}$$

$$\sigma_{IJ} = \frac{1}{M_1 M_2 - 1} \sum_{x=1}^{M_1} \sum_{y=1}^{M_2} \bigl(I(x, y) - m_I\bigr)\bigl(J(x, y) - m_J\bigr) \qquad (4.49)$$

For the rest of the experiments, as the 'ground truth' image is not available, two Image Fusion performance indexes will be used: one proposed by Piella [24] and one proposed by Petrovic and Xydeas [25]. Both indexes are widely used by the image fusion community to benchmark the performance of fusion algorithms. They both attempt at quantifying the amount of 'interesting' information (edge information) that has been conveyed from the input images to the fused image. In addition, as Piella's index employs the Image Quality Index Q_0 to quantify the quality of information transfer between each of the input images and the fused image, it is bounded between -1 and 1.

The ICA and the topographic ICA bases were trained using $10\,000$ 8×8 image patches that were randomly selected from 10 images of similar content to the ground truth or the observed scene. We used 40 out of the 64 possible bases to perform the transformation in either case. The local means of the fused image were reconstructed using an average of the means of the input sensor images. We compared the performance of the ICA and Topographic ICA transforms (TopoICA) with a Wavelet Packet decomposition[1] and the Dual-Tree Wavelet Transform.[2] For the Wavelet Packet decomposition (WP), we used Symmlet-7 (Sym7) bases, with 5-level decomposition using Coifman–Wickerhauser entropy. For the Dual-Tree Wavelet Transform (DTWT), we used 4 levels of decomposition and the filters included in the package. In the next pages, we will present some of the resulting fusion images. However, the visual differences between the fused images may not be very clear in the printed version of this chapter, due to limitation in space. Consequently, the reader is prompted to acquire the whole set either by download[3] or via email to us.

4.7.1 Experiment 1: Artificially distorted images

In the first experiment, we have created three images of an 'airplane' using different localised artificial distortions. The introduced distortions can model several different types of degradation that may occur in visual sensor imaging, such as motion blur, out-of-focus blur and finally pixelate or shape distortion, due to low bit-rate transmission or channel errors. This synthetic example can be a good starting point for evaluation, as there are no registration errors between the input images and we can perform numerical evaluation, as we have the ground truth image. We applied all possible combinations of transforms and the fusion rules (the 'Weighted' and 'Regional' fusion rules can not be applied in the described form for the WP and DTWT transforms). Some results are depicted in Figure 4.7, whereas the full numerical evaluation is presented in Table 4.1.

We can see that using the ICA and the TopoICA bases, we can get better fusion results both in visual quality and metric quality (PSNR, Q_0). We observe the ICA bases provide

---

[1] We used WaveLab v8.02, as available at http://www-stat.stanford.edu/~wavelab/.

[2] DT-WT code available online by the Polytechnic University of Brooklyn, NY at http://taco.poly.edu/WaveletSoftware/.

[3] http://www.commsp.ee.ic.ac.uk/~nikolao/BookElsevierImages.zip.

Table 4.1 *Performance comparison of several combinations of transforms and fusion rules in terms of PSNR (dB)/Q_0 using the 'airplane' example.*

	WP (Sym7)	DT-WT	ICA	TopoICA
Max-abs	14.03/0.8245	13.77/0.8175	16.28/0.9191	17.49/0.9354
Mean	23.19/0.9854	23.19/0.9854	20.99/0.9734	21.21/0.9752
Weighted	–	–	21.18/0.9747	21.41/0.9763
Regional	–	–	21.17/0.9746	21.42/0.9764
Laplacian	–	–	20.99/0.9734	21.73/0.9782

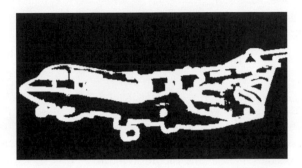

Figure 4.6 *Region mask created for the region-based image fusion scheme. The white areas represent 'active' segments and the black areas 'non-active' segments.*

an improvement of ∼2–4 dB, compared to the wavelet transforms, using the 'max-abs' rule. The TopoICA bases seem to score slightly better than the normal ICA bases, mainly due to better adaptation to local features. In terms of the various fusion schemes, the 'max-abs' rule seems to give very low performance in this example using visual sensors. This can be explained, due to the fact that this scheme seems to highlight the important features of the images, however, it tends to lose some constant background information. On the other hand, the 'mean' rule gives the best performance (especially for the wavelet coefficient), as it seems to balance the high detail with the low-detail information. However, the 'fused' image in this case seems quite 'blurry,' as the fusion rule has oversmoothed the image details. Therefore, the high SNR has to be cross-checked with the actual visual quality and image perception, where we can clearly that the salient features have been filtered. The 'weighted combination' rule seems to balance the pros and cons of the two previous approaches, as the results feature high PSNR and Q_0 (inferior to the 'mean' rule), but the 'fused' images seem sharper with correct constant background information. In Figure 4.6, we can see the segmentation map created by (4.18) and (4.19). The proposed region-based scheme manages to capture most of the salient areas of the input images. It performs reasonably well as an edge detector, however, it produces thicker edges, as the objective is to identify areas around the edges, not the edges themselves. The region-based fusion scheme produces similar results to the 'Weighted' fusion scheme. However, it seems to produce better visual quality in constant background areas, as the 'mean' rule is more suitable for the 'non-active' regions. The adaptive system based on the Laplacian prior seems to achieve the maximum performance in the case of Topographic ICA bases, but not on the trained ICA bases, where it matches the 'mean' rule performance.

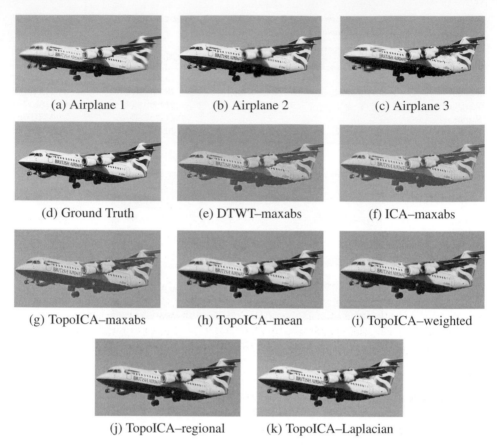

(a) Airplane 1   (b) Airplane 2   (c) Airplane 3

(d) Ground Truth   (e) DTWT–maxabs   (f) ICA–maxabs

(g) TopoICA–maxabs   (h) TopoICA–mean   (i) TopoICA–weighted

(j) TopoICA–regional   (k) TopoICA–Laplacian

Figure 4.7 *Three artificially distorted input images and various fusion results using various transforms and fusion rules.*

4.7.2 Experiment 2: Out-of-focus image fusion

In the second experiment, we use the 'Clocks' and the 'Disk' examples, which are real visual sensor example provided by Lehigh Image Fusion group [26]. In these examples, there are two registered images with different focus points, observing two complicated scenes. In the first image of each set, the focus is on left part and in the second image the focus is on the right part of the image. The ground truth image is not available, which is common in many multi-focus examples. Therefore, SNR-type measurements are not available in this case. Instead, the Piella fusion index [24] and the Petrovic fusion index [25] were used and are depicted in Table 4.2 for various combinations of fusion rules and transform domains. In Figures 4.8 and 4.9 the resulting fused images for different configurations of the two experiments are depicted.

Here, we can see that the ICA and TopoICA bases perform slightly better than wavelet-based approaches in the first example and a lot better in the second example. Also, we can see that the 'max-abs' rule performs slightly better than any other approach, with almost similar performance from the 'Weighted' scheme. The reason is that the three

Table 4.2 *Performance comparison of several combinations of transforms and fusion rules for out-of-focus datasets, in terms of the Piella/Petrovic indexes.*

	WP (Sym7)	DT-WT	ICA	TopoICA
	Clocks dataset			
Max-abs	0.8727/0.6080	0.8910/0.6445	0.8876/0.6530	0.8916/0.6505
Mean	0.8747/0.5782	0.8747/0.5782	0.8523/0.5583	0.8560/0.5615
Weighted	–	–	0.8678/0.6339	0.8743/0.6347
Regional	–	–	0.8583/0.5995	0.8662/0.5954
Laplacian	–	–	0.8521/0.5598	0.8563/0.5624
	Disk dataset			
Max-abs	0.8850/0.6069	0.8881/0.6284	0.9109/0.6521	0.9111/0.6477
Mean	0.8661/0.5500	0.8661/0.5500	0.8639/0.5470	0.8639/0.5459
Weighted	–	–	0.9134/0.6426	0.9134/0.6381
Regional	–	–	0.9069/0.6105	0.9084/0.6068
Laplacian	–	–	0.8679/0.5541	0.8655/0.5489

images have the same colour information, however, most parts of each image are blurred. Therefore, the 'max-abs' that identifies the greatest activity, in terms of edge information, seems more suitable for a multi-focus example. The 'max-abs' simply strengthens the existence of edges in the fused image and can therefore in an out-of-focus situation can excel in restoring these blurred parts of the input images.

4.7.3 Experiment 3: Multi-modal image fusion

In the third experiment, we explore the performance in multi-modal image fusion. In this case, the input images are acquired from different modality sensors to unveil different components in the observed scene. We have used some surveillance images from TNO Human Factors, provided by Toet [27]. More of these can be found in the Image Fusion Server [28]. The images are acquired by three kayaks approaching the viewing location from far away. As a result, their corresponding image size varies from less than 1 pixel to almost the entire field of view, i.e. they are minimal registration errors. The first sensor (AMB) is a Radiance HS IR camera (Raytheon), the second (AIM) is an AIM 256 microLW camera and the third is a Philips LTC500 CCD camera. Consequently, we get three different modality inputs for the same observed scene. The third example is taken from the 'UN Camp' dataset available from the Image Fusion Server [28]. In this case, the inputs consist of a greyscale visual sensor and an infrared sensor. The Piella fusion index [24] and the Petrovic fusion index [25] are measured and are depicted in Table 4.3 for various combinations of fusion rules and transform domains.

In this example, we can witness some minor effects of misregistration in the fused image. We can see that all four transforms seem to have included most salient information from the input sensor images, especially in the 'max-abs' and 'weighted' schemes. However, the ICA and the TopoICA bases approaches seem to excel in comparison to the dual-tree wavelet transform and the wavelet packet approaches. The 'fused image' constructed using the proposed framework seems to be sharper and less blurry compared to the other approaches, especially in the case of the 'max-abs' and 'weighted' schemes. These ob-

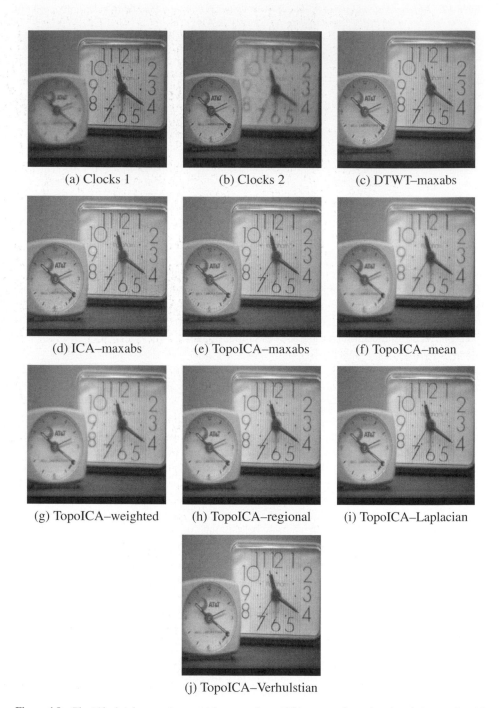

(a) Clocks 1   (b) Clocks 2   (c) DTWT–maxabs

(d) ICA–maxabs   (e) TopoICA–maxabs   (f) TopoICA–mean

(g) TopoICA–weighted   (h) TopoICA–regional   (i) TopoICA–Laplacian

(j) TopoICA–Verhulstian

Figure 4.8 *The 'Clocks' data-set demonstrating several out-of-focus examples and various fusion results with various transforms and fusion rules.*

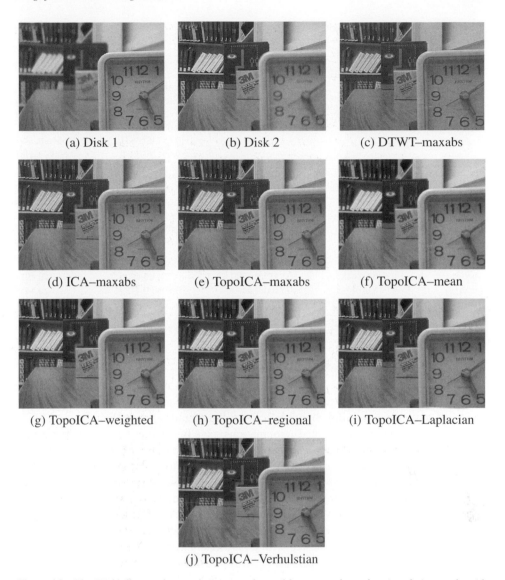

Figure 4.9 *The 'Disk' data-set demonstrating several out-of-focus examples and various fusion results with various transforms and fusion rules.*

servations can be verified in Figures 4.10, 4.11 and 4.12, where some of the produced fused images are depicted for various configurations. The other proposed schemes offer reasonable performance in all multi-modal examples, but not the optimal.

4.8 Conclusion

The authors have introduced the use of ICA and Topographical ICA bases for image fusion applications. These bases seem to construct very efficient tools, which can com-

Table 4.3 *Performance comparison of several combinations of transforms and fusion rules for multimodal datasets, in terms of the Piella/Petrovic indexes.*

	WP (Sym7)	DT-WT	ICA	TopoICA
		Multimodal-1 dataset		
Max-abs	0.6198/0.4163	0.6399/0.4455	0.6592/0.4507	0.6646/0.4551
Mean	0.6609/0.3986	0.6609/0.3986	0.6591/0.3965	0.6593/0.3967
Weighted	–	–	0.6832/0.4487	0.6861/0.4528
Regional	–	–	0.6523/0.3885	0.6566/0.3871
Laplacian	–	–	0.6612/0.3980	0.6608/0.3983
		Multimodal-2 dataset		
Max-abs	0.5170/0.4192	0.58022/0.4683	0.6081/0.4759	0.6092/0.4767
Mean	0.6028/0.420	0.6028/0.4207	0.6056/0.4265	0.6061/0.4274
Weighted	–	–	0.6252/0.4576	0.6286/0.4632
Regional	–	–	0.5989/0.4148	0.5992/0.4133
Laplacian	–	–	0.6071/0.4277	0.6068/0.4279
		'UN Camp' dataset		
Max-abs	0.6864/0.4488	0.7317/0.4780	0.7543/0.4906	0.7540/0.4921
Mean	0.7104/0.4443	0.7104/0.4443	0.7080/0.4459	0.7081/0.4459
Weighted	–	–	0.7361/0.4735	0.7429/0.4801
Regional	–	–	0.7263/0.4485	0.7321/0.4508
Laplacian	–	–	0.7101/0.4475	0.7094/0.4473

pliment common techniques used in image fusion, such as the Dual-Tree Wavelet Transform. The proposed method can outperform wavelet approaches. The Topographical ICA bases offer more accurate directional selectivity, thus capturing the salient features of the image more accurately. A weighted combination image fusion rule seemed to improve the fusion quality over traditional fusion rules in several cases. In addition, a region-based approach was introduced. At first, segmentation into 'active' and 'non-active' areas is performed. The 'active' areas are fused using the pixel-based weighted combination rule and the 'non-active' areas are fused using the pixel-based 'mean' rule. An adaptive fusion rule based on the sparsity of the coefficients in the ICA-domain was also introduced. Sparsity was modelled using either Laplacian or Verhulstian prior with promising results. The proposed framework was tested with an artificial example, two out-of-focus examples and three multi-modal, outperforming current state-of-the-art approaches based on the wavelet transform.

The proposed schemes seem to increase the computational complexity of the image fusion framework. The extra computational cost is not necessarily introduced by the estimation of the ICA bases, as this task is performed only once. The bases can be trained offline using selected image samples and then employed constantly by the fusion applications. The increase in complexity comes from the 'sliding window' technique that is introduced to achieve shift invariance. Implementing this fusion scheme in a more computationally efficient framework than MATLAB will decrease the time required for the image analysis and synthesis part of the algorithm.

For future work, the authors would be looking at evolving to a more autonomous fusion system, exploring the nature of 'topography,' as introduced by Hyvärinen et al., and

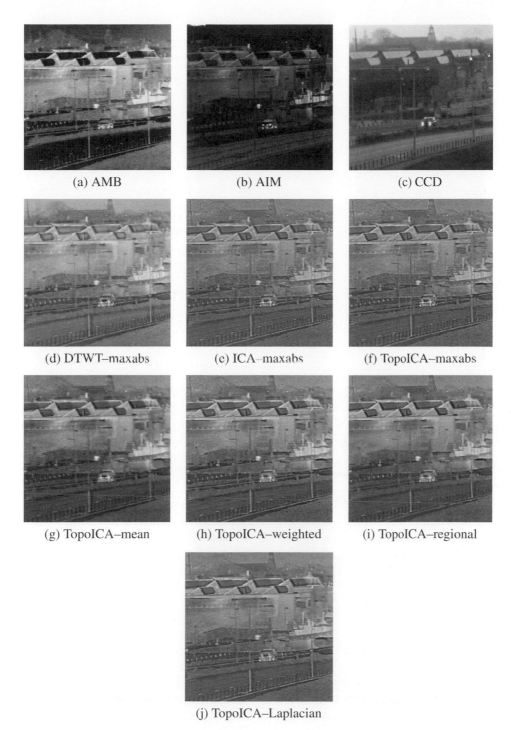

Figure 4.10 *Multi-modal image fusion: Three images acquired through different modality sensors and various fusion results with various transforms and fusion rules.*

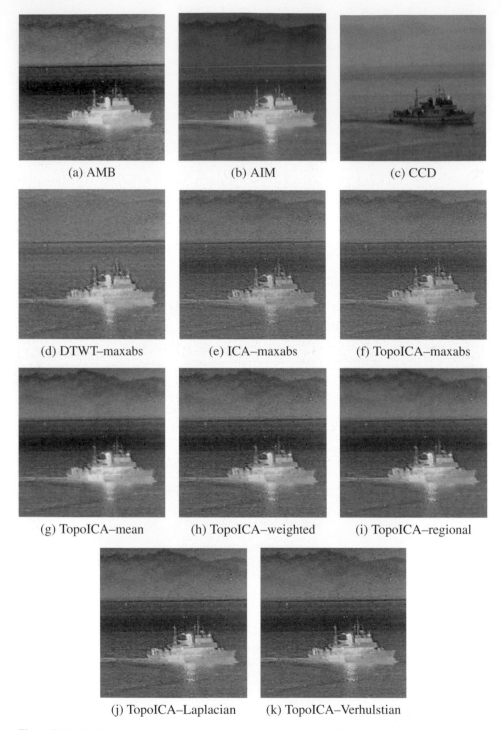

Figure 4.11 *Multi-modal image fusion: Three images acquired through different modality sensors and various fusion results with various transforms and fusion rules.*

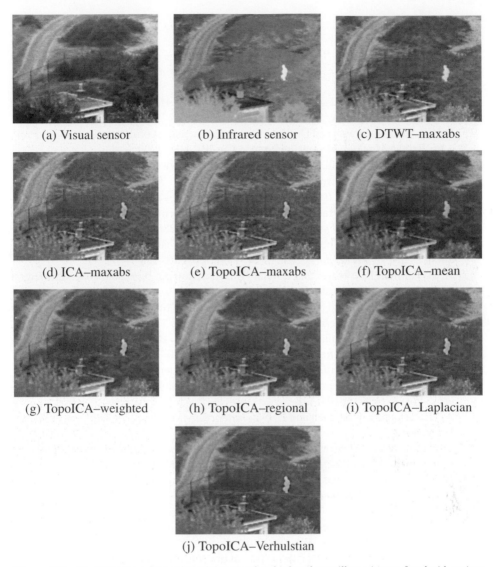

(a) Visual sensor   (b) Infrared sensor   (c) DTWT–maxabs

(d) ICA–maxabs   (e) TopoICA–maxabs   (f) TopoICA–mean

(g) TopoICA–weighted   (h) TopoICA–regional   (i) TopoICA–Laplacian

(j) TopoICA–Verhulstian

Figure 4.12 *The 'UN camp' dataset containing visual and infrared surveillance images fused with various transforms and fusion rules.*

form more efficient activity detectors, based on topographic information. In addition, they would be looking at more sophisticated methods for the selection of intensity or colour range of the fused image in the case of multi-modal or colour image fusion.

Acknowledgements

This work was supported by the Data Information Fusion Phase-I project 6.4 and the Phase-II AMDF cluster project of the Defence Technology Centre, UK.

References

[1] I. Bloch and H. Maitre, 'Data fusion in 2d and 3d image processing: An overview', in *Proc. X Brazilian Symposium on Computer Graphics and Image Processing*, 1997, pp. 122–134.

[2] A. Goshtasby, *2-D and 3-D Image Registration: For Medical, Remote Sensing, and Industrial Applications*, John Wiley & Sons, 2005.

[3] P. Hill, N. Canagarajah and D. Bull, 'Image fusion using complex wavelets', in *Proc. 13th British Machine Vision Conference*, Cardiff, UK, 2002.

[4] S.G. Nikolov, D.R. Bull, C.N. Canagarajah, M. Halliwell and P.N.T. Wells, 'Image fusion using a 3-d wavelet transform', in *Proc. 7th International Conference on Image Processing and Its Applications*, 1999, pp. 235–239.

[5] G. Piella, 'A general framework for multiresolution image fusion: From pixels to regions', *Information Fusion*, Vol. 4, 2003, pp. 259–280.

[6] O. Rockinger and T. Fechner, 'Pixel-level image fusion: The case of image sequences', *SPIE Proceedings*, Vol. 3374, 1998, pp. 378–388.

[7] A. Hyvärinen, P.O. Hoyer and E. Oja, 'Image denoising by sparse code shrinkage', in S. Haykin and B. Kosko (eds.), *Intelligent Signal Processing*, IEEE Press, 2001.

[8] H. Li, S. Manjunath and S. Mitra, 'Multisensor image fusion using the wavelet transform', *Graphical Models and Image Processing*, Vol. 57, No. 3, 1995, pp. 235–245.

[9] J.J. Lewis, R.J. O'Callaghan, S.G. Nikolov, D.R. Bull and C.N. Canagarajah, 'Region-based image fusion using complex wavelets', in *Proc. 7th International Conference on Information Fusion*, Stockholm, Sweden, 2004, pp. 555–562.

[10] A. Hyvärinen, 'Survey on independent component analysis', *Neural Computing Surveys*, Vol. 2, 1999, pp. 94–128.

[11] A. Hyvärinen, J. Karhunen and E. Oja, *Independent Component Analysis*, John Wiley & Sons, 2001.

[12] A. Cichocki and S.I. Amari, *Adaptive Blind Signal and Image Processing. Learning Algorithms and Applications*, John Wiley & Sons, 2002.

[13] A. Hyvärinen, 'Fast and robust fixed-point algorithms for independent component analysis', *IEEE Transactions on Neural Networks*, Vol. 10, No. 3, 1999, pp. 626–634.

[14] A. Hyvärinen, P.O. Hoyer and M. Inki, 'Topographic independent component analysis', *Neural Computation*, Vol. 13, 2001.

[15] R.R. Coifman and D.L. Donoho, 'Translation-invariant de-noising', Technical report, Department of Statistics, Stanford University, Stanford, CA, 1995.

[16] N. Mitianoudis and T. Stathaki, 'Pixel-based and region-based image fusion schemes using ICA bases', *Elsevier Information Fusion*, Vol. 8, No. 2, 2007, pp. 131–142.

[17] N. Mitianoudis and T. Stathaki, 'Adaptive image fusion using ICA bases', in *Proceedings of the International Conference on Acoustics, Speech and Signal Processing*, Toulouse, France, May 2006.

[18] M. Sonka, V. Hlavac and R. Boyle, *Image Processing Analysis and Machine Vision*, second ed., Brooks/Cole Publishing Company, 1999.

[19] B.A. Olshausen, 'Sparse codes and spikes', in R.P.N. Rao, B.A. Olshausen and M.S. Lewicki (eds.), *Probabilistic Models of the Brain: Perception and Neural Function*, MIT Press, 2002.

[20] Z. Xue and R.S. Blum 'Concealed weapon detection using color image fusion', in *Proc. Int. Conf. on Information Fusion*, 2003, pp. 622–627.

[21] A.M. Waxman, M. Aguilar, D.A. Fay, D.B. Ireland, J.P. Racamato Jr., W.D. Ross, J.E. Carrick, A.N. Gove, M.C. Seibert, E.D. Savoye, R.K. Reich, B.E. Burke, W.H. McGonagle and D.M. Craig, 'Solid-state color night vision: Fusion of low-light visible and thermal infrared imagery', *Lincoln Laboratory Journal*, Vol. 11, No. 1, 1998, pp. 41–60.

[22] L. Bogoni, M. Hansen and P. Burt, 'Image enhancement using pattern-selective color image fusion', in *Proc. Int. Conf. on Image Analysis and Processing*, 1999, pp. 44–49.

[23] Z. Wang and A.C. Bovik, 'A universal image quality index', *IEEE Signal Processing Letters*, Vol. 9, No. 3, 2002, pp. 81–84.

[24] G. Piella, 'New quality measures for image fusion', in *7th International Conference on Information Fusion*, Stockholm, Sweden, 2004.

[25] C. Xydeas and V. Petrovic, 'Objective pixel-level image fusion performance measure', in *Sensor Fusion IV: Architectures, Algorithms and Applications*, in *Proc. SPIE*, Vol. 4051, 2000, pp. 88–99.

[26] Lehigh fusion test examples, http://www.eecs.lehigh.edu/spcrl/if/toy.htm.

[27] A. Toet, *Targets and Backgrounds: Characterization and Representation VIII*, The International Society for Optical Engineering, 2002, pp. 118–129.

[28] The Image fusion server, http://www.imagefusion.org/.

5

Statistical modelling for wavelet-domain image fusion

Alin Achim, Artur Łoza, David Bull and Nishan Canagarajah

Department of Electrical and Electronic Engineering, University of Bristol, Bristol, UK

This chapter describes a new methodology for multimodal image fusion based on non-Gaussian statistical modelling of wavelet coefficients of the input images. The use of families of generalised Gaussian and alpha-stable distributions for modelling image wavelet coefficients is investigated and methods for estimating distribution parameters are proposed. Improved techniques for image fusion are developed, by incorporating these models into the weighted average image fusion algorithm. The superior performance of the proposed methods is demonstrated using visual and infrared light image datasets.

5.1 Introduction

The purpose of image fusion is to combine information from multiple images of the same scene into a single image that ideally contains all the important features from each of the original images. The resulting fused image will be thus more suitable for human and machine perception or for further image processing tasks. Many image fusion schemes have been developed in the past. In general, these schemes can be roughly classified into pixel-based and region-based methods. In [1] it has been shown that comparable results can be achieved using both types of methods with added advantages for the region based approaches, mostly in terms of the possibility of implementing more intelligent fusion rules. On the other hand, pixel based algorithms are simpler and thus easier to implement.

The majority of pixel based image fusion approaches, although effective, have not been developed based on strict mathematical foundations. Only in recent years more rigorous approaches have been proposed, including those based on estimation theory [2]. In [3] a Bayesian fusion method has been proposed, allowing to adaptively estimate the relationships between the multiple image sensors in order to generate a single enhanced display. The multiresolution image representation was assumed to follow a Gaussian distribution. This limiting assumption was relaxed in [4], where a generalisation has been presented, allowing modelling both Gaussian and non-Gaussian distortions to the input images. In [4–6], an Expectation Maximisation algorithm was used to estimate model parameters and the fused image. This approach was further refined in [7,8], by using hidden Markov models in order to describe the correlations between the wavelet coefficients across decomposition scales. In [9], a combined Wavelet and Cosine Packets image fusion algorithm was presented. Counts of input image coefficients contributing to the fused image were modelled by a log-linear distribution. The final fused image was constructed based on the confidence intervals for the probabilities of a fused pixel coming from a nth input image being selected within a packet from multiple hypotheses testing.

Recent work on non-Gaussian modelling for image fusion has been proposed in [10]. Specifically, an image fusion prototype method, originally proposed in [11], has been modified to account for non-Gaussianity of the image distributions. Since there exist strong evidence that wavelet coefficient of images are very well modelled by Symmetric Alpha-Stable (SαS) distributions, second-order statistics used in [11], have been replaced by Fractional Lower-Order Moments (FLOMs) of SαS distributions. This novel approach to image fusion resulted in improved performance compared to earlier pixel level fusion techniques.

In this chapter we extend the work presented in [10]. We discuss different possibilities of reformulating and modifying the original Weighted Average (WA) method in order to cope with more appropriate statistical model assumptions like the generalised Gaussian and the alpha-stable. We use a relatively novel framework, that of Mellin transform theory, in order to estimate all statistical parameters involved in the fusion algorithms.

As is the case with many of the recently proposed techniques, our developments are made using the wavelet transform, which constitutes a powerful framework for implementing image fusion algorithms [1,11–13]. Specifically, methods based on multiscale decompositions consist of three main steps: first, the set of images to be fused is analysed by means of the wavelet transform, then the resulting wavelet coefficients are fused through an appropriately designed rule, and finally, the fused image is synthesised from the processed wavelet coefficients through the inverse wavelet transform. This process is depicted in Figure 5.1. In implementing the algorithms described in this chapter we make use of the Dual-Tree Complex Wavelet Transform (DT-CWT) that has been shown to offer near shift invariance and improved directional selectivity compared to the standard wavelet transform [14,15]. Due to these properties, image fusion methods implemented in the complex wavelet domain have been shown to outperform those implemented using the discrete wavelet transform [13].

The chapter is organised as follows: In Section 5.2, we provide some necessary preliminaries on generalised Gaussian and alpha-stable processes and present results on the modelling of subband coefficients of VIsible light (VI) and Infra-Red (IR) images. Section 5.3 describes the modified WA algorithms for wavelet-domain image fusion, which are based on heavy-tailed models. Special emphasis is put on data-driven parameter estimation of the distributions and their use in the fusion framework. Section 5.4 compares the performance of the new algorithms with the performance of the conventional weighted average fusion technique applied to sequences of multimodal test images. Finally, in Section 5.5 we conclude the paper with a short summary and suggest areas of future research.

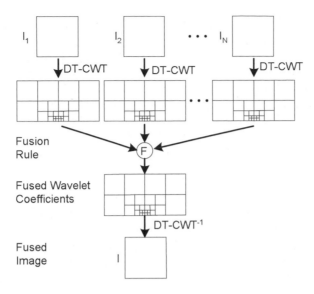

Figure 5.1 *Pixel-based image fusion scheme using the DT-CWT.*

5.2 Statistical modelling of multimodal images wavelet coefficients

5.2.1 Heavy-tailed distributions

In the past, several authors have pointed out that, in a subband representation of images, histograms of wavelet coefficients have heavier tails and more sharply peaked modes at zero than what is assumed by the Gaussian distribution [16–18]. Most commonly, the distributions of image wavelet coefficients are modelled as Generalised Gaussian Distribution (GGD) [16,18] or SαS [19,20]. Both models are families of distributions that are in general non-Gaussian and heavy-tailed. SαS processes include the Cauchy and the Gaussian distributions as limiting cases, whereas the GGD covers both Laplacian and Gaussian cases. The advantage of GGD models consists in the availability of analytical expressions for their probability density functions (pdf) as well as of simple and efficient parameter estimators. On the other hand, SαS distributions are much more flexible and rich. For example, they are also able to capture skewed characteristics.

5.2.1.1 The generalised Gaussian distribution
The generalised Gaussian density function proposed by Mallat [18] and also used by Simoncelli [16] is given by

$$f_{s,p}(x) = \frac{1}{Z(s,p)} \cdot e^{-|x/s|^p} \qquad (5.1)$$

where $Z(s, p)$ is a normalisation constant

$$Z(s, p) = 2\frac{s}{p}\Gamma\left(\frac{1}{p}\right)$$

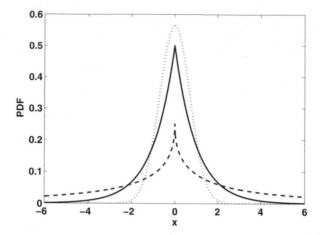

Figure 5.2 *Example of generalised Gaussian probability density functions for $p = 0.5$ (dashed line), 1.0 (Laplacian – solid line), and 2.0 (Gaussian – dotted line). The scale parameter is kept constant at $s = 1$.*

and $\Gamma(\cdot)$ is the well-known Gamma function

$$\Gamma(t) = \int_0^\infty e^{-u} u^{t-1} \, du$$

In (5.1), s models the width of the pdf peak (standard deviation), while p is inversely proportional to the decreasing rate of the peak. Usually, s is referred to as the scale parameter, while p as the shape parameter. The smaller the parameter p is, the heavier the tails of the corresponding GG density. The GGD model contains the Gaussian and Laplacian pdfs as special cases, corresponding to $p = 2$ and $p = 1$, respectively. Examples of Generalised Gaussian family of distributions are shown in Figure 5.2. Note different cusp and tail behaviours depending on the value of the shape parameter p. As we will see later, it is sometimes convenient to express s in terms of the standard deviation σ:

$$s = \sigma \left(\frac{\Gamma\left(\frac{1}{p}\right)}{\Gamma\left(\frac{3}{p}\right)} \right)^{1/2} \tag{5.2}$$

In fact, one of the simplest ways of estimating the parameters s and p of the GGD is using the second and fourth moments of the data [16,18]:

$$\sigma^2 = \frac{s^2 \Gamma\left(\frac{3}{p}\right)}{\Gamma\left(\frac{1}{p}\right)}, \qquad k = \frac{\Gamma\left(\frac{1}{p}\right)\Gamma\left(\frac{5}{p}\right)}{\Gamma^2\left(\frac{3}{p}\right)} \tag{5.3}$$

where k is the kurtosis.

5.2.1.2 Alpha-stable distributions

The appeal of symmetric alpha-stable (SαS) distributions as a statistical model for signals derives from two important theoretical reasons [21,22]. First, stable random variables satisfy the stability property which states that linear combinations of jointly stable variables are indeed stable. Second, stable processes arise as limiting processes of sums of independent and identically distributed (i.i.d.) random variables via the generalised central limit theorem. Actually, the *only* possible non-trivial limit of normalised sums of i.i.d. terms is stable.

The SαS distribution is best defined by its characteristic function

$$\varphi(\omega) = \exp\left(\jmath \delta \omega - \gamma |\omega|^\alpha\right) \tag{5.4}$$

In (5.4) α is the *characteristic exponent*, taking values $0 < \alpha \leqslant 2$, δ $(-\infty < \delta < \infty)$ is the *location parameter*, and γ $(\gamma > 0)$ is the *dispersion* of the distribution. For values of α in the interval $(1, 2]$, the location parameter δ corresponds to the mean of the SαS distribution, while for $0 < \alpha \leqslant 1$, δ corresponds to its median. The dispersion parameter γ determines the spread of the distribution around its location parameter δ, similar to the variance of the Gaussian distribution. An interesting observation is that on analysing Equations (5.1) and (5.4) it can be seen that the expressions corresponding to the GG pdf and to the SαS characteristic function are identical for $\delta = 0$.

The characteristic exponent α is the most important parameter of the SαS distribution and it determines the shape of the distribution. The smaller the characteristic exponent α is, the heavier the tails of the SαS density. This implies that random variables following SαS distributions with small characteristic exponents are highly impulsive. Gaussian processes are stable processes with $\alpha = 2$, while Cauchy processes result when $\alpha = 1$. In fact, no closed-form expressions for the general SαS pdf are known except for the Gaussian and the Cauchy members. Although the SαS density behaves approximately like a Gaussian density near the origin, its tails decay at a lower rate than the Gaussian density tails [21]. Specifically, while the Gaussian density (and generalised Gaussian as well) has exponential tails, the stable densities have algebraic tails. Figure 5.3 shows the tail behaviour of several SαS densities including the Cauchy and the Gaussian.

One consequence of heavy tails is that only moments of order less than α exist for the non-Gaussian alpha-stable family members, i.e.,

$$E|X|^p < \infty \quad \text{for } p < \alpha \tag{5.5}$$

However, FLOMs of SαS random variables can be defined and are given by [22]

$$E|X|^p = C(p, \alpha)\gamma^{p/\alpha} \quad \text{for } -1 < p < \alpha \tag{5.6}$$

where

$$C(p, \alpha) = \frac{2^{p+1}\Gamma\left(\frac{p+1}{2}\right)\Gamma\left(-\frac{p}{\alpha}\right)}{\alpha\sqrt{\pi}\,\Gamma\left(-\frac{p}{2}\right)} \tag{5.7}$$

and $\Gamma(\cdot)$ is the Gamma function. In the past, this infinite variance property of the SαS family has caused sceptics to dismiss the stable model. With the same reasoning, one could argue that the routinely used Gaussian distribution, which has infinite support,

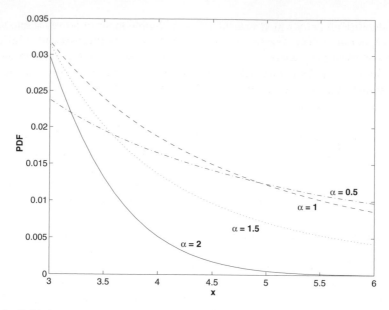

Figure 5.3 *Tail behaviour of the SαS probability density functions for α = 0.5, 1.0 (Cauchy), 1.5, and 2.0 (Gaussian). The dispersion parameter is kept constant at γ = 1.*

should also be dismissed as a model of bounded measurements. In practice, one should remember that it is important to capture the shape of the distribution and that the variance is only one measure of the spread of a density [23].

5.2.2 Modelling results of wavelet subband coefficients

In this section, we study whether the stable and generalised Gaussian families of distributions provide flexible and appropriate tools for modelling the coefficients within the framework of multiscale wavelet analysis of images.

Two sets of surveillance images[1] kindly provided by Dr Lex Toet, from TNO Human Factors, have been used in this research. They include two different image modalities: VI or CCD images and IR (thermal 3–5 μm) images. The image sequences, which we prefer to refer to as *UN camp*, has 32 frames. More information about how this data was gathered can be found in [24].

We proceed in two steps: first, we assess whether the data deviate from the normal distribution and if they have heavy tails. To determine that, we make use of normal probability plots. Then, we check if the data is in the stable or generalised Gaussian domains of attraction by estimating the characteristic exponent α, and shape parameter p, respectively, directly from the data. Several methods have been proposed for estimating stable parameters. Here, we use the maximum likelihood (ML) method described by Nolan [23], which

---

[1] All the images used in this study are available online at http://www.imagefusion.org.

gives reliable estimates and provides the most tight confidence intervals. In order to perform a fair comparison between the two candidate models, we also employ a ML method for estimating GG parameters. Specifically, we make use of the algorithm proposed by Do and Vetterli [25]. As further stability diagnostics, we employ logarithmic probability density plots that give a good indication of whether a statistical model fit matches the data near the mode and on the tails of the distribution.

For every image we iterated three times the DT-CWT decomposition [14] and we studied the coefficients of each subband. The first and third rows in Figure 5.4 depict the normal probability plots corresponding to the first three levels of decomposition and the same orientation for a VI and IR image, respectively. The plots provide strong evidence that the underlying distribution is not normal. The circles in the plots show the empirical probability versus the data values for each point in the sample. The circles are in curves that do not follow the straight Gaussian lines and thus, the normality assumption is violated for all these data. While both stable and generalised Gaussian densities are heavy-tailed, not all heavy-tailed distributions are either stable or generalised Gaussian. Hence, the second and fourth rows of Figure 5.4 assess the appropriateness of using these two heavy-tailed models. The second row corresponds to the VI image while the fourth to the IR. The characteristic exponent and the shape parameter are estimated and the data samples are fitted with the corresponding distributions. For the particular cases shown here, the distribution parameters that best fit the data are shown above each graph. It can be observed that for both models the shape parameters depend on the particular level of decomposition. The shape parameters become smaller as the level increases since the corresponding subbands become more and more sparse.

Naturally, the real question is which of the two candidate models describe the data more accurately. On analysing Figure 5.4 one can observe that the SαS distribution is in general superior to the generalised Gaussian distribution because it provides a better fit to both the mode and the tails of the empirical density of the actual data. Nevertheless, the figure demonstrates that the coefficients of different subbands and decomposition levels exhibit various degrees of non-Gaussianity. The important observation is that all subbands exhibit distinctly non-Gaussian characteristics, with values of α varying between 1.3 and 1.9 and p between 0.6 and 1.4, away from the Gaussian points of α, $p = 2$. Our modelling results clearly point to the need for the design of fusion rules that take into consideration the non-Gaussian heavy-tailed character of the data to achieve close to optimal image fusion performance.

5.3 Model-based weighted average schemes

After more than ten years since its inception, the weighted average method proposed in [11] remains one of the most effective, yet simple and easy to implement image fusion algorithm. However, the method is based on a local-Gaussianity assumption for wavelet subbands, while in the previous section we have shown than wavelet coefficients clearly exhibit non-Gaussian characteristics. In this section we show how the WA method can be reformulated and modified in order to cope with more appropriate statistical model

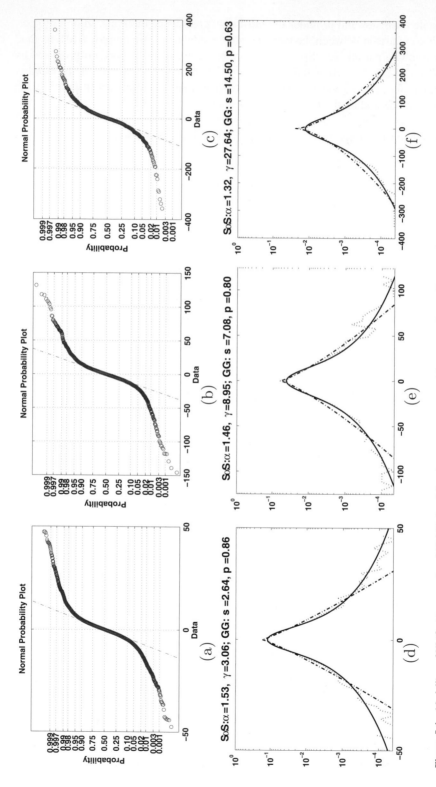

Figure 5.4 *Modelling of V1 (first two rows) and IR (third and fourth rows, see next page) image wavelet coefficients. The second and fourth rows show the fit of the empirical distribution with the SαS and the generalised Gaussian density functions, depicted in solid and dashed lines, respectively. The dotted line denotes the empirical pdf. Note that the SαS pdf provides a better fit to both the mode and the tails of the empirical density of the actual data.*

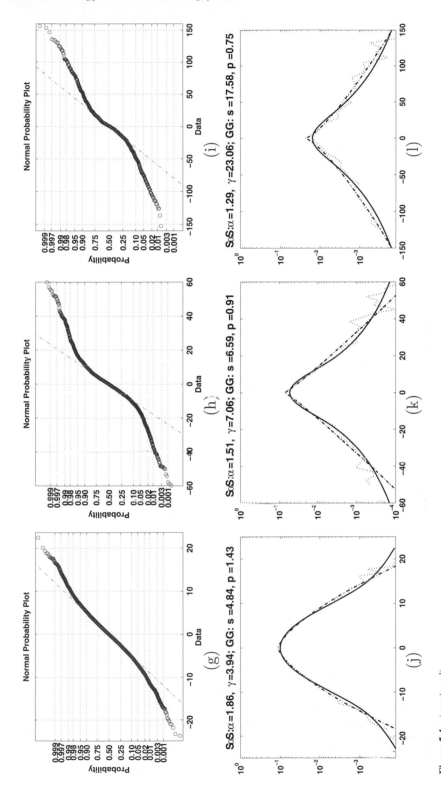

Figure 5.4 *(continued)*

assumptions like the generalised Gaussian and the alpha-stable. For the completeness of the presentation we first recall the original method (based on [11] and [10]):

---

1. Decompose each input image into subbands.
2. For each subband pair X, Y, except the lowpass residuals:
   (a) Compute neighbourhood saliency measures, σ_X and σ_Y.
   (b) Compute neighbourhood matching coefficient

$$M = \frac{2\sigma_{XY}}{\sigma_X^2 + \sigma_Y^2} \qquad (5.8)$$

   where σ_{XY} stands for covariance between X and Y.
   (c) Calculate the fused coefficients using the formula
      $Z = W_X X + W_Y Y$ as follows:
      - if $M > T$ ($T = 0.75$) then $W_{min} = 0.5\left(1 - \frac{1-M}{1-T}\right)$ & $W_{max} = 1 - W_{min}$
        (weighted average mode, including mean mode for $M = 1$),
      - else $W_{min} = 0$ & $W_{max} = 1$ (selection mode),
      - if $\sigma_X > \sigma_Y$, $W_X = W_{max}$ and $W_Y = W_{min}$,
      - else $W_X = W_{min}$ and $W_Y = W_{max}$.
3. Average coefficients in lowpass residuals.
4. Reconstruct the fused image from the processed subbands and the lowpass residual.

Essentially, the algorithm shown above considers two different modes for fusion: selection and averaging. The overall fusion rule is determined by two measures: a match measure that determines which of the two modes is to be employed and a saliency measure that determines which wavelet coefficient in the pair will be copied in the fused subband (selection mode), or which coefficient will be assigned the larger weight (weighted average mode). Adapting the above algorithm to the case of GGDs is relatively easy since it simply presupposes appropriate estimation of the saliency measure (which is also a term in the match measure). For the case of SαS distributions classical second-order moments and correlation cannot be used and consequently new match and saliency measures need to be defined. In [10] we proposed the use of dispersion γ, estimated in a neighbourhood around the reference wavelet coefficient, as a saliency measure. The *symmetric coefficient of covariation* was also introduced in order to account for the similarities between corresponding patterns in the two subbands to be fused. In the following we show how all these quantities can be estimated adaptively, in the context of Mellin transform theory, for both GG and SαS distributions.

5.3.1 Saliency estimation using Mellin transform

Following the arguments in [26] and [27], Nicolas has recently proposed the use of the Mellin Transform (MT) as a powerful tool for deriving novel parameter estimation methods based on log-cumulants [28]. The technique has also been successfully used in [29] for signal parameters estimation in the case of multiplicative noise contamination. In the following, we briefly review the MT and its main properties that we used in our derivations.

5.3.1.1 Mellin transform

Let f be a function defined over $\Re^+$. The MT of f is defined as

$$\Phi(z) = \mathbf{M}[f(x)](z) = \int_0^{+\infty} x^{z-1} f(x)\,\mathrm{d}x \tag{5.9}$$

where z is the complex variable of the transform. The inverse transform is given by

$$f(x) = \mathbf{M}^{-1}[\Phi(z)](x) = \frac{1}{2\pi j} \int_{c-j\infty}^{c+j\infty} x^{-z} \Phi(z)\,\mathrm{d}z \tag{5.10}$$

The transform $\Phi(z)$ exists if the integral $\int_0^{+\infty} |f(x)| x^{k-1}\,\mathrm{d}x$ is bounded for some $k > 0$, in which case the inverse $f(x)$ exists with $c > k$. The functions $\Phi(z)$ and $f(x)$ are called a MT pair, and either can be computed if the other is known.

By analogy with the way in which common statistics are deducted based on the Fourier Transform, the following second-kind statistic functions can be defined, based on the MT [28], for a random variable $x \sim p(x)$:

- Second-kind first characteristic function

$$\Phi(z) = \int_0^{|\infty} x^{z-1} p(x)\,\mathrm{d}x \tag{5.11}$$

- Second-kind second characteristic function

$$\Psi(z) = \log\big(\Phi(z)\big) \tag{5.12}$$

- rth-order second-kind cumulants

$$\tilde{k}_r = \left.\frac{\mathrm{d}^r \Psi(z)}{\mathrm{d}z^r}\right|_{z=1} \tag{5.13}$$

The first two second-kind cumulants can be estimated empirically from N samples y_i as follows:

$$\hat{\tilde{k}}_1 = \frac{1}{N} \sum_{i=1}^{N} \big[\log(x_i)\big],$$

$$\hat{\tilde{k}}_2 = \frac{1}{N} \sum_{i=1}^{N} \big[\big(\log(x_i) - \hat{\tilde{k}}_1\big)^2\big] \tag{5.14}$$

5.3.1.2 Log-moment estimation of the GG model

In this section we show how the saliency and match measures (5.8) can be computed for samples coming from GG distributions. Specifically, the variance terms appearing in the denominator of (5.8) need to be estimated differently, depending on which member of the

GG family is considered. By plugging the expression of the GG pdf given by (5.1) into (5.11) and after some straightforward manipulations, one gets

$$\Phi(z) = \frac{s^z}{p} \Gamma\left(\frac{z}{p}\right)$$

and

$$\Psi(z) = \ln \Phi(z) = z \ln s + \ln \Gamma(z) - \ln p$$

which is the second-kind second characteristic function of a GG density. Calculating the first- and second-order second-kind cumulants (5.13) gives

$$\tilde{k}_1 = \left.\frac{d\Psi(z)}{dz}\right|_{z=1} = \ln s + \frac{\psi_0\left(\frac{1}{p}\right)}{p}$$

and

$$\tilde{k}_2 = F(p) = \left.\frac{d\Psi^2(z)}{dz^2}\right|_{z=1} = \frac{\psi_1\left(\frac{1}{p}\right)}{p^2}$$

respectively, where

$$\psi_n(t) = \frac{d^{n+1}}{dt^{n+1}} \ln \Gamma(t)$$

is the polygamma function. The shape parameter p is estimated by computing the inverse of the function F. Then, p can be substituted back into the equation for $\tilde{k}_1$ in order to find s.

In order to stress the importance of using different estimates of saliency for different members of the GG family of distributions let us do the following analysis. After estimating p as above, a ML estimate of s could be computed from data x as

$$s = \left(\frac{p}{N} \sum_{k=1}^{N} |x_k|^p\right)^{1/p} \tag{5.15}$$

Substituting (5.2) into (5.15) gives a general expression for the ML estimate of variance:

$$\sigma^2 = \frac{\Gamma\left(\frac{3}{p}\right)}{\Gamma\left(\frac{1}{p}\right)} \left(\frac{p}{N} \sum_{k=1}^{N} |x_k|^p\right)^{2/p} \tag{5.16}$$

In particular, if x is assumed to be Gaussian ($p = 2$) the familiar estimate of variance is obtained

$$\sigma_x^2 = \frac{1}{N} \sum_{n=1}^{N} x_k^2$$

while for the Laplacian distribution ($p = 1$) one gets

$$\sigma_x = \frac{\sqrt{2}}{N} \sum_{n=1}^{N} |x_k|$$

To further illustrate the influence of the parameter p, let us now consider two vectors $x = [-a, -b, a, b]$ and $y = [-(a + b), 0, a + b, 0]$, where a, b are positive numbers. Depending on which distribution is assumed, different variances are obtained, and thus it can be shown that for Gaussian $\sigma_x < \sigma_y$, for Laplacian $\sigma_x = \sigma_y$ and for $p = 0.5$ $\sigma_x > \sigma_y$. In general, for high values of p, sparser vectors will tend to have higher variances. This property influences the fusion results significantly.

5.3.1.3 Log-moment estimation of the SαS model

In [27], Zolotarev established the following relationship between the Fourier and Mellin transforms:

$$\Phi(z) = \frac{\Gamma(1 + z)}{2\pi j} \int (j\omega)^{-z} \left(\varphi(\omega) - \frac{1}{1 + j\omega} \right) \frac{d\omega}{\omega} \tag{5.17}$$

By using (5.17) together with the characteristic function $\varphi(\omega)$ of a SαS density in (5.4) with location parameter $\delta = 0$, we obtain

$$\Phi(z) = \frac{\gamma^{(z-1)/\alpha} 2^z \Gamma\left(\frac{z}{2}\right) \Gamma\left(-\frac{z-1}{\alpha}\right)}{\alpha \sqrt{\pi} \Gamma\left(-\frac{z-1}{2}\right)} \tag{5.18}$$

which is the second-kind first characteristic function of a SαS density. Next, by plugging the above expression in (5.12) and subsequently in (5.13), we obtain the following results for the second-kind cumulants of the SαS model

$$\tilde{k}_1 = \frac{\alpha - 1}{\alpha} \psi(1) + \frac{\log \gamma}{\alpha} \tag{5.19}$$

and

$$\tilde{k}_2 = \frac{\pi^2}{12} \frac{\alpha^2 + 2}{\alpha^2} \tag{5.20}$$

The estimation process simply involves now solving (5.20) for α and substituting back in (5.19) to find the value of the dispersion parameter γ (the saliency measure). We should note that this method of estimating SαS parameters was first proposed in [30]. Here, we have only provided an alternative derivation of the method, based on the MT properties.

5.3.2 Match measure for SαS random variables: The symmetric covariation coefficient

The notion of covariance between two random variables plays an important role in the second-order moment theory. However, due to the lack of finite variance, covariances do not exist either on the space of SαS random variables. Instead, quantities like covariations or codifferences, which under certain circumstances play analogous roles for SαS random variables to the one played by covariance for Gaussian random variables have been introduced [21]. Specifically, let X and Y be jointly SαS random variables with $\alpha > 1$, zero location parameters and dispersions γ_X and γ_Y, respectively. The covariation of X with Y is defined in terms of the previously introduced FLOMs by Samorodnitsky and Taqqu [21]:

$$[X, Y]_\alpha = \frac{E(XY^{\langle p-1 \rangle})}{E(|Y|^p)} \gamma_Y \tag{5.21}$$

where the p-order moment is defined as $x^p = |x|^p \operatorname{sign}(x)$. Moreover, the covariation coefficient of X with Y, is the quantity

$$\lambda_{X,Y} = \frac{[X, Y]_\alpha}{[Y, Y]_\alpha} \quad \text{for any } 1 \leqslant p < \alpha \tag{5.22}$$

Unfortunately, it is a well-known fact that the covariation coefficient in (5.22) is neither symmetric nor bounded [21]. Therefore, in [10] we proposed the use of a symmetrised and normalised version of the above quantity, which enables us to define a new match measure for SαS random vectors. The symmetric covariation coefficient that we used for this purpose can be simply defined as

$$\operatorname{Corr}_\alpha(X, Y) = \lambda_{X,Y} \lambda_{Y,X} = \frac{[X, Y]_\alpha [Y, X]_\alpha}{[X, X]_\alpha [Y, Y]_\alpha} \tag{5.23}$$

Garel et al. [31] have shown that the symmetric covariation coefficient is bounded, taking values between -1 and 1. In our implementation, the above similarity measure is computed in a square-shaped neighbourhood of size 7×7 around each reference coefficient.

5.4 Results

In this section, we show results obtained using the model-based approach to image fusion described in this chapter. Appropriate methodology for comparing different image fusion algorithms generally depends on the application. For example, in applications like medical image fusion, the ultimate goal is to combine perceptually salient image elements such as edges and high contrast regions. Evaluation of fusion techniques in such situation can only be effective based on visual assessment. There are also applications (e.g. multifocus image fusion) when computational measures could be employed. Several qualitative measures have been proposed for this purpose. For example, in [32] Zhang and Blum have used a mutual information criterion, the root mean square error as well as a measure representing the percentage of correct decisions. Unfortunately, all these measures involve the existence of a reference image for their computation, which in practice is generally not available. Moreover, the problem with the mutual information measure, or with any other metric, is their connection to the visual interpretation of an human observer. On analysing an image, a human observer does not compute any such measure. Hence, in order to study the merit of the proposed fusion methodology, we chose different images, applied the different algorithms, and visually evaluated the fused images.

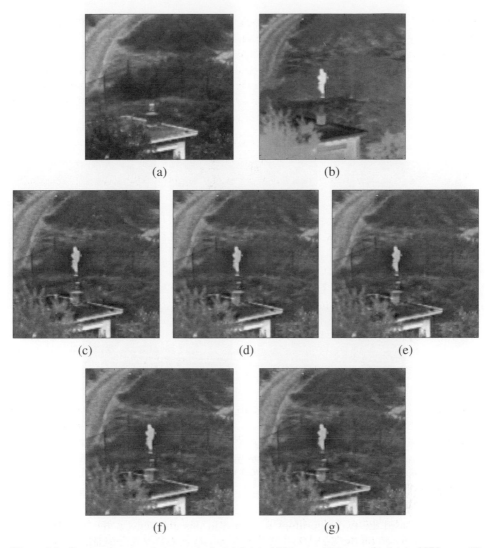

Figure 5.5 *Results of various fusion methods. (a) Original* UN camp *VI image. (b) Original* UN camp *IR image. (c) Image fused using WA. (d) Image fused using GGD-WA. (e) Image fused using SαS-WA. (f) Image fused using Laplacian-WA. (g) Image fused using Cauchy-WA.*

We were interested in performing experiments on images with various content in order to be able to obtain results, which we could claim to be general enough. Thus, the first example shows one image pair from the sequences that we used in the modelling part of this chapter (Section 5.2.2). As a second example, we chose to illustrate the fusion of images from our *Tropical* dataset. These are still images extracted from video sequences recorded during a data gathering exercise organised by the University of Bristol at the Eden Project [33].

The experimental results are shown in Figures 5.5 and 5.6. Results are obtained using five different methods based on the WA scheme. Apart from the original WA method [11],

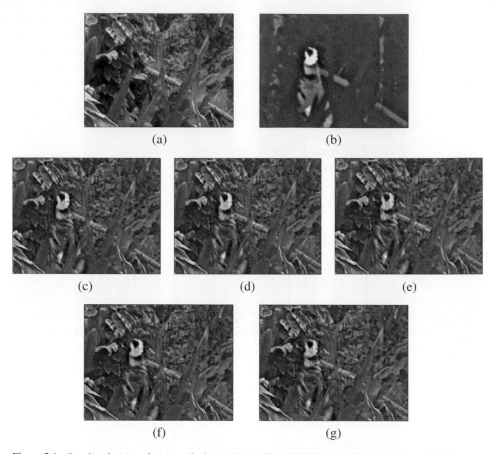

Figure 5.6 *Results of various fusion methods. (a) Original* Tropical *VI image. (b) Original* Tropical *IR image. (c) Image fused using WA. (d) Image fused using GGD-WA. (e) Image fused using SαS-WA. (f) Image fused using Laplacian-WA. (g) Image fused using Cauchy-WA.*

we have included the two algorithms described in this chapter, i.e. the weighted average schemes based on the SαS (SαS-WA) and on GG (GGD-WA) modelling of wavelet coefficients, as well as two particular cases of these, corresponding to the Cauchy and Laplacian densities, respectively. It is important to note at this point that the original WA represents in fact a particular case of any of the algorithms described here. Although further qualitative evaluation in this way is highly subjective, it seems that the best results are achieved by the GG based WA technique for the first pair of images, while both GG and SαS performed equally well in the second example. It appears that our systems, the GGD-WA in particular, perform like feature detectors, retaining the features that are clearly distinguishable in each of the input images.

Admittedly, especially for the SαS based algorithm, the fusion results are not as spectacular as one would expect after the excellent modelling results obtained in Section 5.2. This could be due in part to the fact that SαS parameter estimation methods are relatively sensitive to the sample size. Remember that in Section 5.2 we have shown results obtained by modelling globally whole wavelet subbands, while in the actual fusion algorithms the scale parameters are estimated locally in square windows centred around the reference coefficients. Most importantly, the SαS match measure (i.e. the symmetric covariation coefficient) should be improved or replaced by a more powerful measure. The match measure plays an essential role in the fusion process as it is the one on which the decision to select or to average is based. Hopefully, a measure based on codifferences could represent a viable alternative and constitutes one of our present areas of investigation. Nevertheless, the results shown in this chapter demonstrate that it is feasible to implement more flexible fusion rules by taking into account the non-Gaussian characteristics inherent in image data. They pave the way towards the design of improved fusion algorithm based on the actual data statistics in order to achieve close to optimal data fusion performance.

5.5 Conclusions and future work

In this chapter, we discussed new approaches to image fusion for the case when the data to be fused exhibit heavy tails and sharp cusp of the pdf at the origin. We have shown through extensive modelling that typical VI/IR images and their corresponding wavelet coefficients have highly non-Gaussian characteristics that can be accurately described by GGD or SαS statistical models. Our modelling results show evidence that wavelet decomposition coefficients of images are best characterised by SαS distributions. However, the use of such models has been long time hampered by the fact that no convergent second- or higher-order moments exist. On the other hand, the less accurate fit offered by GGDs is compensated by the availability of analytical expressions for their pdf as well as that of simple parameter estimators.

We proposed new statistical model-based image fusion methods by reformulating the well-known WA scheme in order to account for the heavy-tailed nature of data. Due to the lack of second or higher order moments for the general alpha-stable family members, we introduced new saliency and match measures, which are derived based on Mellin transform properties. In the multiscale domain, we employed the local dispersion of wavelet coefficients as saliency measure, while *symmetric covariation coefficients* were computed in order to account for the similarities between corresponding patterns in the pair of subbands to be fused. A similar approach has been applied to GGD parameters estimation, resulting in a novel estimator based on the variance of the logarithmically scaled random variable.

The fusion results show that in general the best performance is achieved by the GG-WA technique followed by the SαS-WA, which has also the potential for further improvement. The fusion algorithms described in this chapter facilitate efficient feature detection, retaining the salient features present in each of the input images.

An interesting direction in which this work could be extended is the development of algorithms that will additionally capture the inherent dependencies of wavelet coefficients across scales. This could be achieved by the use of multivariate statistical models. Research in this direction is under way and will be presented in a future communication.

Acknowledgements

The authors are grateful for the financial support offered to project 2.1 'Image and video sensor fusion' by the UK MOD Data and Information Fusion Defence Technology Centre.

References

[1] J.J. Lewis, R.J. O'Callaghan, S.G. Nikolov, D.R. Bull and N.C. Canagarajah, 'Region-based image fusion using complex wavelets', in *Proc. Seventh Int. Conf. on Information Fusion*, 2004, pp. 555–562.

[2] R.S. Blum, 'On multisensor image fusion performance limits from an estimation theory perspective', *Information Fusion*, Vol. 7, No. 3, 2006, pp. 250–263.

[3] R. Sharma and M. Pavel, 'Adaptive and statistical image fusion', *Society for Information Display Digest*, Vol. 17, No. 5, 1996, pp. 969–972.

[4] J. Yang and R. Blum, 'A statistical signal processing approach to image fusion for concealed weapon detection', in *Proceedings of 2002 International Conference on Image Processing*, Vol. 1, 2002, pp. I-513–I-516.

[5] Z. Zhang and R. Blum, 'Region-based image fusion scheme for concealed weapon detection', in *Proc. 31th Annual Conference on Information Sciences and Systems*, Baltimore, MD, 1997, pp. 168–173.

[6] R.S. Blum and J. Yang, 'Image fusion using the expectation-maximization algorithm and a Gaussian mixture model', in G.L. Foresti, C.S. Regazzoni and P. Varnshey (eds.), *Advanced Video-Based Surveillance Systems*, Kluwer Academic Publishers, 2003.

[7] J. Yang and R.S. Blum, 'Image fusion using the expectation-maximization algorithm and the hidden Markov models', in *Vehicular Technology Conference (VTC)*, 2004.

[8] J. Yang and R.S. Blum, 'A statistical signal processing approach to image fusion using hidden Markov models', in *Multi-Sensor Image Fusion and Its Applications*, Marcel Dekker/CRC, 2005.

[9] A. Cardinali and G.P. Nason, 'A statistical multiscale approach to image segmentation and fusion', in *Proceedings of the 8th International Conference on Information Fusion (Fusion 2005)*, Philadelphia, PA, USA, 25–29 July 2005.

[10] A. Achim, C.N. Canagarajah and D.R. Bull, 'Complex wavelet domain image fusion based on fractional lower order moments', in *Proceedings of the 8th International Conference on Information Fusion (Fusion 2005)*, Philadelphia, PA, USA, 25–29 July 2005.

[11] P.J. Burt and R.J. Kolczynski, 'Enhanced image capture through fusion', in *Proc. Fourth Int. Conf. on Computer Vision*, 1993, pp. 173–182.

[12] O. Rockinger, 'Image sequence fusion using a shift-invariant wavelet transform', in *Proc. IEEE Int. Conf. on Image Processing*, 1997, pp. 288–291.

[13] P. Hill, N.C. Canagarajah and D.R. Bull, 'Image fusion using complex wavelets', in *Proc. 13th British Machine Vis. Conf. (BMVC-2002)*, 2002, pp. 487–496.

[14] N.G. Kingsbury, 'Image processing with complex wavelets', *Philosophical Transactions of the Royal Society of London A*, Vol. 357, 1999, pp. 2543–2560.

[15] N.G. Kingsbury, 'Complex wavelets for shift invariant analysis and filtering', *Applied and Computational Harmonic Analysis*, Vol. 10, 2001, pp. 234–253.

[16] E.P. Simoncelli, 'Bayesian denoising of visual images in the wavelet domain', in P. Muller and B. Vidakovic (eds.), *Bayesian Inference in Wavelet Based Models*, Springer-Verlag, New York, 1999, Chapter 18, pp. 291–308.

[17] P. Tsakalides, P. Reveliotis and C.L. Nikias, 'Scalar quantization of heavy-tailed signals', *IEE Proceedings – Vision, Image and Signal Processing*, Vol. 147, No. 5, 2000, pp. 475–484.

[18] S.G. Mallat, 'A theory for multiresolution signal decomposition: The wavelet representation', *IEEE Transactions on Pattern Analysis and Machine Intelligence*, Vol. 11, 1989, pp. 674–692.

[19] A. Achim, P. Tsakalides and A. Bezerianos, 'SAR image denoising via Bayesian wavelet shrinkage based on heavy-tailed modeling', *IEEE Transactions on Geoscience and Remote Sensing*, Vol. 41, 2003, pp. 1773–1784.

[20] A. Achim and E.E. Kuruoglu, 'Image denoising using bivariate α-stable distributions in the complex wavelet domain', *IEEE Signal Processing Letters*, Vol. 12, 2005, pp. 17–20.

[21] G. Samorodnitsky and M.S. Taqqu, *Stable Non-Gaussian Random Processes: Stochastic Models with Infinite Variance*, Chapman and Hall, New York, 1994.

[22] C.L. Nikias and M. Shao, *Signal Processing with Alpha-Stable Distributions and Applications*, John Wiley and Sons, New York, 1995.

[23] J.P. Nolan, 'Maximum likelihood estimation and diagnostics for stable distributions', Technical report, Department of Mathematics and Statistics, American University, 1999.

[24] A. Toet, J.K. IJspeert, A.M. Waxman and M. Aguilar, 'Fusion of visible and thermal imagery improves situational awareness', *Displays*, Vol. 24, 1997, pp. 85–95.

[25] M.N. Do and M. Vetterli, 'Wavelet-based texture retrieval using generalized Gaussian density and Kullback–Leibler distance', *IEEE Transactions on Image Processing*, Vol. 11, 2002, pp. 146–158.

[26] B. Epstein, 'Some applications of the Mellin transform in statistics', *The Annals of Mathematical Statistics*, Vol. 19, 1948, pp. 370–379.

[27] V.M. Zolotarev, 'Mellin–Stieltjes transforms in probability theory', *Theory of Probability and Its Applications*, Vol. 2, No. 4, 1957, pp. 432–460.

[28] J.M. Nicolas, 'Introduction aux statistiques de deuxième espèce: Applications des log-momentset des log-cumulants à l'analyse des lois d'images radar', *Traitement du Signal*, Vol. 19, 2002, pp. 139–167.

[29] A. Achim, E.E. Kuruoglu and J. Zerubia, 'SAR image filtering based on the heavy-tailed Rayleigh model', *IEEE Transactions on Image Processing*, Vol. 15, 2006, pp. 2686–2693.

[30] X. Ma and C.L. Nikias, 'Parameter estimation and blind channel identification in impulsive signal environment', *IEEE Transactions on Signal Processing*, Vol. 43, No. 12, 1995, pp. 2884–2897.

[31] B. Garel, L. d'Estampes and D. Tjostheim, 'Revealing some unexpected dependence properties of linear combinations of stable random variables using symmetric covariation', *Communications in Statistics – Theory and Methods*, Vol. 33, No. 4, 2004, pp. 769–786.

[32] Z. Zhang and R. Blum, 'A categorization of multiscale-decomposition-based image fusion schemes with a performance study for a digital camera application', *Proceedings of the IEEE*, Vol. 87, No. 8, 1999, pp. 1315–1326.

[33] J.J. Lewis, S.G. Nikolov, A. Łoza, E.F. Canga, N. Cvejic, J. Li, A. Cardinali, C.N. Canagarajah, D.R. Bull, T. Riley, D. Hickman and M.I. Smith 'The Eden Project Multi-Sensor Data Set', Technical Report TR-UoB-WS-Eden-Project-Data-Set, 2006; available at http://www.imagefusion.org.

6

Theory and implementation of image fusion methods based on the *á trous* algorithm

Xavier Otazu

Computer Vision Center, Universitat Autònoma de Barcelona,
Cerdanyola del Vallès, Barcelona, Spain

In this work, we present an introductory and detailed explanation of several image fusion methods based on the *á trous* multiresolution wavelet decomposition algorithm. As shown by many authors in the literature, from the many existing wavelet decomposition schemes the *á trous* algorithm produces the best results when used for image fusion tasks. We present a short introduction to the different wavelet-based image fusion approaches, and we describe some practical advices and how to implement several image fusion algorithms that use this particular wavelet decomposition.

6.1 Introduction

In remote sensing spaceborne imagery there exists a trade-off between spatial and spectral resolution. The combination of a set of observational constraints imposed by the acquisition system, detector specifications, and satellite motion, among others, are the reasons for this problem. Hence, spaceborne imagery is usually offered to the community as two separate products: a High Resolution Panchromatic (HRP) image and a Low Resolution Multispectral (LRM) image. In addition, an increasing number of applications, such as feature detection, change monitoring and land cover classification often demand the highest spatial and spectral resolution for the best accomplishment of their objectives. In response to those needs, image fusion has become a powerful solution providing a single image which contains simultaneously the multispectral content of the original LRM image and an enhanced spatial resolution.

A large collection of fusion methods developed over the last two decades can be found in the literature. Initial efforts based on component substitution on the Intensity–Hue–Saturation colour space (depending on the particular definition of the intensity, component may be IHS or LHS transforms – see next sections) [1–3], Principal Component Substitution (PCS) [4–6], or relative spectral contribution (Intensity Modulation – IM) [7], PX + S on SPOT Handbook [8], Brovey transform [9,10] are mainly focused on enhancing spatial resolution for easing tasks of human photo-interpretation. However, it is not possible to undertake quantitative analysis of fused images obtained by those methods in a systematic way and with high degree of reliability, since the original multispectral content of the LRM image is greatly distorted.

In an attempt to overcome this limitation, another family of methods was developed. These operate on the basis of the injection of high-frequency components from the HRP image into the LRM image. This family of methods was at the beginning initiated by the High-Pass Filtering (HPF) method [4], which provides far less spectral distortion with respect to its predecessors [11]. However, it was not until a second more recent stage that, with the upcoming of methods based on multiresolution analysis, fused products accomplish good enough results which can be employed for quantitative studies of their multispectral content (land-cover mapping and urban areas mapping [12]). In the following, we outline the different fusion methods whose decomposition algorithms are inscribed inside this category and briefly discuss their distinctive nature.

6.1.1 Multiresolution-based algorithms

The basic idea of all fusion methods based on wavelets is to extract the spatial detail information from the HRP image not present in the LRM to inject it into the latter, usually using a multiresolution framework [13–15]. In the next subsection, we explain how to implement some of these methods.

Different multiscale wavelet-based image fusion methods can be distinguished by the algorithm used to perform the detail extraction of the HRP image, for example:

- Based on decimated wavelet transform algorithms, used by Ranchin and Wald [16], Garguet-Duport et al. [17], Yocky [18], and Zhou et al. [19] amongst others.
- Based on undecimated or redundant wavelet transform algorithms, used by Aiazzi et al. [20], González-Audícana et al. [21], and Núñez et al. [22].

The main difference between the decimated and undecimated fusion algorithms is the presence or absence of subsampling when the wavelet decomposition is performed. This subsampling causes a loss of linear continuity in spatial features such as edges and the appearance of artifacts in those structures with neither horizontal nor vertical directions [23].

As shown in previous works [20,21,23], the undecimated algorithms are more appropriate for image fusion purposes than the decimated ones.

Besides, different wavelet-based image fusion methods can be distinguished by the method used to inject the spatial detail of the HRP image into the LRM one:

- Injection directly into each LRM band [16–18].
- Injection into each LRM band through an IHS transform [22–24].
- Injection into each LRM band through a Principal Component Analysis [21].

When the first methods are applied, the spatial detail of the HRP image is injected into each LRM band, so that the spatial information introduced into the whole LRM image is n times that of the HRM image, n being the number of spectral bands of the LRM sensor. Consequently, redundant spatial detail incorporation may appear when using these methods.

When methods based on Intensity–Hue–Saturation (IHS) and Principal Component Analysis (PCA) are used, the quality of the resulting merged images depends on the bandwidth of the HRP sensor. The best performance for IHS-based methods occurs when it covers the entire range of bandwidths of all the LRM bands, as with the Ikonos and QuickBird panchromatic sensor [25], for example. The use of a very narrow HRP band, as that of the SPOT 4 M mode, is more favourable to PCA than to IHS.

In order to preserve to the highest extent the spectral information of the multispectral image, different transformation models could be applied to the spatial detail information extracted from the HRP image before its injection into the LRM one. The simplest one is the identity model proposed by Mangolini [26]. Recently, more complex models have been proposed by Aiazzi et al. [20].

The combination of different wavelet decomposition algorithms, different spatial detail transformation models, and different spatial detail injection procedures results in many different image fusion methods. In the following, we show in detail how to implement some of these methods, focusing on the ones that use the *à trous* wavelet decomposition algorithm.

6.2 Image fusion algorithms

In this section we explain in detail how to implement some image fusion methods that use the wavelet transform in order to extract the spatial detail from the PAN image. We show a detailed explanation of the *à trous* algorithm and several detailed injection methods.

6.2.1 Energy matching

Depending on the format and the information available about the image data sets, they can be supplied in digital counts or physical radiance units. Data sets supplied in digital counts are obtained while preprocessing original data in order to describe it in a compact notation using 8 or 16 bit numbers. Furthermore, digital counts from different bands cannot be compared because every band is processed independently from the others. However, if we know how to convert these counts into radiance units, we can obtain this original physical magnitude. In this case, we can compare values from different bands. Therefore, when this information is available, we should always convert digital counts into radiance units.

6.2.1.1 Histogram equalisation

If we cannot describe data sets in radiance units, we may preprocess them before the image fusion process in order to obtain image data sets with comparable values.

The usual process to obtain comparable data sets is to perform histogram matching which allows obtaining data sets with comparable dynamic range and statistical distribution. Since we work with a panchromatic and several multispectral bands, the usual procedure is to modify the histogram of the panchromatic image to match the one from the intensity I calculated from the set of the several multispectral ones. The mathematical expression to obtain I depends on both the image fusion algorithm we are going to use and the numerical dynamic range of the multispectral bands.

6.2.1.2 Radiance

When all data sets are described in radiance units, no preprocessing is needed because the data from different data sets is directly comparable.

6.2.2 Spatial detail extraction. The à trous algorithm

Wavelet Transform (also called wavelet decomposition) is a frequency transform. Fourier Transform is also a frequency transform, but there are some important differences with the Wavelet Transform. The most important difference is that while the Fourier Transform is defined on the spatial frequency domain, the Wavelet Transform is defined in both the spatial frequency and spatial location. That is, the Fourier Transform depends only on the spatial frequency, i.e. $FT(\nu)$, whereas the Wavelet Transform depends on frequency ν and location t, i.e. it can be written as a function of the form $WT(\nu, t)$. It means that Fourier Transform tells us about the spatial frequencies present in our image, but the Wavelet Transform tells us about them and also where they are located in our image. Several Wavelet Transform algorithms exist, but for image fusion tasks the *à trous* algorithm is one of the most widely used [22].

The implementation of the *à trous* is one of the simplest wavelet decomposition algorithms. Most of the wavelet decomposition algorithms usually decompose the original image into a sequence of new images (usually called *wavelet planes*) of decreasing size. In this sequence, every wavelet plane ω_i has a half of the number of rows and columns of ω_{i-1}, i.e. the number of pixels of ω_i is $N_i^2 = (N_{i-1}/2)^2$. These are called pyramidal wavelet decomposition algorithms [15]. In contrast, in the *à trous* algorithm the number of pixels is the same for all wavelet planes and it is the same as in the original image, i.e. $N_i = N, \forall i$. One of the interesting properties of this decomposition is that it is translation invariant, i.e. the wavelet planes resulting from the *à trous* decomposition of a translated image are just a translation of the wavelet coefficients. Pyramidal wavelet decomposition algorithms do not fulfil this property and the final wavelet coefficients may be completely different. Furthermore, the higher the i value, the lower the spatial frequency features contained in the ω_i wavelet plane. That is, the wavelet planes with lower i values contain the image features of higher spatial frequency, and the ones with the higher values contain the lower spatial frequency.

The *à trous* wavelet decomposition algorithm can be described in the following way:

- Define a convolution kernel as

$$
h_i(j,k) = \frac{1}{256}
\begin{pmatrix}
1 & 4 & 6 & 4 & 1 \\
4 & 16 & 24 & 16 & 4 \\
6 & 24 & 36 & 24 & 6 \\
4 & 16 & 24 & 16 & 4 \\
1 & 4 & 6 & 4 & 1
\end{pmatrix}
$$

- Define $c_0 = I$, I being the original image.
- Define n as the number of wavelet planes into which decompose the image.
- Take $i = 1$.
- Step 1: Convolve c_{i-1} with the kernel h_i to obtain the image c_i, i.e. $c_i = c_{i-1} \otimes h_i$. (See the description below for details on how to correctly perform this convolution.)
- Step 2: Obtain the wavelet plane ω_i by taking the difference between c_{i-1} and c_i, i.e. $\omega_i = c_{i-1} - c_i$.
- Step 3: Define $i = i + 1$.
- Step 4: If $i \leqslant n$ go to Step 1.

The spatial convolution performed by kernel h_i in Step 1 is a very important process that usually leads to confusion and wrong results if it is not performed in the correct way. In contrast to what we may initially think, this convolution is not performed in the same way during all iterations. The kernel h_i is modified during every iteration by doubling its size and inserting null values between the original ones, but ignoring these intermediate new values when performing the convolution with the enlarged kernel. This procedure is graphically explained in Figure 6.1.

In Figure 6.1(a), we show the pixels used when performing the convolution of the kernel h_1 with the original image $I = c_0$. In this case, the convolution process is performed in the usual way, i.e. by multiplying the pixels in the local neighbourhood with the corresponding value of the kernel h_1 and adding these values to store the final value in the central black pixel of the new resulting image. But during the next iteration $i + 1$, the size of kernel h_{i+1} is doubled. In Figure 6.1(b) we show how the convolution is performed using the new h_{i+1} kernel. In contrast to the convolution with the h_i kernel, in the present case the used pixels are not the immediate neighbours of the central pixel, but a few further ones (the dashed squares in Figure 6.1(b)). That is, the values of the h_{i+1} kernel are the same as h_i kernel, but they are applied to a different set of neighbour pixels. The

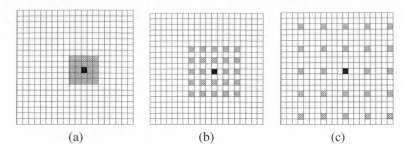

(a)   (b)   (c)

Figure 6.1 *Convolution of kernel h_i is performed doubling its size during every new iteration and ignoring the new values (see text for details). White squares are image pixels, the black square is the central pixel of the kernel h_i and the image pixel where final result will be stored. Dashed squares are the only pixels used to perform the convolution.*

white pixels placed between the central black pixel and the dashed ones are not used to perform the convolution, hence the name *à trous* (with holes).

Since wavelet planes obtained by the *à trous* algorithm have the same size, recovery of the original image is easily performed by just an addition of all the wavelet planes and the residual plane:

$$I = \sum_{i=1}^{n} \omega_i + c_n$$

6.2.3 Spatial detail injection

The general idea of spatial detail injection image fusion methods is to extract from the PAN image the spatial details not present in the LRM bands and inject these details into them (Figure 6.2). Several methods can be used in order to inject the spatial details extracted from the PAN image into the several m LRM bands [14,18,20,22]. Depending on the information we have about the PAN image and LRM bands, some methods are more suitable than others. Furthermore, some of them can only be used when information about the spectral response function (SRF) of both the PAN and the LRM sensors is available [25].

The summation of the n wavelet planes ω_i, $i = 1, \ldots, n$, obtained from the decomposition of the PAN data set contains the spatial detail we will inject into the LRM bands. The value n is chosen according to the relation of the spatial resolution between the PAN and the LRM bands. A property of the wavelet transform is that the ith wavelet plane contains features with spatial frequency $v = 2^{-i}$ [15]. Hence, since the spatial resolution (or spatial period) of the ith wavelet plane is 2^i pixels, the number of wavelet planes that contain the spatial detail present in the PAN image but not present in the LRM bands is $n = \log_2(s_{\text{LRM}}/s_{\text{PAN}})$, where s_{LRM} is the spatial resolution of the LRM bands and s_{PAN} is the spatial resolution of the PAN band. When this value is not an exact integer, we take the closest integer value.

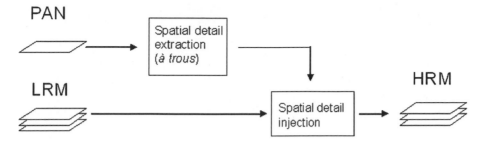

Figure 6.2 *General scheme of spatial detail injection image fusion algorithms. Spatial detail is extracted from the PAN image and injected into the LRM bands. Different injection methods produce different results.*

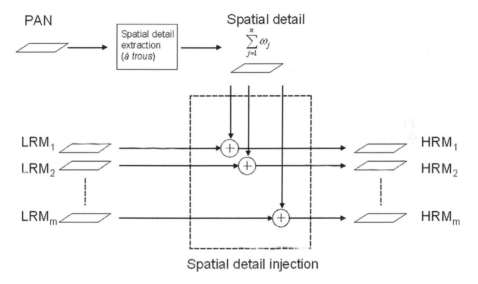

Figure 6.3 *AWRGB image fusion method. Spatial detail is directly injected into every LRM band.*

6.2.3.1 Direct injection

The simplest way to inject a spatial detail into the LRM bands is to directly add it into every LRM band (Figure 6.3) [22]. The mathematical expression is

$$\mathrm{HRM}_i = \mathrm{LRM}_i + \sum_{j=1}^{n} \omega_j$$

where HRM_i is the ith merged High Resolution Multispectral band. In this method, the same amount of spatial detail is added into all the LRM_i, $i = 1, \ldots, m$, bands. It implies that all LRM bands are equally treated and no information about them is used during the spatial injection procedure. Since the same amount of spatial detail is injected into all the LRM bands, a grey feature (i.e. a feature with the same value for all the spectral bands) is added into the LRM data set. It implies that this method biases towards grey values the areas where spatial detail is added. It clearly modifies the multispectral information present in the original LRM bands.

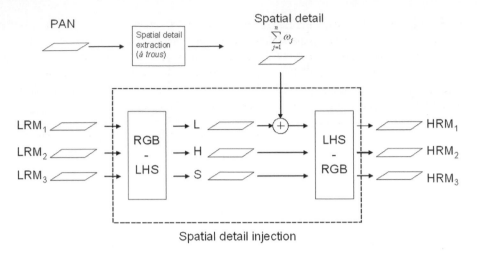

Figure 6.4 *AWL image fusion method. Spatial detail is injected into the intensity component obtained from the three LRM bands.*

This method is called AWRGB (Additive Wavelet RGB) and is only suitable when no information about PAN and LRM Spectral Response Functions (SRFs) is available.

6.2.3.2 Intensity-based (LHS)

Intensity-based methods are amongst the first developed approaches that try to preserve the original multispectral information of the LRM data sets [1]. Similarly to the classical LHS image fusion methods, these methods add the spatial detail into the intensity component of the LRM bands:

$$\mathrm{HRL} = \mathrm{LRL} + \sum_{j=0}^{n} \omega_j$$

where LRL is the intensity component of the LRM bands and HRL is the obtained high resolution merged intensity band (Figure 6.4). The intensity band L is usually defined as $L = (R + G + B)/3$ [27].

Combining this new HRL intensity component and the H and S components from the original LRM image, we define the LHS values of the merged image (Figure 6.4). Performing a LHS–RGB transformation [27,28] we obtain a merged RGB image.

Since the spectral information of an image is mainly included into the hue (H) and saturation (S) components of the LHS representation, the merged HRM image preserves the spectral information of the original LRM in a certain degree. But one of the drawbacks of this algorithm is that the LRM image has to contain only three bands, i.e. $m = 3$, because the RGB–LHS transform is defined on a three-dimensional colour space. It reduces the general applicability of this algorithm to very few cases but, fortunately, some of the most widely used data sets in remote sensing are on this category. One of these cases is the LANDSAT and SPOT images. This multispectral sensor has three bands on the visible spectrum and a fourth band on the infrared. Since the infrared sensor has a spatial resolution much lower than the three visible sensors, it is usually dismissed for tasks related to production of visual products. This implies that the above mentioned LHS-based algorithm is perfectly suited to these types of data sets and situations.

The above described method is called AWL (Additive Wavelet L-band).

6.2.3.3 Principal component

When the number of LRM bands is different from 3, the previous LHS-based procedures cannot be directly used because we cannot apply the RGB–LHS transform, which implies that we cannot obtain the L intensity component. In this situation, we need an alternative procedure in order to obtain a data set which can be considered as an approximation to the intensity component of the LRM bands. An available approach is to perform a principal component analysis (PCA) on the LRM bands, taking the first principal component as an approximation to the intensity component of the LRM bands and considering the remaining principal components as the ones containing the multispectral information of the scene.

In this approach, the spatial detail is added to the first principal component of the PCA decomposition of the LRM bands, denoted as LRP_1, keeping unmodified the remaining principal components (Figure 6.5):

$$HRP_1 = LRP_1 + \sum_{j=0}^{n} \omega_j$$

Taking this new HRP_1 first principal component and the remaining unmodified principal components in order to perform the inverse PCA procedure, we recover the merged HRM bands.

This algorithm is called AWPC and can be used when any number m of multispectral bands are available.

6.2.3.4 Proportional injection

The multispectral information of the LRM image is described by the relative information between their LRM bands, that is, the multispectral content on a certain pixel is described by the relative values between all the m LRM data sets. For example, suppose we have three LRM bands equivalent to the RGB channels. In this case, when in a certain pixel the values of the different $m = 3$ LRM data sets are very similar, we have a grey pixel; if the LRM data set equivalent to the R channel is greater than the other two channels, we have a reddish pixel. Therefore, if we want to maintain the multispectral information of

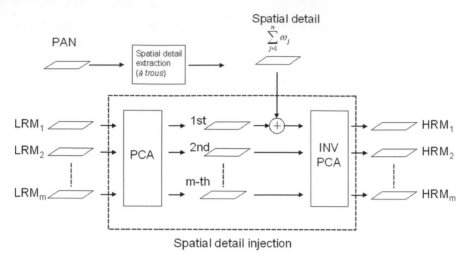

Figure 6.5   *AWPC image fusion method. Spatial detail is injected into the first principal component of the PCA transform of the LRM bands.*

the LRM bands when performing the image fusion procedure, we have to maintain the ratios between the values of the LRM data sets.

In order to maintain this multispectral information, instead of injecting the spatial detail into an intensity component derived from the LRM bands, we can inject the spatial detail into the LRM bands trying to preserve this relative multispectral information in the following way:

$$\text{HRM}_i = \text{LRM}_i + \frac{\text{LRM}_i}{(1/m)\sum_{i=1}^{m}\text{LRM}_i}\sum_{j=0}^{n}\omega_j$$

This equation is only valid when data sets are described in digital counts and the dynamic range of the PAN is similar to the LRM bands. In case the data set of the PAN is approximately m times the dynamic range of the LRM bands or the data sets of both PAN and LRM are described in radiance units, the $1/m$ factor has to be eliminated.

In this expression, we add the spatial detail into every LRM band applying a weighting factor to the spatial details in order to maintain the original proportional relative values between the LRM bands. In order to obtain this weighting factor, we calculate the ratio between the LRM$_i$ data set and the mean value of all the m LRM data sets. It allows injecting the spatial detail into the LRM bands proportionally to their original values (Figure 6.6).

This method is called AWLP and is suitable for all types of PAN and LRM data sets.

6.2.3.5 SRF-based methods

When information about SRF (spectral response function) of the sensors is available, the complexity of the resulting image fusion algorithms highly increases [25], but the ob-

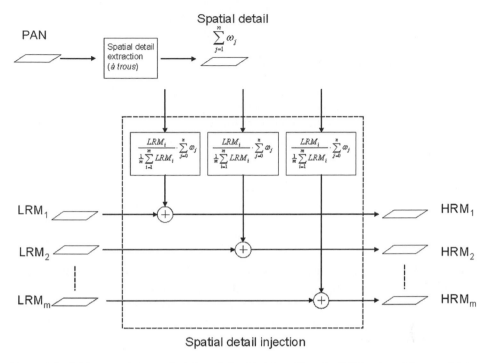

Figure 6.6 *AWLP image fusion method. Spatial detail is injected into the LRM bands applying a weighting factor which depends on their values.*

tained results are much better compared to those obtained by the previous algorithms. These algorithms take into account the physical characteristics of the sensors, e.g. quantum efficiency, wavelength sensibility, etc. A representative example of these methods is [25].

6.2.3.6 Substitution methods

In all of the presented methods, the spatial detail is added into the LR bands, e.g. the LR· terms in the previous equations. But instead of adding the spatial detail into these original LR· data sets, we could substitute some features of the LR bands by the injected spatial detail (a familiar example of this procedure is the classical LHS image fusion method where the intensity component obtained from the LRM data sets is substituted by the PAN data set). In this case, the n first wavelet planes of the LR· data set can be substituted by the n first wavelet planes of the PAN data set. This procedure has been used in the first versions of wavelet-based image fusion algorithms, but it was abandoned, since the addition of the spatial details to the LR· data sets have been shown to produce better results. One explanation is that the substitution methods eliminate some information from the LR bands, in contrast to the addition methods that maintain all the original information from the LR bands and just add the spatial information needed to improve its spatial resolution.

6.3 Results

In this section we demonstrate several results obtained with the image fusion methods presented in the previous section and compare them by focusing on the visual differences. In the literature it is also common to compare results obtained with different image fusion methods using statistical numerical estimators. The goal of this work is not to compare image fusion methods, but to explain how to implement them and illustrate some results obtained by these methods. The reader interested in the numerical comparisons of different image fusion methods can easily find these in the literature.

To illustrate the results obtained by the image fusion methods explained in the previous section, we have used two QuickBird images kindly supplied by Eurimage (http://www.eurimage.com). The first is a 4-band multispectral image (blue, green, red, and near infrared channels) with 2.8 meters per pixel. The other is a panchromatic image with 0.7 meters per pixel. In order to compare the merged images with the ideal result, we degraded these two images to 11.2 and 2.8 meters per pixel, respectively. It allowed us to obtain a high resolution multispectral merged image (HRM) with 2.8 meters per pixel, which could be directly compared with the original multispectral QuickBird image.

In Figure 6.7 we show a subimage of both the downsampled multispectral and panchromatic images, e.g. the ones with 11.2 and 2.8 meters per pixel, respectively.

(a)   (b)

Figure 6.7   *(a) False colour subimage obtained using red, green, and blue channels from the four channels multispectral QuickBird image. Spatial resolution is 11.2 meters per pixel. It is obtained from a downgraded original multispectral QuickBird image with 2.8 meters per pixel, shown in (c). (b) Panchromatic image with 2.8 meters per pixel. (c) Original multispectral QuickBird image with 2.8 meters per pixel. This is the ideal image that image fusion methods should obtain when merging images from (a) and (b). (This image fusion example is a synthetic case. We used real multispectral and panchromatic images and we spatially degraded them in order to obtain out LRM and PAN images. This procedure allows performing an image fusion task using these degraded images, knowing that the ideal solution is the original multispectral image.). Images obtained by (d) LHS method, (e) AWRGB method, (f) AWL method, (g) AWPC method, (h) AWLP method and (i) WiS-peR method. (j) False colour subimage obtained using red, green, and blue channels from the four channels multispectral QuickBird image. Spatial resolution is 11.2 meters per pixel. It is obtained from a downgraded original multispectral QuickBird image with 2.8 meters per pixel, shown in (c). (k) Panchromatic image with 2.8 meters per pixel. (l) Original multispectral QuickBird image with 2.8 meters per pixel. This is the ideal image that image fusion methods should obtain. Images obtained by (m) LHS method, (n) AWRGB method, (o) AWL method, (p) AWPC method, (q) AWLP method and (r) WiSpeR method.*

In Figure 6.7(c) we show the ideal result we should obtain when merging the images shown in Figures 6.7(a) and 6.7(b). The merged images obtained by the several image fusion methods are shown in Figures 6.7(d)–6.7(h).

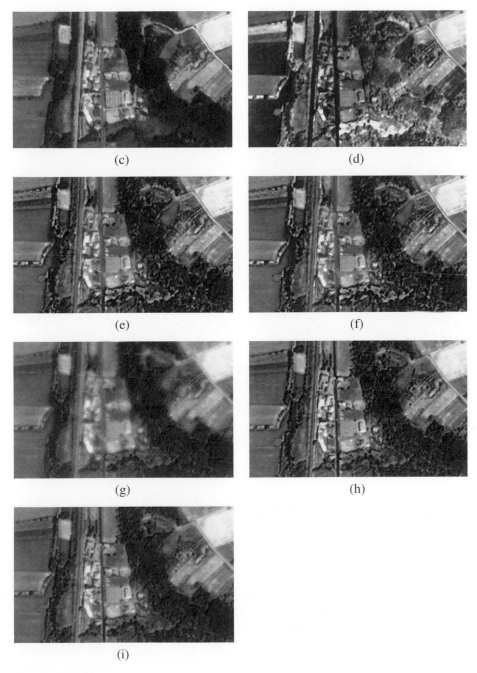

(c)　　　　　　　　　　　　　　　　(d)

(e)　　　　　　　　　　　　　　　　(f)

(g)　　　　　　　　　　　　　　　　(h)

(i)

Figure 6.7　*(continued)*

We can see that the classical method LHS in Figure 6.7(d) greatly distorts the multispectral information of the original image. The spatial detail is the same as the ideal solution
in Figure 6.7(c), but the colours are very different. When using wavelet based methods,

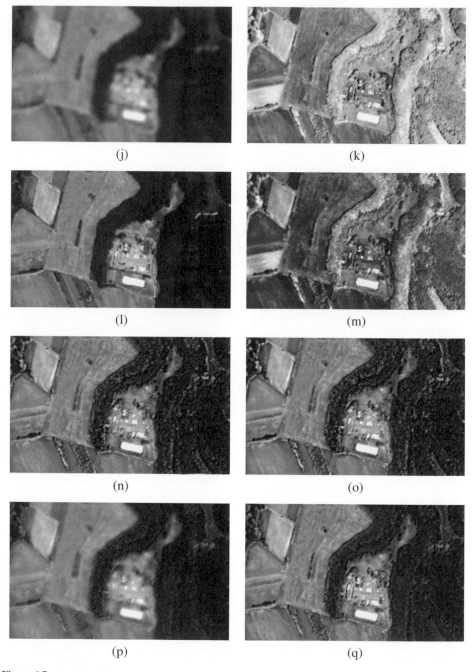

Figure 6.7   *(continued)*

(r)

Figure 6.7 *(continued)*

results greatly improve. In Figure 6.7(e) we show the AWRGB method. The global colour of this image is more close to the ideal result, but some particular defects can be observed. One of the more clearly visible defects is the presence of some pixels on the forest areas with an intense bluish colour. The results obtained by the AWL method and shown in Figure 6.7(f) are not visually different, although several studies in the literature show that the results are numerically better. The APWC method, shown in Figure 6.7(g), presents a global colour of the image more similar to the ideal one, but spatial detail is poor. But the AWLP method in Figure 6.7(h) has all the detail of ideal image and a global colour closer to the ideal one.

The best result is obtained with the WiSpeR method [25], shown in Figure 6.7(i). In this method we introduced the information about the spectral sensitivity of the sensor supplied by the manufacturer. It allows injecting the spatial detail on every multispectral channel in a more accurate way. We can see that the spatial detail is almost equal to the ideal image, and the colour of the image is very close to the ideal one.

We can see similar results in images of Figures 6.7(l)–6.7(r), where images in Figures 6.7(j) and 6.7(k) are the original images to be merged.

Acknowledgements

Author is partially supported by the programme Ramon y Cajal, founded by Spanish Science Ministry. Author wants to thanks Eurimage (http://www.eurimage.com) for kindly supplying QuickBird images.

References

[1] R. Haydn, G.W. Dalke, J. Henkel and J.E. Bare, 'Application of HIS color transform to the processing of multisensor data and image enhancement', in *Proc. Int. Symp. on Remote Sensing of Arid and Semi-Arid Lands*, Cairo, Egypt, 1982, pp. 599–616.
[2] J.W. Carper, T.M. Lillesand and R.W. Kiefer, 'The use of intensity–hue–saturation transformations for merging SPOT panchromatic and multispectral image data', *Photogrammetric Engineering and Remote Sensing*, Vol. 56, 1990, pp. 459–467.

[3] C. Pohl and J.L. van Genderen, 'Multisensor image fusion in remote sensing: Concepts, methods and applications', *International Journal of Remote Sensing*, Vol. 19, 1998, pp. 823–854.

[4] P.S. Chavez Jr., S.C. Sides and J.A. Anderson, 'Comparison of three different methods to merge multiresolution and multispectral data: Landsat TM and SPOT panchromatic', *Photogrammetric Engineering and Remote Sensing*, Vol. 57, No. 3, 1991, pp. 295–303.

[5] M. Ehlers, 'Multisensor image fusion techniques in remote sensing', *ISPRS Journal of Photogrammetry and Remote Sensing*, Vol. 51, 1991, pp. 311–316.

[6] V.K. Shettigara, 'A generalized component substitution technique for spatial enhancement of multispectral images using a higher resolution data set', *Photogrammetric Engineering and Remote Sensing*, Vol. 58, 1992, pp. 561–567.

[7] G. Cliche and F. Bonn, 'Integration of SPOT panchromatic channel into multispectral mode for image sharpness enhancement', *Photogrammetric Engineering and Remote Sensing*, Vol. 51, No. 3, 1985, pp. 811–816.

[8] SPOT Image, *SPOT User's Handbook, Vols. 1–3*, Centre National d'Etude Spatiale (CNES) and SPOT Image, France, 1988.

[9] A.R. Gillespie, A.B. Kahle and R.E. Walker, 'Color enhancement of highly correlated images II. Channel ratio and chromaticity transformation technique', *Remote Sensing of Environment*, Vol. 22, No. 3, 1987, pp. 343–365.

[10] C. Pohl, 'Geometric aspects of multisensor image fusion for topographic map updating in the humid tropics', ITC Publication No. 39, Enschede, The Netherlands, 1996.

[11] L. Wald, T. Ranchin and M. Mangolini, 'Fusion of satellite images of different spatial resolutions: Assessing the quality of resulting images', *Photogrammetric Engineering and Remote Sensing*, Vol. 63, No. 6, 1997, pp. 691–699.

[12] I. Couloigner, T. Ranchin, V.P. Valtonen and L. Wald, 'Benefit of the future SPOT 5 and of data fusion to urban mapping', *International Journal of Remote Sensing*, Vol. 19, No. 8, 1998, pp. 1519–1532.

[13] I. Daubechies, *Ten Lectures on Wavelets*, CBMS-NSR Regional Conference Series in Applied Mathematics, Society for Industrial and Applied Mathematics, 1992.

[14] C.H. Chui, *An Introduction to Wavelets*, Academic Press, Boston, 1992.

[15] S. Mallat, *A Wavelet Tour of Signal Processing*, second ed., Academic Press, 1999.

[16] T. Ranchin and L. Wald, 'Fusion of high spatial and spectral resolution images: The ARSIS concept and its implementation', *Photogrammetric Engineering and Remote Sensing*, Vol. 66, 2000, pp. 49–61.

[17] B. Garguet-Duport, J. Girel, J.M. Chasseny and G. Pautou, 'The use of multiresolution analysis and wavelet transform for merging SPOT panchromatic and multispectral image data', *Photogrammetric Engineering and Remote Sensing*, Vol. 62, 1996, pp. 1057–1066.

[18] D.A. Yocky, 'Image merging and data fusion by means of the discrete two-dimensional wavelet transform', *Journal of the Optical Society of America A*, Vol. 12, 1995, pp. 1834–1841.

[19] J. Zhou, D.L. Civco and J.A. Silander, 'A wavelet transform method to merge Landsat TM and SPOT panchromatic data', *International Journal of Remote Sensing*, Vol. 19, No. 4, 1998, pp. 743–757.

[20] B. Aiazzi, L. Alparone, S. Baronti and A. Garzelli, 'Context-driven fusion of high spatial and spectral resolution images based on oversampled multiresolution

analysis', *IEEE Transactions on Geoscience and Remote Sensing*, Vol. 40, 2002, pp. 2300–2312.

[21] M. González-Audícana, J.L. Saleta, O.G. Catalán and R. García, 'Fusion of multispectral and panchromatic images using improved IHS and PCA mergers based on wavelet decomposition', *IEEE Transactions on Geoscience and Remote Sensing*, Vol. 42, 2004, pp. 1291–1299.

[22] J. Nuñez, X. Otazu, O. Fors, A. Prades, V. Pala and R. Arbiol, 'Multiresolution-based image fusion with additive wavelet decomposition', *IEEE Transactions on Geoscience and Remote Sensing*, Vol. 37, 1999, pp. 1204–1211.

[23] M. González-Audícana, X. Otazu, O. Fors and A. Seco, 'Comparison between the Mallat's and the "à trous" discrete wavelet transform based algorithms for the fusion of multispectral and panchromatic images', *International Journal of Remote Sensing*, Vol. 26, No. 3, 2005, pp. 597–616.

[24] Y. Chibani and A. Houacine, 'The joint use of IHS transform and redundant wavelet decomposition for fusing multispectral and panchromatic images', *International Journal of Remote Sensing*, Vol. 23, 2002, pp. 3821–3833.

[25] X. Otazu, M. González-Audícana, O. Fors and J. Núñez, 'Introduction of sensor spectral response into image fusion methods. Application to wavelet-based methods', *IEEE Transactions on Geoscience and Remote Sensing*, Vol. 43, 2005, pp. 2376–2385.

[26] M. Mangolini, 'Apport de la fusion d'images satellitaires multicapteurs au niveau pixel en télédétection et photo-interprétation', Eng. Doctoral Thesis, Univesité de Nice–Sophia Antipolis, France, 1994.

[27] D.H. Ballard and C.M. Brown, *Computer Vision*, Prentice–Hall, New Jersey, 1982.

[28] R.C. Gonzalez and R.E. Woods, *Digital Image Processing*, Addison–Wesley, 1992.

7

Bayesian methods for image fusion

Jürgen Beyerer [a,b], Michael Heizmann [a], Jennifer Sander [b] and Ioana Gheţa [b]

[a] *Fraunhofer-Institut für Informations- und Datenverarbeitung IITB, Karlsruhe, Germany*

[b] *Universität Karlsruhe (TH), Institut für Technische Informatik, Lehrstuhl für Interaktive Echtzeitsysteme, Karlsruhe, Germany*

The Bayesian fusion methodology bases upon a solid mathematical theory, provides a rich ensemble of methods and allows an intuitive interpretation of the fusion process. It is applicable independently of the goal pursued by image fusion, at different abstraction levels, and also if different kinds of image data have to be fused. It allows transformation, fusion, and focusing, i.e. it fulfils the basic requirements that a reasonable fusion methodology has to satisfy.

Within the Bayesian methodology, image processing and image fusion problems are representable naturally as inverse problems which are usually ill-posed. We review how the Bayesian approach handles inverse problems probabilistically and relate these techniques to the classical regularisation methods. Within the Bayesian framework, the final fusion result is extracted from the Bayesian posterior distribution using an adequate Bayes estimator from decision theory. Prior knowledge as well as artificial constraints on the fusion result can be incorporated via the prior distribution. If a non-informative prior distribution is used, all information contained in the posterior distribution results from the image data. By Bayesian multistage models, complex structures can be modelled as combinations of simpler structures in multiple hierarchical levels. Generalising Gaussian assumptions, we describe the concept of conjugate families by which the computational costs of Bayesian fusion tasks are drastically reduced.

Via Gibbs' distributions, Bayesian methods are linked to energy formalisms which base upon the formulation of energy functionals. An optimal fusion result is obtained by minimising the global energy, i.e. the weighted sum of energy terms into which known and desired properties of the fusion result are formalised in an easy and flexible manner. An overview of the most important classes of energy terms according to the modelled requirements for the fusion result is given: we consider data terms, quality terms, and constraint terms. Some techniques for the exact and approximative minimisation of the global energy are considered: direct minimisation, successive optimisation, graph cuts, and dynamic programming.

The disadvantage of the Bayesian fusion methodology is its global perspective on the fusion problem that also causes high computational costs. We present a local fusion approach within the Bayesian framework: by this, global modelling is avoided and local Bayesian approximations are computed. Local Bayesian fusion is closely related with the concept of an agent-based architecture that is inspired by criminal investigations.

7.1 Introduction: fusion using Bayes' theorem

7.1.1 Why image fusion?

One image of a complex scene usually does not contain enough information to solve a problem at hand. However, several images concerning the same ground truth[1] acquired under different conditions can comprise the information necessary to solve a given problem. On the one hand, such images can form a series according to a systematic variation of the acquisition constellation with respect to geometrical as well as optical parameters, e.g. a focus series where the focal plane is shifted systematically through the scenery with an intent to produce a fused image with quasi infinite depth of field. On the other hand, these images can stem from acquisition devices which are different with respect to their physical principles, e.g. an X-ray image and a nuclear magnetic resonance (NMR) tomography image from a human body, or photographic images and synthetic aperture radar images for reconnaissance tasks. A further generalisation considers also information sources other than images. This leads to the problem of fusion of heterogeneous information sources [1].

In the following, images are denoted by $d(x)$ (like data) and the fusion result is denoted by $r(x)$. In general, we have S different images $d_i(x)$, $i = 1, \ldots, S$, that are to be fused to a fusion result:

$$\{d_1(x), d_2(x), \ldots, d_S(x)\}, \quad \text{with } d_i(x): \Omega_i \to \Delta_i,$$

$$\Omega_i := \text{supp}\{d_i(x)\} \subset \mathbb{R}^2, \ \Delta_i := \text{range}\{d_i(x)\} \tag{7.1}$$

The fusion result $r(x)$ is specified by $N \in \mathbb{N}$ properties of interest $r_j(x)$, $j = 1, \ldots, N$, about which we want to infer when we fuse the images $d_i(x)$, $i = 1, \ldots, S$. Therefore,

---

[1]Information from different sources that is to be fused must be caused by a common underlying truth. It is a necessary prerequisite for the fusion to make sense at all. This is tacitly assumed throughout this chapter.

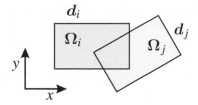

Figure 7.1 *Images to be fused and their spatial supports.*

we have

$$r(x) = \{r_1(x), \dots, r_N(x)\}, \quad \text{with } r(x) : \Omega \to R,$$

$$\Omega := \text{supp}\{r(x)\} = \text{supp}\{r_j(x)\} \tag{7.2}$$

$$R = R_1 \times \cdots \times R_N, \quad \text{with } R_j := \text{range}\{r_j(x)\} \tag{7.3}$$

Depending on the problem at hand fusion can pursue different goals:

- Reduction of noise, i.e. improvement of Signal-to-Noise-Ratio (SNR) by averaging over several images; see Figure 7.1 at $\Omega_i \cap \Omega_j$.
- Improvement of spatial resolution (super-resolution); see Figure 7.1 at $\Omega_i \cap \Omega_j$.
- Extension of the spatial domain; e.g. mosaicking [2]; see Figure 7.1 at $\Omega_i \cup \Omega_j$.
- Improvement of the value resolution and extension of the value range (Dynamic Range Increase (DRI) [3]); see Figure 7.1 at $\Omega_i \cap \Omega_j$.
- Qualitative extension of the image values, e.g. registration of images for different spectral bands to a vector-valued (multi-spectral) image.
- Visualisation of high-dimensional images (multi- and hyper-spectral) as false-colour images.
- Creating designed fusion results in the case that images are generated from different physical principles, e.g. sonar image and X-ray image, etc.

Fusion can take place on different abstraction levels:

- *Fusion on data level*: the combination takes place directly on the image data – intensities like grey or RGB values. The precondition for such an operation is that the images have been acquired by homogeneous sensors, such that the images reproduce similar or comparable physical properties of the scene. The fusion result is an image.
- *Fusion on feature level*:
  - Direct fusion of features: the combination of information is performed by using features which are extracted from the set of images. The features used may be calculated separately from each image $d_i(x)$, or they may be obtained by simultaneous processing of all images.
  - Object based fusion: features extracted from images are assigned to objects that describe the scene in a symbolic manner; features from different images are used to determine attributes of and relations between objects.

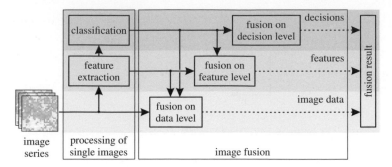

Figure 7.2 *Fusion on different abstraction levels: to determine the abstraction level of the fusion, the processing of the single images accomplished prior to the fusion is decisive. Even if the fusion takes place on a rather low abstraction level, information from higher levels may be beneficial or necessary.*

- *Fusion on decision level*: decisions – such as classification results obtained from single images – are combined to the fusion result. For such a fusion, inhomogeneous sensors may be used, which allow to generate decisions that are compatible on a decision level.

Of course, fusion on different abstraction levels can be coupled to construct powerful algorithms; see Figure 7.2.

7.1.2 Three basic requirements for a fusion methodology

Before we take a closer look on Bayesian fusion, three basic requirements are discussed that a methodology for fusion should fulfil. If we have different information sources, e.g. an X-ray image and an NMR-image, due to their different physical origin, the data are not compatible from a semantic point of view. In order to fuse such different information sources, a suitable mathematical framework for fusion needs to allow:

1. **Transformation** of information stemming from heterogeneous sources into a common mathematical description; this step makes them compatible. Depending on the abstraction level of the information, this transform can mean an abstraction (with minimal loss of information) or a specialisation (with minimal generation of artefacts).
2. **Fusion**: superposition of the transformed information.
3. **Focusing**: concentrating the pooled information in order to derive specific statements for the problem at hand.

Bayesian statistics is based on a certain admissible interpretation of Kolmogorov's axioms of probability theory. Whereas in classical statistics, probability is interpreted as a properly defined limit of the frequency of events when an experiment is repeated infinitely often, the Bayesian point of view allows interpreting probability also as a Degree of Belief (DoB). That is a measure for uncertainty, or to say it conversely, a probabilistic measure to describe what is known. In the following, probabilities are always interpreted as DoBs.

The Bayesian methodology fulfils all three requirements: The transformation can be accomplished by the Maximum Entropy (ME) principle. As known from Shannon's information theory, entropy is a measure of DoB concentration. The idea is to take everything known as constraints and then to find the ME DoB distribution fulfilling those constraints. Thus, this principle delivers a DoB distribution that incorporates the given knowledge with the minimal concentration of DoB.

For example, if we have knowledge about $r(x)$, the transformation can be stated as:

$$p_{\text{ME}}\big(r(x)\big) = \arg \max_{p(r(x)) \in \Pi} \big\{ \mathbf{E}\big[-\log p\big(r(x)\big)\big] \big\} \qquad (7.4)$$

$$\Pi = \bigg\{ p\big(r(x)\big) \,\bigg|\, p\big(r(x)\big) \geqslant 0 \wedge \oint\!\!\!\!\!\sum p\big(r(x)\big)\, \mathrm{d}r(x) = 1$$

$$\wedge\, p\big(r(x)\big) \text{ conforms with given knowledge} \bigg\}^2 \qquad (7.5)$$

Three examples show the application of the ME approach: In the case when the knowledge is a finite restriction of the range R of $r(x)$, the ME DoB is the uniform distribution on R. If knowledge is given in the form of a value ρ which can be interpreted as the expectation of $r(x)$ together with an uncertainty σ about ρ that can be interpreted as the standard deviation, the ME DoB distribution is the $N(\rho, \sigma^2)$ distribution. For the important case where knowledge is given in the form of expectations

$$\mathbf{E}\big[k_l\big(r(x)\big)\big] = \kappa_l, \quad l = 1, \ldots, L \qquad (7.6)$$

the well-known closed solution for the ME DoB distribution is obtained as

$$p_{\text{ME}}(r) = \lambda_0 \exp\bigg\{ \sum_{l=1}^{L} \lambda_l k_l\big(r(x)\big) \bigg\} \qquad (7.7)$$

where the constants λ_i have to be determined by evaluating the constraints of Equation (7.6) [4].

Even though the DoB interpretation has by definition a subjective nature, the ME approach is an objective procedure, since two different individuals would derive the same DoB distributions given the same knowledge.

It follows from $\mathbf{E}[-\log p(d(x), r(x))] = \mathbf{E}[-\log p(d(x)|r(x))] + \mathbf{E}[-\log p(r(x))]$ that if $p(r(x))$ and $p(d(x)|r(x))$ are ME, then the mutual distribution $p(d(x), r(x))$ is ME,

---

[2]Note that the Bayesian framework can treat continuous, discrete, as well as mixtures of both types of variables simultaneously. The same holds also for qualitative and quantitative properties [1]. For discrete variables, $p(.)$ has to be interpreted as a probability mass function and for continuous variables it has to be interpreted as a probability density function. The symbol $\oint\!\!\!\!\!\sum$ means an integral with respect to continuous and a sum with respect to discrete variables.

too. Therefore, knowledge about how $d(x)$ depends on $r(x)$ can also be transformed into a conditional DoB distribution $p(d(x)|r(x))$ using the ME principle.

If Bayesian inference about $r(x)$ based on two observed data $d_1(x)$ and $d_2(x)$ is to be performed and if no Shannon information should be lost, the inference result is equal to the posterior distribution $p(r(x)|d_1(x), d_2(x))$ [5]. It can be written as:

$$p\big(r(x)|d_1(x), d_2(x)\big) \propto p\big(d_1(x), d_2(x)|r(x)\big) \cdot p\big(r(x)\big) \qquad (7.8)$$

$$= p\big(d_2(x)|d_1(x), r(x)\big) \cdot p\big(d_1(x)|r(x)\big) \cdot p\big(r(x)\big) \quad (7.9)$$

$$\overset{(A)}{=} p\big(d_2(x)|r(x)\big) \cdot p\big(d_1(x)|r(x)\big) \cdot p\big(r(x)\big) \qquad (7.10)$$

where equality (A) holds, if $d_1(x)$ and $d_2(x)$ are independent given the cause $r(x)$. In practice, this is often fulfilled, e.g. if $d_1(x)$ and $d_2(x)$ are based on different sensor principles and the noise processes disturbing the two sensors are independent. According to Equation (7.10), the fusion of both sources $d_1(x)$ and $d_2(x)$ is essentially accomplished by multiplying their likelihood functions and the prior distribution. The generalisation to more than two sources is straightforward.

Bayesian methodology also has natural mechanisms for focusing. Assume that we are interested only in some components of $r(x)$, say of $r_I(x) := \{r_i(x) \mid i \in I\}$ and not interested in $r_J(x) := \{r_i(x) \mid i \in J\}$. Here, the index sets I and J constitute a partition of the index set $\{1, \ldots, N\}$ by which we subscript the components of the original fusion result $r(x)$, i.e. $I \cup J = \{1, \ldots, N\}$, $I \cap J = \varnothing$. Then we can use the original fusion scheme by treating $r_J(x)$ as so-called nuisance parameters and integrating them out of the posterior distribution (marginalisation):

$$p\big(r_I(x)|d_1(x), \ldots, d_S(x)\big) = \int_{\times_{j \in J} R_j} p\big(r(x)|d_1(x), \ldots, d_S(x)\big) \, \mathrm{d}r_J(x) \qquad (7.11)$$

A second way to focus is calculating posterior expectations of functions $f(r(x))$ specific to the problem at hand:

$$\mathbf{E}\big[f(r(x))\big] = \int_R f(r(x)) \cdot p\big(r(x)|d_1(x), \ldots, d_S(x)\big) \, \mathrm{d}r(x) \qquad (7.12)$$

If $f(r(x))$ does not depend on all components of $r(x)$, this integration implicitly involves also a marginalisation.

7.1.3 Why Bayesian fusion?

So far it has been shown that the Bayesian framework fulfils all three basic requirements for a fusion methodology stated above.

Moreover, Bayesian statistics has been developed over a long time to a very mature methodology that supports a rich ensemble of methods together with an intuitive interpretation.

Other methodologies, like Fuzzy theory or Dempster–Shafer theory [6], do not possess such a lean concept as does the Bayesian approach, where only one single measure is used to describe DoB. From a theoretical point of view, Occam's razor should be cited here: 'The simplest explanation is best' [7]. From a practical point of view, it is hard to find real world (non-pathologic) problems, where the Bayesian approach does not outperform the above mentioned methodologies.

Since within the Bayesian framework the probability measure is interpreted as a DoB, a further generalisation is rather appealing. Instead of interpreting $r(x)$ as the 'real' cause of the data $d(x)$, $r(x)$ could be seen as a matter of design. From this perspective, $p(r(x))$ is no longer only a mathematical representation of prior knowledge about $r(x)$, but also describes the designers wishes concerning an artificial fusion result $r(x)$. Analogously, $p(d(x)|r(x))$ can be interpreted as the intended dependency between data $d(x)$ and the fusion result $r(x)$.

7.2 Direct application of Bayes' theorem to image fusion problems

In the next two sections, we describe the direct application of the Bayesian approach to image fusion mainly exemplarily on the basis of some commonly used modelling assumptions, namely additive linear models and Gaussian distributions. Despite the associated restrictions, these assumptions can be shown to be approximatively valid for a large class of image fusion problems. Beside the motivation of Bayesian image fusion, the aim of Section 7.2.1 is to illustrate that the Bayesian methodology is predestinated for handling inverse problems that generally underly fusion tasks [8]. In this context, the connection between the Bayesian approach and classical methods for solving inverse problems, namely regularisation methods, are reviewed. For Tikhonov regularisation, this connection is described in more detail later in this paragraph. In Section 7.2.2, a practical demonstration of Bayesian fusion is developed for one image as well as for image sequences in the case of Gaussian densities, an important conjugate family (cf. Section 7.2.5.1). Obtaining an estimate for the solution of a fusion task is a substantial step. This topic is discussed in Section 7.2.3. As described in Section 7.2.4, several kinds of multi-stage models represent important tools for modelling and evaluating more complex structures via Bayesian techniques. Finally, Section 7.2.5 depicts some basic possibilities for prior modelling.

7.2.1 Bayesian solution of inverse problems in imaging

Conforming with the definition given in Equation (7.1), an image $d_i(x)$ of a scene is, in principle, a continuous function of two spatial variables. $d_i(x)$ may be vector valued, e.g. an RGB colour image is composed of the continuous intensity functions of the three basis colours: red, green, and blue; and hence $\Delta_i \subseteq \mathbb{R}^3$.

Image processing systems can handle only a discrete (digitalised) approximation of $d_i(x)$ over a finite lattice $L_i \subset \Omega_i$. Thus, $d_i(x)$ is commonly identified with its approximation by an image matrix. However, for mathematical convenience, we understand in the following a discretised image as a vector $\mathbf{d}_i \in \mathbb{R}^{m_i}$ that results from concatenating the entries of the corresponding image matrix in a suitable manner pertaining to the subsequent probabilistic modelling task. Obviously, m_i is equal to the number of entries of the image matrix.

Similarly, the vector $\mathbf{d} \in \mathbb{R}^m$ denotes the suitable concatenation of all image vectors $\mathbf{d}_i \in \mathbb{R}^{m_i}$ that are given in the fusion task. According to the notation of Section 7.1.1, we have $i = 1, \ldots, S$, i.e. $m = \sum_{i=1}^{S} m_i$. To illustrate the situation underlying inverse problems in imaging, we first restrict ourself to the case $S = 1$.

Additional to discretisation errors, there are usually other factors that make an image $\mathbf{d}_1$ imperfect. The sensing process by which $\mathbf{d}_1$ has been acquired comes along with a loss of information, e.g. due to projections or filtering operations. $\mathbf{d}_1$ may also be incomplete, e.g. in case of occlusions, imprecise due to measurement errors, and non-deterministic by random influences like sensor noise and non-predictable effects outside the sensing system caused for instance by atmospheric influences. Inferring features of the scene from $\mathbf{d}_1$ or obtaining an image that is qualitatively better than $\mathbf{d}_1$ (image restoration) means solving an inverse problem. Usually, the forward model that relates an image to the unknown quantity of interest has been formulated independently of observing the actual image $\mathbf{d}_1$. Additionally, simplifying assumptions concerning the image acquisition process of the sensing system are necessary to keep the corresponding system modelling task manageable.

In the following, we will use the common assumption that the forward model is given by a linear measurement equation

$$\mathbf{d}_1 = \mathbf{A}_1 \mathbf{r} \qquad (7.13)$$

Here – in analogy to the definition of $\mathbf{d}_1$ – the unknown quantity of interest is also represented by an appropriate vector $\mathbf{r} \in \mathbb{R}^n$. $\mathbf{A}_1 \in \mathbb{R}^{m_1 \times n}$ denotes the linear mapping by which $\mathbf{r}$ has been transformed onto the image $\mathbf{d}_1$.

As defined by Hadamard [9], a problem is well-posed if its solution exists, is unique, and depends continuously on the observed data. An ill-posed problem violates at least one of these demands. Obtaining the image $\mathbf{d}_1$ from $\mathbf{r}$ is generally a well-posed problem. However, due to the non-ideality and incompleteness of both $\mathbf{d}_1$ and $\mathbf{A}_1$, its inverse, i.e. calculating $\mathbf{r}$ from $\mathbf{d}_1$, is ill-posed [10]. We point out that ill-posedness is not restricted to the case $m_1 < n$. Dealing with an ill-conditioned $\mathbf{A}_1$, the apparent solution $\mathbf{r} = \mathbf{A}_1^{-1} \mathbf{d}_1$ and $\mathbf{r} = \mathbf{A}_1^{+} \mathbf{d}_1$, respectively (with $\mathbf{A}_1^{+}$ denoting the pseudo-inverse of $\mathbf{A}_1$, i.e. $\mathbf{A}_1^{+} = (\mathbf{A}_1^{\mathsf{T}} \mathbf{A}_1)^{-1} \mathbf{A}_1^{\mathsf{T}}$), is a useless approximation to $\mathbf{r}$ even in the case of existence [11].

Regularisation methods [8,11,12] transform an ill-posed problem into a well-posed one by imposing additional prior knowledge (cf. Section 7.2.2.1) and heuristic constraints. Thereby, a unique stable solution of a well-posed but modified problem is obtained. In Tikhonov regularisation for instance, the additional information gets included via the image of an operator $\mathbf{A}_0$ into a normed space. The Tikhonov regularised solution of (7.13) with regularisation parameter $\theta > 0$ is then obtained by minimising the Tikhonov func-

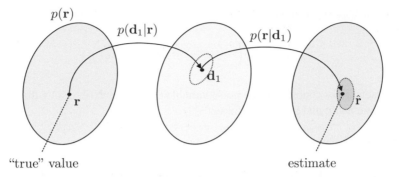

"true" value   estimate

Figure 7.3 $\mathbf{r}$ *has an unknown 'true' value. Additional prior knowledge is described by the prior* $p(\mathbf{r})$. $\mathbf{r}$ *causes the observation* $\mathbf{d}_1$. *Uncertainty about* $\mathbf{d}_1$ *for a given* $\mathbf{r}$ *is described via the Likelihood* $p(\mathbf{d}_1|\mathbf{r})$. *Solving an inverse problem by the Bayesian approach means inferring the posterior* $p(\mathbf{r}|\mathbf{d}_1)$. *From the posterior, different estimates* $\hat{\mathbf{r}}$ *of* $\mathbf{r}$ *can be made.*

tional

$$T_{\theta}(\mathbf{r}) := \|\mathbf{d}_1 - \mathbf{A}_1\mathbf{r}\|^2 + \theta\|\mathbf{A}_0\mathbf{r}\|^2 \tag{7.14}$$

The Bayesian approach to inverse problems handles an ill-posed problem by embedding it in a more comprehensive probabilistic framework [1,8,13]. Solving an inverse problem means inferring the posterior distribution $p(\mathbf{r}|\mathbf{d}_1)$ of the unknown quantity of interest given the observation that communicates the final DoB with respect to the unknown, see Figure 7.3.

To describe the image formation by a more suitable forward model within the Bayesian framework, we add a non-deterministic noise term $\mathbf{e}$ to the right side of Equation (7.13) and get

$$\mathbf{d}_1 = \mathbf{A}_1\mathbf{r} + \mathbf{e} \tag{7.15}$$

7.2.2 Bayesian image fusion exemplified for Gaussian distributions

7.2.2.1 Fusion of one image with prior knowledge

In the Bayesian methodology, additional prior knowledge concerning $\mathbf{r}$ that is independent of observing $\mathbf{d}_1$ should get incorporated into the stochastic model via the prior distribution $p(\mathbf{r})$. By additional prior knowledge we mean prior knowledge that is not already incorporated in form of the forward model. We denote that because information that is included via the latter, of course, influences the Likelihood instead of the prior [14].

In the following, we assume the validity of the linear forward model (7.15) and, additionally, that our information concerning $\mathbf{r}$ is restricted to the knowledge of the first two moments, that are the mean $\bar{\mathbf{r}}_0 := \mathbf{E}[\mathbf{r}]$ and the covariance matrix $\bar{\mathbf{R}}_0 := \mathbf{Cov}[\mathbf{r}] = \mathbf{E}[(\mathbf{r} - \bar{\mathbf{r}}_0)(\mathbf{r} - \bar{\mathbf{r}}_0)^{\mathsf{T}}]$. Respecting the ME principle, we inevitably obtain for the prior the $\mathrm{N}(\bar{\mathbf{r}}_0, \bar{\mathbf{R}}_0)$-distribution, that is the multivariate Gaussian distribution with mean $\bar{\mathbf{r}}_0$ and covariance matrix $\bar{\mathbf{R}}_0$. Thus, we have

$$p(\mathbf{r}) \propto \exp\left(-\frac{1}{2}(\mathbf{r} - \bar{\mathbf{r}}_0)^{\mathrm{T}}\bar{\mathbf{R}}_0^{-1}(\mathbf{r} - \bar{\mathbf{r}}_0)\right)$$

Similar assumptions concerning the noise variable deliver $\mathbf{e} \sim \mathrm{N}(\bar{\mathbf{e}}_1, \bar{\mathbf{E}}_1)$ with $\bar{\mathbf{E}}_1 > 0$, i.e. $\bar{\mathbf{E}}_1$ being symmetric and positive definite.

If $\mathbf{e}$ and the unknown $\mathbf{r}$ are furthermore mutually independent, the stochastic forward model (7.15) yields for the Likelihood $p(\mathbf{d}_1|\mathbf{r})$ a Gaussian distribution with mean vector $\mathbf{A}_1\mathbf{r}$ and covariance matrix $\bar{\mathbf{E}}_1$. Within the Bayesian methodology, Gaussian distributions constitute an important class of parametric families for several reasons. One reason results, of course, from the central limit theorem. Another reason is the closed formulation of the posterior distribution

$$p(\mathbf{r}|\mathbf{d}_1) \propto \exp\left(-\frac{1}{2}(\mathbf{d}_1 - \mathbf{A}_1\mathbf{r} - \bar{\mathbf{e}}_1)^{\mathrm{T}}\bar{\mathbf{E}}_1^{-1}(\mathbf{d}_1 - \mathbf{A}_1\mathbf{r} - \bar{\mathbf{e}}_1)\right.$$

$$\left. -\frac{1}{2}(\mathbf{r} - \bar{\mathbf{r}}_0)^{\mathrm{T}}\bar{\mathbf{R}}_0^{-1}(\mathbf{r} - \bar{\mathbf{r}}_0)\right) \tag{7.16}$$

in the Gaussian case, because the product of two Gaussian densities belongs again to the family of Gaussian densities. Hence, the posterior $p(\mathbf{r}|\mathbf{d}_1)$ of $\mathbf{r}$ given the observation $\mathbf{d}_1$ is the $\mathrm{N}(\bar{\mathbf{r}}_1, \bar{\mathbf{R}}_1)$-distribution with

$$\bar{\mathbf{R}}_1 = \left(\bar{\mathbf{R}}_0^{-1} + \mathbf{A}_1^{\mathrm{T}}\bar{\mathbf{E}}_1^{-1}\mathbf{A}_1\right)^{-1}, \qquad \bar{\mathbf{r}}_1 = \bar{\mathbf{R}}_1\left(\mathbf{A}_1^{\mathrm{T}}\bar{\mathbf{E}}_1^{-1}(\mathbf{d}_1 - \bar{\mathbf{e}}_1) + \bar{\mathbf{R}}_0^{-1}\bar{\mathbf{r}}_0\right)$$

These results are obtained from Equation (7.16), e.g. by a completing the square procedure.

7.2.2.2 Extension to image sequences and a review of conditional independence

The presented approach to image fusion can be simply extended to the case where more than one image observation is available. To show this, we denote by $\mathbf{d}_1, \ldots, \mathbf{d}_S$ a (registered) image sequence of the scene. Assuming the structure of the linear additive forward model (7.15) to be valid for each single image leads to the system of equations

$$\mathbf{d}_k = \mathbf{A}_k\mathbf{r} + \mathbf{e}_k, \quad k = 1, \ldots, S \tag{7.17}$$

in which $\mathbf{A}_k$ and $\mathbf{e}_k$ denote the transformation operator and the noise term, respectively, that correspond to the forward model for the kth image.

If the noise variables $\mathbf{e}_1, \ldots, \mathbf{e}_S$ are mutually independent, the images are conditionally independent given $\mathbf{r}$. In this case, we can fuse the information which is provided by $\mathbf{d}_1, \ldots, \mathbf{d}_S$ sequentially by applying the Bayesian theorem repeatedly. Thereby, the posterior at a certain step becomes the prior for the next one:

$$p(\mathbf{r}|\mathbf{d}_1) \propto p(\mathbf{d}_1|\mathbf{r})p(\mathbf{r})$$

$$p(\mathbf{r}|\mathbf{d}_1, \mathbf{d}_2) \propto p(\mathbf{d}_2|\mathbf{r}) p(\mathbf{r}|\mathbf{d}_1)$$

$$\vdots$$

$$p(\mathbf{r}|\mathbf{d}_1, \ldots, \mathbf{d}_S) \propto p(\mathbf{d}_S|\mathbf{r}) p(\mathbf{r}|\mathbf{d}_1, \ldots, \mathbf{d}_{S-1}) \tag{7.18}$$

In the Gaussian case, the means $\bar{\mathbf{r}}_k$ and the covariance matrices $\bar{\mathbf{R}}_k$ of the posterior distributions $p(\mathbf{r}|\mathbf{d}_1, \ldots, \mathbf{d}_k)$, $k = 1, \ldots, S$, are then computable via the iterative scheme

$$\bar{\mathbf{R}}_k = \left(\bar{\mathbf{R}}_{k-1}^{-1} + \mathbf{A}_k^{\mathrm{T}}\bar{\mathbf{E}}_k^{-1}\mathbf{A}_k\right)^{-1},$$

$$\bar{\mathbf{r}}_k = \bar{\mathbf{R}}_k\left(\mathbf{A}_k^{\mathrm{T}}\bar{\mathbf{E}}_k^{-1}(\mathbf{d}_k - \bar{\mathbf{e}}_k) + \bar{\mathbf{R}}_{k-1}^{-1}\bar{\mathbf{r}}_{k-1}\right) \tag{7.19}$$

If the whole image sequence $\mathbf{d}_1, \ldots, \mathbf{d}_S$ has been gathered by the same sensing process, we can assume $\mathbf{A} := \mathbf{A}_k$, $\mathbf{e} := \bar{\mathbf{e}}_k$ and $\mathbf{E} := \bar{\mathbf{E}}_k$ for all $k \in \{1, \ldots, S\}$. Hence, Equation (7.19) delivers the simple algebraic equation

$$\bar{\mathbf{R}}_S = \left(\bar{\mathbf{R}}_0^{-1} + S\mathbf{A}^{\mathrm{T}}\mathbf{E}^{-1}\mathbf{A}\right)^{-1},$$

$$\bar{\mathbf{r}}_S = \bar{\mathbf{R}}_S\left(\mathbf{A}^{\mathrm{T}}\mathbf{E}^{-1}\left(\sum_{k=1}^{S}\mathbf{d}_k - S\mathbf{e}\right) + \bar{\mathbf{R}}_0^{-1}\bar{\mathbf{r}}_0\right) \tag{7.20}$$

for the calculation of the final posterior $p(\mathbf{r}|\mathbf{d}_1, \ldots, \mathbf{d}_S)$.

We have seen that conditional independence of the data $\mathbf{d}_1, \ldots, \mathbf{d}_S$ given the unknown $\mathbf{r}$ significantly simplifies Bayesian fusion. It is usually a realistic assumption if the images have been obtained by different sensing mechanisms.

Often, Bayesian fusion has to be performed for highly heterogeneous kinds of information, e.g. information delivered by IMINT, SIGINT, ACINT, ..., HUMINT.[3] In such cases, the uniform description of the information via DoB distributions (cf. Section 7.1) is generally interpretable as a uniform symbolic encoding of the information in the form of (possibly high-dimensional) image data. Additionally, one can assume that these image data have been delivered via such kind of conspicuously different sensing schemes.

7.2.3 Bayes estimators

Having inferred the posterior distribution that communicates all the information which is provided by prior knowledge and the observations $\mathbf{d}_1, \ldots, \mathbf{d}_S$, often a single point estimate $\hat{\mathbf{r}}$ is extracted from $p(\mathbf{r}|\mathbf{d}_1, \ldots, \mathbf{d}_S)$: one usually chooses a single image that in a sense represents a good approximation of the unknown $\mathbf{r}$. Thereby, some of the information that is encoded in the posterior distribution gets lost.

---

[3]IMINT: image intelligence, SIGINT: signal intelligence, ACINT: acoustic intelligence, HUMINT: human intelligence.

Using decision theoretic concepts, the loss that arises from focusing on a special estimate $\hat{\mathbf{r}}$ is quantifiable. This quantification is performed by a loss function l which is appropriate with regard to the intention that has to be accomplished by the fusion task. $l(\hat{\mathbf{r}}, \mathbf{r})$ states the loss that is caused by selecting an estimate $\hat{\mathbf{r}}$ for the unknown provided that its 'true' value is $\mathbf{r}$. Of course, one always constitutes $l(\mathbf{r}, \mathbf{r}) = 0$. With a fixed image sequence $\mathbf{d}_1, \ldots, \mathbf{d}_S$ given, the expected loss that results from the selection of a particular $\hat{\mathbf{r}}$ is given by

$$c(\hat{\mathbf{r}}|\mathbf{d}_1, \ldots, \mathbf{d}_S) = \sum_{\mathbf{r}} l(\hat{\mathbf{r}}, \mathbf{r}) p(\mathbf{r}|\mathbf{d}_1, \ldots, \mathbf{d}_S) \tag{7.21}$$

where c is called the Bayes cost function. Minimising c with respect to $\hat{\mathbf{r}}$ delivers the Bayes estimator, i.e. the best estimator for $\mathbf{r}$ – of course, subject to the selected loss function. Needless to say, the uniqueness of the resulting Bayesian estimate is usually not warranted, since c may possess several global minima. Below, we give some examples for useful loss functions. The corresponding theoretical background as well as various other examples can be found in the respective literature [8,14–17].

EXAMPLE 7.1. The loss function

$$l(\hat{\mathbf{r}}, \mathbf{r}) = \begin{cases} 0 & \text{if } \hat{\mathbf{r}} = \mathbf{r} \\ 1 & \text{if } \hat{\mathbf{r}} \neq \mathbf{r} \end{cases} = 1 - \delta_{\mathbf{r}}^{\hat{\mathbf{r}}} \tag{7.22}$$

where δ_a^b denotes the Kronecker delta function, which takes the value 1 if $a = b$ and 0 otherwise, requires exactly the 'true' value of the unknown because it appraises all wrong values as equally bad. In this situation, the resulting cost function is minimised by the Maximum A Posteriori (MAP) estimator

$$\hat{\mathbf{r}}_{\text{MAP}} = \arg\max_{\mathbf{r}} p(\mathbf{r}|\mathbf{d}_1, \ldots, \mathbf{d}_S)$$

that is the most common estimator. It delivers the most probable value with respect to the posterior distribution. $\square$

EXAMPLE 7.2. Another useful loss function is given by $l(\hat{\mathbf{r}}, \mathbf{r}) = \|\hat{\mathbf{r}} - \mathbf{r}\|^2$. Here, the loss is assumed to be the lower the less $\hat{\mathbf{r}}$ deviates from the 'true' value. In that case, the Bayes costs are minimised by the minimum mean squares estimator that coincides with the posterior expectation of $\mathbf{r}$ given $\mathbf{d}_1, \ldots, \mathbf{d}_S$

$$\hat{\mathbf{r}}_{\text{MMS}} = \mathbf{E}_{p(\mathbf{r}|\mathbf{d}_1, \ldots, \mathbf{d}_S)}[\mathbf{r}] \qquad \square$$

EXAMPLE 7.3. A pixel based version of (7.22) is given by $l(\hat{\mathbf{r}}, \mathbf{r}) = \sum_{i=1}^{n} (1 - \delta_{\mathbf{r}\langle i\rangle}^{\hat{\mathbf{r}}\langle i\rangle})$ where $\hat{\mathbf{r}}\langle i\rangle$ and $\mathbf{r}\langle i\rangle$ denote the i-entries of the image vectors $\hat{\mathbf{r}}$ and $\mathbf{r}$, respectively. The corresponding minimiser of the cost function $\hat{\mathbf{r}}_{\text{MPM}}$ is the marginal posterior mode. It is determined by the requirements

$$\hat{\mathbf{r}}\langle i\rangle_{\text{MPM}} = \arg\max_{\mathbf{r}} \langle i\rangle p(\mathbf{r}\langle i\rangle|\mathbf{d}_1, \ldots, \mathbf{d}_S), \quad i = 1, \ldots, n \qquad \square$$

Obviously, for a Gaussian posterior distribution (cf. Section 7.2.2), each of the Bayes estimators proposed here delivers the posterior mean. This results simply from knowledge concerning the structure of this parametric family. However, the calculation of Bayes estimates by a complete search or a global integration with respect to $\mathbf{r}$ is a costly task in general. Being based on the whole posterior distribution, the entailed computational costs grow exponentially with the dimension n of the unknown $\mathbf{r}$. Usually, greedy approaches to obtain the desired estimate in a more efficient manner are not applicable because of the existence of multiple local extremal values of the posterior – especially in the non-parametric case. To facilitate the analysis of the posterior in these situations, usually Markov Chain Monte Carlo (MCMC) methods [5] are applied (cf. Section 7.2.5.1). MCMC methods are basically dynamic sampling methods that have been developed from the theory of Markov chains. Introducing the concept of local Bayesian fusion, a promising alternative approach has been developed by the authors (cf. Section 7.4).

Many deterministic regularisation techniques are interrelated to Bayesian estimation [8,11]. The connection between Tikhonov regularisation and Bayesian estimation appears by reviewing the following example.

EXAMPLE 7.4. One has observed an image $\mathbf{d}_1$ that is related to $\mathbf{r}$ via the additive linear model (7.15) with a zero mean Gaussian prior and white Gaussian noise, that means $\mathbf{r} \sim N(\mathbf{0}, \bar{\mathbf{R}}_0)$ and $\mathbf{e} \sim N(\mathbf{0}, \sigma_{\mathbf{e}}^2 \mathbf{I})$ with $\sigma_{\mathbf{e}}^2 > 0$ and $\bar{\mathbf{R}}_0 > 0$. $\mathbf{0}$ and $\mathbf{I}$ denote the zero vector and the identity matrix, both of appropriate dimension. Hence, the MAP estimator is given by

$$\hat{\mathbf{r}}_{\text{MAP}} = \max_{\mathbf{r}} \left\{ \exp\left(-\frac{1}{2\sigma_{\mathbf{e}}^2}(\mathbf{d}_1 - \mathbf{A}_1\mathbf{r})^{\mathsf{T}}(\mathbf{d}_1 - \mathbf{A}_1\mathbf{r}) - \frac{1}{2}\mathbf{r}^{\mathsf{T}}\bar{\mathbf{R}}_0^{-1}\mathbf{r} \right) \right\}$$

or equivalently,

$$\hat{\mathbf{r}}_{\text{MAP}} = \min_{\mathbf{r}} \left\{ \frac{1}{\sigma_{\mathbf{e}}^2}(\mathbf{d}_1 - \mathbf{A}_1\mathbf{r})^{\mathsf{T}}(\mathbf{d}_1 - \mathbf{A}_1\mathbf{r}) + \mathbf{r}^{\mathsf{T}}\bar{\mathbf{R}}_0^{-1}\mathbf{r} \right\}$$

For $\bar{\mathbf{R}}_0 = (\mathbf{A}_0^{\mathsf{T}}\mathbf{A}_0)^{-1}$ and $\sigma_{\mathbf{e}}^2 = \theta$, the MAP estimator coincides with the Tikhonov regularised solution (7.14) of Equation (7.13). In this sense, one can state that Tikhonov regularisation of a linear model is a special case of Bayesian point estimation. Now assume additionally that $\bar{\mathbf{R}}_0 = \sigma_{\mathbf{r}}^2 \mathbf{I}$ for a $\sigma_{\mathbf{r}}^2 > 0$, i.e. we have a white Gaussian prior. Then we get $\mathbf{A}_0 = \mathbf{I}$ and $\theta = \sigma_{\mathbf{e}}^2/\sigma_{\mathbf{r}}^2$. Hence, minimising the resulting Tikhonov functional (7.14) is equivalent to the minimisation of the weighted sum

$$\frac{1}{\sigma_{\mathbf{r}}^2} T_{\sigma_{\mathbf{e}}^2/\sigma_{\mathbf{r}}^2}(\mathbf{r}) = \frac{1}{\sigma_{\mathbf{e}}^2} \|\mathbf{A}\mathbf{r} - \mathbf{d}_1\|^2 + \frac{1}{\sigma_{\mathbf{r}}^2} \|\mathbf{r}\|^2 = \frac{1}{\sigma_{\mathbf{e}}^2} \|\mathbf{e}\|^2 + \frac{1}{\sigma_{\mathbf{r}}^2} \|\mathbf{r}\|^2 \qquad \square$$

A huge research topic concerning regularisation techniques deals with the selection of the regularisation parameters. Within the Bayesian framework, this question is treated in conjunction with hierarchical modelling.

7.2.4 Multi-stage models

7.2.4.1 Hierarchical models

Using hierarchical modelling [17,18], complex structures can be formalised by combining several simpler models in hierarchical levels. An unknown parameter at a certain level demands the specification of one or more parameters at the next level. The hierarchy has not been build up completely until all parameters in the last level are known. In the following, the set H_j denotes the set of the unknowns in level j.

EXAMPLE 7.5. If (with the usual notations) $\mathbf{r} \sim N(\bar{\mathbf{r}}_0, \sigma_{\mathbf{r}}^2 \mathbf{I})$ (level 1) with both $\bar{\mathbf{r}}_0$ and $\sigma_{\mathbf{r}}^2$ unknown, we should add a second level in that hyperpriors $p(\bar{\mathbf{r}}_0)$, $p(\sigma_{\mathbf{r}})$ are defined. If the hyperpriors contain also unknown parameters, at least a third level is necessary. We have $H_1 = \{\mathbf{r}\}$, $H_2 = \{\bar{\mathbf{r}}_0, \sigma_{\mathbf{r}}^2\}$, etc.   □

The advantage of hierarchical modelling is induced by conditional independence across several levels. Hence, the basic model remains unchanged. Assuming one observation $\mathbf{d}_1$ and L levels given, we have

$$p(\mathbf{r}|\mathbf{d}_1) \propto p(\mathbf{d}_1|\mathbf{r})p(\mathbf{r})$$

with

$$p(\mathbf{r}) = p(H_1) = \sum\!\!\!\!\!\int p(H_1|H_2)p(H_2|H_3)\ldots p(H_{L-1}|H_L)\,\mathrm{d}H_2\ldots\mathrm{d}H_L$$

REMARK. One of the foundations of the Bayesian approach is that all quantities in a model that are not certainly known have to be treated as random. At first sight, this paradigm could lead into a dilemma. Generally, most of the parameters in a model are specified by measurements or experience values resulting from former measurements. Another common paradigm [19] originated from measurement engineering states that a measurement process delivers solely an estimate for the measured quantity. Respecting both paradigms simultaneously would obviously lead to unfeasible high-dimensional hierarchical models. However, Occam's razor prevents the choice of unnecessarily complex models.

7.2.4.2 Compound models

Compound models are convenient in image fusion problems in which heterogeneous images are obtained, i.e. images delivering information concerning different properties of the unknown $\mathbf{r}$ [20,21]. Assuming linear additive forward models, one can formulate the imaging processes by

$$\mathbf{d}_1 = \mathbf{A}_1\mathbf{r}_1 + \mathbf{e}_1, \qquad \mathbf{d}_2 = \mathbf{A}_2\mathbf{r}_2 + \mathbf{e}_2 \tag{7.23}$$

and relate them by a the decomposition of $\mathbf{r}$ into $\mathbf{r}_1$, $\mathbf{r}_2$ with $\mathbf{r}_2 = \mathbf{F}\mathbf{r}_1$, where $\mathbf{F}$ denotes a linear operator. Therefore, the posterior distribution at Bayesian fusion is given by

$$p(\mathbf{r}|\mathbf{d}_1, \mathbf{d}_2) = p(\mathbf{r}_1, \mathbf{r}_2|\mathbf{d}_1, \mathbf{d}_2) \propto p(\mathbf{d}_1, \mathbf{d}_2|\mathbf{r}_1, \mathbf{r}_2)p(\mathbf{r}_1, \mathbf{r}_2)$$

$$= p(\mathbf{d}_1, \mathbf{d}_2|\mathbf{r}_1, \mathbf{r}_2)p(\mathbf{r}_2|\mathbf{r}_1)p(\mathbf{r}_1)$$

Assuming further conditional independence of the images, this leads to

$$p(\mathbf{r}|\mathbf{d}_1, \mathbf{d}_2) = p(\mathbf{d}_1|\mathbf{r}_1)p(\mathbf{d}_2|\mathbf{r}_2)p(\mathbf{r}_2|\mathbf{r}_1)p(\mathbf{r}_1)$$

Thus, if an additional linear measurement equation between the resulting components of the unknown $\mathbf{r}$ is valid, a kind of hierarchical model has been induced via introducing a compound model. Thereby, the fusion problem has been basically reduced to the more homogeneous case (7.17). Concerning the more complicated situation in which an operator like $\mathbf{F}$ is not determinable, a similar hierarchical approach based on a compound Markov model with a hidden variable can be made [20,21]. This model and also the question how the unknown $\mathbf{r}$ can be finally estimated from the posterior are comprehensively discussed in the corresponding publications.

Dealing with heterogeneous image data that have been obtained by different sensing mechanisms, the formulation of the forward models should be performed carefully. To avoid poor fusion results, their compatibility has to be ensured. Incompatibility can arise due to exclusive prior information (in the form of heuristics or constraints) which has been incorporated into one of the forward models. Therefore, a strong coupling approach may be unavoidable if the forward models have been formulated independently without regard to the fusion task [14].

Beside the proposed multi-stage modelling techniques, modelling also via Bayesian networks may be a helpful method for simplifying Bayesian image fusion tasks [22,23]. Bayesian networks represent, in a sense, a generalisation of the proposed multi-stage models. We skip this topic because it would go beyond the scope of this publication.

7.2.5 Prior modelling

The aim of this section is to describe some helpful concepts of the Bayesian approach for choosing an appropriate prior, not to give an overview over common prior distributions.

Often, prior knowledge, information concerning the forward model, as well as additional constraints are formulated via energy functionals [16,24,25] that are connected to Bayesian techniques via Gibbs' densities. This concept will be treated in Section 7.3.

The ME principle has been discussed already in Section 7.1. Its application guarantees that all kinds of additional prior information, e.g. expertise knowledge, maps of the scene, knowledge concerning the scene geometry or material properties, are transformed in a loss-, error- and artefact-free manner. Therefore, it ensures that the prior is mainly concentrated on the a priori highly probable images.

Two other important concepts concerning also the choice of a suitable prior are discussed in the following. Firstly, the theory of conjugate priors is introduced. This can sometimes guarantee the manageability of computationally intensive fusion problems. Afterwards, the concept of non-informative priors is described. It makes the Bayesian approach also applicable if no prior knowledge is given.

7.2.5.1 Conjugate priors

A class P_1 of prior distributions is defined to be a conjugate family for a class P_2 of Likelihoods, if for all $p_1 \in P_1$ and $p_2 \in P_2$ the resulting posterior distribution is again contained in P_1.

EXAMPLE 7.6. We have seen, that the class of Gaussian densities represents a conjugate family for itself. ☐

Prior to the introduction of MCMC Methods, conjugate families have been of immense importance within the Bayesian approach. Non-conjugate priors could lead to posterior distributions that are mathematically not manageable. Due to the advantages that MCMC methods provide, conjugate priors, in principle, have become less important. However, despite the availability of dynamic sampling based techniques, the handling of extremely high-dimensional distributions is computationally still problematic, e.g. if one has to deal with huge Bayesian networks or to fuse a larger amount of sensing information that concerns high-dimensional unknowns. A serious problem concerning MCMC methods arises also from lacking and insufficient results regarding convergence rates of the corresponding sampling procedures. Hence, the concept of the conjugate priors may be still promising because – as exemplified for the Gaussian case (cf. Section 7.2.2) – thereby a Bayesian fusion task can be reduced to the problem of solving some algebraic manipulations. Consequently, conjugate prior theory and adjacent topics still represent an active research field. Actual research addresses, e.g. the approximation of arbitrary, maybe nonparametric, distributions by conjugate families.

7.2.5.2 Non-informative priors

The Bayesian approach works also for image fusion problems, where no prior knowledge is available. In this situation, all information concerning $\mathbf{r}$ that is encoded in the posterior $p(\mathbf{r}|\mathbf{d})$ should originate from the observations $\mathbf{d}_1, \ldots, \mathbf{d}_S$.

EXAMPLE 7.7. We assume that the image sequence $\mathbf{d}_1, \ldots, \mathbf{d}_S$ originates from a (local stationary) camera and is degenerated by additive white Gaussian noise with variance $\sigma_e^2 > 0$. The aim is fusing $\mathbf{d}_1, \ldots, \mathbf{d}_S$ to obtain a qualitatively better image. Despite of its extreme simplicity, this model is of great practical importance.

We obtain for $p(\mathbf{d}_k|\mathbf{r})$ the $N(\mathbf{r}, \sigma_e^2 \mathbf{I})$ distribution. If no prior knowledge is given, the ME principle delivers for the prior distribution the uniform distribution over the range of values of $\mathbf{r}$. Hence, $p(\mathbf{r}|\mathbf{d}_1, \ldots, \mathbf{d}_S)$ is the $N(\bar{\mathbf{r}}_S, \bar{\mathbf{R}}_S)$ distribution with

$$\bar{\mathbf{r}}_S = \frac{1}{S} \sum_{k=1}^{S} \mathbf{d}_k, \qquad \bar{\mathbf{R}}_S = \frac{\sigma_e^2}{S} \mathbf{I}$$

The MAP estimate is $\bar{\mathbf{r}}_S$. Generally, the MAP estimator delivers the estimate $\hat{\mathbf{r}}$ for $\mathbf{r}$ which is the most probable value after weighting the prior distribution and the Likelihood of the data via Bayes' theorem. However, due to the lack of prior knowledge in this example, it delivers here the same estimate as the Maximum Likelihood estimator. Note that the latter generally delivers the estimate which fits the observations $\mathbf{d}_1, \ldots, \mathbf{d}_S$ in the sense that they give the highest Degree of Evidence for it.

We remark that by the common assumption of independent, identically distributed random variables the overall probability over the (common) domain of the images $\mathbf{d}_k$ results in the product of the separate probabilities over each single pixel.

In this example, the ad hoc method of averaging pixel-wise over the images $\mathbf{d}_1, \ldots, \mathbf{d}_S$ forms a satisfactory alternative. It also delivers $\bar{\mathbf{r}}_S$ and reduces the variance of the resulting image about the factor S, too.

However, caused by the lack of flexibility, averaging generally can produce unwanted results, e.g. if the information contained in corresponding pixels of two images is complementary, it gets erased by averaging the images. The Bayesian approach provides clearly broad benefits in such situations. $\qquad \square$

The kind of non-informative prior that we have used in the preceding example is called the Laplace prior. It is not invariant under reparametrisation – a fact that causes problems. Due to the insufficiency of Laplace priors, a lot of effort has been exerted by statisticians to find distributions by which lack of knowledge can be expressed in a better manner. The approaches are based on sampling, the ME principle, information theoretic concepts, etc., and lead to (not perfect but mainly realistic) models for non-informative priors [15,18].

7.3 Formulation by energy functionals

Energy functionals are a powerful tool to formulate a processing task on given data $\mathbf{d}$. The central idea is to formalise all known or desired properties of the final result $\mathbf{r}$ or some intermediate results together with constraints and prior knowledge into I so-called energy terms $E_i \in \mathbb{R}, i = 1, \ldots, I, I \in \mathbb{N}^+$, which decrease monotonously when the result adopts a more desired value or when imposed constraints are better met. The global energy is obtained by weighting and summing the energy terms:

$$E(\mathbf{r}, \mathbf{d}) := \sum_i \lambda_i E_i(\mathbf{r}, \mathbf{d}), \quad \lambda_i > 0$$

Without loss of generality, one weighting coefficient can be set equal to 1.

By minimising the global energy with respect to $\mathbf{r}$, the optimum result $\hat{\mathbf{r}}$ with respect to the properties, constraints, and prior knowledge expressed in the energy terms is achieved:

$$\hat{\mathbf{r}} := \arg \min_{\mathbf{r}} \{ E(\mathbf{r}, \mathbf{d}) \}$$

The energy formalism has several advantages:

- The number of energy terms is arbitrary and not limited. This way, all available information can be integrated into the energy formulation. If new information becomes available, a suitable energy term is simply added.
- Energy terms can refer to the original data, to intermediate results, or to the final result. This way, the formulation of desired or expected properties offers many degrees of freedom.
- The relative relevance of desired and constraining properties can be represented by means of choosing appropriate values for the weighting coefficients λ_i.

It is obvious that energy functionals fulfil all requirements for fusing data and images: The original data $\{\mathbf{d}_i\}$ are the images to fuse, the fusion result $\hat{\mathbf{r}}$ is obtained by minimising the global energy, and the energies embody the fusion rules that model the desired and required properties of the final and intermediate results.

7.3.1 Energy terms

The formulation of energy terms should be done so that each desired or required property of the final result or any intermediate result as well as each constraint imposed is transformed into only one term. The overall specification of the fusion task is then obtained by weighting and combining these terms. Energy terms can base on the input data $\mathbf{d}$, on the final result $\mathbf{r}$, on some intermediate results, or on combinations of them.

Depending on the nature of the properties and constraints modelled in the energy terms, they can be classified into several groups which however are not always strictly separable.

7.3.1.1 Data terms

These terms serve to ensure a desired, a reasonable, or a given connection between the input data and the fusion result. A common formulation is given by

$$E_d(\mathbf{r}, \mathbf{d}) = \sum_{M_d \subseteq (D \times R)} D\{F\{\mathbf{r}\}, F\{\mathbf{d}\}\} + \sum_{O_d \subseteq (D \times R) \setminus M_d} \beta_d$$

where $D\{.\,,\,.\}$ is based on a distance measure (ideally a metric) of the two operands, $F\{.\}$ is a feature operator which defines the fusion-relevant property the result has to match with respect to the data; $M_d \subseteq (D \times R)$ denotes the part of the combined domains of $\mathbf{d}$ and $\mathbf{r}$ which is used for the data matching, and $\beta_d \geqslant 0$ serves as penalty for image points in a part of or in the entire remaining area $O_d \subseteq (D \times R) \setminus M_d$. Note that, since the data term assesses some kind of similarity between the input data and the result, it is a function of both signals $\mathbf{d}$ and $\mathbf{r}$.

EXAMPLE 7.8. A very simple example is obtained by using the Euclidean metric to $\mathbf{d}$ and $\mathbf{r}$ for $D = R$ in a region $M_d \subseteq R$, while deviations are allowed at positions $O_d = R \setminus M_d$:

$$E_d(\mathbf{r}, \mathbf{d}) = \sum_{M_d \subseteq R} (\mathbf{r} - \mathbf{d})^2 + \sum_{O_d = R \setminus M_d} \beta_d$$

thus demanding that the result keeps close to the input signal within the regions M_d. Deviations in O_d are rated with the penalty β_d. □

7.3.1.2 Quality terms

In such terms, the desired properties of the fusion result are expressed by means of an assessment of meaningful features:

$$E_q(\mathbf{r}) = \sum_{M_q \subseteq R} Q\{\mathbf{r}\} + \sum_{O_q \subseteq R \setminus M_q} \beta_q$$

where $Q\{.\}$ is an operator that acts as a measure for the quality of the fusion result $\mathbf{r}$ within the part of the domain $M_q \subseteq R$ that is used for the quality assessment, and β_q acts as a penalty in a part of or in the entire remaining area $O_q \subseteq R \setminus M_q$. Depending on the objective of the fusion, an appropriate term can model various characteristics of the fusion result. Since the quality term only rates the quality of the fusion result, it is only a function of $\mathbf{r}$.

EXAMPLE 7.9. Assume that a locally high contrast is the relevant property of the fusion result $r(x)$ which is for convenience formulated in terms of Equation (7.2). The quality term can be formulated, e.g., by means of its local signal variance, or equivalently, by its high-frequency components:

$$E_q(r(x)) = \sum_{M_q \subseteq R} -(r(x) - LP\{r(x)\})^2 + \sum_{O_q = R \setminus M_q} \beta_q$$

$$= \sum_{M_q \subseteq R} -HP^2\{r(x)\} + \sum_{O_q = R \setminus M_q} \beta_q$$

where $LP\{.\}$ is a low-pass filter that smoothes the function of interest by keeping homogeneous and slowly varying signal components, and $HP\{.\}$ is the corresponding high-pass filter such that for any function $s(x)$: $s(x) = LP\{s(x)\} + HP\{s(x)\}$. Although the first formulation needs an additional subtraction, it reveals the effect of the assessment better: Only deviations of $r(x)$ from the local average $LP\{r(x)\}$ are counted. Thus, the cutoff frequency f_{cutoff} of $LP\{.\}$ or $HP\{.\}$, respectively, determines the local neighbourhood that is used for calculating the local average. For $f_{cutoff} \to 0$, i.e. $LP\{r(x)\} = (1/|M_q|) \sum_{M_q} r(x)$, the energy term assesses global variance within the region M_q. The low-pass and high-pass filtering can be efficiently accomplished in the frequency domain, e.g. by using fast Fourier transform or fast wavelet transform. □

7.3.1.3 Constraint terms

Such terms are used, when certain conditions are imposed on the fusion result:

$$E_c(\mathbf{r}) = \sum_{M_c \subseteq R} C\{\mathbf{r}\} + \sum_{O_c \subseteq R \setminus M_c} \beta_c$$

where $C\{.\}$ is an operator that assesses how a certain condition is met within the part of the domain $M_c \subseteq R$. β_c is a penalty that is assigned to points in a part of or in the entire remaining area $O_c \subseteq R \setminus M_c$.

Depending on the degree to which the constraint must be fulfilled, an appropriate energy term must be chosen: If the constraint is soft, the respective term should use a continuous function for $C\{.\}$. However, when a certain fusion result is known to be impossible or strongly undesirable (i.e. a hard constraint), it is sensible to assign infinity to such a result, which causes the global energy also to become infinite, and therefore prevents the energy minimisation from choosing this result.

Since constraints often are not only applicable to the final result but rather to an intermediate result, constraint terms may contain assessments of intermediate results.

EXAMPLE 7.10. Depending on the objective of the fusion, an appropriate term can model, e.g., the n-continuity (the smoothness) of the fusion result or some intermediate result. Note that in the context of image series, smoothness does not need be restricted to the spatial dimensions. A common term used for assessing the n-continuity of the fusion result is obtained by

$$E_c(\mathbf{r}) = \sum_{M_c \subseteq R} L\{\mathbf{r}\} + \sum_{O_c = R \setminus M_c} \beta_c$$

where $L\{.\}$ is an operator which assesses the desired smoothness property within the region M_c. In the case that the first derivative should be small, $L\{\mathbf{r}\}$ could be chosen as a difference quotient. Regions O_c which are not included in the smoothness rating could be, e.g., image borders or visible edges which are known to belong to the scene and should not be smoothed.   □

The distinction of quality and constraint terms is not always strict: Depending on the fusion task, the smoothness constraint of the fusion result mentioned above can be interpreted as a constraint imposed to the final or to an intermediate result (in case that, e.g., some local contrast measure is to be optimised under some smoothness constraint) or as a quality property (in case that the processing serves to locally smooth the input data).

In practice, the formulation of such terms often involves evaluating local neighbourhood structures (e.g., the above mentioned smoothness terms). These terms then adopt the form of a Markov Random Field [26].

EXAMPLE 7.11. Consider the common task of stereo fusion: given a stereo pair $\{d_1(x), d_2(x)\}$, the 3-D reconstruction of the observed scene is required. For this purpose, the essential step is to precisely determine the disparity map $r(x)$, which here is the fusion target. A pixel x_1 of the first image and a pixel x_2 of the second image correspond if $x_2 = x_1 + r(x_1)$.

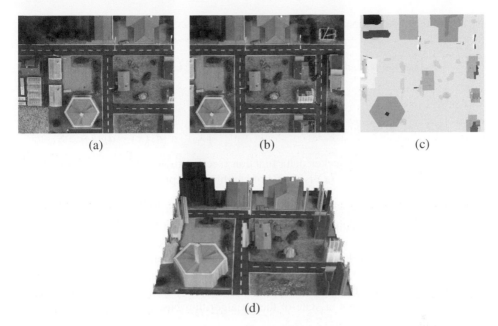

(a) (b) (c)

(d)

Figure 7.4 *(a, b) Stereo images of a landscape. (c) Disparity map: $r(x)$. (d) Landscape reconstruction.*

The fusion problem is modelled using the energy functional:

$$E_{\text{stereo}}\big(r(x), d_1(x), d_2(x)\big) = E_d\big(r(x), d_1(x), d_2(x)\big)$$
$$+ E_q\big(r(x), d_1(x), d_2(x)\big)$$
$$+ E_c\big(r(x), d_1(x), d_2(x)\big) \qquad (7.24)$$

The energy functional E_{stereo} consists of three terms, which in practice ensure that (1) the fusion result relates to the input signal (expressed by the data term E_d); (2) that the fusion result has the desired properties (expressed by the quality term E_q); and (3) that the fusion result satisfies previously known constraints (expressed by the constraint term E_c).

The *data term* $E_d(r(x), d_1(x), d_2(x))$: it ensures photo-consistency, i.e. corresponding pixels must have similar grey values. For this, a cost function based on a pixel dissimilarity measure, e.g. the squared intensity difference $D(x_1, x_2) := (d_1(x_1) - d_2(x_2))^2$, is defined:

$$E_d\big(r(x), d_1(x), d_2(x)\big) = \sum_{(x_1, x_2) \in C} \min\big\{0, D(x_1, x_2) - K\big\}$$

where K is a threshold and C is the set of corresponding pixels.

The *quality term* $E_q(r(x), d_1(x), d_2(x))$: as the reconstruction problem is underconstrained, additional knowledge is required to obtain a unique solution. The assumption used here is that the disparity values vary significantly only at intensity edges. The quality term models this property:

$$E_q\big(r(x), d_1(x), d_2(x)\big) = \sum_{k=1}^{2} \sum_{x \in \Omega_k} \sum_{\xi \in \mathcal{N}(x)} Q\big(d_k(x), d_k(\xi), r(x), r(\xi)\big)$$

$$Q\big(d_k(x), d_k(\xi), r(x), r(\xi)\big) := \big(1 - \delta_{r(x)}^{r(\xi)}\big) \cdot \begin{cases} \lambda_1 & \text{if } |d_k(x) - d_k(\xi)| < S, \\ \lambda_2 & \text{if } |d_k(x) - d_k(\xi)| \geqslant S \end{cases}$$

where δ_a^b denotes the Kronecker delta function (see Example 7.1). S is a threshold for detecting intensity edges, $\mathcal{N}(x)$ is the set of neighbouring pixels of x and $\lambda_1 > \lambda_2 > 0$.

The *constraint term* $E_c(r(x), d_1(x), d_2(x))$: this term assesses the visibility of the scene and excludes some physically impossible disparity configurations by assigning them infinite energy. The reader is referred to [27–29] for more details on the visibility constraint and on the appropriate selection of the parameters S, λ_1, λ_2, and K.

The solution of the depth estimation problem is found by minimising the energy functional (7.24) [29]. For this, a state-of-the-art algorithm based on graph cuts has been employed [30,31], see Section 7.3.4.3.   □

7.3.2 Connection with Bayes' methodology via Gibbs' distributions

By means of Gibbs' densities, the energy formalism can be directly related to the Bayes' methodology. Gibbs' distribution using the global energy is defined by

$$\pi(\mathbf{r}, \mathbf{d}) := \frac{1}{Z} e^{-\gamma E(\mathbf{r}, \mathbf{d})}$$

$$= \frac{1}{Z} \prod_i e^{-\gamma \lambda_i E_i(\mathbf{r}, \mathbf{d})}, \quad \gamma, \lambda_i > 0 \tag{7.25}$$

with a normalisation constant Z ensuring that $\pi(\mathbf{r}, \mathbf{d})$ is a DoB distribution. By means of the exponential function, the summation of the energy terms turns into a multiplication of the respective Gibbs' densities.

To establish a connection to the Bayes' formalism, the Gibbs' distribution of Equation (7.25) can be interpreted as proportional to the posterior distribution of $\mathbf{r}$ with given input data $\mathbf{d}$:

$$\pi(\mathbf{r}, \mathbf{d}) \propto p(\mathbf{r}|\mathbf{d})$$

On the other hand, the energy terms form the conditional DoB $p(\mathbf{d}|\mathbf{r})$ and the prior DoB $p(\mathbf{r})$:

$$p(\mathbf{r}|\mathbf{d}) \propto p(\mathbf{d}|\mathbf{r}) \cdot p(\mathbf{r}) \propto \underbrace{\prod_j e^{-\gamma \lambda_j E_j(\mathbf{r}, \mathbf{d})}}_{\propto\, p(\mathbf{d}|\mathbf{r})} \underbrace{\prod_k e^{-\gamma \lambda_k E_k(\mathbf{r})}}_{\propto\, p(\mathbf{r})}, \quad \gamma, \lambda_j, \lambda_k > 0 \tag{7.26}$$

where the energy terms have been assigned according to their arguments and $j = 1, \ldots, J, k = 1, \ldots, K, K + J = I$. The normalisation constant Z plays the role of $p(\mathbf{d})$ and can be omitted, if only the maximum of $p(\mathbf{r}|\mathbf{d})$ and the corresponding argument $\hat{\mathbf{r}}$ is of interest. The assignment of the energy terms to the conditional DoB or to the prior DoB depends on the individual energy formulation and the arguments of the respective energy terms. However, since data terms $E_d(\mathbf{r}, \mathbf{d})$ always describe the connection between the input data $\mathbf{d}$ and the fusion result $\mathbf{r}$, they can always be related to conditional DoBs. Quality terms $E_q(\mathbf{r})$ assess the fusion result and can therefore be read as prior information inducing a prior DoB distribution.

Note that the transition from (subjectively defined) energy functionals to Gibbs' densities is defined arbitrarily. The problem statement using Equation (7.26) consists only formally of probability distributions, which are subject to the underlying energy formulations. The advantage of the transition, however, consists in the multitude of methods from statistics (including Bayesian methods), which are now applicable to the optimisation problem.

Since the exponential function $e^{-\alpha}$ is strictly decreasing with α, the energy minimisation turns into a maximisation of the Gibbs' distribution. The optimum is independent of the constant γ, which therefore can be neglected:

$$\hat{\mathbf{r}} := \arg\min_{\mathbf{r}} \{ E(\mathbf{r}, \mathbf{d}) \}$$
$$= \arg\max_{\mathbf{r}} \{ \pi(\mathbf{r}, \mathbf{d}) \} = \arg\max_{\mathbf{r}} \{ e^{-E(\mathbf{r}, \mathbf{d})} \}$$
$$= \arg\max_{\mathbf{r}} \{ p(\mathbf{r}|\mathbf{d}) \}$$

Thus, maximising the Gibbs' distribution with respect to $\mathbf{r}$ and consequently minimising the corresponding global energy leads to the MAP estimate for $\mathbf{r}$, see Section 7.2.3. However, the interpretation of the Gibbs' distribution in the Bayesian sense offers many other methods to calculate an estimate for $\mathbf{r}$, such as other Bayesian estimators (see Section 7.2.3) or marginalisation to obtain a statement on parts of $\mathbf{r}$:

$$p(\mathbf{r} \in R_a|\mathbf{d}) \propto \sum_{\mathbf{r} \in R \setminus R_a} p(\mathbf{r}|\mathbf{d})$$

Note that only the MAP estimate of the posterior distribution yields the very same solution as the energy minimisation. However, the use of integral estimates such as the MMS estimate (Example 7.2) is restricted to measures that can be interpreted in the sense of DoB distributions and therefore justifies the transition from the energy formulation to Gibbs' densities.

When only quadratic energy terms are used, the Gibbs' distribution turns into a Gaussian distribution. As a consequence, the resulting distributions are especially easy to analyse since they are fully determined by their expectations and covariances, see Section 7.2.2. In this case, an objective choice for the weighting coefficients λ_i can be made.

EXAMPLE 7.12. Assume the task of unidirectionally smoothing an input function $d(x)$ defined on a discrete support $D = R$ with spacing 1 by using the energy terms [32,33]

$$E_d\big(r(x), d(x)\big) = \sum_R \big(r(x) - d(x)\big)^2$$

$$E_q\big(r(x)\big) = \sum_R \big(r(x + \delta) - r(x)\big)^2, \quad \delta = (1, 0)^T$$

Energy minimisation leads to the problem

$$\hat{r}(x) := \arg\min_{r(x)} \Big\{ \lambda_d \sum_R \big(r(x) - d(x)\big)^2 + \lambda_q \sum_R \big(r(x + \delta) - r(x)\big)^2 \Big\}, \quad \lambda_d, \lambda_q > 0$$

Inserting the global energy into Gibbs' distribution yields

$$p\big(r(x)|d(x)\big) \propto \pi\big(r(x), d(x)\big)$$

$$\propto e^{\left(-\lambda_d \sum_R (r(x)-d(x))^2 - \lambda_q \sum_R (r(x+\delta)-r(x))^2\right)}$$

$$= \underbrace{\prod_R e^{-\lambda_d(r(x)-d(x))^2}}_{\propto\, p(d(x)|r(x))} \underbrace{\prod_R e^{-\lambda_q(r(x+\delta)-r(x))^2}}_{\propto\, p(r(x))}, \quad \lambda_d, \lambda_q > 0$$

The first part is the desired conditional DoB and weighs for each point of the domain R how far the result $r(x)$ deviates from the input data $d(x)$. By means of the Gibbs' distribution, the Likelihood $p(d(x)|r(x))$ is modelled as the Gaussian distribution $\mathcal{N}(d(x), 1/2\lambda_d)$. The second part plays the role of prior information that models the previously defined favourable property of the fusion result with $p(r(x)) = \mathcal{N}(r(x + \delta),$ $1/2\lambda_q)$. This term is closely connected to the classical regularisation used in image restoration, where a comparable term is used to impose a smoothness constraint to the final result.

The definition of the global energy causes that $p(r(x)|d(x))$ has the structure of a Markov Random Field (MRF) [5,33]. An MRF is a probability distribution over a set of random variables (here the values $r(x)$ on the discrete domain R with spacing 1), where the probability of a single random variable $r(x_i)$, $x_i \in R$, given the values of all other random variables $r(x_j)$, $x_i \neq x_j \in R$, only depends on some neighbours N_i of x_i, i.e. $p(r(x_i)|r(x_j), x_j \neq x_i) = p(r(x_i)|r(x_j), x_j \in N_i)$. In this case, we will see in the solution of the direct minimisation (Example 7.13) that the neighbourhood is given by $N_i = \{x_i - \delta, x_i + \delta\}$. □

7.3.3 Connection with regularisation

Energy functionals are also closely related to regularisation problems where the task is to conclude from some (observation) data $\mathbf{d}$ on the signal of interest $\mathbf{r}$ under the assumption of a linear measurement equation, see Section 7.2.1.

The regularisation issue can be interpreted in a sense of an energy representation with the data term $E_d(\mathbf{r}, \mathbf{d}) := \|\mathbf{d} - \mathbf{A}\mathbf{r}\|^2$ and the prior term $E_p(\mathbf{r}) := \|\mathbf{A}_0\mathbf{r}\|^2$.

7.3.4 Energy minimisation

Although we made it obvious that the energy formulation of a fusion task offers a lot of freedom, the minimisation of the global energy poses heavy problems. The global energy is usually a non-convex function in a high-dimensional space spanned by the components of $\mathbf{r}$. A possible approach is to use a universal optimisation technique such as simulated annealing [33] which, however, requires much computational expense. Indeed, there is no fast and simultaneously exact general purpose solution that could be applied in any case to this problem. Instead, there exists a number of approximative approaches to energy minimisation which all have specific advantages and drawbacks.

7.3.4.1 Direct minimisation

In some special cases, the global energy functional $E(\mathbf{r}, \mathbf{d})$ can be minimised directly with respect to $\mathbf{r}$, e.g. by solving the pertaining Euler–Lagrange equations. The direct minimisation is especially feasible if the conditions for the components of $\mathbf{r}$ are separable as in the following example:

EXAMPLE 7.13. Consider the smoothing task of Example 7.12. The minimisation of the global energy

$$E\big(r(x), d(x)\big) = \lambda_d \sum_R \big(r(x) - d(x)\big)^2 + \lambda_q \sum_R \big(r(x + \delta) - r(x)\big)^2,$$

$$\delta = (1, 0)^\mathsf{T}, \ \lambda_d, \lambda_q > 0$$

leads to the Euler–Lagrange equations (neglecting the boundaries of R)

$$\frac{\partial E(r(x), d(x))}{\partial r(x_i)} = 2\lambda_d\big(r(x_i) - d(x_i)\big)$$

$$+ 2\lambda_q\big(2r(x_i) - r(x_i + \delta) - r(x_i - \delta)\big) \overset{!}{=} 0 \quad \forall i$$

Writing the signals in vector notation $\mathbf{d}$ and $\mathbf{r}$ according to Section 7.2.1 (with the components $\mathbf{d}\langle i \rangle := d(x_i)$ and $\mathbf{r}\langle i \rangle := r(x_i)$, respectively), the matrix equation $\mathbf{r} = \mathbf{L}^{-1}\mathbf{d}$ with the symmetric matrix ($\lambda = \lambda_q/\lambda_d$)

$$\mathbf{L} = \begin{pmatrix} \ddots & \ddots & & \ddots & \\ & -\lambda & 1 + 2\lambda & -\lambda & \\ & \ddots & & \ddots & \ddots \end{pmatrix}$$

is obtained, which is always invertible since $\det(\mathbf{L}) > 0$. Note that if the prior term is suppressed ($\lambda_q = 0$), the minimisation of the metric distance term $E_d(r(x), d(x))$ leads

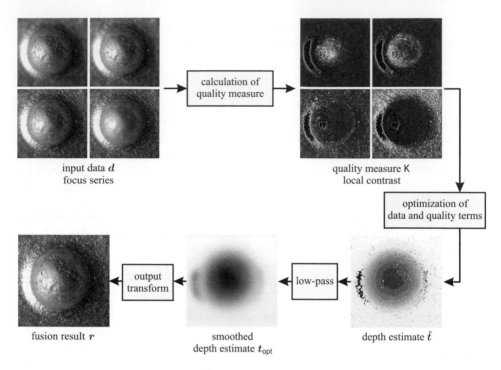

Figure 7.5 *Approximative solution by successive optimisation.*

to $\mathbf{L} = \mathbf{I}$ and the identity $r(x) = d(x)$. The MRF character of the pertaining posterior distribution $p(r(x)|d(x))$ with the neighbourhood $N_i = \{x_i - \delta, x_i + \delta\}$ can be recognised in the diagonal shape of $\mathbf{L}$.   □

7.3.4.2 Approximative solution by successive optimisation

Although finding the optimal solution to energy minimisation requires simultaneous optimisation of the global energy in principle, a practical alternative to solve energy minimisation tasks approximatively is established by successive optimisation of the energy terms. The main idea is to optimise the energy terms consecutively, starting with the most important one. Even if the solution obtained is not the optimal one in most cases, its application is often justified by the reduced computational effort and useful results.

EXAMPLE 7.14.   Consider the task of fusing a focus series $\{d_1(x), \ldots, d_S(x)\}$ of S images which have been recorded at equally spaced focus positions to a fusion result $r(x)$ with synthetically enhanced depth of focus [34], see Figure 7.5. The fusion task can be formulated by means of the energy functional

$$E\big(r(x), d(x)\big) = \lambda_{\mathrm{d}} E_{\mathrm{d}}\big(r(x), d(x), t(x)\big) + \lambda_{\mathrm{q}} E_{\mathrm{q}}\big(r(x)\big) + \lambda_{\mathrm{c}} E_{\mathrm{c}}\big(t(x)\big)$$

$$= \lambda_{\mathrm{d}} \sum_{\substack{D, R \\ t(x) = i}} \big(r(x) - d_i(x)\big)^2 + \lambda_{\mathrm{q}} \sum_{R} -K\{r(x)\} + \lambda_{\mathrm{c}} \sum_{D} -LP\{t(x)\}$$

where the intermediate result $t(x)$ with range$\{t(x)\} = \{1, \ldots, S\}$ contains the information whether an intensity value of a specific point x_j in an image of the series $d_i(x_j)$ is transferred in the fusion result $(t(x_j) = i)$ or not $(t(x_j) \neq i)$, i.e. $r(x_j) = d_i(x_j) \Leftrightarrow i = t(x_j)$, and can be interpreted as the depth estimate for that image point x_j. The data term describes the difference between the input signal and the fusion result, but only for the subset $d_i(x_j)$ of the input signal that is actually used for the fusion result, i.e. $t(x_j) = i$. The quality term seeks to find locally focused images by maximising a local contrast measure $K\{.\}$. Finally, the constraint term ensures that neighbouring image points have neighbouring depth estimates by means of a low-pass operator $LP\{.\}$ that checks for locally 'smooth' variation of the depth estimate $t(x)$, thus avoiding artifacts in the fusion result.

A closed minimisation of the global energy is very complex, since the optimisation has to consider both the intermediate result $t(x)$ and the final result $r(x)$. Instead, an efficient successive strategy initially calculates the quality measure $K\{.\}$ for each input image and then optimises the data and the quality term simultaneously in order to obtain a first estimate $\hat{t}(x)$:

$$\hat{t}(x) = \arg\max_{t(x)} \left\{ \lambda_d E_d\big(r(x), d(x), t(x)\big) + \lambda_q E_q\big(r(x)\big) \right\}$$

In a second step, the constraint term is incorporated by means of a smoothing of $\hat{t}(x)$:

$$t_{opt}(x) = LP\{\hat{t}(x)\}$$

and the fusion result is finally obtained by the output transform

$$\hat{r}(x) = d_i(x), \quad i = t_{opt}(x) \qquad \square$$

7.3.4.3 Optimisation using graph cuts

Graph cuts are means to solve optimisation tasks and have been originally developed for binary pixel labelling problems [35–37]. They define the optimisation task by means of a graph $\mathcal{G} = \langle \mathcal{V}, \mathcal{E} \rangle$ consisting of a set of vertices $\mathcal{V} = \{s, t\} \cup \mathcal{P}$ and a set of directed edges $\mathcal{E}$; see Figure 7.6. The special vertices s and t are the source and sink, respectively, which are both connected to any other vertex $p_i \in \mathcal{P}$ by so-called t-edges. Edges between other vertices $p_i, p_j \in \mathcal{P}$ are called n-edges. Each directed edge $(p_i, p_j) \in \mathcal{E}$, $p_i, p_j \in \mathcal{G}$ obtains a cost or weight $c(p_i, p_j)$, which is interpretable as individual energy of the edge.

An s/t-cut (or just cut) $\mathcal{C} = \{\mathcal{S}, \mathcal{T}\}$ is a partitioning of the set of vertices $\mathcal{V}$ into two disjoint subsets $\mathcal{S}$ and $\mathcal{T}$ such that $s \in \mathcal{S}$ and $t \in \mathcal{T}$. The global cost (i.e. the energy) of the cut $c(\mathcal{C})$ is then the sum of all severed edges:

$$c(\mathcal{C}) = \sum_{\substack{p_s \in \mathcal{S}, \, p_t \in \mathcal{T} \\ (p_s, p_t) \in \mathcal{E}}} c(p_s, p_t) \tag{7.27}$$

The minimum cut problem is then to find the cut $\mathcal{C}$ which minimises the cost $c(\mathcal{C})$. This problem is equivalent to finding the maximum flow from s to the t, when the graph

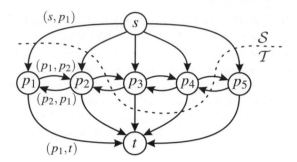

Figure 7.6 *Graph modelling a binary labelling with a possible cut.*

edges are interpreted as pipes and the weights are their capacity [38]. What makes the use of graph cuts so interesting is that a large number of algorithms exists to compute the maximum flow in a graph, and they can be applied to efficiently solve the optimisation problem [30,36,39,40].

Since a cut must sever exactly one t-edge (s, p_i) or (p_i, t), $p_i \in \mathcal{P}$, per vertex, each vertex $p_i \in \mathcal{P}$ contributes either with $c(s, p_i)$ or $c(p_i, t)$ to the global cost. The cut can be interpreted as an assignment $r : \mathcal{P} \rightarrow \{0, 1\}$ of a binary label to p_i: $r_i := r(p_i) = 0$ if $p_i \in \mathcal{S}$ and $r_i = 1$ if $p_i \in \mathcal{T}$. Thus, the cost per vertex is $D(r_i) = c(s, p_i)$ if $r_i = 1$ and $D(r_i) = c(p_i, t)$ if $r_i = 0$. Cuts between connected vertices $p_i, p_j \in \mathcal{P}$ with $r_i \neq r_j$ contribute with $V(r_i, r_j) = c(p_i, p_j) + c(p_j, p_i)$, whereas $V(r_i, r_j) = 0$ for $r_i = r_j$. Equation (7.27) can then be formulated as

$$E(\mathbf{r}, \mathbf{d}) := c(\mathcal{C}) = \underbrace{\sum_{p \in \mathcal{P}} D(r_i)}_{= E_\mathrm{d}(\mathbf{r}, \mathbf{d})} + \lambda \underbrace{\sum_{p_i, p_j \in \mathcal{N}} V(r_i, r_j)}_{= E_\mathrm{q,c}(\mathbf{r})}, \quad \lambda > 0 \qquad (7.28)$$

with a neighbourhood system $\mathcal{N}$ which reproduces the n-edges.

The connection to energy functionals is obtained by interpreting the terms in Equation (7.28): When the costs referring to a vertex r_i are defined so that they are a function of the respective observation d_i, i.e. $c(s, r_i), c(r_i, t) = f(d_i)$, the t-edges play the role of the data term in a related energy formulation. The costs for severed n-edges on the other side play the role of quality or constraint terms and assess cuts, i.e. discontinuities between neighbouring vertices.

Although not every energy formulation can be transferred in a graph representation, it has been shown that graphs can be constructed for certain classes of energy functions [31]. In the above case of $V(r_i, r_j)$ being a function of two variables, it is necessary and sufficient that the terms $V(r_i, r_j)$ satisfy the so-called regularity condition $V(0, 0) + V(1, 1) \leqslant V(0, 1) + V(1, 0)$. In order to model multi-label tasks instead of binary labels, additional rows of vertices can be inserted in the graph. In most of such generalised cases, graph cuts yield an approximative optimisation [35]. Graph cuts have been successfully applied for image fusion, e.g., in the context of stereo fusion [29,41,42] and fusion of multivariate series [27].

An interesting link of graph cuts to level sets [43,44] can be established by interpreting the cut in a way of the zero level of a level set function $\phi(.)$ [45]. The labels of the vertices $p_i \in \mathcal{P}$ are then the values $\phi(p_i)$. Both technologies include an implicit surface representation. However, there are some major differences:

- The level set optimisation allows a better localisation of the zero level in between the vertices which permits a subpixel accuracy.
- For the iterative optimisation, level sets need derivatives of the level set function $\phi(.)$. Graph cut optimisation, in contrast, does not use derivatives.
- In contrast to level sets, graph cuts are able to find the global minimum of the optimisation problem and are therefore insensitive to initialisation [46].

7.3.4.4 Optimisation using dynamic programming

A further approach to minimise energy functionals is provided by dynamic programming [47]. The main idea is to break the global optimisation problem into several stages which can be treated on their own. The problem is thus split into a recursive structure that is easier to solve. The simplification usually comprises both computation time and memory requirements.

The applicability of dynamic programming to energy minimisation problems strongly depends on the particular problem structure, i.e. the requirement that the optimisation can be separated into several stages must be fulfilled. It may be sensible to combine dynamic programming with a successive optimisation scheme [24].

7.4 Agent based architecture for local Bayesian fusion

Since the computational complexity for getting the optimal estimate of **r** is prohibitive even for moderate dimensions of the domain R of **r**, a direct calculation is often impossible (see Section 7.2.3). The underlying problem of the Bayesian approach is that everything is based on attaining and evaluating the whole posterior distribution: performing the requested fusion task and extracting estimates, a global view is retained.

The aim of the local Bayesian fusion approach is obtaining estimates with justifiable complexity. Therefore, Bayesian inference and fusion should be performed only in task relevant regions of R. This fundamental idea inspired an agent-based fusion architecture. Consistent with the local approach, the architecture establishes an analogy to criminal investigations [1,13,48–50].

In an initialisation step, clues, i.e. suspicious points in R, are detected with respect to each of the information sources using source specific operators. The clues are given to fusion agents [51] that try to confirm their DoBs with respect to the corresponding initial hypotheses by incorporating the information contributions of the other information sources in a local manner.

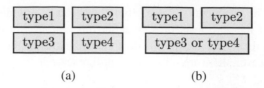

(a)   (b)

Figure 7.7 *Example:* $\mathbf{r} \in \mathbf{R} = \{\text{type1, type2, type3, type4}\}$ *is a qualitative feature.* $\mathbf{R}$ *becomes coarser by merging the values with low DoB, given the information of the analysed source. The DoB of the merged values results from cumulating their single DoBs. The significant values are exactly those that are contained in the local environment. Their DoBs are communicated separately. Here, three (b) instead of four values (a) have to be specified for communicating the local DoBs. In realistic applications, usually a large domain – e.g. a huge number of types – has to be considered. In this case, the decrease of computational complexity resulting from the local approach is considerable.*

7.4.1 Local Bayesian fusion

For realising a local view, every significant clue $\mathbf{c} \in \mathbf{R}$ is described locally within an appropriate local environment $U(\mathbf{c})$. Bayesian fusion with respect to $U(\mathbf{c})$ has to be still feasible.

In general, $U(\mathbf{c})$ is specified as a neighbourhood of $\mathbf{c}$ with respect to an appropriate norm, i.e.

$$U(\mathbf{c}) = \left\{ \mathbf{r} \in \mathbf{R} \mid \|\mathbf{c} - \mathbf{r}\| < \text{const}(\mathbf{c}) \right\}$$

For components for which such a specification of neighbourship is not meaningful (that are the most qualitative features), $U(\mathbf{c})$ is defined as the set of values whose DoB, given the evaluated information source, exceeds a certain threshold.

By local DoBs we mean DoBs that are specified with respect to a partition that is coarser and possibly more restricted than the domain $\mathbf{R}$ [48]. Local Bayesian fusion means performing Bayesian fusion using the local DoBs: exact Bayesian fusion is performed only within the local environments. This way, the computational complexity gets reduced significantly, see Figure 7.7.

7.4.2 Agent-based architecture

7.4.2.1 Initialisation

Taking into account the given problem, the prior information, and the nature of the unknown property of interest $\mathbf{r}$, source specific operators are performed on each of the information sources. In doing so, all available expertise as how to evaluate special kinds of information should be used. Values of $\mathbf{r}$ that appear conspicuous are stored in source specific clue lists together with the corresponding local DoBs, see Figure 7.8. The lists get ordered by falling significance which is measured by the local DoBs. For each clue, a fusion agent should be initialised, in principle. However, in the case of restricted computational resources, a smaller number of fusion agents for the most significant clues can also be initialised.

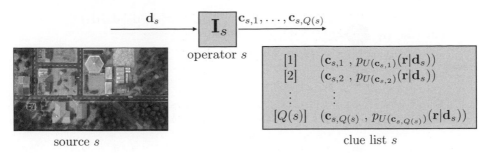

$$\mathbf{d}_s \longrightarrow \boxed{\mathbf{I}_s} \xrightarrow{\mathbf{c}_{s,1}, \ldots, \mathbf{c}_{s,Q(s)}}$$

operator s

$$
\begin{array}{ll}
[1] & (\mathbf{c}_{s,1} , p_{U(\mathbf{c}_{s,1})}(\mathbf{r}|\mathbf{d}_s)) \\
[2] & (\mathbf{c}_{s,2} , p_{U(\mathbf{c}_{s,2})}(\mathbf{r}|\mathbf{d}_s)) \\
\vdots & \vdots \\
[Q(s)] & (\mathbf{c}_{s,Q(s)} , p_{U(\mathbf{c}_{s,Q(s)})}(\mathbf{r}|\mathbf{d}_s))
\end{array}
$$

source s clue list s

Figure 7.8 *The initialisation phase exemplified for source number s, $s \in \{1, \ldots, S\}$, an IMINT (image intelligence) source: a source specific operator $\mathbf{I}_s$ delivers clues $\mathbf{c}_{s,q(s)}$ ($q(s) = 1, \ldots, Q(s)$ with $Q(s)$ being the number of significant clues with respect to $\mathbf{d}_s$). The clues are stored in the clue list s together with the corresponding local DoBs $p_{U(\mathbf{c}_{s,q(s)})}$. For each of them, a fusion agent is initialised.*

7.4.2.2 Investigation

Subsequent to the initialisation, each of the fusion agents starts to elaborate his initial local DoB, see Figure 7.9. For this purpose, he visits other information sources whose information has not yet been incorporated into his current local DoB. He can seek support of expert agents if he is not capable of accessing an information source or of evaluating the pertinent information contribution that is delivered by this source. If conditional independence of the information sources given $\mathbf{r}$ can be assumed (see Section 7.2.2.2), the expert can perform a local Bayesian fusion in a simple manner: he uses the local DoB of the fusion agent as prior distribution and transforms it by a simple multiplication into a posterior distribution with respect to the additional information of the corresponding source.

7.4.2.3 Basic competencies of the agents

A fusion agent has to visit all available information sources. To accomplish this task, he may have to migrate. Additionally, he has to administrate his clue, i.e. to memorise which information contributions have been already incorporated into his local DoB and to adjust the size of the local neighbourhood, if necessary. The fusion agents communicate with others to find out if the clues of some of them refer to the same cause: in this case, the corresponding fusion agents merge. The investigations of the individual fusion agents could be also implemented in a parallel manner. If one fusion agent was capable to duplicate himself after the initialisation, several copies of him could evaluate different information sources in an parallel manner. At integrating this possibility within the system, ad hoc methods for combining the resultant DoBs of the different copies, e.g. the imprudent application of a independent opinion pool scheme [15], have to be avoided.

An expert agent should be capable of evaluating some kind of information best and to possess a communication interface to the fusion agents. Also a human expert could constitute an expert agent. A realistic scenario for such kind of modelling arises if for some of the given information sources no automatic evaluation mechanisms exist.

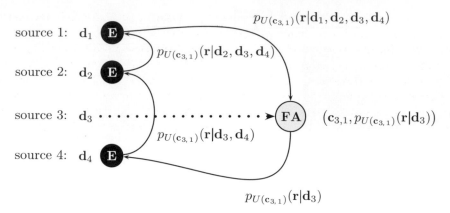

Figure 7.9   *Here, the investigation of one fusion agent at fusing four information sources is illustrated. For example, this fusion agent is responsible for the clue $\mathbf{c}_{3,1}$ (clue number 1 found in source 3) that has been delivered by the operator $\mathbf{I}_3$.*

7.4.3 The high potential of the proposed conception

The high computational costs caused by the Bayesian approach can be circumvented by realising a local Bayesian approach: in analogy to the local view of human criminal investigators, fusion agents evaluate and relate clues and evidence in a local manner.

The agent based architecture is scalable according to the given computational resources: as drafted in Section 7.4.2.1, only an affordable number of fusion agents is initialised. Doing so, only the hot traces on the first positions of the clue lists get investigated. Because of the modular structure of the architecture, all kinds of source specific operators and experts are replaceable in an uncomplicated manner. This option should be used if better evaluation mechanisms for specific information sources become available.

Additional sources can be added to the given ones in a simple manner. Alternatively, the fusion agents could also access sources that are additional to those used for the initialisation if the given ones do not provide enough information.

Obviously, the approximation of a given DoB by a simpler DoB distribution tends to be less difficult if instead of the global DoB only local DoBs have to be approximated. Hence, the application of the concept of conjugate families (see Section 7.2.5.1), which is very attractive from a computational point of view, is much more promising to be a good approximation for local Bayesian inference.

7.5 Summary

Bayesian image fusion is a powerful methodology for fusing images based on a probabilistic interpretation of images. It meets the major requirements for a useful fusion methodology: the capabilities to transform the information about the fusion result $r(x)$

that is available by prior knowledge, imposed by constraints or contained in the images $d_i(x)$ into probabilistic descriptions $p(r(x))$ and $p(d_i(x)|r(x))$, to combine the available information into a posterior DoB $p(r(x)|d_i(x))$, and to focus on some interesting detail of the fusion result $r(x)$ using conventional methods of probabilistic calculus. Bayesian methods are linked to other optimisation techniques such as regularisation techniques and energy minimisation which enable handling of many practical fusion tasks. In case a global perspective of the fusion problem is not manageable or desirable, local approaches offer the opportunity to transfer the idea of Bayesian image fusion to such tasks.

References

[1] J. Beyerer, J. Sander and S. Werling, 'Bayes'sche Methodik zur lokalen Fusion heterogener Informationsquellen', *Technisches Messen*, Vol. 74, No. 3, 2007, pp. 103–111.

[2] M. Müller, W. Krüger and G. Saur, 'Robust image registration for fusion', *Information Fusion*, Vol. 8, No. 4, 2007, pp. 347–353.

[3] J. Beyerer and F. Puente León, 'Bildoptimierung durch kontrolliertes Aktives Sehen und Bildfusion', *Automatisierungstechnik*, Vol. 53, No. 10, 2005, pp. 493–502.

[4] J.N. Kapur, *Maximum Entropy Models in Science and Engineering*, John Wiley & Sons, New York, 1989.

[5] G. Winkler, *Image Analysis, Random Fields and Dynamic Monte Carlo Methods: A Mathematical Introduction*, Springer, Berlin/Heidelberg, 1995.

[6] G.J. Klir and B. Yuan, *Fuzzy Sets and Fuzzy Logic: Theory and Applications*, Prentice Hall, Upper Saddle River, NJ, 1995.

[7] T.M. Cover and J.A. Thomas, *Elements of Information Theory*, John Wiley & Sons, New York, 1991.

[8] J. Kaipio and E. Somersalo, *Statistical and Computational Inverse Problems*, Springer, New York/Heidelberg, 2005.

[9] J. Hadamard, 'Sur les problèmes aux dérivées partielles et leur signification physique', *Princeton University Bulletin*, Vol. 13, 1902, pp. 49–52.

[10] G. Demoment, 'Image reconstruction and restoration: Overview of common estimation problems', *IEEE Transactions on Acoustics, Speech, and Signal Processing*, Vol. 37, No. 12, 1989, pp. 2024–2036.

[11] A. Neumaier, 'Solving ill-conditioned and singular linear systems: A tutorial on regularization', *SIAM Review*, Vol. 40, No. 3, 1998, pp. 636–666.

[12] H.W. Engl, M. Hanke and A. Neubauer, *Regularization of Inverse Problems*, Kluwer Academic Publishers, Dordrecht, 2000.

[13] J. Beyerer, M. Heizmann and J. Sander, 'Fuselets – an agent based architecture for fusion of heterogeneous information and data', in B.V. Dasarathy (ed.), *Multisensor, Multisource Information Fusion: Architectures, Algorithms, and Applications 2006*, in *Proceedings of SPIE*, Vol. 6242, 2006, pp. 235–243.

[14] A.L. Yuille and H.H. Bülthoff, 'Bayesian decision theory and psychophysics', in D.C. Knill and W. Richards (eds.), *Perception as Bayesian Inference*, Cambridge University Press, Cambridge, 1996, pp. 123–161.

[15] J.O. Berger, *Statistical Decision Theory and Bayesian Analysis*, second ed., Springer, New York, 1993.

[16] B. Chalmond, *Modeling and Inverse Problems in Image Analysis*, Springer, New York, 2003.

[17] S.J. Press, *Subjective and Objective Bayesian Statistics: Principles, Models, and Applications*, second ed., John Wiley & Sons, Hoboken, NJ, 2003.

[18] C.P. Robert, *The Bayesian Choice: From Decision-Theoretic Foundations to Computational Implementation*, second ed., Springer, New York/Berlin/Heidelberg, 2001.

[19] *Guide to the Expression of Uncertainty in Measurement*, International Organization for Standardization (ISO), Genf, 1995.

[20] A. Mohammad-Djafari, 'Probabilistic methods for data fusion', in G.J. Erickson, J.T. Rychert and C.R. Smith (eds.), *Maximum Entropy and Bayesian Methods*, Kluwer Academic Publishers, 1998, pp. 57–69.

[21] A. Mohammad-Djafari, 'Bayesian approach with hierarchical Markov modelling for data fusion in image reconstruction applications', in *Proceedings of the Fifth International Conference on Information Fusion (FUSION)*, Vol. 1, 2002, pp. 440–447.

[22] D. Kersten and P.R. Schrater, 'Pattern inference theory: A probabilistic approach to vision', in D. Heyer and R. Mausfeld (eds.), *Perception and the Physical World: Psychological and Philosophical Issues in Perception*, John Wiley & Sons, 2002, pp. 191–228.

[23] D. Kersten and A.L. Yuille, 'Bayesian models of object perception', *Current Opinion in Neurobiology*, Vol. 13, No. 2, 2003, pp. 150–158.

[24] P.N. Belhumeur, 'A Bayesian approach to binocular stereopsis', *International Journal of Computer Vision*, Vol. 19, No. 3, 1996, pp. 237–260.

[25] J.L. Marroquin, S.K. Mitter and T. Poggio, 'Probabilistic solution of ill-posed problems in computational vision', *Journal of the American Statistical Association*, Vol. 82, No. 397, 1987, pp. 76–89.

[26] S.Z. Li, *Markov Random Field Modeling in Computer Vision*, Springer, Tokyo/Heidelberg, 1995.

[27] C. Frese and I. Gheţa, 'Robust depth estimation by fusion of stereo and focus series acquired with a camera array', in *IEEE International Conference on Multisensor Fusion and Integration for Intelligent Systems (MFI)*, Heidelberg, Germany, September 2006, pp. 243–248.

[28] I. Gheţa, C. Frese and M. Heizmann, 'Fusion of combined stereo and focus series for depth estimation', in C. Hochberger and R. Liskowsky (eds.), *INFORMATIK 2006: Informatik für Menschen – Beiträge der 36. Jahrestagung der Gesellschaft für Informatik*, Vol. 1, Dresden, September 2006, pp. 359–363.

[29] V. Kolmogorov and R. Zabih, 'Multi-camera scene reconstruction via graph cuts', in *Proceedings of the Seventh European Conference on Computer Vision (ECCV) – Part III*, in *Lecture Notes in Computer Science*, Vol. 2352, Springer, 2002, pp. 82–96.

[30] Y. Boykov and V. Kolmogorov, 'An experimental comparison of min-cut/max-flow algorithms for energy minimization in vision', *IEEE Transactions on Pattern Analysis and Machine Intelligence*, Vol. 26, No. 9, 2004, pp. 1124–1137.

[31] V. Kolmogorov and R. Zabih, 'What energy functions can be minimized via graph cuts?', *IEEE Transactions on Pattern Analysis and Machine Intelligence*, Vol. 26, No. 2, 2004, pp. 147–159.

[32] J.J. Clark and A.L. Yuille, *Data Fusion for Sensory Information Processing Systems*, Kluwer Academic Publishers, Boston/Dordrecht/London, 1990.

[33] S. Geman and D. Geman, 'Stochastic relaxation, Gibbs distributions, and the Bayesian restoration of images', *IEEE Transactions on Pattern Analysis and Machine Intelligence*, Vol. 6, No. 6, 1984, pp. 721–741.

[34] F. Puente León and J. Beyerer, 'Datenfusion zur Gewinnung hochwertiger Bilder in der automatischen Sichtprüfung', *Automatisierungstechnik*, Vol. 45, No. 10, 1997, pp. 480–489.

[35] Y. Boykov and O. Veksler, 'Graph cuts in vision and graphics: Theories and applications', in N. Paragios, Y. Chen and O. Faugeras (eds.), *Handbook of Mathematical Models in Computer Vision*, Springer, New York, 2006.

[36] W.J. Cook, W.H. Cunningham, W.R. Pulleyblank and A. Schrijver, *Combinatorial Optimization*, John Wiley & Sons, New York, 1998.

[37] D.M. Greig, B.T. Porteous and A.H. Seheult, 'Exact maximum a posteriori estimation for binary images', *Journal of the Royal Statistical Society, Series B (Methodological)*, Vol. 51, No. 2, 1989, pp. 271–279.

[38] L.R. Ford and D.R. Fulkerson, *Flows in Networks*, Princeton University Press, Princeton, NJ, 1962.

[39] R.K. Ahuja, T.L. Magnanti and J.B. Orlin, *Network Flows: Theory, Algorithms, and Applications*, Prentice Hall, Upper Saddle River, NJ, 1993.

[40] Y. Boykov, O. Veksler and R. Zabih, 'Fast approximate energy minimization via graph cuts', *IEEE Transactions on Pattern Analysis and Machine Intelligence*, Vol. 23, No. 11, 2001, pp. 1222–1239.

[41] V. Kolmogorov and R. Zabih, 'Graph cut algorithms for binocular stereo with occlusions', in N. Paragios, Y. Chen and O. Faugeras (eds.), *Handbook of Mathematical Models in Computer Vision*, Springer, New York, 2006.

[42] D. Scharstein and R. Szeliski, 'A taxonomy and evaluation of dense two-frame stereo correspondence algorithms', *International Journal of Computer Vision*, Vol. 47, No. 1–3, 2002, pp. 7–42.

[43] S. Osher and R. Fedkiw, *Level Set Methods and Dynamic Implicit Surfaces*, Springer, New York/Berlin/Heidelberg, 2003.

[44] J.A. Sethian, *Level Set Methods and Fast Marching Methods: Evolving Interfaces in Computational Geometry, Fluid Mechanics, Computer Vision, and Materials Sciences*, second ed., Cambridge University Press, Cambridge, 2005.

[45] Y. Boykov, D. Cremers and V. Kolmogorov, 'Graph-cuts versus level-sets', Full-day tutorial at the *Ninth European Conference on Computer Vision (ECCV)*, Graz, Austria, May 6, 2006.

[46] Y. Boykov and V. Kolmogorov, 'Computing geodesics and minimal surfaces via graph cuts', in *Proceedings of the Ninth IEEE International Conference on Computer Vision (ICCV)*, Vol. 1, 2003, pp. 26–33.

[47] D.P. Bertsekas, *Dynamic Programming and Optimal Control*, Athena Scientific, Belmont, MA, 1995.

[48] J. Sander and J. Beyerer, 'Fusion agents – realizing Bayesian fusion via a local approach', in *IEEE International Conference on Multisensor Fusion and Integration for Intelligent Systems (MFI)*, Heidelberg, Germany, September 2006, pp. 243–248.

[49] J. Sander and J. Beyerer, 'A local approach for Bayesian fusion: Mathematical analysis and agent based conception', *Journal of Robotics and Autonomous Systems*, submitted for publication.

[50] J. Sander and J. Beyerer, 'Local Bayesian fusion realized via an agent based architecture', in *INFORMATIK 2007: Informatik trifft Logistik – Beiträge der 37. Jahrestagung der Gesellschaft für Informatik*, Vol. 2, Bremen, September 2007, pp. 95–99.

[51] S.J. Russell and P. Norvig, *Artificial Intelligence: A Modern Approach*, second ed., Prentice Hall, Upper Saddle River, NJ, 2003.

8

Multidimensional fusion by image mosaics

Yoav Y. Schechner[a] and Shree K. Nayar[b]

[a] *Department of Electrical Engineering, Technion – Israel Institute of Technology, Haifa, Israel*

[b] *Department of Computer Science, Columbia University, New York, USA*

Image mosaicing creates a wide field of view image of a scene by fusing data from narrow field images. As a camera moves, each scene point is typically sensed multiple times during frame acquisition. Here we describe *generalised mosaicing*, which is an approach that enhances this process. An optical component with spatially varying properties is rigidly attached to the camera. This way, the multiple measurements corresponding to any scene point are made under different optical settings. Fusing the data captured by the multiple frames yields an image mosaic that includes additional information about the scene. This information can come in the form of extended dynamic range, high spectral quality, polarisation sensitivity or extended depth of field (focus). For instance, suppose the state of best focus in the camera is spatially varying. This can be achieved by placing a transparent dielectric on the detector array. As the camera rigidly moves to enlarge the field of view, it senses each scene point multiple times, each time in a different focus setting. This yields a wide depth of field, wide field of view image, and a rough depth map of the scene.

8.1 Introduction

Image mosaicing[1] is a common method to obtain a wide field of view (FOV) image of a scene [8–10]. The basic idea is to capture frames of different parts of a scene, and then fuse the data from these frames, to obtain a larger image. The data is acquired by a relative motion between the camera and the scene: this way, each frame captures different

---

[1] In different communities the terms *mosaicing* [1,2] and *mosaicking* [3–7] are used.

scene parts. Image mosaicing has long been used in a variety of fields, such as optical observational astronomy [11,12], radio astronomy [13], and remote sensing [3,7,14–18], optically or by synthetic aperture radar (SAR). It is also used in underwater research [5, 6,19–21]. Moreover, image mosaicing has found applications in consumer photography [1,4,9,22–29].

As depicted in Figure 8.1, image mosaicing mainly addressed the extension of the FOV. However, there are other imaging dimensions that require enhanced information by fusing multiple measurements. In the following, we show how this can be done, within a unified framework that includes mosaicing. The framework is termed *generalised mosaicing*. It extracts significantly more information about the scene, given an amount of acquired data similar to that acquired in traditional mosaicing.

A typical video sequence acquired during mosaicing has great redundancy in terms of the data it contains. The reason is that there is typically a significant overlap between frames acquired for the mosaic, thus each point is observed multiple times. Now, let us rigidly attach to the camera a fixed filter with spatially varying properties, as in the setup shown in Figure 8.2. As the camera moves (or simply rotates), each scene point is measured under different optical settings. This significantly reduces the redundancy in the captured video stream. In return, the filtering embeds in the acquired data more information about each point in the mosaic FOV. Except for mounting the fixed filter, the image acquisition in generalised mosaicing is identical to traditional mosaicing.

In the following sections we describe several realisations of this principle. Specifically, when a filter with spatially varying transmittance is attached to the camera, each scene point is effectively measured with different exposures as the camera moves. These measurements are then fused to a high dynamic range (HDR) mosaic. Similarly, if the filter transmits a spatially varying spectral band, multispectral information is obtained for each scene point. In another implementation, a spatially varying polarisation filter is used, yielding wide FOV polarimetric imaging. Such systems were described in [30–34].

A particular realisation, which we describe in more detail is one having spatially varying focus settings. Fusion of image data acquired by such a sensor can yield an all-focused image in a wide FOV, as well as a rough depth map of the scene.

8.2 Panoramic focus

8.2.1 Background on focus

Focusing is required in most cameras. Let the camera view an object at a distance s_{object} from the first (front) principal plane of the lens. A focused image of this object is formed at a distance s_{image} behind the second (back) principal plane of the lens, as illustrated in Figure 8.3. For simplicity, consider first an aberration-free flat-field camera, having an effective focal length f. Then,

$$\frac{1}{s_{\text{image}}} = \frac{1}{f} - \frac{1}{s_{\text{object}}} \tag{8.1}$$

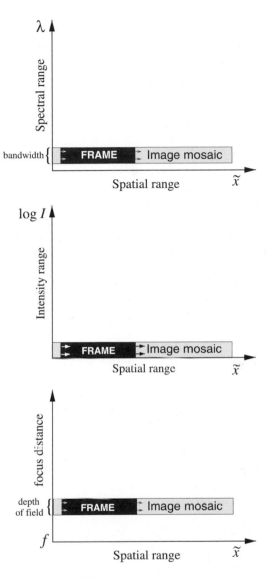

Figure 8.1 *An image frame has a limited FOV of the scene, i.e., it has a limited extent spatially ($\tilde{x}$ coordinates). By fusing partly overlapping frames, an image mosaic extends the FOV of any camera. However, there is a need to enhance additional imaging dimensions, such as the dynamic range [33] of intensity I, the hyperspectral [32] quality (sensitivity to the wavelength λ), and depth of field. The latter refers to the need to view objects in-focus at a distance extending from f (front focal plane) to infinity. ((The top part) Y.Y. Schechner and S.K. Nayar, 'Generalized mosaicing: Wide field of view multispectral imaging', IEEE Transactions on Pattern Analysis and Machine Intelligence, Vol. 24, 2002, pp. 1334–1348. © 2005 IEEE.)*

Hence, s_{image} is equivalent to s_{object}. The image is sensed by a detector array (e.g., a CCD), situated at a distance s_{detector} from the back principal plane. If $s_{\text{detector}} = s_{\text{image}}$, then the detector array senses the focused image. Generally, however, $s_{\text{detector}} \neq s_{\text{image}}$. If $|s_{\text{image}} - s_{\text{detector}}|$ is sufficiently large, then the detector senses an image which is *defocus blurred*.

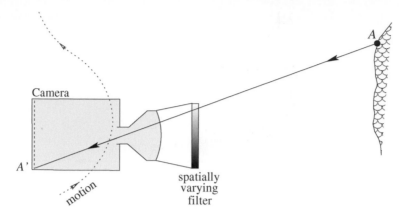

Figure 8.2   *Scene point A is imaged on the detector at A′ through a spatially varying filter attached to the camera. As the imaging system moves [31], each scene point is sensed through different portions of the filter, thus multiple measurements are obtained under different optical settings. (Y.Y. Schechner and S.K. Nayar, 'Generalized mosaicing', in Proc. IEEE International Conference on Computer Vision, Vol. I, 2001, pp. 17–24. © 2005 IEEE.)*

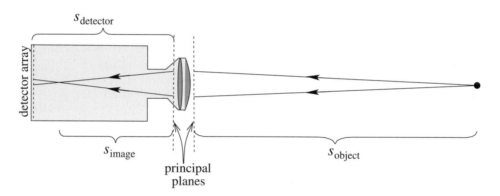

Figure 8.3   *Geometry of a simple camera system.*

For a given s_{detector}, there is a small range of s_{image} values for which the defocus blur is insignificant. This range is the *depth of focus*. Due to the equivalence of s_{image} to s_{object}, this corresponds to a range of object distances which are imaged sharply on the detector array. This range is the *depth of field* (DOF). Hence, a single frame can generally capture in focus objects that are in this limited span. However, typically, different objects or points in the FOV have different distances s_{object}, extending beyond the DOF. Hence, while some objects in a frame are in focus, others are defocus blurred.

There is a common method to capture each object point in focus, using a stationary camera. In this method, the FOV is fixed, while K frames of the scene are acquired. In each frame, indexed $k \in [1, K]$, the *focus settings* of the system change relative to the previous frame. Change of the settings can be achieved by varying s_{detector}, or f, or s_{object}, or any combination of them. This way, for any specific object point (x, y), there is a frame

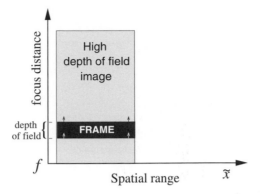

Figure 8.4 *An image frame has a limited FOV of the scene (marked by $\tilde{x}$) and a limited DOF. By fusing differently focused images, the DOF can be extended by image post processing, but the FOV remains limited.*

$k(x, y)$ for which Equation (8.1) is approximated as

$$\frac{1}{s_{\text{detector}}^{(k)}} \approx \frac{1}{f^{(k)}} - \frac{1}{s_{\text{object}}^{(k)}(x, y)} \tag{8.2}$$

i.e. $s_{\text{detector}}^{(k)} \approx s_{\text{image}}^{(k)}$, bringing the image of this object point into focus. This is the *focusing* process. Since each point (x, y) is acquired in focus at some frame k, then fusing the information from all K frames yields an image in which all points appear in focus. This principle is sketched in Figure 8.4. The result of this image fusion is effectively a high DOF image. However, the FOV remains limited, since the camera is static while the frames are acquired. In the subsequent sections, we will show that *focusing* and *extension of the FOV* can be obtained in a single, efficient scan.

The surface of least confusion

The object distance $s_{\text{object}}(x, y)$ is a function of the transversal coordinates. Following Equation (8.1), this function is equivalent to a surface $s_{\text{image}}(x, y)$ inside the camera chamber.[2] On this surface, the image is at best focus (least blur). This is the *surface of least confusion* (SLC) [35]. Apparently, for a flat object having a spatially invariant distance, the SLC is flat as well. In such a case, the entire object can be focused in a single frame, since the detector array is flat.

However, the SLC is generally not flat, even if s_{object} is constant. Typically, it is curved [35] radially from the centre of the camera FOV. Since it no longer obeys Equation (8.1), we denote the SLC as $s_{\text{image}}^{\text{effective}}$. This effect is caused by lens aberrations, which have been considered as a hindering effect. Thus, optical engineering makes an effort to

---

[2]The transversal coordinates of the image (x, y) are a scaled version of the object coordinates. The scale is the magnification of the camera. Since the magnification is fixed for given camera settings, we do not make explicit use of this magnification. Thus for simplicity, we do not scale the coordinates, and thus (x, y) are used for both the image domain and the object.

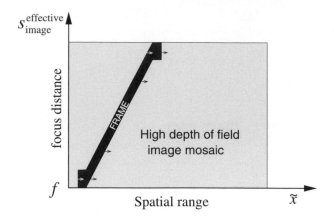

Figure 8.5 *Each frame has a spatially varying focus. Generalised mosaicing of such frames extends both the FOV and the DOF.*

flat-field optical systems [35], i.e. to minimise the departure of the SLC from a flat surface normal to the optical axis.

8.2.2 Intentional aberration

We now describe a unified way for expanding both the FOV and the DOF. The key is an intentionally aberrated imaging system, in which the distance between the SLC and the flat detector array spatially varies significantly. At a given frame, thus, different object points on a flat frontal plane are focused or defocused differently. This concept is depicted in Figure 8.5, by the support of the frame: across the FOV of a frame, the focus distance changes.

Now, the camera scans the scene *transversally*, in order to increase the FOV using mosaicing, as in Figure 8.1. However, due to the spatially varying focus of this system, during the transversal scan any object becomes focused at *some* frame k, as seen in Figure 8.5. By use of computational analysis of the acquired images, information about focus is extracted for each object.

A focused state may be obtained for all the pixels in a wide FOV image mosaic. In addition, information becomes available about the periphery of the central region of interest. The periphery has a gradually narrower DOF, but at least as wide as the inherent DOF of the camera (Figure 8.6). Such a gradual variation is analogous to foveated imaging systems, in which the acquisition quality improves from the periphery towards the centre of the FOV. The periphery is at most one frame wide, and is eliminated in 360° panoramic mosaics.

Optical implementation
Panoramic focusing based on this principle had been obtained by a system in which the CCD array was tilted relative to the optical axis [36]. We now describe an alternative

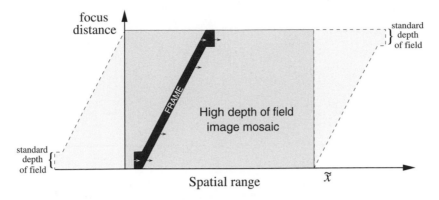

Figure 8.6 *Outside the main region of interest, the mosaic provides additional information about the scene periphery, whose quality gradually coincides with that of a single frame.*

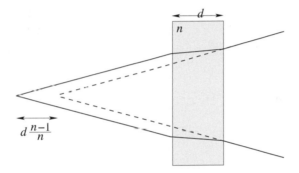

Figure 8.7 *A dielectric with refraction index n and thickness d. If it is placed before the detector, it shifts the focused image point away from the imaging optics.*

implementation, that can be more flexible. It can be based on mounting a transparent dielectric object on the detector array. By letting this transparent object have spatially varying characteristics, the SLC can be deformed to suit our needs.

First, we describe a simpler setup. Suppose a thin transparent slab is inserted between the imaging optics and the plane of best image focus. Now, let a light beam be focused through the slab. As seen in Figure 8.7, light refracts at the slab interfaces. Hence, the plane of best focus is shifted. The shift effectively increases s_{image} by approximately

$$\Delta s \approx d(n-1)/n \tag{8.3}$$

where d is the thickness of the slab and n is its refractive index relative to air.[3] The SLC is then

$$s_{\text{image}}^{\text{effective}} = s_{\text{image}} + \Delta s \tag{8.4}$$

---

[3]This is based on the paraxial approximation.

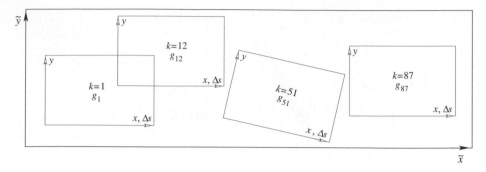

Figure 8.8 *The coordinates $(\tilde{x}, \tilde{y})$ denote the position of a scene point in the global coordinate system of the mosaic [32]. The mosaic is composed of raw frames indexed by k. The intensity in a raw frame is $g_k(x, y)$, where (x, y) are the internal coordinates in the frame. The axial shift Δs of the SLC is a function of (x, y). (Y.Y. Schechner and S.K. Nayar, 'Generalized mosaicing: Wide field of view multispectral imaging', IEEE Transactions on Pattern Analysis and Machine Intelligence, Vol. 24, 2002, pp. 1334–1348. © 2005 IEEE.)*

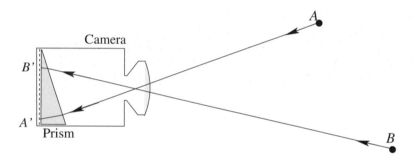

Figure 8.9 *The focus settings can change across the frame's FOV by placing a transparent object with variable thickness on the detector. Here, a wedge-prism is placed on the detector array, such that both objects A and B are focused even-though they are at different distances from the camera.*

where s_{image} is given by Equation (8.1). The distance between the focused image and the detector is thus $|s_{\text{image}}^{\text{effective}} - s_{\text{detector}}|$. It can be affected by setting d or n.

To obtain a spatially varying SLC, we may vary n or d across the camera FOV. Let us better define the spatial coordinates we use. The coordinates $(\tilde{x}, \tilde{y})$ denote a scene point, in the global coordinate system of the mosaic, as depicted in Figure 8.8. In each frame, the internal coordinates of a pixel are (x, y), equivalent to a position on the detector array. As an example for spatial variation of d, let the transparent object be a wedge prism, as illustrated in Figure 8.9. In the paraxial approximation (small angles), d and Δs change linearly along the detector's x-axis.

$$d = \gamma x \qquad (8.5)$$

where γ encapsulates the wedge-prism slope. Following Equations (8.3) and (8.5),

$$\Delta s = \gamma x (n - 1)/n \qquad (8.6)$$

The defocus blur changes accordingly.

Figure 8.10 *The frame in which an object is focused depends on its position in the frame. The teddy bear is focused only when it appears on the right-hand side of the frame's FOV. The words 'don't panic' are focused only when they appear in the left-hand side of the frame's FOV.*

To demonstrate this, we placed a wedge-prism on the CCD array of a Sony monochrome machine vision camera, and then mounted a C-mount lens over it. The system was then positioned on a tripod, and panned to obtain video data for a wide FOV mosaic. Four sample raw frames extracted from the video sequence are shown in Figure 8.10. Consider two objects in the scene: a teddy bear and the words 'don't panic' written on a newspaper. The two objects have a significantly different distance from the camera. The teddy bear is defocus blurred when it appears on the left-hand side of an image frame, while focused on the right-hand side of another frame. In contrast, the words 'don't panic' focus at the opposite frame part.

A wedge-prism is a special case. More generally, the transparent object placed on the detector array can have other shapes, leading to more general SLCs. Actually, even in the absence of any transparent object attachment, the focal distance is somewhat spatially varying (curved radially) by default, unless an effort is made to flat field the system [35]. Thus, one can use simpler lenses in which no effort is made to minimise the Petzval curvature [35]. In any case, the raw images need to have spatially varying focusing characteristics. While this makes the raw images look strange, the spatially varying focus is compensated for and even exploited using fusion algorithms, as described in the next sections.

8.2.3 Data fusion

Thanks to algorithms, a wide FOV and a wide DOF can be obtained from the raw images. The algorithms combine the principles of image mosaicing (registration; image fusion that reduces artifacts) with the principles of wide DOF imaging (focus search and image fusion). Therefore, mosaicing includes three consecutive stages: image registration, focus sensing for the registered images, and image fusion that reduces the appearance of artifacts.

First of all, the frames should be registered, assuming the motion between frames is unknown. A scene point has different coordinates in each frame. The measurements corresponding to this point should be identified before they can be fused. Registration algorithms for image mosaicing are well developed, and we refer the readers to [1,4,9,27,29, 32,33] for details. We registered the images, samples of which are shown in Figure 8.10, by minimising the mean squared difference between the frames.

Recall that $(\tilde{x}, \tilde{y})$ are the coordinates of a scene point in the global coordinate system of the mosaic (see Figure 8.8). Let k be the index of the individual frames that compose the mosaic. Since the raw frames are registered, we have for each scene point in the mosaic FOV a set of intensity measurements $\{g_k(\tilde{x}, \tilde{y})\}_{k=1}^{K}$. Now we may analyse the focus or defocus blur.

8.2.3.1 The best focus

Once the frames are registered, we may find which one of them is focused at each mosaic coordinate $(\tilde{x}, \tilde{y})$. The focused state of an object point is detected by maximising some focus criterion [37–45]. Since focus is associated with high sharpness, we maximise a sharpness criterion, say the image Laplacian

$$\hat{k}(\tilde{x}, \tilde{y}) = \arg\max_{k} \left| \nabla^2_{\tilde{x}, \tilde{y}} \, g_k(\tilde{x}, \tilde{y}) \right| \tag{8.7}$$

where $\nabla^2_{\tilde{x}, \tilde{y}}$ is the Laplacian over the mosaic's spatial domain.

Based on $\hat{k}$, a depth map of the scene can be estimated, as in standard methods for *depth from focus* [38–41,43,46–49]. As illustrated in Figure 8.8, once we know $\hat{k}$ at $(\tilde{x}, \tilde{y})$, we can retrieve the corresponding (x, y) coordinates in the $\hat{k}$th frame. Denote them as $(x_{\hat{k}}, y_{\hat{k}})$. The SLC $s_{\text{image}}^{\text{effective}}(x, y)$ of the imaging system can be known, e.g., by pre-calibration of the system. Hence the retrieved position $(x_{\hat{k}}, y_{\hat{k}})$ immediately indicates $s_{\text{image}}(x_{\hat{k}}, y_{\hat{k}})$. As explained in Section 8.2.1, s_{image} is equivalent to the object distance, i.e., the depth map.

For example, consider again our experiment using the wedge-prism. At focus, $s_{\text{image}}^{\text{effective}} \approx s_{\text{detector}}$. Thus, following Equations (8.4) and (8.6),

$$s_{\text{image}} = s_{\text{detector}} - \gamma x_{\hat{k}} \frac{n-1}{n} \tag{8.8}$$

Left periphery   Right periphery

Figure 8.11 *A map equivalent to a rough object distance, estimated in an experiment. Brighter pixels correspond to objects at larger object distances.*

This was applied to our image data, samples of which were shown in Figure 8.10. The resulting depth-equivalent map s_{image} is shown in Figure 8.11, where brighter pixels correspond to objects at larger object distances. This demonstrates the feasibility of obtaining rough depth information. The depth map was smoothed by median filtering. As mentioned in Section 8.2.2, the peripheral scene parts are measured in fewer frames, thus with less focus states. Therefore, the dynamic range of depths gradually decreases in the mosaic periphery, leaving partial knowledge of depth from focus.[4]

8.2.3.2 Wide DOF mosaic by fusion

In Section 8.2.3.1, Equation (8.7) led to a rough estimate of the object depth. In this section, Equation (8.7) is the basis for forming a sharp image of the objects in view (as if they are all in focus), no matter their depth. The output image is $\hat{I}$. Let us assign an output pixel $(\tilde{x}, \tilde{y})$ the value

$$\hat{I}(\tilde{x}, \tilde{y}) = g_{\hat{k}(\tilde{x}, \tilde{y})}(\tilde{x}, \tilde{y}) \tag{8.9}$$

Here, each object point is extracted at its sharpest state. This principle is common in techniques that create high DOF images from a stack of differently focused frames, in a stationary camera [37,44], as described in Section 8.2.1.

While Equations (8.7) and (8.9) are a simple recipe, the result may not be visually pleasing. Two kinds of artifacts are created. One of them is a noisy appearance, stemming from a noisy estimation of the state of best-focus. The other artifact is seam-lines in the mosaic, corresponding to the boundaries of raw frames. It stems from the slight inconsistencies of the exposure and illumination settings between frames. Both artifacts are inherent to image fusion and image mosaicing, and have been observed consistently in past work about either mosaicing or focusing. In the following, we detail the reasons for these two visual artifacts, and describe a way to overcome both. This is not just a cosmetic technicality: overcoming visual artifacts in elegant ways had been a topic for research in the mosaicing and fusion communities.

---

[4]Even if no frame measures the focus state at an area, as occurs in the periphery, depth may still be estimated there. A method that enables this is *depth from defocus* [50–53], which requires as few as two differently defocused measurements in order to estimate depth.

Focus artifacts Focus measurement as in Equation (8.7) is not an ideal process. Image noise may cause a random shift in the maximum focus measure. In other words, there is randomness in the value of $\hat{k}(\tilde{x}, \tilde{y})$ about the value that would have been obtained in the absence of noise. This results in a noisy looking mosaic. This randomness can be attenuated if the focus measure (e.g., Equation (8.7)) is calculated over a wide patch. This would have been fine if the object was equidistant from the camera. However, $s_{\text{object}}(\tilde{x}, \tilde{y})$ may have significant spatial variations (depth edges). In this case, large patches blur the estimated $\hat{k}$ map, degrading the resulting fused image.

This artifact is bypassed by an elegant approach which yields fusion results that are more perceptually pleasing, as has been shown in various fusion studies [54–57]. It is based on calculations performed in *multiple scales* [54–57], as described next. For each frame $g_k(\tilde{x}, \tilde{y})$, derive Laplacian and Gaussian pyramids [58] having M pyramid levels. Denote the image at level $m \in [0, M-1]$ of the Laplacian pyramid as $\mathbf{l}_k^{(m)}$, with a corresponding Gaussian pyramid image $\mathbf{g}_k^{(m)}$. As m increases, the image $\mathbf{l}_k^{(m)}$ expresses lower spatial frequencies. Moreover, a pixel of $\mathbf{l}_k^{(m)}$, denoted as $l_k^{(m)}(\tilde{x}^{(m)}, \tilde{y}^{(m)})$ represents an equivalent image area in the full image domain $(\tilde{x}, \tilde{y})$ that increases with m. The lowest spatial frequencies are represented by another image in the Gaussian pyramid, denoted as $\mathbf{g}_k^{(M)}$. Based on $\mathbf{g}_k^{(M)}$ and $\{\mathbf{l}_k^{(m)}\}_{m=0}^{M-1}$, the raw image $g_k(\tilde{x}, \tilde{y})$ can be reconstructed [58].

Now, instead of Equation (8.7), the relation

$$\hat{k}\big(\tilde{x}^{(m)}, \tilde{y}^{(m)}\big) = \arg\max_k \big|l_k^{(m)}\big(\tilde{x}^{(m)}, \tilde{y}^{(m)}\big)\big| \tag{8.10}$$

determines which is the sharpest frame $\hat{k}$, in each level m of the pyramid, and in each pixel $(\tilde{x}^{(m)}, \tilde{y}^{(m)})$. Define an image

$$l_{\text{best}}^{(m)}\big(\tilde{x}^{(m)}, \tilde{y}^{(m)}\big) = l_{\hat{k}}^{(m)}\big(\tilde{x}^{(m)}, \tilde{y}^{(m)}\big) \tag{8.11}$$

where $\hat{k}$ is given by Equation (8.10). The image $\mathbf{l}_{\text{best}}^{(m)}$ defined in Equation (8.11) is analogous to Equation (8.9), but it fuses the images in each scale of the Laplacian pyramid, as in [55,56]. Effectively, Equation (8.11) can be interpreted as using large areas of the full image domain $(\tilde{x}, \tilde{y})$ to fuse rough image components (low frequencies), while using small effective areas to fuse small features (high frequencies).

The images $\{\mathbf{g}_k^{(M)}\}_{k=1}^K$ represent the lowest spatial frequencies. These components are least affected by defocus blur, as defocus is essentially a low-pass filter. Therefore, there is not much use in trying to select the sharpest $\mathbf{g}_k^{(M)}$ among all K frames. Hence, this component is fused by a linear superposition of $\{\mathbf{g}_k^{(M)}\}_{k=1}^K$, yielding a new image, which we term $\mathbf{g}_{\text{best}}^{(M)}$. Finally, using all the frequency components (pyramid levels) $\mathbf{g}_{\text{best}}^{(M)}$ and $\{\mathbf{l}_{\text{best}}^{(m)}\}_{m=0}^{M-1}$, we reconstruct the full size, complete fused image, $g_{\text{best}}(\tilde{x}, \tilde{y})$. This is in analogy to the operation mentioned above, of reconstructing $g_k(\tilde{x}, \tilde{y})$ from its pyramid components.

Left periphery   Right periphery

Figure 8.12 *A wide DOF mosaic. It is composed of a sequence of frames, each having a narrow FOV and a spatially varying focus. Samples of the sequence are shown in Figure 8.10.*

Mosaicing artifacts Image mosaicing creates artifacts in the form of *seams* in lines that correspond to the boundaries of the raw frames [4,9,16,17,25,28,54]. One reason for seams is spatial variability of exposure, created when an object is seen through different parts of the camera's FOV. For example, it may be caused by vignetting. Other reasons include slight temporal variations of illumination or camera gain between frames. This problem has been easily solved in traditional image mosaicing using image feathering. There, fusion of frames $\{\mathbf{g}_k\}_{k=1}^{K}$ is obtained by a weighted linear superposition of the raw pixels. The superposition weight of a pixel (x, y) in frame k decreases the closer the pixel is to the boundary of this frame [28,54]. Seam removal by the feathering operation is particularly effective if done in low-frequency components [17], since exposure variations (which cause seams) change very smoothly across the camera FOV.

We easily adapted this principle to our problem. Recall that we create the low frequency component $\mathbf{g}_{\text{best}}^{(M)}$ of the fused image by a linear superposition of $\{\mathbf{g}_k^{(M)}\}_{k=1}^{K}$. Hence, we set the weights of the superposition of $\{\mathbf{g}_k^{(M)}\}_{k=1}^{K}$ according to the described feathering principle. To conclude, the multi-scale fusion approach handles both kinds of artifacts: those associated with non-ideal focusing and those stemming from non-ideal mosaicing.

As an example, we analysed a sequence, samples of which are shown in Figure 8.10. The resulting wide DOF mosaic appears in Figure 8.12. The leftmost part of the image contains defocused objects: as explained in Figure 8.6, the peripheral parts are seen in fewer frames, thus with less focus states. As said, the periphery is at most one frame wide.

8.3 Panorama with intensity high dynamic range

8.3.1 Image acquisition

In many scenarios, object radiance changes by orders of magnitude across the FOV. For this reason, there has recently been an upsurge of interest in obtaining HDR image data and in their representation [59–63]. On the other hand, raw images have a limited optical dynamic range [64], set by the limited dynamic range of the camera detector. Above a certain detector irradiance, the images become saturated and their content at the saturated region is lost. Attenuating the irradiance by a shorter exposure time, a smaller aperture, or a neutral (space invariant) density filter can ensure that saturation is avoided. However, at

Figure 8.13 *Two generalised mosaicing systems [31–33]. (Left) A system composed of a Sony black/white video camera and an extended arm which holds the filter. (Right) A system that includes a Canon Optura digital video camera and a cylindrical attachment that holds the filter. In both cases, the camera moves with the attached filter as a rigid system. (With kind permission from Springer Science and Business Media.)*

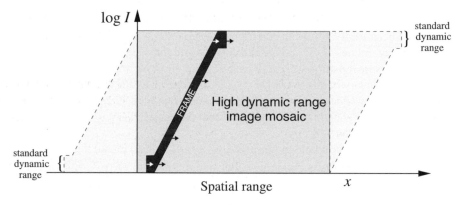

Figure 8.14 *Image mosaicing coupled with vignetting effects yields HDR image mosaics [33]. Besides the FOV, it also extends the intensity dynamic range of the system. Outside the main region of interest, the mosaic provides additional information about the scene periphery, whose quality gradually coincides with that of a single frame. (With kind permission from Springer Science and Business Media.)*

the same time other information is lost since light may be below the detection threshold in regions of low image irradiance.

Using generalised mosaicing, extension of both the dynamic range and the FOV are done in a unified framework [33]. We mount a fixed filter on the camera, as in Figures 8.2 and 8.13. The intensity transmittance varies across the filter's extent. This causes an *intentional vignetting*. Including vignetting effects originating from the lens, the overall effect is equivalent to spatially attenuating the image by a mask $A(x, y)$.

Now, as in Section 8.2.2, the scene is scanned by the motion of the camera. The moving system attenuates the light from any scene point differently in each frame. Effectively, the camera captures each point with different exposures during the sequence. Therefore, the system acquires both dark and bright areas with high quality while extending the FOV. It may be viewed as introducing a new dimension to the mosaicing process (Figure 8.14). This dimension leads to the introduction of the concept of the *spatio-intensity space*. In

Figure 8.14, the spatio-intensity support of a single frame occupies a diagonal region in the spatio-intensity space. This occurs if $\log[A(x, y)]$ varies linearly with x. The spatial frame motion then covers the intensity dynamic range, as a by product. High definition intensity may be obtained for all the pixels in a wide FOV image mosaic. In addition, information becomes available about the periphery of the central region of interest: the periphery has a smaller dynamic range, but at least the standard dynamic range of the detector (Figure 8.14).

To demonstrate the appearance of frames taken this way, we used [33] a linear variable density filter, 3 cm long, rigidly attached to an 8-bit CCD camera system, ≈ 30 cm in front of its 25 mm lens. The filter has a maximum attenuation of 1:100. The camera was rotated about its centre of projection so that each point was imaged 14 times across the camera FOV. Some images of this sequence of 36 frames are presented in Figure 8.15.

8.3.2 Data fusion

We now describe the method we used [33] to estimate the intensity at each mosaic point, given its multiple corresponding measurements. As in Section 8.2.3, this is done after the images have been registered. Let a measured intensity readout at a point be g_k with uncertainty Δg_k, and the estimated mask be $\hat{A}$ with uncertainty $\Delta \hat{A}$. Compensating the readout for the mask, the scene point's intensity is

$$I_k = \frac{g_k}{\hat{A}} \tag{8.12}$$

and its uncertainty is

$$\Delta I_k = \sqrt{\left(\frac{\partial I_k}{\partial g_k} \Delta g_k\right)^2 + \left(\frac{\partial I_k}{\partial \hat{A}} \Delta \hat{A}\right)^2} \tag{8.13}$$

For instance, we may set the readout uncertainty to be $\Delta g_k = 0.5$, since the intensity readout values are integers. Any image pixel considered to be saturated (g_k close to 255 for an 8-bit detector) is treated as having a high uncertainty. Thus, its corresponding Δg_k is set to be a very large number.

Assuming the measurements I_k to be Gaussian and independent, the log-likelihood for a value I behaves as $-E^2$, where

$$E^2 \equiv \sum_k \left(\frac{I - I_k}{\Delta I_k}\right)^2 \tag{8.14}$$

The maximum likelihood (ML) solution for the intensity I in this scene point is the one that minimises E^2:

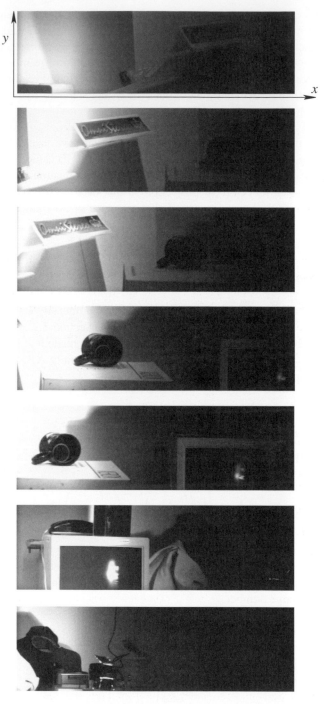

Figure 8.15   *Frames 4, 9, 11, 15, 17, 23, and 31 from a sequence taken with a linear variable density filter [33]. Scene features become brighter as they move leftwards in the frame. Bright scene points gradually reach saturation. Dim scene points, which are not visible in the right-hand side of the frames, become visible when they appear on the left. (With kind permission from Springer Science and Business Media.)*

$$\hat{I} = \widehat{\Delta I}^2 \sum_k \frac{I_k}{\Delta I_k^2} \tag{8.15}$$

where

$$\widehat{\Delta I} = \left(0.5 \cdot \frac{d^2 E^2}{dI^2}\right)^{-1/2} = \left(\sqrt{\sum_k \frac{1}{\Delta I_k^2}}\right)^{-1} \tag{8.16}$$

Although Equation (8.15) suffices to determine the value at each point, annoying seams may appear at the boundaries of the frames that compose the mosaic. At these boundaries there is a transition between points that have been estimated using somewhat different sources of data. Seams appear also at the boundaries of saturated areas, where there is an abrupt change in the uncertainty Δg, while the change in g is usually small. These seams are removed by feathering techniques: see [33].

Images from a sequence, of which samples are shown in Figure 8.15, were fused into a mosaic using this method. The histogram equalised version of $\log \hat{I}$ is shown in Figure 8.16. Contrast stretching of $\hat{I}$ in selected regions shows that the mosaic is not saturated anywhere, and details are seen wherever $I \geqslant 1$. The HDR of this mosaic is evident. The periphery parts of the mosaic are left of the $\mathcal{L}$ mark and right of the $\mathcal{R}$ mark, having a width of a single frame. These parts were not exposed at the full range of attenuation. HDR is observed there as well, but the range gradually decreases to that of the native detector.

8.4 Multispectral wide field of view imaging

Multispectral imaging is very useful in numerous imaging applications, including object and material recognition [65], colour analysis and constancy [66–68], remote sensing [65,69–71], and astronomy [72]. The applications for multispectral imaging are expanding [73], and include for example, medical imaging, agriculture, archaeology and art.

In this section we describe the use of a spatially varying interference filter in the generalised mosaicing framework [32]. In particular, a linear interference filter passes a narrow wavelength band, and its central wavelength λ varies linearly with x. Such a filter can be used in a system as depicted in Figures 8.2 and 8.13. In this case, the spectral information in each raw frame is multiplexed with the spatial features which appear in ordinary images. This is seen, for example, in frames shown in Figure 8.17, acquired using a monochrome camera. This scene was acquired under incandescent lighting. The spatial details of the scene are clearly recognisable (e.g., the computer monitor). The spatial features are clear because, as depicted in Figure 8.2, the system is an imaging device, thus each of its frames captures an area of the scene.

Once the raw frames are registered, we have for each scene point in the mosaic FOV a set of wavelength samples $\{\lambda_k(\tilde{x}, \tilde{y})\}$, and corresponding intensity measurements $\{g_k(\tilde{x}, \tilde{y})\}$, where k is the index of the individual frames that compose the mosaic. This raw data structure is converted [32] to a multispectral *image cube*, denoted $g(\tilde{x}, \tilde{y}, \lambda)$. The multispectral data is now available at each scene point, and can be used in multispectral

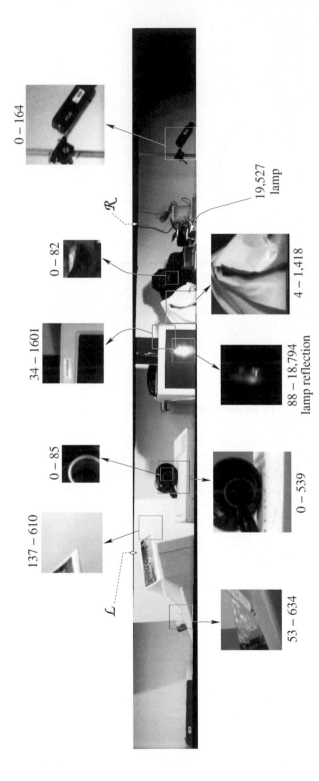

Figure 8.16 *An image created using a generalised mosaicing system [33]. It is based on a single rotation about the centre of projection of an 8-bit video camera. Contrast stretching in the selected squares reveals the details that reside within the computed mosaic. The numbers near the squares are the actual (unstretched) brightness ranges within the squares. Note the shape of the filament of the lamp in its reflection from the computer monitor. The periphery regions are left of the L mark and right of the R mark. (With kind permission from Springer Science and Business Media.)*

Figure 8.17 *Frames 29, 32, 38, 40, 60, and 64 of a sequence [32] taken through the linear variable interference filter. The left of the FOV senses the energy density at 700 nm, while the right senses it at 400 nm. The spatial features of the scene are clearly seen. (Y.Y. Schechner and S.K. Nayar, 'Generalized mosaicing: Wide field of view multispectral imaging', IEEE Transactions on Pattern Analysis and Machine Intelligence, Vol. 24, 2002, pp. 1334–1348. © 2005 IEEE.)*

imaging applications. For display, it is possible to convert the spectrum in each point to Red–Green–Blue values, as described in [32].

In an experiment corresponding to the images in Figure 8.17, about 21 spectral samples were acquired for each scene point. The grabbed images were compensated for cam-

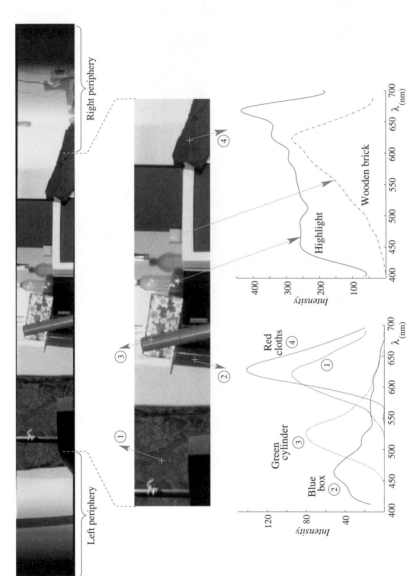

Figure 8.18 (*Top*) *A colour image mosaic [32] rendered using the spectral data acquired at each point in its FOV, based on a single pass (rotation about the centre of projection) of an ordinary black/white camera with a single fixed filter. The scene was illuminated by incandescent lamps. In the mosaic periphery the spectral range becomes narrower towards the outer boundaries, thus gradually deteriorating the colour rendering. (Middle) The mosaic's central region of interest contains the full spectral range the filter can scan. (Bottom) The spectrum is plotted for selected points.* (Y.Y. Schechner and S.K. Nayar, 'Generalized mosaicing: Wide field of view multispectral imaging', *IEEE Transactions on Pattern Analysis and Machine Intelligence, Vol. 24, 2002, pp. 1334–1348.* © *2005 IEEE.*)

era vignetting effects that were calibrated beforehand. We registered the images using a method discussed in [32]. The registration yielded a wide multispectral image mosaic, where the spectrum can be computed for each point. The multispectral mosaic was then converted to a colour mosaic, shown in Figure 8.18. Similarly to Sections 8.2 and 8.3, a full range spectrum is obtained in the central region of interest. This region seems yellowish in Figure 8.18 because the light coming from the incandescent lamps is rather yellow.[5] Other than that, the estimated colours were consistent with the appearance of the objects.

Information is obtained also about the periphery, though with decreasing spectral range. The top of Figure 8.18 indeed shows the periphery regions, one frame wide, on both sides of the mosaic central part. Since the spectral range in these regions changes gradually from the central part, there is no abrupt decrease of quality in the periphery. However, towards the outer boundaries of the left and right periphery, the image becomes red and blue respectively. This is due to the absence of data on the complimentary wavelengths in these regions [32]. Even there, substantial information can still be available for algorithms that make do with partial spectral data, or that do not rely on colour but on spatial features. For example, the objects in the right periphery clearly appear in the raw frame shown at the last photograph in Figure 8.17. It shows [32] loose dark cables hanging down through the frame, and their shadows on the wall behind. This can indicate the number and spatial distribution of the light sources in the scene, as in [74]. In addition, other objects (shaped bricks) can be recognised in this region. Therefore, the peripheral regions are not wasted data, but can be useful for computer vision.

8.5 Polarisation as well

Polarimetric imaging has been used in numerous imaging applications [75,76], including object and material recognition [77,78], shape recovery [79–81], and removal and analysis of specular reflection in photography and computer vision [82–85]. It has also been used for removal of scattering effects [86,87], e.g. in haze [88–90], underwater [78,88, 91–94] and tissue [95].

The polarisation state has several parameters. In linear polarisation these parameters are the intensity, the degree of polarisation, and the orientation of the plane of polarisation. To recover the polarisation parameters corresponding to each scene point, it is usually sufficient to measure the scene several times, each time with different polarisation settings [34]. Typically, man-made systems achieve this by filtering the light through a linear polariser, oriented differently in different images. Biological systems, however, use other mechanisms to capture polarisation images. Specifically, the retina of the mantis shrimp [96,97] has several distinct regions, each having different optical properties. In order to capture all the information in high quality, the shrimp moves its eyes to scan the FOV, thereby sequentially measuring each scene point using different optical settings [96]. This gives the shrimp vision high quality colour, polarisation and spatial information in a wide

---

[5]The human visual system adapts when it is embedded in such a coloured illumination (colour constancy).

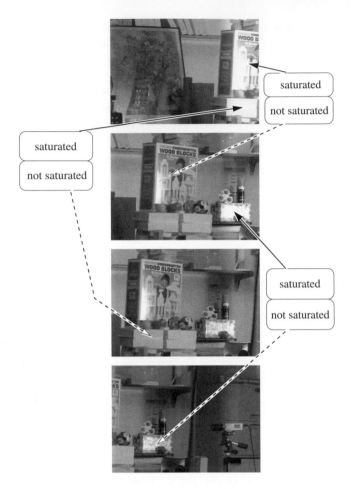

Figure 8.19 *A few frames from a sequence [34] taken through a mounted filter. The filter has spatially vary-ing attenuation and polarisation characteristics. The effective filtering variations are gradual and not easily visible, due to defocus blur of the filter. Nevertheless, the spatial variations make objects appear brighter on the right-hand side. Thus saturated pixels on the right appear unsaturated in other frames, in corresponding left-hand side pixels. (Y.Y. Schechner and S.K. Nayar, 'Generalized mosaicing: Polarization panorama', IEEE Transactions on Pattern Analysis and Machine Intelligence, Vol. 27, 2005, pp. 631–636. © 2005 IEEE.)*

FOV. This biological evidence motivates the application of generalised mosaicing to po-larisation sensing.

This principle can be demonstrated by attaching a spatially varying polarisation filter in front of the camera [34]. As the camera moves, a given scene point is measured multiple times, each through a filter part with different polarisation characteristics and/or orien-tation. Computational algorithms tailored to this kind of a system extract the required polarisation parameters, in a wide FOV. To demonstrate this, a camera was mounted, as in Figure 8.13, with a filter that varies the polarisation filtering across it. Furthermore, the system was built such that its transmittance varied across the camera FOV as well (see details in Ref. [34]). This enhances the dynamic range of the mosaic, as in Section 8.3.

For instance, let scene points appear saturated in a frame, when viewed through a filter part having high transmittance. These points may become unsaturated in other frames, when the points are viewed through darker portions of the filter. Furthermore, the polarisation sensitivity of the mounted spatially varying filter can significantly reduce specular highlights, thus aiding in the dynamic range extension. This is seen in sample frames taken by the system [34], shown in Figure 8.19. Note that objects are brighter (and even saturated) when they appear on the right-hand side of the frame.

Defocus blur affects the filtering properties of the mounted filter. Thus Ref. [34] describes a single framework that handles analysis of data acquired by non-ideal polarisation filters (partial polarisers), variable exposures and saturation. As the images are automatically registered, they can be analysed by algorithms described in Ref. [34], to yield a panoramic, polarimetric seamless image mosaic [34].

8.6 Conclusions

Generalised mosaicing is a framework for capturing information in multiple imaging dimensions. In Section 8.2 we described how the DOF can be extended while enlarging the FOV. Contrary to common optical designs, an SLC that is not flat or not normal to the optical axis can be beneficial, as it enables the extraction of depth information when the scene is scanned. This has implications for several aspects of computer vision, such as image-based rendering [98]. It may also be applied to machine vision systems (e.g., microscopic ones) used for industrialised inspection.

In addition to focus, we demonstrated this framework by deriving mosaics having HDR, multispectral or polarimetric outputs. Nevertheless, generalised mosaicing permits simultaneous enhancement of multiple dimensions. One example for this was described in Section 8.5, where both an extended intensity dynamic range and polarisation information were extracted using a single spatially varying filter. Other possibilities may include simultaneous extraction of polarisation and spectral information in a wide field, as the mantis shrimp [96,97]; simultaneous expansion of focus and intensity range, or other combinations. Furthermore, generalised mosaicing can be used for self-calibration of simultaneous radiometric effects that occur in the camera, including vignetting, automatic-gain control (AGC) and radiometric nonlinearity [99,100]. This is achieved by exploiting the redundancy encapsulated in multiple overlapping frames to retrieve these radiometric parameters.

Acknowledgements

The studies described in this work were supported in parts by a National Science Foundation ITR Award, IIS-00-85864, a David and Lucile Packard Fellowship, the Louis Morin Fellowship, the US–Israel Binational Science Foundation (BSF), and the Ollendorff Minerva Center in the Electrical Engineering Department at the Technion. Minerva is funded through the BMBF. Yoav Schechner is a Landau Fellow – supported by the Taub Foundation, and an Alon Fellow.

References

[1] D. Capel and A. Zisserman, 'Automated mosaicing with super-resolution zoom', in *Proc. IEEE Computer Society Conference on Computer Vision and Pattern Recognition*, 1998, pp. 885–891.

[2] S. Peleg, M. Ben-Ezra and Y. Pritch, 'Omnistereo: Panoramic stereo imaging', *IEEE Transactions on Pattern Analysis and Machine Intelligence*, Vol. 23, 2001, pp. 279–290.

[3] R.M. Batson, 'Digital cartography of the planets: New methods, its status, and its future', *Photogrammetric Engineering and Remote Sensing*, Vol. 53, 1987, pp. 1211–1218.

[4] M.L. Duplaquet, 'Building large image mosaics with invisible seam lines', in *Visual Information Processing VII*, in *Proc. of the SPIE*, Vol. 3387, 1998, pp. 369–377.

[5] R. Eustice, O. Pizarro, H. Singh and J. Howland, 'UWIT: Underwater Image Toolbox for optical image processing and mosaicking in MATLAB', in *Proc. IEEE International Symposium on Underwater Technology*, 2002, pp. 141–145.

[6] R. Garcia, J. Batlle, X. Cufi and J. Amat, 'Positioning an underwater vehicle through image mosaicking', in *Proc. IEEE International Conference on Robotics and Automation*, Part 3, 2001, pp. 2779–2784.

[7] R. Kwok, J.C. Curlander and S. Pang, 'An automated system for mosaicking spaceborne SAR imagery', *International Journal of Remote Sensing*, Vol. 11, 1990, pp. 209–223.

[8] S. Hsu, H.S. Sawhney and R. Kumar, 'Automated mosaics via topology inference', *IEEE Computer Graphics and Application*, Vol. 22, No. 2, 2002, pp. 44–54.

[9] M. Irani, P. Anandan, J. Bergen, R. Kumar and S. Hsu, 'Efficient representations of video sequences and their application', *Signal Processing: Image Communication*, Vol. 8, 1996, pp. 327–351.

[10] A. Smolić and T. Wiegand, 'High-resolution image mosaicing', in *Proc. International Conference on Image Processing*, Vol. 3, 2001, pp. 872–875.

[11] C.J. Lada, D.L. DePoy, K.M. Merrill and I. Gatley, 'Infrared images of M17', *The Astronomical Journal*, Vol. 374, 1991, pp. 533–539.

[12] J.M. Uson, S.P. Boughn and J.R. Kuhn, 'The central galaxy in Abel 2029: An old supergiant', *Science*, Vol. 250, 1990, pp. 539–540.

[13] E.M. Reynoso, G.M. Dubncr, W.M. Goss and E.M. Arnal, 'VLA observations of neutral hydrogen in the direction of Puppis A', *The Astronomical Journal*, Vol. 110, 1995, pp. 318–324.

[14] R. Bernstein, 'Digital image processing of earth observation sensor data', *IBM Journal of Research and Development*, Vol. 20, No. 1, 1976, pp. 40–57.

[15] M. Hansen, P. Anandan, K. Dana, G. van der Wal and P. Burt, 'Real-time scene stabilization and mosaic construction', in *Proc. IEEE Workshop on Applications of Computer Vision*, 1994, pp. 54–62.

[16] E. Fernandez, R. Garfinkel and R. Arbiol, 'Mosaicking of aerial photographic maps via seams defined by bottleneck shortest paths', *Operations Research*, Vol. 46, No. 3, 1998, pp. 293–304.

[17] L.A. Soderblom, K. Edwards, E.M. Eliason, E.M. Sanchez and M.P. Charette, 'Global color variations on the Martian surface', *Icarus*, Vol. 34, 1978, pp. 446–464.

[18] A.R. Vasavada, A.P. Ingersoll, D. Banfield, M. Bell, P.J. Gierasch and M.J.S. Belton, 'Galileo imaging of Jupiter's atmosphere: The great red spot, equatorial region, and white ovals', *Icarus*, Vol. 135, 1998, pp. 265–275.

[19] R.D. Ballard, 'Ancient Ashkelon', *National Geographic Magazine*, Vol. 199, No. 1, 2001, pp. 61–93.

[20] D. Drollette, 'Photonics defies the depths', *Photonics Spectra*, Vol. 34, No. 11, 2000, pp. 80–90.

[21] S. Negahdaripour, X. Xu, A. Khemene and Z. Awan, '3-D motion and depth estimation from sea-floor images for mosaic-based station-keeping and navigation of ROV's/AUV's and high-resolution sea-floor mapping', in *Proc. IEEE Workshop on Autonomous Underwater Vehicles*, 1998, pp. 191–200.

[22] S. Coorg, N. Master and S. Teller, 'Acquisition of a large pose-mosaic dataset', in *Proc. IEEE Computer Society Conference on Computer Vision and Pattern Recognition*, 1998, pp. 872–878.

[23] S. Gumustekin and R.W. Hall, 'Mosaic image generation on a flattened sphere', in *Proc. IEEE Workshop on Applications of Computer Vision*, 1996, pp. 50–55.

[24] S. Mann, 'Joint parameter estimation in both domain and range of functions in same orbit of the projective-Wyckoff group', in *Proc. International Conference on Image Processing*, 1996, pp. 193–196.

[25] S. Peleg and J. Herman, 'Panoramic mosaics by manifold projection', in *Proc. IEEE Computer Society Conference on Computer Vision and Pattern Recognition*, 1997, pp. 338–343.

[26] H.S. Sawhney and S. Ayer, 'Compact representations of videos through dominant and multiple motion estimation', *IEEE Transactions on Pattern Analysis and Machine Intelligence*, Vol. 18, 1996, pp. 814–830.

[27] H.S. Sawhney, R. Kumar, G. Gendel, J. Bergen, D. Dixon and V. Paragano, 'VideoBrush^{TM}: Experiences with consumer video mosaicing', in *Proc. IEEE Workshop on Applications of Computer Vision*, 1998, pp. 52–62.

[28] H.Y. Shum and R. Szeliski, 'Systems and experiment paper: Construction of panoramic image mosaics with global and local alignment', *International Journal of Computer Vision*, Vol. 36, 2000, pp. 101–130.

[29] R. Szeliski, 'Image mosaicing for telereality applications', in *Proc. IEEE Workshop on Applications of Computer Vision*, 1994, pp. 44–53.

[30] M. Aggarwal and N. Ahuja, 'High dynamic range panoramic imaging', in *Proc. IEEE International Conference on Computer Vision*, Vol. I, 2001, pp. 2–9.

[31] Y.Y. Schechner and S.K. Nayar, 'Generalized mosaicing', in *Proc. IEEE International Conference on Computer Vision*, Vol. I, 2001, pp. 17–24.

[32] Y.Y. Schechner and S.K. Nayar, 'Generalized mosaicing: Wide field of view multispectral imaging', *IEEE Transactions on Pattern Analysis and Machine Intelligence*, Vol. 24, 2002, pp. 1334–1348.

[33] Y.Y. Schechner and S.K. Nayar, 'Generalized mosaicing: High dynamic range in a wide field of view', *International Journal of Computer Vision*, Vol. 53, 2003, pp. 245–267.

[34] Y.Y. Schechner and S.K. Nayar, 'Generalized mosaicing: Polarization panorama', *IEEE Transactions on Pattern Analysis and Machine Intelligence*, Vol. 27, 2005, pp. 631–636.

[35] E. Hecht, *Optics*, fourth ed., Addison–Wesley, New York, 2002, pp. 264–266.

[36] A. Krishnan and N. Ahuja, 'Panoramic image acquisition', in *Proc. IEEE Computer Society Conference on Computer Vision and Pattern Recognition*, 1996, pp. 379–384.

[37] K. Itoh, A. Hayashi and Y. Ichioka, 'Digitized optical microscopy with extended depth of field', *Applied Optics*, Vol. 28, No. 16, 1989, pp. 3487–3493.

[38] R.A. Jarvis, 'A perspective on range-finding techniques for computer vision', *IEEE Transactions on Pattern Analysis Machine Intelligence*, Vol. 5, No. 2, 1983, pp. 122–139.

[39] H.N. Nair and C.V. Stewart, 'Robust focus ranging', in *Proc. IEEE Computer Society Conference on Computer Vision and Pattern Recognition*, 1992, pp. 309–314.

[40] S.K. Nayar, 'Shape from focus system', in *Proc. IEEE Computer Society Conference on Computer Vision and Pattern Recognition*, 1992, pp. 302–308.

[41] M. Noguchi and S.K. Nayar, 'Microscopic shape from focus using active illumination', in *Proc. International Conference on Image Processing*, Vol. 1, 1994, pp. 147–152.

[42] Y.Y. Schechner, N. Kiryati and R. Basri, 'Separation of transparent layers using focus', *International Journal of Computer Vision*, Vol. 89, 2000, pp. 25–39.

[43] M. Subbarao and J.K. Tyan, 'The optimal focus measure for passive autofocusing', in *Videometrics VI*, in *Proc. of the SPIE*, Vol. 2598, 1995, pp. 88–99.

[44] P. Torroba, N. Cap and H. Rabal, 'Defocus detection using a visibility criterion', *Journal of Modern Optics*, Vol. 41, No. 1, 1994, pp. 111–117.

[45] T.T.E. Yeo, S.H. Ong, Jayasooriah and R. Sinniah, 'Autofocusing for tissue microscopy', *Image and Vision Computing*, Vol. 11, No. 10, 1993, pp. 629–639.

[46] T. Darrell and K. Wohn, 'Pyramid based depth from focus', in *Proc. IEEE Computer Society Conference on Computer Vision and Pattern Recognition*, 1988, pp. 504–509.

[47] K. Engelhardt and G. Hausler, 'Acquisition of 3-D data by focus sensing', *Applied Optics*, Vol. 27, No. 22, 1988, pp. 4684–4689.

[48] S.A. Sugimoto and Y. Ichioka, 'Digital composition of images with increased depth of focus considering depth information', *Applied Optics*, Vol. 24, No. 14, 1985, pp. 2076–2080.

[49] Y. Xiong and S.A. Shafer, 'Depth from focusing and defocusing', in *Proc. IEEE Computer Society Conference on Computer Vision and Pattern Recognition*, 1993, pp. 68–73.

[50] P. Favaro and S. Soatto, 'A geometric approach to shape from defocus', *IEEE Transactions on Pattern Analysis and Machine Intelligence*, Vol. 27, 2006, pp. 406–417.

[51] A. Pentland, S. Scherock, T. Darrell and B. Girod, 'Simple range camera based on focal error', *Journal of the Optical Society of America A*, Vol. 11, 1994, pp. 2925–2934.

[52] A.N. Rajan and S. Chaudhuri, 'Simultaneous estimation of super-resolved scene and depth map from low resolution defocused observations', *IEEE Transactions on Pattern Analysis and Machine Intelligence*, Vol. 25, 2003, pp. 1102–1117.

[53] Y.Y. Schechner and N. Kiryati, 'Depth from defocus vs. stereo: How different really are they?', *International Journal of Computer Vision*, Vol. 89, 2000, pp. 141–162.

[54] P.J. Burt and E.H. Adelson, 'A multiresolution spline with application to image mosaics', *ACM Transactions on Graphics*, Vol. 2, 1983, pp. 217–236.

[55] P.J. Burt and R.J. Kolczynski, 'Enhanced image capture through fusion', in *Proc. IEEE International Conference on Computer Vision*, 1993, pp. 173–182.

[56] J.M. Odgen, E.H. Adelson, J.R. Bergen and P.J. Burt, 'Pyramid-based computer graphics', *RCA Engineer*, Vol. 30, No. 5, 1985, pp. 4–15.

[57] A. Toet, 'Hierarchical image fusion', *Machine Vision and Applications*, Vol. 3, 1990, pp. 1–11.

[58] P.J. Burt and E.H. Adelson, 'The Laplacian pyramid as a compact image code', *IEEE Transactions on Communications*, Vol. 31, 1983, pp. 532–540.

[59] F. Durand and J. Dorsey, 'Fast bilateral filtering for the display of high-dynamic-range images', *ACM Transactions on Graphics*, Vol. 21, No. 3, 2002, pp. 257–266.

[60] R. Fattal, D. Lischinski and M. Werman, 'Gradient domain high dynamic range compression', *ACM Transactions on Graphics*, Vol. 21, No. 3, 2002, pp. 249–256.

[61] G.W. Larson, H. Rushmeier and C. Piatko, 'A visibility matching tone reproduction operator for high dynamic range scenes', *IEEE Transactions on Visualization and Computer Graphics*, Vol. 3, 1997, pp. 291–306.

[62] A. Pardo and G. Sapiro, 'Visualization of high dynamic range images', in *Proc. International Conference on Image Processing*, Vol. 1, 2002, pp. 633–636.

[63] D.A. Socolinsky, 'Dynamic range constraints in image fusion and realization', in *Proc. IASTED International Conference on Signal and Image Processing*, 2000, pp. 349–354.

[64] W. Ogiers, 'Survey of CMOS imagers', IMEC Report P60280-MS-RP-002, Issue 1.1, Part 1, 1997.

[65] D. Slater and G. Healey, 'Material classification for 3D objects in aerial hyperspectral images', in *Proc. IEEE Computer Society Conference on Computer Vision and Pattern Recognition*, Vol. 2, 1999, pp. 268–273.

[66] A. Abrardo, L. Alparone, V. Cappellini and A. Prosperi, 'Color constancy from multispectral images', in *Proc. International Conference on Image Processing*, Vol. 3, 1999, pp. 570–574.

[67] M. Hauta-Kasari, K. Miyazawa, S. Toyooka and J. Parkkinen, 'A prototype of the spectral vision system,' in *Proc. Scandinavian Conference on Image Analysis*, Vol. 1, 1999, pp. 79–86.

[68] H.M.G. Stokman, T. Gevers and J.J. Koenderink, 'Color measurement by imaging spectrometry', *Computer Vision and Image Understanding*, Vol. 79, No. 2, 2000, pp. 236–249.

[69] A.M. Mika, 'Linear-wedge spectrometer', in *Imaging Spectroscopy of the Terrestrial Environment*, in *Proc. of the SPIE*, Vol. 1298, 1990, pp. 127–131.

[70] X. Sun and J.M. Anderson, 'A spatially variable light-frequency-selective component-based, airborne pushbroom imaging spectrometer for the water environment', *Photogrammetric Engineering and Remote Sensing*, Vol. 59, No. 3, 1993, pp. 399–406.

[71] J.B. Wellman, 'Multispectral mapper: Imaging spectroscopy as applied to the mapping of earth resources', in *Imaging Spectroscopy*, in *Proc. of the SPIE*, Vol. 268, 1981, pp. 64–73.

[72] G. Monnet, '3D spectroscopy with large telescopes: Past, present and prospects', in *Tridimensional Optical Spectroscopic Methods in Astronomy*, in *ASP Conference Series*, Vol. 71, 1995, pp. 2–17.

[73] N. Gat, 'Imaging spectroscopy using tunable filters: A review', in *Wavelet Applications VII*, in *Proc. of the SPIE*, Vol. 4056, 2000, pp. 50–64.

[74] I. Sato, Y. Sato and K. Ikeuchi, 'Illumination distribution from brightness in shadows: Adaptive estimation of illumination distribution with unknown reflectance properties in shadow regions', in *Proc. IEEE International Conference on Computer Vision*, 1999, pp. 875–883.

[75] G.P. Können, *Polarized Light in Nature*, Cambridge University Press, Cambridge, 1985.

[76] W.A. Shurcliff and S.S. Ballard, *Polarized Light*, Van Nostrand, Princeton, 1964.

[77] L.B. Wolff, 'Polarization camera for computer vision with a beam splitter', *Journal of the Optical Society of America A*, Vol. 11, 1994, pp. 2935–2945.

[78] L.B. Wolff, 'Polarization vision: A new sensory approach to image understanding', *Image and Vision Computing*, Vol. 15, 1997, pp. 81–93.

[79] D. Miyazaki, M. Masataka and K. Ikeuchi, 'Transparent surface modeling from a pair of polarization images', *IEEE Transactions on Pattern Analysis and Machine Intelligence*, Vol. 26, 2004, pp. 73–82.

[80] S. Rahmann and N. Canterakis, 'Reconstruction of specular surfaces using polarization imaging', in *Proc. IEEE Computer Society Conference on Computer Vision and Pattern Recognition*, Vol. 1, 2001, pp. 149–155.

[81] A.M. Wallace, B. Laing, E. Trucco and J. Clark, 'Improving depth image acquisition using polarized light', *International Journal of Computer Vision*, Vol. 32, 1999, pp. 87–109.

[82] H. Farid and E.H. Adelson, 'Separating reflections from images by use of independent component analysis', *Journal of the Optical Society of America A*, Vol. 16, 1999, pp. 2136–2145.

[83] Hermanto, A.K. Barros, T. Yamamura and N. Ohnishi, 'Separating virtual and real objects using independent component analysis', *IEICE Transactions on Information & Systems*, Vol. E84-D, No. 9, 2001, pp. 1241–1248.

[84] S. Lin and S.W. Lee, 'Detection of specularity using stereo in color and polarization space', *Computer Vision and Image Understanding*, Vol. 65, 1997, pp. 336–346.

[85] Y.Y. Schechner, J. Shamir and N. Kiryati, 'Polarization and statistical analysis of scenes containing a semireflector', *Journal of the Optical Society of America A*, Vol. 17, 2000, pp. 276–284.

[86] L.J. Denes, M. Gottlieb, B. Kaminsky and P. Metes, 'AOTF polarization difference imaging', *Proc. of the SPIE*, Vol. 3584, 1999, pp. 106–115.

[87] J.S. Tyo, M.P. Rowe, E.N. Pugh Jr. and N. Engheta, 'Target detection in optically scattering media by polarization-difference imaging', *Applied Optics*, Vol. 35, 1996, pp. 1855–1870.

[88] Y.Y. Schechner and Y. Averbuch, 'Regularized image recovery in scattering media', *IEEE Transactions on Pattern Analysis and Machine Intelligence*, Vol. 29, 2007, pp. 1655–1660.

[89] Y.Y. Schechner, S.G. Narasimhan and S.K. Nayar, 'Polarization-based vision through haze', *Applied Optics*, Vol. 42, No. 3, 2003, pp. 511–525.

[90] S. Shwartz, E. Namer and Y.Y. Schechner, 'Blind haze separation', in *Proc. IEEE Computer Society Conference on Computer Vision and Pattern Recognition*, Vol. 2, 2006, pp. 1984–1991.

[91] S. Harsdorf, R. Reuter and S. Tönebön, 'Contrast-enhanced optical imaging of submersible targets', *Proc. of the SPIE*, Vol. 3821, 1999, pp. 378–383.

[92] Y.Y. Schechner and N. Karpel, 'Recovery of underwater visibility and structure by polarization analysis', *IEEE Journal of Oceanic Engineering*, Vol. 30, 2005, pp. 570–587.

[93] J.S. Talyor Jr. and L.B. Wolff, 'Partial polarization signature results from the field testing of the shallow water real-time imaging polarimeter (SHRIMP)', in *Proc. MTS/IEEE Oceans*, Vol. 1, 2001, pp. 107–116.

[94] T. Treibitz and Y.Y. Schechner, 'Instant 3Descatter', in *Proc. IEEE Computer Society Conference on Computer Vision and Pattern Recognition*, Vol. 2, 2006, pp. 1861–1868.

[95] S.G. Demos and R.R. Alfano, 'Optical polarization imaging', *Applied Optics*, Vol. 36, 1997, pp. 150–155.

[96] T.W. Cronin and J. Marshall, 'Parallel processing and image analysis in the eyes of the mantis shrimp', *The Biological Bulletin*, Vol. 200, 2001, pp. 177–183.

[97] T.W. Cronin, N. Shashar, R. Caldwell, J. Marshall, A.G. Cheroske and T.H. Chiou, 'Polarization vision and its role in biological signaling', *Integrative and Comparative Biology*, Vol. 43, 2003, pp. 549–558.

[98] K. Aizawa, K. Kodama and A. Kubota, 'Producing object-based special effects by fusing multiple differently focused images', *IEEE Transactions on Circuits and Systems for Video Technology*, Vol. 10, 2000, pp. 323–330.

[99] A. Litvinov and Y.Y. Schechner, 'Radiometric framework for image mosaicking', *Journal of the Optical Society of America A*, Vol. 22, 2005, pp. 839–848.

[100] A. Litvinov and Y.Y. Schechner, 'Addressing radiometric nonidealities: A unified framework', in *Proc. IEEE Computer Society Conference on Computer Vision and Pattern Recognition*, Vol. 2, 2005, pp. 52 59.

9

Fusion of multispectral and panchromatic images as an optimisation problem

Andrea Garzelli, Luca Capobianco and Filippo Nencini

Department of Information Engineering, University of Siena, Siena, Italy

In this chapter, different approaches to image fusion for pan-sharpening of multispectral images are presented and critically compared. Particular emphasis is devoted to the advantages resulting from defining pan-sharpening as an optimisation problem. Implementation issues are also considered and extensive results in terms of quality of the fused products, both visual and objective, and computational time comparisons are presented for classical, state-of-the-art, and innovative solutions.

9.1 Introduction

Multisensor data fusion has nowadays become a discipline to which more and more general formal solutions to a number of application cases are demanded. In remote sensing applications, the increasing availability of spaceborne sensors, imaging in a variety of ground scales and spectral bands undoubtedly provides strong motivations. Due to the physical constraint of a tradeoff between spatial and spectral resolutions, spatial enhancement of multispectral (MS) data is desirable. This objective is equivalent, from another point of view, to the spectral enhancement (through MS) of data collected with adequate ground resolution but poor spectral selection (panchromatic image).

Spaceborne imaging sensors routinely allow a global coverage of the Earth surface. MS observations, however, may exhibit limited ground resolutions that may be inadequate to specific identification tasks. Data fusion techniques have been designed not only to allow integration of different information sources, but also to take advantage of complementary spatial and spectral resolution characteristics. In fact, the panchromatic (PAN) band is transmitted with the maximum resolution allowed by the imaging sensor, while the MS bands are usually acquired and transmitted with coarser resolutions, e.g. two or four times lower. At the receiving station, the PAN image may be merged with the MS bands to enhance the spatial resolution of the latter.

Since the pioneering high-pass filtering (HPF) technique [1], fusion methods based on injecting high-frequency components into resampled versions of the MS data have demonstrated superior performance [2]. By following this approach, three main issues arise:

(1) How to extract the spatial detail information from the panchromatic image?
(2) How to inject details into resampled multispectral data?
(3) How to assess the quality of the spatially enhanced MS images and possibly drive the fusion process?

On the one hand, multiresolution analysis (MRA) employing the discrete wavelet transform, namely, DWT [3–5], or wavelet frames [6], or the Laplacian pyramid [7] can be successfully adopted for spatial detail extraction from Pan image. According to the basic DWT fusion scheme [8], couples of sub-bands of corresponding frequency content are merged together. The fused image is synthesised by taking the inverse transform. Wavelet frames are more appropriate for pan-sharpening: the missing decimation of the 'à trous' wavelet transform (ATWT) allows an image to be decomposed into nearly disjointed bandpass channels in the spatial frequency domain without losing the spatial connectivity of its highpass details, e.g. edges and textures. On the other hand, MRA approaches present one main critical point: filtering operations may produce ringing artifacts, e.g. when high frequency details are extracted from the panchromatic image. This problem does not decrease significantly any global quality index, but it may locally reduce the visual quality of the fused product in a considerable way. To avoid this problem, different pan-sharpening algorithms which do not make use of MRA have been proposed in the last years [9–11].

Regardless of how the spatial details are extracted from the panchromatic image, data fusion methods require the definition of a model establishing how the missing high-pass information is injected into the resampled MS bands [12]. The goal is to make the fused bands the most similar to what the narrow-band MS sensor would image if it had the same resolution as the broad-band one (PAN). Some examples of injection models are additive combination of 'à trous' wavelet frames as in the Additive Wavelet to the Luminance component (AWL) technique [6], injection of wavelet details after applying Intensity–Hue–Saturation transformation (IHS) or Principal Component Analysis (PCA) [13], Spectral Distortion Minimization (SDM) with respect to the resampled MS data [14], or spatially adaptive injection as in the Context-Based Decision (CBD) algorithm [16] and in the Ranchin–Wald–Mangolini (RWM) method [17]. More efficient schemes can be obtained by incorporating the Modulation Transfer Functions (MTFs) of the multispectral scanner to design the MRA reduction filters. As a consequence, the interband structure model (IBSM), which is calculated at a coarser scale, where both MS and PAN data are available, can be extended to a finer scale, without the drawback of the poor enhancement occurring when MTFs are assumed to be ideal filters [18]. Theoretical considerations on injection models and experimental comparisons among MRA-based pan-sharpening methods can be found in [19].

Concerning the problem of assessing the quality of spatially enhanced MS images, unique quality indexes have been proposed for multiband images. ERGAS, after its name in French, which means relative dimensionless global error in synthesis, is a very efficient index for radiometric distortion evaluation [2], while $Q4$, designed for four-band MS data, such as QuickBird and Ikonos MS images, simultaneously accounts for local mean bias, changes in contrast, and loss of correlation of individual bands, together with spectral distortion [15]. An efficient quality index may be selected to drive the fusion process [11] in order to contextually

- find the optimal way to extract the spatial detail information from PAN;
- determine the optimal model parameters defining the injection model;
- maximise the score index (or a combination of different quality indexes) specifically designed to objectively assess the quality of the fused product.

From all above considerations it is clear that approaching pan-sharpening as an optimisation problem is very attractive, since it allows to avoid all those problems deriving from improper modelling of the injection process or inaccurate estimation of model parameters from image data.

This chapter is structured as follows. Section 9.2 introduces two of the most appropriate methodologies for image fusion: the 'à trous' wavelet transform and the generalised IHS (GIHS) transform. In Section 9.3, a linear injection model is defined and a framework for optimal computation of the model parameters is presented, suitable for both ATWT and GIHS-based fusion schemes which are recalled and referenced in Section 9.2.2. Section 9.4 illustrates different methodologies to perform the optimisation of the parameters defining the fusion process. Section 9.5 introduces the quality indexes which may be adopted as cost functions (or fitness functions) driving the optimisation process. In Section 9.6, a novel algorithm is presented, which substantially reduces the computational complexity of the optimisation-based pan-sharpening methods. Experimental results and comparisons for classical, state-of-the-art, commercial solutions, and the proposed fast algorithm for pan-sharpening of multispectral images are reported in Section 9.7.

9.2 Image fusion methodologies

9.2.1 'À trous' wavelet transform

The octave multiresolution analysis introduced by Mallat [20] for digital images does not preserve the translation invariance property. In other words, a translation of the original signal does not necessarily imply a translation of the corresponding wavelet coefficient. This property is essential in image processing. This *non-stationarity* in the representation is a direct consequence of the down-sampling operation following each filtering stage.

In order to preserve the translation invariance property, the down-sampling operation is suppressed, but filters are up-sampled by 2^j, i.e. dilated by inserting $2^j - 1$ zeroes between any couple of consecutive coefficients. An interesting property of the undecimated domain [16] is that at the jth decomposition level, the sequences of approximation, $c_j(k, m)$, and detail, $d_j(k, m)$, coefficients are straightforwardly obtained by filtering the original signal through a bank of equivalent filters, given by the convolution of recursively up-sampled versions of the lowpass filter h and the highpass filter g of the analysis bank:

$$h_j^* = \bigotimes_{m=0}^{j-1}(h \uparrow 2^m) \tag{9.1}$$

$$g_j^* = \left[\bigotimes_{m=0}^{j-2}(h \uparrow 2^m)\right] \otimes (g \uparrow 2^{j-1}) = h_{j-1}^* \otimes (g \uparrow 2^{j-1}) \tag{9.2}$$

In Equations (9.1) and (9.2), $\bigotimes_{n=1}^{N} f_n$ denotes the discrete convolution of the sequences $f_1, f_2, \ldots, f_N$.

The 'à trous' wavelet transform (ATWT) [21] is an undecimated non-orthogonal multiresolution decomposition defined by a filter bank $\{h_i\}$ and $\{g_i = \delta_i - h_i\}$, with the Kronecker operator δ_i denoting an allpass filter. In the absence of decimation, the lowpass filter is up-sampled by 2^j, before processing the jth level; hence the name 'à trous' which means 'with holes.' In two dimensions, the filter bank becomes $\{h_i h_j\}$ and $\{\delta_i \delta_j - h_i h_j\}$, which means that the 2-D detail signal is given by the pixel difference between two successive approximations, which have all the same scale 2^0, i.e. 1. The jth level of ATWT, $j = 0, \ldots, J - 1$, is obtained by filtering the original image with a separable 2-D version of the jth equivalent filter 1.

For a J-level decomposition, the ATWT accommodates a number of coefficients $J + 1$ times greater than the number of pixels. Due to the absence of decimation, as well as to the zero phase and -6 dB amplitude cutoff of the filter, the synthesis is simply obtained by summing all detail levels to the approximation:

$$\tilde{x}(k, m) = \sum_{j=0}^{J-1} d_j(k, m) + c_J(k, m) \tag{9.3}$$

in which $c_J(k, m)$ and $d_j(k, m)$, $j = 0, \ldots, J - 1$, are obtained through 2-D separable linear convolution with h_j^* and g_j^*, $j = 0, \ldots, J - 1$, shown in (9.1) and (9.2), respectively. Equivalently, they can be calculated by means of a tree-split algorithm, i.e. by taking pixel differences between convolutions of the original signal with progressively up-sampled versions of the lowpass filter.

For pan-sharpening purposes, the ATWT allows to obtain fused MS images by simply adding the high frequency details of the PAN image, $d_j(k, m)$ (conveniently scaled by a gain factor), with the MS approximation, $c_J(k, m)$ [6,16,19].

9.2.2 Generalised Intensity–Hue–Saturation transform

The Intensity–Hue–Saturation transform (IHS) [22] is one of the most widespread image fusion methods in the remote sensing community which has been used as a standard procedure in many commercial software solutions. Most literature recognises IHS as a third-order method because it employs a 3×3 matrix as its transform kernel in the RGB-IHS conversion model. The linear RGB-IHS conversion system is

$$\begin{bmatrix} I \\ v_1 \\ v_2 \end{bmatrix} = \begin{bmatrix} 1/3 & 1/3 & 1/3 \\ -\sqrt{2}/6 & -\sqrt{2}/6 & \sqrt{2}/6 \\ 1/\sqrt{2} & -1/\sqrt{2} & 0 \end{bmatrix} \cdot \begin{bmatrix} R \\ G \\ B \end{bmatrix} \tag{9.4}$$

and

$$\begin{bmatrix} R \\ G \\ B \end{bmatrix} = \begin{bmatrix} 1 & -1/\sqrt{2} & 1/\sqrt{2} \\ 1 & -1/\sqrt{2} & -1/\sqrt{2} \\ 1 & \sqrt{2} & 0 \end{bmatrix} \cdot \begin{bmatrix} I \\ v_1 \\ v_2 \end{bmatrix} \tag{9.5}$$

Variables v_1 and v_2 in (9.5) can be considered as x and y axes in the Cartesian coordinate system, while intensity I corresponds to the z axis. In this way, the hue (H) and the saturation (S) can be represented by

$$H = \tan^{-1}\left(\frac{v_2}{v_1}\right), \qquad S = \sqrt{v_1^2 + v_2^2} \tag{9.6}$$

More recently, the Generalised IHS algorithm, capable to extend the traditional third-order transformation to an arbitrary order N, has been proposed by Te-Ming et al. [23]. By using GIHS, the low resolution intensity component in the IHS space is replaced by a grey-level image with higher spatial resolution and transformed back into the original RGB space. Directly implementing the GIHS method requires many operations, making it computationally inefficient. To develop a low-complexity method for GIHS-based pan-sharpening, Equations (9.4) and (9.5) can be written as

$$\widehat{B}_l = \widetilde{B}_l + D \tag{9.7}$$

with $\widetilde{B}_l$ ($l = 1, 2, \ldots, N$) denoting the MS images up-sampled to the PAN scale, $\widehat{B}_l$ the fusion products, and

$$D = P - \frac{1}{N}\sum_{k=1}^{N} \widetilde{B}_k \tag{9.8}$$

the spatial detail extracted from the panchromatic image P.

Image fusion techniques with an arbitrary order can easily handle multispectral images with more than three bands, or even hyperspectral data. However, the fusion products may exhibit significant drifts from the true means. Spectral distortion is often severe for true-colour fusion products, while it is moderately noticeable in a false colour combination of bands. GIHS-based fusion schemes may preserve spectral information when a convenient injection model is adopted.

9.3 Injection model and optimum parameters computation

A spectral-preserving injection model may be obtained by adopting a simple linear model in which the coefficients that regulate the injection of the PAN image are derived globally from coarser scales but not *a priori* defined on image statistics, e.g. variance, mean, correlation coefficient, etc. The fused *l*th MS band can be computed, similarly to Equation (9.7), by

$$\widehat{B}_l = \widetilde{B}_l + g_l \cdot D \tag{9.9}$$

where the difference image D is equalised by a gain parameter, g_l, to reduce spectral distortions. When the fusion method is based on ATWT, the D image is simply the detail of the panchromatic image P, while for GIHS-based methods the D image is given by

$$D = P - \sum_{k=1}^{N} \alpha_k \cdot \widetilde{B}_k \tag{9.10}$$

similar to Equation (9.8) with coefficients α_k to be determined, as well as the gain parameters g_l.

An optimisation algorithm (OA) (different OAs are described in Section 9.4) can be applied to determine the g_l and α_k coefficients of Equations (9.9) and (9.10) that maximise an image quality score index (two quality indexes are described in Section 9.5). The goal of the OA is to find the best combination of the N real coefficients g_l, $l = 1, \ldots, N$, in the ATWT fusion method and the α_k and g_l, $l, k = 1, \ldots, N$, coefficients in the GIHS fusion method, according to an objective criterion that evaluates the quality of the spatially enhanced MS images.

The pan-sharpening scheme in Figure 9.1 represents a general framework for optimal spatial detail injection into MS images, suitable for both ATWT and GIHS-based fusion schemes.

The parameters driving the fusion process are optimised at coarser resolution, i.e. at a resolution degraded by a factor equal to the scale ratio between MS and PAN spatial resolutions (e.g., 4 for Ikonos and QuickBird data sets). The cost (or fitness) function of the OA is an image quality index. The OA simulates the injection process at degraded scale and computes the optimal parameters by using the original MS images. The same parameters are successively used to perform the injection at full resolution.

9.4 Functional optimisation algorithms

This section describes two different approaches to optimise the fusion parameters which drive the pan-sharpening process modelled as shown in Figure 9.1. First, unconstrained optimisation is recalled in Section 9.4.1, then the genetic algorithm approach is resumed in Section 9.4.2 and a possible set of fusion parameters is defined for efficient pan-sharpening. The quality index acting as the objective – or fitness – function of the optimisation algorithm can be selected between the two score indexes described in Section 9.5.

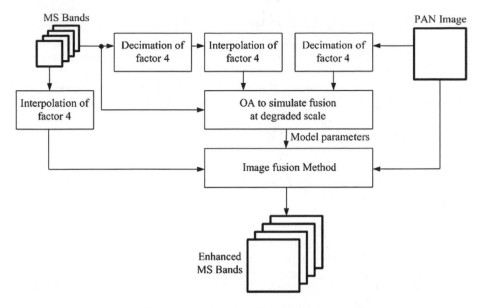

Figure 9.1 *Flowchart of pan-sharpening based on OA with 1:4 scale ratio.*

9.4.1 Unconstrained optimisation

Unconstrained optimisation [24] deals with the search for a local minimiser of a positive real-valued objective (or 'cost') function, $f(x)$, where x is a vector of real variables, $x \in \mathbb{R}^n$. In other words, we seek a vector, $\bar{x}$, such that

$$f(\bar{x}) \leqslant f(x) \quad \forall x \text{ close to } \bar{x}$$

Due to a large number of variables or to the complex expression of the objective function, it is not always possible to find a solution for the optimisation problem in a closed form. In this case, the problem may be approached by using an iterative algorithm.

The main idea of such an algorithm is actually simple: the values of the function $f(x)$ and its gradient $\nabla f(x)$ are evaluated for an initial point $x_0 \in \mathbb{R}$: if the null vector is obtained, the algorithm is stopped. Otherwise, another value of x is chosen, $x_1 \in \mathbb{R}$, by using a reasonable criterion. Since the minimising value of x for the function $f(x)$ is being searched, the simplest choice for this criterion is to move by one *step* in the descent direction, searching for a new value of x. Once more, if the null vector is found, the algorithm is stopped, otherwise a new value for x is evaluated. For each iteration k, the updated value may be written as

$$x_{k+1} := x_k + \alpha_k d_k \tag{9.11}$$

obtaining a sequence $x_0, x_1, x_2, \ldots, x_k, \ldots$, where $\alpha_k \in \mathbb{R}$ are the coefficients to be estimated, and the directions d_k are those search directions such that

$$f(x_{k+1}) < f(x_k)$$

Of course, the differences between one implementation of the method and one other, are determined by the descent direction and the length of the step $\alpha_k d_k$; moreover, the convergence of the methods and the speed of the convergence depend on these parameters.

Newton's method is one of the most general way to approach the unconstrained optimisation. It gives rise to a wide and important class of algorithms that require computation of the gradient vector

$$\nabla f(x) = \left(\partial_1 f(x), \partial_2 f(x), \ldots, \partial_n f(x) \right)^{\mathrm{T}}$$

and the Hessian matrix

$$\nabla^2 f(x) = \left(\partial_i \partial_j f(x) \right)$$

The exact estimation or approximation of the Hessian could have high computational costs; we describe Newton's algorithm in which the Hessian is explicitly evaluated, and then move to a discussion of an algorithm that does not require Hessian calculation. Both methods at each iteration use a *model* of the objective function evaluated using Taylor polynomial *approximation* and employ information about the slope of the function to search for a direction where the minimum is supposed to be located.

9.4.1.1 Newton's method

Newton's method uses the Taylor approximation of the objective function around the current iterate x_k. Given the search direction d, the model function is defined by

$$f(x_k + d) \approx f(x_k) + \nabla f(x_k)^{\mathrm{T}} d + \frac{1}{2} d^{\mathrm{T}} \nabla^2 f(x_k) d, \quad \text{as } \|d\| \to 0$$

where the symbol $\| \cdot \|$ indicates the Euclidean distance. Then, the objective function is

$$\psi_k(d) = f(x_k) + \nabla f(x_k)^{\mathrm{T}} d + \frac{1}{2} d^{\mathrm{T}} \nabla^2 f(x_k) d$$

In the basic Newton's method, the next iterate is obtained by assessing the minimiser of ψ_k: setting the $\nabla \psi$ equal to zero, we obtain the expression

$$\nabla f(x_k) + \nabla^2 f(x_k) d = 0$$

When the Hessian matrix, $\nabla^2 f(x_k)$, is positive definite, the quadratic model has a unique minimiser that can be obtained by solving the symmetric $n \times n$ linear system

$$\nabla^2 f(x_k) d_k = -\nabla f(x_k)$$

The solution d^* can be written as

$$d^* = -\left(\nabla^2 f(x_k) \right)^{-1} \nabla f(x_k) \tag{9.12}$$

Once the solution is found, expression (9.11) for the next iterate can be rewritten as

$$x_{k+1} = x_k + d^* = x_k - \left(\nabla^2 f(x_k) \right)^{-1} \nabla f(x_k)$$

9.4.1.2 The Gradient Method

The Gradient Method is the simplest among descent methods. It uses a *linear approximation* of $f(x_k + d)$, as a function of the vector d, and the search is performed in a direction $-\nabla f(x)$, where $\nabla f(x)$ is the gradient of the objective function.

Let us consider the first-order Taylor polynomial

$$f(x_k + d) \approx f(x_k) + \nabla f(x_k)^\mathrm{T} d, \quad \text{as } \|d\| \to 0 \tag{9.13}$$

The idea of the method is to approximate $f(x_k + d)$ with the function

$$\psi_k(d) = f(x_k) + \nabla f(x_k)^\mathrm{T} d$$

and to choose as the descent direction the vector d_k that minimises $\psi_k(d)$ for which

$$\|d\| = 1 \quad \text{(sphere of unitary radius)} \tag{9.14}$$

According to the Hölder inequality and condition (9.14), the second term in Equation (9.13) is rewritten as

$$\left| \nabla f(x_k)^\mathrm{T} d \right| \leqslant \left\| \nabla f(x_k)^\mathrm{T} \right\| \|d\| = \left\| \nabla f(x_k)^\mathrm{T} \right\|$$

The minimum is obtained by choosing the value verifying the equality, i.e.

$$d = -\nabla f(x_k) / \left\| \nabla f(x_k)^\mathrm{T} \right\|$$

Since the proposed algorithm relies on minimising the directional derivative of f, $-\nabla f(x_k)$, this method is called the *steepest descent* method. The algorithm moves in the direction where the minimum of the objective function is found. Expression (9.11) is rewritten as

$$x_{k+1} := x_k - \alpha_k \nabla f(x_k) \tag{9.15}$$

9.4.1.3 Line Search examples by using Matlab

In general, the problem of searching for the coefficient α_k is called *Line Search*, since it takes place along a line, having the direction of d_k.

This section describes one algorithm used to evaluate α_k and x_{k+1} at each iteration: the code of a simple Matlab implementation of the Line Search in the Steepest Descent, shown in Appendix A, uses the same algorithm.

The issue is how we may use the information given by the descent direction (i.e. such that $\nabla f(x_k)^\mathrm{T} d < 0$) to evaluate the next vector x_{k+1}. The main idea of the *Line Search* algorithm is the following: Since the function $\phi(\alpha) = f(x_k + \alpha d_k)$ expresses the behaviour near the value x_k in the direction d_k, it seems to be logical to search the value of α^* that minimises the function $\phi(\alpha)$ (with $\alpha > 0$). This condition may be expressed as

$$\phi'(\alpha^*) = \nabla f(x_k + \alpha^* d_k)^{\mathrm{T}} d_k \tag{9.16}$$

and has an interesting geometrical interpretation. For each iteration, the gradient is orthogonal to the direction of the previous step. To find the minimum of the function ϕ, the algorithm moves in the direction of d_k and stops where one x_k minimising the ϕ is found, among the set of x_k for which the gradient and d_k are orthogonal.

Unfortunately, this approach has heavy computational costs: it is usual to replace it with an iterative algorithm, called *Backtracking technique*, minimising the function ϕ by using its quadratic or cubic approximation.

Backtracking and Armijo's rule

In the *Backtracking* iterative approach, a set of values $\{\alpha_1, \alpha_2, \ldots, \alpha_i, \ldots\}$ is generated at each iteration, until the condition

$$f(x_k + \alpha d_k) \leqslant f(x_k) + \gamma \alpha \nabla f(x_k)^{\mathrm{T}} d_k, \quad 0 < \gamma < 1 \tag{9.17}$$

is satisfied. At each step, if α_i is not the value which is being searched, it is *multiplied* by a stretch factor $0 < \sigma < 1/2$ (Armijo's rule).

A better and a more efficient method may be built by considering a second- or third-order polynomial approximation of ϕ on the interval of interest, where $0 < \alpha < \alpha_i$, with α_i being the current value. In this interval, the value $\hat{\alpha}$ minimising the function is searched.

In the following a simple approximation method is shown. The first step is to verify whether the initial α_0 satisfies the condition expressed by (9.17) and can be rewritten as

$$\phi(\alpha_0) \leqslant \phi(0) + \gamma \alpha_0 \phi'(0) \tag{9.18}$$

If not, we evaluate a quadratic approximation ϕ_q of ϕ:

$$\phi_q(\alpha) = a\alpha^2 + b\alpha + c$$

by choosing the parameters so that

(i) $\phi_q(0) = \phi(0) = f(x_k)$,
(ii) $\phi_q'(0) = \phi'(0) = -\nabla f(x_k)^{\mathrm{T}} \nabla f(x_k)$,
(iii) $\phi_q(\alpha_0) = \phi(\alpha_0)$.

As a consequence of this approximation, the expression of α_1 can be written as

$$\alpha_1 = -\frac{\phi'(0)\alpha_0^2}{2[\phi(\alpha_0) - \phi(0) - \alpha_0 \phi'(0)]} \tag{9.19}$$

If this value does not satisfy condition (9.17), a cubic approximation may be chosen, since a new value of the function, $\phi(\alpha_1)$, is available:

$$\phi_c(\alpha) = a\alpha^3 + b\alpha^2 + \alpha \phi'(0) + \phi(0)$$

where a and b values may be evaluated using the constraints $(\alpha_0, \phi(\alpha_0))$ and $(\alpha_1, \phi(\alpha_1))$, thus obtaining

$$\begin{bmatrix} a \\ b \end{bmatrix} = \frac{1}{\alpha_0^2\alpha_1^2(\alpha_1 - \alpha_0)} \begin{bmatrix} \alpha_0^2 & -\alpha_1^2 \\ -\alpha_0^3 & \alpha_1^3 \end{bmatrix} \begin{bmatrix} \phi(\alpha_1) - \phi(0) - \phi'(0)\alpha_1 \\ \phi(\alpha_0) - \phi(0) - \phi'(0)\alpha_0 \end{bmatrix} \qquad (9.20)$$

Taking the derivative, we find the minimiser for ϕ_c in the interval $(0, \alpha_1)$:

$$\alpha_2 = -\frac{-b + \sqrt{b^2 - 3a\phi'(0)}}{3a} \qquad (9.21)$$

If the condition is still unsatisfied, the algorithm starts the search for α_3 by using a cubic approximation once more. New values for a and b are evaluated by (9.20), where α_0 and α_1 are replaced with α_1 and α_2. In general, until the condition is satisfied, at the ith iteration, α_i is searched using (9.20) with the values α_{i-2} and α_{i-1}.

9.4.2 Genetic algorithms

Genetic algorithms (GA) [25,26] are inspired by the evolution of populations. In a particular environment, individuals which fit the environment better will be able to survive and hand down chromosomes to their descendants, while less fit individuals will become extinct. The aim of genetic algorithms is to use simple representations to encode complex structures and simple operations to improve these structures. Therefore, genetic algorithms are characterised by their representations and operators. A fitness function is defined which measures the fitness of each individual. The populations are evolved to find good individuals as measured by the fitness function. A GA flow diagram is shown in Figure 9.2, and each of the major components is discussed in the following sections. A GA requires the definition of these fundamental steps: chromosome representation, selection of a function called fitness function, creation of the initial population, reproduction function, mutation and crossover operators, termination criteria, and the evaluation of fitness function. The following subsections describe these issues.

9.4.2.1 Chromosome representation

A chromosome representation is necessary to describe each individual in the GA population. The representation scheme determines how the problem is structured in the GA and also determines the genetic operators that are used. Each chromosome is made up of a sequence of genes from a predefined alphabet. One useful representation of chromosome for function optimisation involves genes from an alphabet of floating point numbers with values limited by an upper and a lower bound. It has been shown by Michalewicz [26] that a real-valued GA is more efficient in terms of *CPU time* and more accurate in terms of precisions for replications than binary GA representations.

9.4.2.2 Reproduction

The selection of parents to produce successive generations plays an important role in a genetic algorithm. The goal is to allow the best individuals to be selected more often to

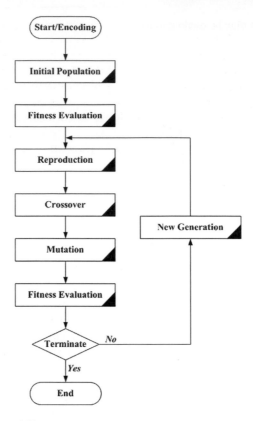

Figure 9.2   *Flow diagram of GA.*

reproduce. However, all individuals in the population have a chance of being selected to reproduce the next generation. Each individual is assigned a probability of being selected, with better individuals having larger probabilities. There are several schemes for determining and assigning the selection probability.

Ranking methods, and in particular the normalised geometric ranking (NGR) scheme, provide best performances. Individuals in the population are ranked from best to worst according to their fitness value. Then, each individual is assigned a probability of selection based upon some distribution, e.g. triangular or geometric. It has been shown in [27] that GAs incorporating ranking methods based upon the geometric distribution outperform those based upon the triangular distribution. It has been also shown that a pure geometric distribution is not appropriate since its range is defined on the interval one to infinity. Thus, for a finite range population size, the normalised geometric distribution shown below is considered

$$\sum_{\text{Population}} \text{Prob[selecting individual of rank } r] = \sum_{i=1}^{P} \frac{q\,(1-q)^{r-1}}{1-(1-q)^{P}} \tag{9.22}$$

where q is the probability of selecting the best individual, P is the overall number of individuals in the population and r is the rank, where 1 is the best.

9.4.2.3 Genetic operators

Genetic operators provide the basic search mechanism of the GA. The operators are used to create new solutions based on existing solutions in the population. There are two basic types of operators: *crossover* and *mutation*. Operators for real-valued representations, i.e. an alphabet of floats, were developed in [26]. Crossover takes two individuals and produces two new individuals while mutation alters one individual to produce a single new solution. The application of these two basic types of operators and their derivatives depends on the chromosome representation used. For real $\overline{X}$ and $\overline{Y}$ m-dimensional vectors representing chromosomes, the following operators are defined: uniform mutation, non-uniform mutation, multi-non-uniform mutation, boundary mutation, simple crossover, arithmetic crossover, and heuristic crossover. Let a_i and b_i be the lower and upper bound, respectively, for each variable i.

Mutation   Uniform mutation randomly selects one variable, j, and sets it equal to a uniform random number bounded by a_i and b_i terms:

$$x_i' = \begin{cases} U(a_i, b_i), & \text{if } i = j, \\ x_i, & \text{otherwise} \end{cases} \tag{9.23}$$

Boundary mutation randomly selects one variable, j, and sets it equal to either its lower or upper bound, where $r = U(0, 1)$:

$$x_i' = \begin{cases} a_i, & \text{if } i = j, \ r < 0.5, \\ b_i, & \text{if } i = j, \ r \geqslant 0.5, \\ x_i, & \text{otherwise} \end{cases} \tag{9.24}$$

Non-uniform mutation randomly selects one variable, j, and sets it equal to a non-uniform random number:

$$x_i' = \begin{cases} x_i + (b_i - x_i)\left(r_2\left(1 - \frac{G}{G_{\max}}\right)\right)^b, & \text{if } r_1 < 0.5, \\ x_i - (x_i + a_i)\left(r_2\left(1 - \frac{G}{G_{\max}}\right)\right)^b, & \text{if } r_1 \geqslant 0.5, \\ x_i, & \text{otherwise} \end{cases} \tag{9.25}$$

where r_1 and r_2 are uniform random numbers between 0 and 1, G and $G_{\max}$ are the current and the maximum number of generations, respectively, b is a shape parameter. The multi-non-uniform mutation operator applies the non-uniform operator to all of the variables in the parent $\overline{X}$.

Crossover   Real-valued simple crossover generates a random number r from a uniform distribution from 1 to m and creates two new individuals ($\overline{X}'$ and $\overline{Y}'$) according to the equation

$$x_i'(y_i') = \begin{cases} x_i(y_i), & \text{if } i < r, \\ y_i(x_i), & \text{otherwise} \end{cases} \tag{9.26}$$

Arithmetic crossover produces two complimentary linear combinations of the parents, where $r = U(0, 1)$:

$$\overline{X}' = r\overline{X} + (1 - r)\overline{Y} \qquad (9.27)$$

$$\overline{Y}' = (1 - r)\overline{X} + r\overline{Y} \qquad (9.28)$$

Heuristic crossover produces a linear extrapolation of the two individuals. This is the only operator that utilises fitness information. A new individual is created when $\overline{X}$ is better than $\overline{Y}$ in terms of fitness. If the new individual is infeasible, i.e. there is at least one new gene smaller than a_i or bigger than b_i, then the algorithm generates a random number r and creates a new solution. After t failures, the process is not repeated and the children are set equal to the parents.

9.4.2.4 Initialisation, termination and fitness function

To start the search of the optimal solution by a GA, it is necessary to provide an initial population as indicated in Figure 9.2. The most common method is to randomly generate solutions for the entire population. The GA moves from generation to generation selecting and reproducing parents until a termination criterion is met. A maximum number of generations is commonly used to stop the GA search. Another termination strategy involves population convergence criteria. Evaluation functions of many forms can be used in a GA, subject to the minimal requirement that the function can map the population into a partially ordered set. The evaluation functions to be optimised are summarised in the next section.

9.4.2.5 Summary

The GA parameters selected for the proposed fusion optimisation algorithm are listed in Table 9.1. Once $Q4$ (or ERGAS, see Section 9.5) has been considered as the fitness function, and all the parameters reported in Table 9.1 are set, a standard GA implementation is sufficient to start the optimisation procedure. The interval of variation for the g_l and α_k parameters is the same for all bands and spans the intervals $[-10, 10]$ and $[0, 10]$, respectively, in order to ensure a wide state space. Each unknown parameter is spatially constant on the corresponding band i.

9.5 Quality evaluation criteria

9.5.1 Q4 quality index

The image quality index $Q4$ for multispectral images having four spectral bands can be used to evaluate pan-sharpened MS images as described in [15]. The index $Q4$ is derived from the theory of hypercomplex numbers, in particular of 'quaternions,' which can be represented in the form $a = a_1 + a_2 i + a_3 j + a_4 k$, where a_1, a_2, a_3, a_4 are real numbers, and $i^2 = j^2 = k^2 = ijk = -1$. Every quaternion is a unique and a real linear combination of the basis quaternions $1, i, j,$ and k. For MS images with four spectral bands, typical

Table 9.1 *GA parameters used for real-valued Q4 function optimisation.*

Operation	Parameters
Initial population	200
Normalised geometric selection	0.05
Uniform mutation	4
Non-uniform mutation	[4, 100, 3]
Multi-non-uniform mutation	[6, 100, 3]
Boundary mutation	4
Simple crossover	2
Arithmetic crossover	2
Heuristic crossover	[2, 3]
Maximum generation	200
Chromosomes bounds (for each band)	g_l: [−10, 10] (only GIHS) α_k: [0, 10]

for the new generation satellite images, a_1, a_2, a_3, a_4 represent the values assumed by a given image pixel in the four bands, acquired in the Blue, Green, Red, and Near Infrared wavelengths. The quality index is a generalisation of the Q index defined in [28] for an original image signal x and a test image signal y, which can be stated as

$$Q = \frac{4 \cos(x, y) \cdot \bar{x} \cdot \bar{y}}{(\text{var}(x) + \text{var}(y)) \cdot [(\bar{x})^2 + (\bar{y})^2]} \tag{9.29}$$

and may be equivalently rewritten as

$$Q_{N \times N} = \frac{\text{cov}(x, y)}{\sigma_x \sigma_y} \cdot \frac{2\bar{x}\bar{y}}{(\bar{x})^2 + (\bar{y})^2} \cdot \frac{2\sigma_x \sigma_y}{\sigma_x^2 + \sigma_y^2} \tag{9.30}$$

where σ_f denotes the standard deviation of f, $\bar{x}$ is the mean of x, and $\text{cov}(x, y)$ is the covariance of x and y, all computed over a given $N \times N$ block. In practice, the first factor is the correlation coefficient (CC), the second one (always $\leqslant 1$ and equal to 1 iff $x = y$) accounts for the mean bias, the third one measures the change in contrast. Eventually, the quality index Q of y is obtained by averaging the values obtained starting from all the $N \times N$ blocks of the images x and y. This quality factor can be applied only to monochrome images.

The unique score index $Q4$ for 4-band MS images, which assumes a real value in the interval [0, 1], is 1 iff the MS image is identical to the reference image. Again, $Q4$ is made up of different components (factors) to take into account for the correlation: the mean of each spectral band, the intra-band local variance, and the spectral angle. The first three factors are also taken into account by Q for each band, while the spectral angle is introduced by $Q4$ by properly defining a CC of multivariate data. In this way, both radiometric and spectral distortions are considered by a single parameter. $Q4$ can be computed from

$$Q4_{N \times N} = \frac{\text{cov}(z_1, z_2^*)}{\sigma_{z_1} \sigma_{z_2}} \cdot \frac{2\bar{z}_1 \bar{z}_2}{(\bar{z}_1)^2 + (\bar{z}_2)^2} \cdot \frac{2\sigma_{z_1} \sigma_{z_2}}{\sigma_{z_1}^2 + \sigma_{z_2}^2} \tag{9.31}$$

where quaternions are indicated in boldface. Finally $Q4$ is obtained by averaging the magnitudes of all $\mathbf{Q4}_{N \times N}$ over the whole image, i.e.

$$Q4 = E\big[\|\, \mathbf{Q4}_{N \times N}\,\|\big] \tag{9.32}$$

where $E[\cdot]$ denotes the average over an $N \times N$ block and $\|x\| = (x_r^2 + x_1^2 + x_2^2 + x_3^2)^{1/2}$ is the magnitude of the quaternion.

9.5.2 Relative dimensionless global error in synthesis

Another score index [2] called ERGAS (erreur relative globale adimensionelle de synthese), i.e. dimensionless global relative error of synthesis, is defined as

$$\text{ERGAS} = \frac{100}{\text{SR}}\sqrt{\frac{1}{N}\sum_{k=1}^{N}\left(\frac{\text{RMSE}_k}{\mu_k}\right)^2} \tag{9.33}$$

with N being the total number of bands, SR the scale ratio of MS and PAN spatial resolutions, and μ_k the average of the kth band. This index is capable of measuring global radiometric distortion of the fused images. An ERGAS value equal to zero denotes absence of radiometric distortion, but possible spectral distortion.

9.6 A fast optimum implementation

The CPU times of the above algorithms rise with the scene dimension and with the minimum tolerance necessary to satisfy the OA requirements. Therefore, a procedure that reduces the computation complexity of the OA would be very useful. Suppose we combine (9.9) and (9.10):

$$\widehat{B}_l = \widetilde{B}_l + g_l\left(P - \sum_{k=1}^{N}\alpha_k \cdot \widetilde{B}_k\right) \tag{9.34}$$

and replace α_l with $\alpha_{l,k}$ to calculate a band-dependent pseudo-intensity image. Equation (9.34) can be rewritten as

$$\widehat{B}_l = \sum_{k=1}^{N}\gamma_{l,k} \cdot \widetilde{B}_k + \gamma_{l,N+1} \cdot P, \quad l = 1, \dots, N \tag{9.35}$$

where $\gamma_{l,N+1} = g_l$ and

$$\gamma_{l,k} = \begin{cases} 1 - g_l\alpha_{l,k}, & k = l, \\ -g_l\alpha_{l,k}, & k \neq l \end{cases} \tag{9.36}$$

The $\gamma_{l,k}$ coefficients are determined simulating the fusion process at a degraded scale applying an OA that minimises the mean square error (MSE) on each band between the reference and the fused data set. Given the linear problem (9.35), the OA could be easily implemented with a multiple linear regression at least squares. The optimum $\gamma_{l,k}$ coefficients are then applied at full resolution to generate the fused images. The main advantage of this algorithm is a tremendous reduction of the CPU time in the calculation of the optimal parameters.

9.7 Experimental results and comparisons

The proposed fusion algorithm has been assessed on two very high-resolution image data sets collected by Ikonos and QuickBird MS and PAN scanners. The first data set has been acquired on the urban area of Toulouse, France, and the second data set has been acquired on the city of Athens, Greece. The four MS bands of the Ikonos data set span the visible and near infrared (NIR) wavelengths and are spectrally disjoint with the exception of the blue and the green band: blue (B1 = 440–530 nm), green (B2 = 520–600 nm), red (B3 = 630–700 nm), and NIR (B4 = 760–850 nm). The PAN band embraces the whole interval (PAN = 450–900 nm). The four MS bands of QuickBird span the visible and NIR wavelengths and are not-overlapping: B1 = 450–520 nm, B2 = 520–600 nm, B3 = 630–690 nm, and B4 = 760–900 nm. The bandwidth of PAN embraces the interval 450–950 nm. Both the data sets have been radiometrically calibrated from digital counts, orthorectified, i.e. resampled to uniform ground resolutions of 4m and 1m for MS and PAN for Ikonos and 2.8m and 0.7m for MS and PAN for QuickBird, respectively, and packed in 16-bit words. The full scale of all the bands is 2047 and is reached in the NIR wavelengths.

A thorough performance comparison was carried out among the proposed optimisation-based methods and five state-of-the-art image fusion methods:

- Eight different optimisation methods obtained by combining two methodologies (ATWT and GIHS), two optimisation approaches (Gradient-Descent, GRA, and Genetic Algorithm, GA), and two objective functions ($Q4$ and ERGAS).
- The fast fusion algorithm defined in Section 9.6, namely Fast-OA.
- *Gram–Schmidt spectral sharpening method* (GS) [9] as implemented in the ENVI® software package, with a low resolution PAN image obtained as the pixel average of the MS bands (Mode 1).
- *Gram–Schmidt spectral sharpening method* (GS) as implemented in ENVI®, with a low-pass PAN image given by preliminary low-pass filtering and decimation of the PAN image (Mode 2).
- *Enhanced–Gram–Schmidt spectral sharpening method* (EGS) as implemented in [29].
- *Synthetic Variable Ratio* (SVR) as proposed by Zhang [30].
- *PCI–Geomatics Fusion Method* (Geomatica®) proposed by Zhang [10].

The fusion methods are evaluated on three different *band-independent* quality indexes. The first two indexes, which account for spectral and radiometric distortion, are described in Section 9.5, while the last index, which provides a measure of spectral quality only, is the Spectral Angle Mapper (SAM) which denotes the absolute value of the spectral angle

Table 9.2 *Band-independent quality/distortion indexes calculated for the Toulouse data set at 4m spatial resolution. Best results shown in boldface.*

			Q4	ERGAS	SAM
MS expanded to the PAN scale			0.649	5.546	4.471
ATWT	GRA	OPT$_{Q4}$	0.932	2.488	2.643
		OPT$_{ERGAS}$	0.931	2.543	2.720
	GA	OPT$_{Q4}$	0.932	2.488	2.643
		OPT$_{ERGAS}$	0.931	2.543	2.720
GIHS	GRA	OPT$_{Q4}$	**0.936**	**2.465**	2.600
		OPT$_{ERGAS}$	0.933	2.540	2.720
	GA	OPT$_{Q4}$	**0.936**	2.471	**2.595**
		OPT$_{ERGAS}$	0.932	2.553	2.720
Fast-OA			0.934	2.520	2.700
GS		Mode 1	0.854	3.782	3.904
		Mode 2	0.857	3.623	3.673
		Enhanced	0.858	3.600	3.675
Other methods		SVR	0.797	3.958	4.471
		Geomatica	0.867	3.531	3.587

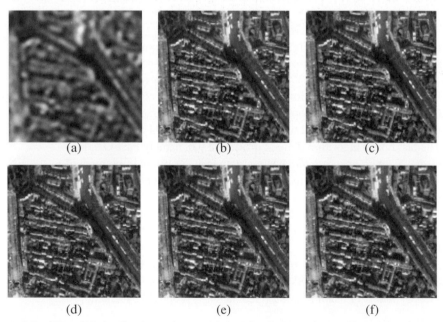

(a) (b) (c)

(d) (e) (f)

Figure 9.3 *128 × 128 details of pan-sharpened MS of the Toulouse dataset (degraded scale). (a) Resampled 16m MS; (b) ATWT-GRA-OPT$_{Q4}$ fusion; (c) ATWT-GRA-OPT$_{ERGAS}$; (d) ATWT-GA-OPT$_{Q4}$; (e) ATWT-GA-OPT$_{ERGAS}$; (f) GIHS-GRA-OPT$_{Q4}$; (g) GIHS-GRA-OPT$_{ERGAS}$ fusion; (h) GIHS-GA-OPT$_{Q4}$; (i) GIHS-GA-OPT$_{ERGAS}$; (j) Fast-OA; (k) GS-Mode 1; (l) GS-Mode 2; (m) EGS; (n) SVR; (o) Geomatica; (p) true 4m MS.*

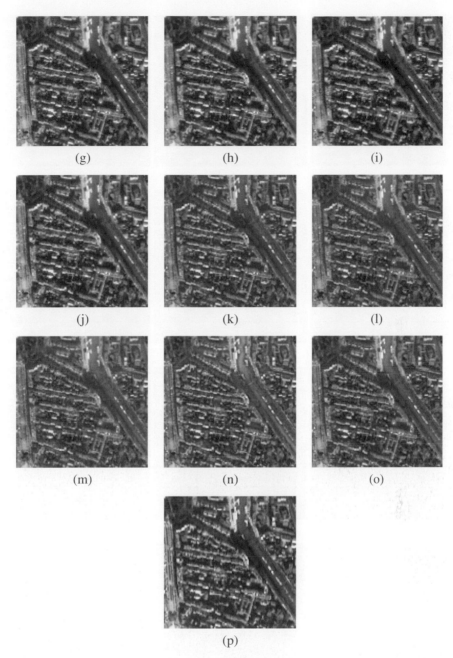

Figure 9.3 *(continued)*

between two pixel vectors, $\boldsymbol{v}$ and $\hat{\boldsymbol{v}}$,

$$\text{SAM} = \arccos\left(\frac{\langle \boldsymbol{v}, \hat{\boldsymbol{v}} \rangle}{\|\boldsymbol{v}\|_2 \cdot \|\hat{\boldsymbol{v}}\|_2}\right) \qquad (9.37)$$

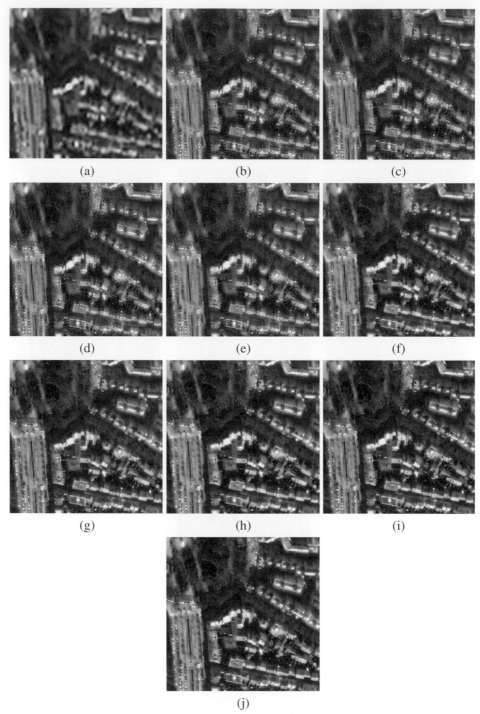

Figure 9.4 *256 × 256 details of pan-sharpened MS of the Toulouse dataset (full resolution). (a) Resampled MS; (b) ATWT-GRA-OPT$_{Q4}$ fusion; (c) ATWT-GRA-OPT$_{ERGAS}$; (d) ATWT-GA-OPT$_{Q4}$; (e) ATWT-GA-OPT$_{ERGAS}$; (f) GIHS-GRA-OPT$_{Q4}$; (g) GIHS-GRA-OPT$_{ERGAS}$ fusion; (h) GIHS-GA-OPT$_{Q4}$; (i) GIHS-GA-OPT$_{ERGAS}$; (j) Fast-OA.*

Table 9.3 *Band-independent quality/distortion indexes calculated for the Athens data set at 2.8m spatial resolution. Best results shown in boldface.*

			$Q4$	ERGAS	SAM
MS expanded to the PAN scale			0.805	5.537	4.111
ATWT	GRA	OPT$_{Q4}$	0.922	3.660	3.561
		OPT$_{ERGAS}$	0.922	3.602	3.568
	GA	OPT$_{Q4}$	0.922	3.660	3.561
		OPT$_{ERGAS}$	0.922	3.602	3.568
GIHS	GRA	OPT$_{Q4}$	**0.927**	3.673	3.531
		OPT$_{ERGAS}$	0.926	3.573	**3.507**
	GA	OPT$_{Q4}$	**0.927**	3.660	3.533
		OPT$_{ERGAS}$	0.924	3.613	3.554
Fast-OA			0.926	**3.572**	3.537
GS		Mode 1	0.886	4.031	3.911
		Mode 2	0.887	4.008	3.885
		Enhanced	0.890	3.924	3.846
Other methods		SVR	0.825	4.414	4.111
		Geomatica	0.891	3.908	3.763

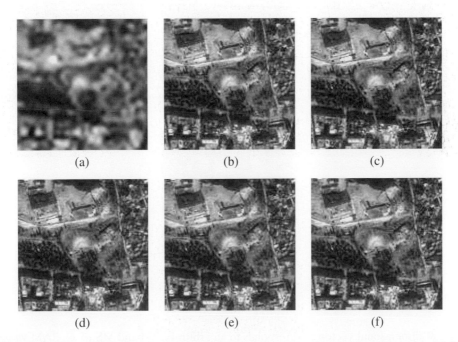

(a) (b) (c)

(d) (e) (f)

Figure 9.5 *128 × 128 details of pan-sharpened MS of the Athens dataset (degraded scale). (a) Re-sampled 11.2m MS; (b) ATWT-GRA-OPT$_{Q4}$ fusion; (c) ATWT-GRA-OPT$_{ERGAS}$; (d) ATWT-GA-OPT$_{Q4}$; (e) ATWT-GA-OPT$_{ERGAS}$; (f) GIHS-GRA-OPT$_{Q4}$; (g) GIHS-GRA-OPT$_{ERGAS}$ fusion; (h) GIHS-GA-OPT$_{Q4}$; (i) GIHS-GA-OPT$_{ERGAS}$; (j) Fast-OA; (k) GS-Mode 1; (l) GS-Mode 2; (m) EGS; (n) SVR; (o) Geomatica; (p) true 2.8m MS.*

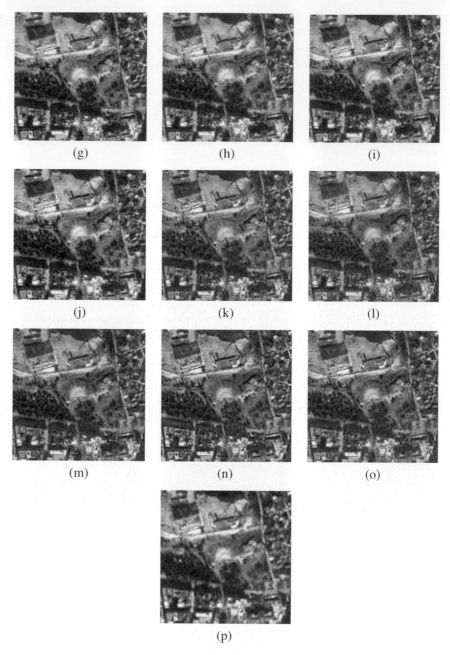

Figure 9.5 *(continued)*

The N-dimensional vector v corresponds to an arbitrary N-band MS pixel. SAM values equal to zero denote the absence of spectral distortion, but possible radiometric distortion. To this purpose, the datasets have been spatially degraded by four, according to the protocol proposed in [2], and statistics have been calculated between fused and original data.

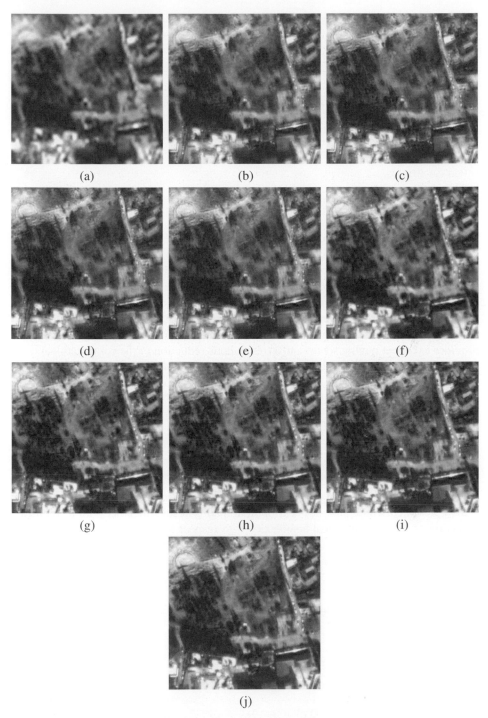

Figure 9.6 *Detail of Athens merged MS image (256 × 256) at full resolution. (a) Resampled 2.8m MS; (b) ATWT-GRA-OPT$_{Q4}$ fusion; (c) ATWT-GRA-OPT$_{\text{ERGAS}}$; (d) ATWT-GA-OPT$_{Q4}$; (e) ATWT-GA-OPT$_{\text{ERGAS}}$; (f) GIHS-GRA-OPT$_{Q4}$; (g) GIHS-GRA-OPT$_{\text{ERGAS}}$ fusion; (h) GIHS-GA-OPT$_{Q4}$; (i) GIHS-GA-OPT$_{\text{ERGAS}}$; (j) Fast-OA.*

Table 9.4 *CPU seconds of the proposed fusion algorithms for the two data sets.*

	CPU time (s)	
	Toulouse	Athens
ATWT-GRA-OPT$_{Q4}$	60	24
ATWT-GRA-OPT$_{ERGAS}$	9	9
ATWT-GA-OPT$_{Q4}$	1350	351
ATWT-GA-OPT$_{ERGAS}$	38	20
GIHS-GRA-OPT$_{Q4}$	136	130
GIHS-GRA-OPT$_{ERGAS}$	25	23
GIHS-GA-OPT$_{Q4}$	1361	290
GIHS-GA-OPT$_{ERGAS}$	38	19
Fast-OA	2	2

Table 9.2 reports the three global indexes calculated for the Toulouse data set at reduced resolution, i.e. by merging 4m-PAN with 16m-MS data to obtain 4m-MS fused images which are compared to the original 4m-MS images. Analogous results for the Athens data set are presented in Table 9.3, where in this case the original spatial resolution of MS data is 2.8m as for QuickBird data products. The fusion methods based on optimisation procedures show better performances (lower ERGAS and SAM, and higher $Q4$) with slight differences among each other. The Fast-OA method is very promising, since it provides excellent quality scores and a tremendous reduction of the computational complexity as demonstrated by Table 9.4 which reports the computing time for each considered optimisation procedure, in the case of Matlab implementation on a Pentium-4 PC platform.

9.8 Conclusions

This chapter reviews several pan-sharpening methods based on optimisation procedures adopting one of two different score indexes as their objective function. A new fast method is also presented which provides near-optimal quality results together with a significant reduction of the computational complexity with respect to genetic or gradient-descent-based algorithms. The experimental results demonstrate that the optimisation approach to pan-sharpening outperforms very efficient state-of-the-art solutions.

Appendix A. Matlab implementation of the Line Search algorithm in the steepest descent

```
function [fout gout] = fdef(x);
fout =  function definition;
gout = [gradient definition];

function [x] = steep(x0,f,tol,iteration_max)

% steepest descent with Armijo rule and Polynomial linesearch
%
% Input: x0 = initial iterate
```

```
%     f = handle of objective function and its gradient (@fdef)
%     tol = termination criterion norm(grad) < tol (opt) default = 1e-6
%     iteration_max = maximum iterations (opt) default = 5000

f_handle = f; gamma = 1e-4;

if nargin < 4 iteration_max = 5000; end

if nargin < 3 tol = 1e-6; end

n_iteration = 0;

xc = x0;

[fc grad_fc] = feval(f_handle,xc);

while(n_iteration <= iteration_max & norm(grad_fc) > tol)

  % Take an Armijo step in -grad_fc direction

    alpha_1 = 1;                              % Stretch factor alpha
    xt = xc - alpha_1 * grad_fc;             % Update x vector
    ft = feval(f_handle,xt);                 % Update the function value

    Armijo_iteration = 0;        % Control for number of expansion steps
    n_iteration = n_iteration+1; % Control for number of iteration
    fgoal = fc - gamma * alpha_1 * (grad_fc'*grad_fc);% Goal function

    %-------- Polynomial Linesearch (Cubic or Quadratic) --------------

    phi_q0 = fc;
    phi_qd0 = -grad_fc' * grad_fc;
    alpha_current = alpha_1;
    qc = ft;                                  % Updated value Stored

    while(ft > fgoal)
        Armijo_iteration = Armijo_iteration + 1
        lleft = alpha_current * (1/10);
        lright = alpha_current * (1/2);
        if Armijo_iteration = 1          % First iteration is quadratic
            alpha_new = -phi_qd0/(2*alpha_current*(qc-phi_q0-phi_qd0));
            if alpha_new < lleft
                alpha_new = lleft;
            elseif alpha_new > lright
                alpha_new = lright;
            end
        else                             % Cubic
            INV_left_mat = [alpha_current^2, alpha_current^3;
                            alpha_previous^2, alpha_previous^3];
            right_mat = [qc; qm]-[phi_q0 + phi_qd0 * alpha_current;
                            phi_q0 + phi_qd0 * alpha_previous];
            [b a] = INV_left_mat \ right_mat;
            alpha_new = (-b + sqrt(b * b - 3 * a * phi_qd0))/(3 * a);
            if alpha_new < lleft
                alpha_new = lleft;
            elseif alpha_new > lright
                alpha_new = lright;
```

```
        end
    end
    qm = qc;
    alpha_previous = alpha_current;
    alpha_current = alpha_new;
    xt = xc - alpha_new * grad_fc;        % Update x vector
    ft = feval(f_handle,xt);          % Update the function value

    if(Armijo_iteration > 10)      % Try up to 10 expansion steps
    disp('Armijo error in steepest descent')
    return;
    end
    fgoal = fc -  gamma * alpha_new * (grad_fc' * grad_fc);
  end
  xc = xt;
  [fc grad_fc] = feval(f_handle,xc);

end x=xc;
```

References

[1] P.S. Chavez, S.C. Sides and J.A. Anderson, 'Comparison of three different methods to merge multiresolution and multispectral data: Landsat tm and spot panchromatic', *Photogrammetric Engineering and Remote Sensing*, Vol. 57, No. 3, 2001, pp. 295–303.

[2] L. Wald, T. Ranchin and M. Mangolini, 'Fusion of satellite images of different spatial resolutions: Assessing the quality of resulting images', *Photogrammetric Engineering and Remote Sensing*, Vol. 63, No. 6, 1997, pp. 691–699.

[3] D.A. Yocky, 'Multiresolution wavelet decomposition image merger of landsat thematic mapper and spot panchromatic data', *Photogrammetric Engineering and Remote Sensing*, Vol. 62, No. 9, 1996, pp. 1067–1074.

[4] J. Zhou, D.L. Civco and J.A. Silander, 'A wavelet transform method to merge landsat tm and spot panchromatic data', *International Journal of Remote Sensing*, Vol. 19, No. 4, 1998, pp. 743–757.

[5] P. Scheunders and S.D. Backer, 'Fusion and merging of multispectral images with use of multiscale fundamental forms', *Journal of the Optical Society of America*, Vol. 18, No. 10, 2001, pp. 2468–2477.

[6] J. Núñez, X. Otazu, O. Fors, A. Prades, V. Palà and R. Arbiol, 'Multiresolution-based image fusion with additive wavelet decomposition', *IEEE Transactions on Geoscience and Remote Sensing*, Vol. 37, No. 3, 1999, pp. 1204–1211.

[7] L. Alparone, V. Cappellini, L. Mortelli, B. Aiazzi, S. Baronti and R. Carla, 'A pyramid-based approach to multisensor image data fusion with preservation of spectral signatures', in P. Gudmandsen (ed.), *Future Trends in Remote Sensing*, Balkema, Rotterdam, 1998.

[8] S. Mallat, 'A theory for multiresolution signal decomposition: The wavelet representation', *IEEE Transactions on Pattern Analysis and Machine Intelligence*, Vol. 11, No. 7, 1989, pp. 674–693.

[9] C.A. Laben and B.V. Brower, 'Process for enhancing the spatial resolution of multispectral imagery using pan-sharpening', US Patent #6,011,875, Eastman Kodak Company, 2000.

[10] Y. Zhang, 'A new automatic approach for effectively fusing landsat 7 images and ikonos images', in *Proc. IEEE Int. Geoscience and Remote Sensing Symposium*, 2002, pp. 2429–2431.

[11] A. Garzelli and F. Nencini, 'PAN-sharpening of very high resolution multispectral images using genetic algorithms', *International Journal of Remote Sensing*, Vol. 27, No. 15, 2006, pp. 3273–3292.

[12] T. Ranchin and L. Wald, 'Fusion of high spatial and spectral resolution images: The arsis concept and its implementation', *Photogrammetric Engineering and Remote Sensing*, Vol. 66, No. 1, 2000, pp. 49–61.

[13] M. Gonzalez-Audicana, J. Saleta, R. Catalan and R. Garcia, 'Fusion of multispectral and panchromatic images using improved IHS and PCA mergers based on wavelet decomposition', *IEEE Transactions on Geoscience and Remote Sensing*, Vol. 42, No. 6, 2004, pp. 1291–1299.

[14] B. Aiazzi, L. Alparone, S. Baronti, I. Pippi and M. Selva, 'Generalised Laplacian pyramid-based fusion of MS + P image data with spectral distortion minimisation', *ISPRS Internat. Archives of Photogrammetry and Remote Sensing*, Vol. 34, No. 3A-W3, 2002, pp. 3–6.

[15] L. Alparone, S. Baronti, A. Garzelli and F. Nencini, 'A global quality measurement of pan-sharpened multispectral imagery', *IEEE Geoscience and Remote Sensing Letters*, Vol. 1, No. 4, 2004, pp. 313–317.

[16] B. Aiazzi, L. Alparone, S. Baronti and A. Garzelli, 'Context-driven fusion of high spatial and spectral resolution data based on oversampled multiresolution analysis', *IEEE Transactions on Geoscience and Remote Sensing*, Vol. 40, No. 10, 2002, pp. 2300–2312.

[17] T. Ranchin, B. Aiazzi, L. Alparone, S. Baronti and L. Wald, 'Image fusion – The arsis concept and some successful implementation schemes', *ISPRS Journal of Photogrammetry and Remote Sensing*, Vol. 58, 2003, pp. 4–18.

[18] B. Aiazzi, L. Alparone, S. Baronti, A. Garzelli and M. Selva, 'MTF-tailored multiscale fusion of high-resolution MS and pan imagery', *Photogrammetric Engineering and Remote Sensing*, Vol. 72, No. 5, 2006, pp. 591–596.

[19] A. Garzelli and F. Nencini, 'Interband structure modeling for pan-sharpening of very high-resolution multispectral images', *Information Fusion*, Vol. 6, No. 3, 2005, pp. 213–224.

[20] S.G. Mallat, *A Wavelet Tour of Signal Processing*, second ed., Academic Press, 1999.

[21] P. Dutillex, 'An implementation of the algorithm "a trous" to compute the wavelet transform', in *Wavelets: Time–Frequency Methods and Phase Space*, 1989.

[22] W. Carper, T. Lillesand and R. Kiefer, 'The use of intensity–hue–saturation transformations for merging spot panchromatic and multispectral image data', *Photogrammetric Engineering and Remote Sensing*, Vol. 56, No. 4, 1990, pp. 459–467.

[23] T. Te-Ming, S. Shun-Chi, S. Hsuen-Chyun and S.H. Ping, 'A new look at HIS-like image fusion methods', *Information Fusion*, Vol. 2, No. 3, 2001, pp. 177–186.

[24] S. Conte and C. De Boor, *Elementary Numerical Analysis: An Algorithmic Approach*, McGraw–Hill Higher Education, 1980.

[25] L. Davis, *The Handbook of Genetic Algorithms*, Van Nostrand Reinhold, New York, 1991.

[26] Z. Michalewicz, *Genetic Algorithms + Data Structures. Evolution Programs*, Springer-Verlag, New York, 1994.

[27] J. Joines and C. Houck, 'On the use of non-stationary penalty functions to solve non-linear constrained optimization problems with GA's', in *Evolutionary Computation, 1994. IEEE World Congress on Computational Intelligence*, 1994, pp. 579–584.

[28] Z. Wang and A.C. Bovik, 'A universal image quality index', *IEEE Signal Processing Letters*, Vol. 9, No. 3, 2002, pp. 81–84.

[29] B. Aiazzi, L. Alparone, S. Baronti and M. Selva, 'Enhanced Gram–Schmidt spectral sharpening based on multivariate regression on MS and PAN data', in *Proc. IEEE Int. Geoscience and Remote Sensing Symposium*, 2006.

[30] Y. Zhang, 'A new merging method and its spectral and spatial effects', *International Journal of Remote Sensing*, Vol. 20, No. 10, 1999, pp. 2003–2014.

10

Image fusion using optimisation of statistical measurements

Laurent Oudre, Tania Stathaki and Nikolaos Mitianoudis

Imperial College London, UK

The purpose of image fusion is to create a perceptually enhanced image from a set of multi-focus or multi-sensors images. In the methods we are about to describe we do not *a priori* know the ground truth image: these are blind fusion methods. There are mainly two groups of fusion methods depending on the signal domain they are applied: spatial domain methods and transform domain methods. The Dispersion Minimisation Fusion (DMF) and Kurtosis Maximisation Fusion (KMF) based techniques we are going to discuss are spatial domain methods that is to say the fusion is simply performed on the image itself. In this work we propose to linearly combine the input images with appropriate weights estimated using specific mathematical performance criteria which evaluate in various ways improvement in visual perception. More specifically, in order to estimate the weights we propose iterative methods which use cost functions based on two statistical parameters, i.e. the dispersion and the kurtosis. The optimisation of the proposed cost functions enables us to obtain a fused image which is less distorted compared to the input ones.

10.1 Introduction

Let us have K source images $X_1, \ldots, X_K$ describing different realisations of the same true scene F. The available images have been acquired from different sensors (multi-sensor scenario) or they are of the same type but exhibit different types of distortion, as for example blurring (multi-focus scenario). Our aim is to create from these images a single image Y which will be perceptually enhanced. The composite image should contain a more useful description of the scene than the one provided by any of the individual sources, and therefore, should be more useful for human visual or machine perception. The task of combining images to form a single improved image is called *image fusion*. Image fusion has been used in many fields such as aerial and satellite imaging, medical imaging, robot vision etc. In recent years image fusion has become an important and useful technique for image analysis, computer vision, concealed weapon detection, autonomous landing guidance and others. Image fusion can be performed either in the spatial or in the transform domain.

As far as the transform domain fusion methods are concerned the input images are first transformed into a new domain, then fused and the result is converted back by an inverse transform. Popular transform domain fusion methods are for example the Dual-Tree Wavelet Transform (DT-WT) method [1] or the Independent Component Analysis (ICA) method [2]. In these methods the fusion coefficients are calculated with either pixel based or region based fusion rules.

The methods we are proposing in this work are spatial domain methods, that is to say, we work on the input images directly. A linear combination of the available source images is used, where the weights are estimated using novel optimisation formulations.

In order to understand the mechanisms behind the proposed fusion rules a thorough mathematical background and some notations are required. Section 10.2 of this chapter will be dedicated to the notations, the definitions and the problem formulation. Sections 10.3 and 10.4 are dedicated to the description of the proposed methods, as well as modified versions of them. The first method we propose is the Distortion Minimisation Fusion (DMF). This spatial domain fusion technique utilises the cost function of one of the most studied and implemented methods, i.e. the Constant Modulus (CM) algorithm, and the concept of signal dispersion [3]. An iterative process updates at every step the weights for the pixels by minimising a function of the dispersion of the unknown original image. An alternative technique is also proposed where we use the Central Limit Theorem and the characteristics of smoothing (blurring) operators to assume that the non-Gaussianity is an indicator of image quality. The statistical parameter we consider to measure non-Gaussianity is the absolute value of kurtosis. This method is called Kurtosis Maximisation Fusion (KMF). The additional methods we shall introduce are improvements of the previous ones. Section 10.5 is a presentation of indicative results we obtain with the proposed methods.

10.2 Mathematical preliminaries

Assume K two-dimensional source digital images $X_1, \ldots, X_K$ of equal size $M \times N$ describing the same true scene F. The images are registered to each other. By scanning the rows sequentially we transfer each image X_k to a row vector $\underline{x}_k$ (lexicographic ordering) with elements $x_k(n)$ where $n \in [1, MN]$. The aim of image fusion is to reconstruct a fused image Y which demonstrates an improved image quality over any individual image X_k. For the fused image we also use its lexicographically ordered version $\underline{y}$ with elements $y(n)$.

To examine a spatially adaptive image fusion scheme we are interested in assigning to the nth pixel $x_k(n)$ a distinct weight $w_k(n)$ that measures the contribution of the pixel $x_k(n)$

to the fused pixel $y(n)$. It is convenient to gather all the weights and intensity values at the nth pixel location together and denote them by single vectors as follows:

$$\underline{w}(n) = \left[w_1(n), \ldots, w_K(n) \right]^T \tag{10.1}$$

and

$$\underline{x}(n) = \left[x_1(n), \ldots, x_K(n) \right]^T \tag{10.2}$$

where $n \in [1, MN]$.

Consequently the nth pixel $y(n)$ in the fused image is obtained as in Equation (10.3) below by linearly combining the pixels $x_k(n)$ at the same location n from the available source images:

$$y(n) = \sum_{k=1}^{K} w_k(n) x_k(n) = \underline{w}^T(n) \underline{x}(n) \tag{10.3}$$

Furthermore, we call $\underline{\underline{x}} = [\underline{x}_1 \ \cdots \ \underline{x}_K]^T$ the $K \times NM$ matrix containing all the source images. The same notation is used for the weights, that is to say, $\underline{\underline{w}} = [\underline{w}_1 \ \cdots \ \underline{w}_K]^T$.

The weights have to be positive and also $\sum_{i=1}^{K} w_i(n) = 1$. The aim of the proposed algorithms is to determine the matrix $\underline{\underline{w}}$.

10.3 Dispersion Minimisation Fusion (DMF) based methods

Recently we introduced a preliminary version of the Dispersion Minimisation based Fusion scheme (DMF) [4]. The concept of dispersion was originally studied in its one-dimensional form and used for blind equalisation of communication signals over dispersive channels [3]. In [4] we investigated the use of two-dimensional dispersion to the problem of image fusion [5].

The dispersion constant of a real-valued image F with its zero-mean version denoted by $\tilde{F}$ is defined as follows:

$$D_F = \frac{E\{\tilde{F}^4\}}{E\{\tilde{F}^2\}} \tag{10.4}$$

where $E\{\cdot\}$ denotes the expectation operator performed along the dimension n defined previously.

In this work we are seeking for fusion weights that minimise the following cost function

$$J_{CM} = E\{(\tilde{y}^2(n) - D_F)^2\} \tag{10.5}$$

where D_F is the dispersion value of the original image F defined as in Equation (10.4) above and $\tilde{y}(n)$ denotes the nth pixel of the zero-mean version of the lexicographically ordered fused image. Cost functions similar to that in (10.5) have been used in communications [3] and the term Constant Modulus (CM) is widely used to refer to them. This term justifies the use of CM as subscript in the notation of the cost function J_{CM}.

It is straightforward from its definition that the cost function in (10.5) penalises the deviations of $\tilde{y}^2(n)$ from the dispersion constant D_F. Since a closed form solution for the minimisation of (10.5) does not exist, iterative approaches, as for example the widely used Gradient Descent (GD) method, are generally used to solve it. The algorithm that performs a stochastic Gradient Descent minimisation of a CM type of cost function is referred to in the existing literature as the Constant Modulus Algorithm or CMA [3]. CMA attempts to minimise the CM cost function by starting with arbitrary values for the unknown parameters and following the trajectory of the steepest descent.

In this work, the particularity of the proposed cost function is that we do not know the value of D_F. Therefore, Equation (10.5) involves the estimation of both the fusion weights $[w_1(n), \ldots, w_K(n)]$, $\forall n$ and the dispersion D_F of the original true scene. Thus, the minimisation of (10.5) is performed using an alternating stochastic Gradient Descent algorithm.

One can notice that in the previous definition we need to deal with zero-mean images. That is why in the rest of the chapter we will use the notation $\tilde{x}(n)$ instead of $x(n)$. However, we will use the non-zero mean version of the source images for the final step, that is to say the reconstruction of the image with the final weights.

From Equation (10.3) we deduct that $\tilde{y}(n) = \underline{w}^{\mathrm{T}}(n)\tilde{\underline{x}}(n)$ and hence, we can rewrite the proposed cost function as follows:

$$J_{\mathrm{CM}}\big(\underline{w}(n), D_F\big) = E\big\{\big[\big(\underline{w}^{\mathrm{T}}(n)\tilde{\underline{x}}(n)\big)^2 - D_F\big]^2\big\} \tag{10.6}$$

In order to minimise the cost function in (10.6), we are going to use a Gradient Descent method with two learning rates μ and η. We need then to calculate the gradient of $J_{\mathrm{CM}}(\underline{w}(n), D_F)$ relative to both $\underline{w}(n)$ and D_F. Throughout the chapter we will often interchange the notations $J_{\mathrm{CM}}(\underline{w}(n), D_F)$ and J_{CM} for simplicity.

Calculation of $\partial J_{\mathrm{CM}}/\partial \underline{w}(n)$

We know that $J_{\mathrm{CM}} = E\{(\tilde{y}^2(n) - D_F)^2\} = E\{\tilde{y}^4(n)\} - 2D_F E\{\tilde{y}^2(n)\} + D_F^2$ and $\tilde{y}(n) = \underline{w}^{\mathrm{T}}(n)\tilde{\underline{x}}(n)$. The expectations calculated along the dimension n are approximated by the sample mean $E\{\tilde{y}^m(n)\} = (1/MN)\sum_{n=1}^{MN}\tilde{y}^m(n)$. As a result, the derivative of these expectations with respect to the specific weight $\underline{w}(n)$ will simply be reduced to the derivative of the sample means' term for the corresponding n. Consequently, the requested derivative can be given by

$$\frac{\partial J_{\mathrm{CM}}}{\partial \underline{w}(n)} = 4\big(\tilde{y}(n)^2 - D_F\big)\tilde{y}(n)\tilde{\underline{x}}(n) \tag{10.7}$$

Calculation of $\partial J_{CM}/\partial D_F$

From the expression of J_{CM} we have $\partial J_{CM}/\partial D_F = -2E\{\tilde{y}^2(n)\}+2D_F$ or, alternatively,

$$\frac{\partial J_{CM}}{\partial D_F} = 2\big(D_F - E\{\tilde{y}^2(n)\}\big) \tag{10.8}$$

10.3.1 The Dispersion Minimisation Fusion method (DMF)

The proposed algorithm is summarised in the steps below.

Initialisation

- Set all the weights $\underline{w}(n)$ at the value K^{-1}. The first estimate of the fused image will then be simply the mean of the K source images.
- Set the initial value of D_F as the mean of the dispersion parameters of the K source images.

Iteration

- Update the values of $\underline{w}(n)$:

$$\underline{w}^+(n) \Leftarrow w(n) - \mu\frac{\partial J_{CM}}{\partial \underline{w}(n)}$$

- Normalise the values of $\underline{w}(n)$:

$$\underline{w}^+(n) \Leftarrow \text{abs}\left(\frac{\underline{w}(n)}{\|\underline{w}(n)\|}\right)$$

- Update the value of D_F:

$$D_F^+ \Leftarrow D_F - \eta\frac{\partial J_{CM}}{\partial D_F}$$

- Check that D_F is positive and if not take its absolute value:

$$D_F^+ \Leftarrow \text{abs}(D_F)$$

The parameters μ and η have a very important role in the convergence of the proposed method. By selecting inappropriate values for these learning rates, the cost function may converge to a local minimum instead of the global minimum. In order to tackle this problem an exhaustive search for optimal combinations of values for μ and η is realised, prior to updating the values for $\underline{w}$ and D_F. By the term optimal we refer to the values that minimise the cost function or yield a value at convergence sufficiently close to the minimum. After a large number of experimental simulations we have concluded that appropriate values for μ lie approximately around 10^{-6} and for η around 0.9. Prior selection of learning rate values enables us to get better results as far as both convergence to the global minimum and speed of convergence are concerned (Figure 10.1). We shall refer to this modified version of the method as the Robust DMF method [4]. Once the matrix $\underline{w}$ has converged, the fused image can be reconstructed using the non-zero mean source images.

10.3.2 The Dispersion Minimisation Fusion method With Neighbourhood (DMF_WN)

In the previous section the update for $\underline{w}(n)$ was essentially estimated using a stochastic update, due to the lack of multiple realisations of the fused image. By assuming that the image signal is locally ergodic we allow the pixels within the $L \times L$ neighbourhood around the pixel of interest n to be treated as multiple realisations of that pixel in the fused image. In addition, we can assume that the weight $\underline{w}(n)$ in this neighbourhood remains constant for each pixel located within the neighbourhood. Consequently, the gradient term can now be calculated via

$$\frac{\partial J_{CM}}{\partial \underline{w}(n)} = E_L\left\{4\big(\tilde{y}(n)^2 - D_F\big)\tilde{y}(n)\underline{\tilde{x}}(n)\right\}$$

$$\Rightarrow \quad \frac{\partial J_{CM}}{\partial \underline{w}(n)} = \begin{bmatrix} E_L\{4(\tilde{y}(n)^2 - D_F)\tilde{y}(n)\tilde{x}_1(n)\} \\ \vdots \\ E_L\{4(\tilde{y}(n)^2 - D_F)\tilde{y}(n)\tilde{x}_K(n)\} \end{bmatrix} \quad (10.9)$$

where the expectations $E_L\{\cdot\}$ are calculated via sample averaging using the pixels located within an $L \times L$ (L odd) neighbourhood placed symmetrically around pixel n. The optimal size of the neighbourhood depends on the particular image and on the type and severity of distortion. We call this method Dispersion Minimisation Fusion method With Neighbourhood (DMF_WN). Regarding the question whether the DMF_WN technique can be combined with the robust version described in Section 10.3.1, the tests have proved that although the combination of the two methods increases the performance it leads to a very long computation time.

10.4 The Kurtosis Maximisation Fusion (KMF) based methods

A possible limitation of the previous fusion scheme based on dispersion minimisation is that it requires a priori some statistical information, namely, the dispersion value of the ground truth image which is unavailable in practical cases. Although we have formulated a framework of alternating minimisation which gives reasonable estimates of the true dispersion value, the instability and bias of the fusion performance could still dominate due to the lack of required information. Therefore, we propose an alternative fusion scheme which is purely based on the available sensor images, and thus, does not require knowledge of the original ground truth image. We refer to this method as Kurtosis Maximisation based Fusion scheme (KMF).

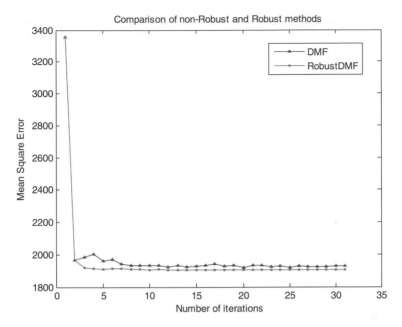

Figure 10.1 *An indicative comparison between DMF and Robust DMF.*

The motivation of using kurtosis maximisation stems from two facts:

- The Central Limit Theorem states that the probability density function of the sum of several independent random variables tends towards a Gaussian distribution [6].
- Due to the physical limitations of the sensors and imperfect observational conditions, the acquired sensor images represent a degraded version of the original scene by smoothing operators and additive noise [5], which is assumed to be independent to the image scene.

A smoothing operator often acts as a low-pass filter which results in a flatter (more Gaussian) distribution of the filtered image, as the high frequency information is suppressed, degraded or missing [5]. In addition, the combination of an image scene and additive noise, which is independent of the image, further increases the Gaussianity of sensor images due to the Central Limit Theorem. Combining these two facts together, we can see that it is likely that the probability distribution of an image is less Gaussian than that of a distorted version of it or of linear combinations of distorted versions of it [5]. We can assume that the fused image is expected to be as close to the original scene as possible, and furthermore, both the fused image and the original image feature a non-Gaussianity property. Such a principle implies that if we find a fused image Y that follows the minimum Gaussian behaviour (or alternatively, maximum non-Gaussianity), then that image will be closer to the original scene F compared to the acquired sensor images. To some extent, non-Gaussianity reflects the quality of the fused image. We can therefore identify the optimal fused image by maximising its non-Gaussianity via updating the fusion weights. To quantify the non-Gaussianity of the image, measurements, such as high-order central moments are frequently used. Here we choose the absolute value of the kurtosis, a normalised fourth-order central moment, to serve for non-Gaussianity maximisation.

Consider an image F. We define the kurtosis K_F of its zero-mean version $\tilde{F}$ as

$$K_F = \frac{\text{cum}_4\{\tilde{F}\}}{E^2\{\tilde{F}^2\}} = \frac{E\{\tilde{F}^4\} - 3E^2\{\tilde{F}^2\}}{E^2\{\tilde{F}^2\}} = \frac{E\{\tilde{F}^4\}}{E\{\tilde{F}^2\}E\{\tilde{F}^2\}} - 3 = \frac{D_F}{\sigma_{\tilde{F}}^2} - 3 \qquad (10.10)$$

where $\text{cum}_4\{\tilde{F}\}$ and $\sigma_{\tilde{F}}^2$ denote the fourth-order cumulant and the standard deviation of $\tilde{F}$, respectively. From a statistical perspective, kurtosis measures the peakedness of a distribution [7]. More specifically, a Gaussian distribution has kurtosis equal to zero ($K_F = 0$). Moreover, it exhibits moderate tails and it is called mesokurtic. A distribution with small tails has negative kurtosis ($K_F < 0$) and is called sub-Gaussian or platykurtic and one with long tails has positive kurtosis ($K_F > 0$) and is called super-Gaussian or leptokurtic. The absolute value of kurtosis is usually used as a measurement of non-Gaussianity as it tends to be zero for a Gaussian distribution and non-zero for any other non-Gaussian distribution. In order to demonstrate the correlation among distortion, non-Gaussianity and the absolute value of kurtosis, we assume the original image Cameraman and a distorted version of it by Gaussian blur. The histograms and the absolute kurtosis of the two images are illustrated in Figure 10.2, in which we observe that when distortion occurs, the corresponding $|K_F|$ value decreases as the image data becomes more Gaussian. In other words, it is safe to state that the actual non-distorted representation of the observed scene, and therefore the fused image that is produced using the available sources have larger values of $|K_F|$, or alternatively follow a more non-Gaussian behaviour and are less distorted. Inspired by the fact that the absolute value of $|K_F|$ (non-Gaussianity) can be a sound criterion to reflect the quality of a fused image, we derive a novel fusion scheme, which solves for optimal fusion weights by maximising a non-quadratic cost function J_K, describing the absolute value of the kurtosis of the fused image Y.

Based on the above analysis, it seems logical to choose for the cost function the absolute value of the kurtosis

$$J_K = |K_F| = \left| \frac{E\{\tilde{y}^4(n)\}}{E^2\{\tilde{y}^2(n)\}} - 3 \right| \qquad (10.11)$$

where $\tilde{y}(n)$ denotes the nth pixel of the zero-mean version of the lexicographically ordered fused image y as already mentioned. Using Equation (10.3) we can rewrite the cost function as follows:

$$J_K\big(w(n)\big) = \left| \frac{E\{(\underline{w}^{\mathrm{T}}(n)\underline{\tilde{x}}(n))^4\}}{E^2\{(\underline{w}^{\mathrm{T}}(n)\underline{\tilde{x}}(n))^2\}} - 3 \right| \qquad (10.12)$$

We therefore need to maximise a cost function depending on one unknown parameter, namely, the vector $\underline{w}(n)$. To solve this problem we will again use a Gradient Descent method with one learning rate λ. We need then to calculate the gradient of $J_K(\underline{w}(n))$ relative to $\underline{w}(n)$. Throughout the chapter we will often interchange the notations $J_K(\underline{w}(n))$ and J_K for simplicity.

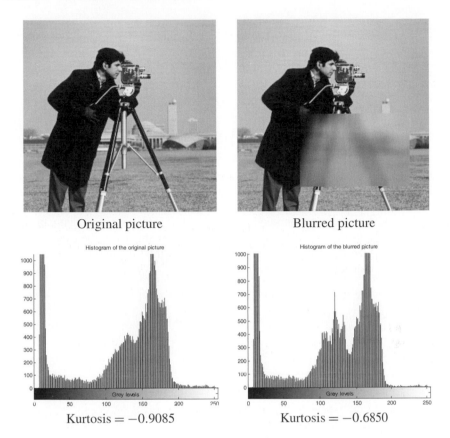

Figure 10.2 *Histograms and kurtosis of two pictures.*

Calculation of $\partial J_K / \partial \underline{w}(n)$

$$J_K = \frac{|E\{\tilde{y}^4(n)\} - 3E^2\{\tilde{y}^2(n)\}|}{E^2\{\tilde{y}^2(n)\}} = \frac{|\mathrm{cum}_4\{\tilde{y}(n)\}|}{E^2\{\tilde{y}^2(n)\}}$$

$$\frac{\partial J_K}{\partial \underline{w}(n)} = \frac{1}{E^4\{\tilde{y}^2(n)\}} \left[\frac{\partial |\mathrm{cum}_4\{\tilde{y}(n)\}|}{\partial \underline{w}(n)} E^2\{\tilde{y}^2(n)\} - \frac{\partial E^2\{\tilde{y}^2(n)\}}{\partial \underline{w}(n)} |\mathrm{cum}_4\{\tilde{y}(n)\}| \right]$$

where

$$\frac{\partial E^2\{\tilde{y}^2(n)\}}{\partial \underline{w}(n)} = 2E\{\tilde{y}^2(n)\} \frac{\partial E\{\tilde{y}^2(n)\}}{\partial \underline{w}(n)} = 4E\{\tilde{y}^2(n)\} E\{\tilde{y}(n)\underline{\tilde{x}}(n)\}$$

and

$$\frac{\partial |\mathrm{cum}_4\{\tilde{y}(n)\}|}{\partial \underline{w}(n)} = \mathrm{sgn}\left(\mathrm{cum}_4\{\tilde{y}(n)\}\right) \left[\frac{\partial E\{\tilde{y}^4(n)\}}{\partial \underline{w}(n)} - 3\frac{\partial E^2\{\tilde{y}^2(n)\}}{\partial \underline{w}(n)} \right] \Rightarrow$$

$$= 4\mathrm{sgn}\left(\mathrm{cum}_4\{\tilde{y}(n)\}\right) \left[E\{\tilde{y}^3(n)\underline{\tilde{x}}(n)\} - 3E\{\tilde{y}^2(n)\} E\{\tilde{y}(n)\underline{\tilde{x}}(n)\} \right]$$

and hence,

$$
\begin{aligned}
\frac{\partial J_K}{\partial \underline{w}(n)} &= 4\frac{\text{sgn}(\text{cum}_4\{\tilde{y}(n)\})}{E^4\{\tilde{y}^2(n)\}}\Big[E^2\{\tilde{y}^2(n)\}E\{\tilde{y}^3(n)\underline{\tilde{x}}(n)\} \\
&\quad - 3E^3\{\tilde{y}^2(n)\}E\{\tilde{y}(n)\underline{\tilde{x}}(n)\} - E\{\tilde{y}^2(n)\}E\{\tilde{y}(n)\underline{\tilde{x}}(n)\}\text{cum}_4\{\tilde{y}\}\Big] \\
&= 4\frac{\text{sgn}(\text{cum}_4\{\tilde{y}(n)\})}{E^3\{\tilde{y}^2(n)\}}\Big[E\{\tilde{y}^2(n)\}E\{\tilde{y}^3(n)\underline{\tilde{x}}(n)\} - E\{\tilde{y}^4(n)\}E\{\tilde{y}(n)\underline{\tilde{x}}(n)\}\Big]
\end{aligned}
$$

As in the dispersion case, the expectation $E\{\cdot\}$ is referring to multiple realisations of the fused image. If we assume that there is only a single realisation, i.e. the image $\tilde{y}(n)$, then the expectation can be dropped for a stochastic update of the gradient. Equally, we can assume that an $L \times L$ neighbourhood around pixel n contains pixels that can serve as multiple realisations of $\tilde{y}(n)$ if local ergodicity exists. In this case, the expectations can be estimated by sample averaging using the pixels in this neighbourhood, assuming a single weight vector $\underline{w}(n)$ for all these pixels.

10.4.1 The Kurtosis Minimisation Fusion method (KMF)

The proposed algorithm is summarised in the steps below.

Initialisation

- Set all the weights at the value K^{-1}. The first iteration of the fused image will then be simply the mean of the K source images.

Iteration

- Update the values of $\underline{w}(n)$:

$$
\underline{w}^+(n) \Leftarrow \underline{w}(n) - \lambda\frac{\partial J_K}{\partial \underline{w}(n)}
$$

- Normalise the values of $\underline{w}(n)$:

$$
\underline{w}^+(n) \Leftarrow \text{abs}\left(\frac{\underline{w}(n)}{\|\underline{w}(n)\|}\right)
$$

Once the matrix $\underline{w}$ has converged, the fused image can be reconstructed using the non-zero mean source images.

10.4.2 The Robust Kurtosis Minimisation Fusion method (Robust KMF)

As with the DMF method we can also use here an optimised learning rate λ. An exhaustive search for optimal values for λ is realised, prior to updating the values of $\underline{w}$. We shall refer to this modified version of the method as the Robust KMF method (Figure 10.3).

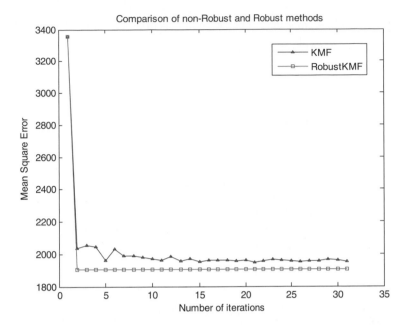

Figure 10.3 *An indicative comparison between KMF and Robust KMF.*

10.5 Experimental results

In order to evaluate the performance of the proposed methods we will compare them to the following well known methods on a selection of sets of images.

- Dual-Tree Wavelet Transform (DT_WT). This is a widely used transform domain method based on wavelet transforms. We use for evaluation this method in conjunction with the so called max-abs fusion rule. One can find further analysis in [1].
- Error Estimation Fusion (EEF). This is a spatial domain iterative method which has been developed very recently and uses the so called robust error estimation theory [8].

In order to provide numerical results we will use the following image fusion performance metrics.

- Q_0 stands for the so called Universal Image Quality Index. This is a measurement that evaluates the quality of an image in general and requires the ground truth in order to be calculated [9].
- MG stands for mean gradient image quality assessment method [10].
- S stands for the Petrovic image fusion metric [6].
- Q, Q_W, Q_e stand for the three variations of the Piella image fusion metric [11].

When we have the choice of robust and non-robust version of a method we will always choose the robust version since it always exhibits improved performance. In these cases as well as in the EEF method 15 iterations approximately are often enough. In the DMF_WN we realise 10 iterations since the computation time is often very long. Therefore, for each set of images we will apply the following techniques.

- Robust KMF
- Robust DMF
- DMF_WN, small neighbourhood 3×3
- DMF_WN, large neighbourhood 9×9 or 15×15
- DT_WT, max-abs fusion rule
- EEF

What follows is a description of the experiments.

10.5.1 Case one: Multi-focus images, small amount of distortion

Example 1: Clocks (see Figures 10.4 and 10.5; Table 10.1)
As shown in Table 10.1, the results obtained using the proposed techniques are not very encouraging compared to the DT_WT method in the context of the image fusion metrics used. The edges of the big clock still remain quite blurred. However, the DMF_WN with a large neighbourhood (9×9) yields acceptable results. Considering that the input images are large (512×512), this is the largest size of local neighbourhood we can take without facing serious computational burden. The numbers shown in bold demonstrate

MG: 2.8434   MG: 1.992
Blurred picture n°1   **Blurred picture n°2**

Figure 10.4

Table 10.1

	Robust KMF	Robust DMF	DMF_WN 3×3	DMF_WN 9×9	DT_WT	EEF
Q	0.8259	0.8258	0.8298	**0.8408**	0.7387	0.8404
Q_W	0.8554	0.8553	0.8492	0.8835	**0.9120**	0.8761
Q_e	0.5839	0.5843	0.5846	0.6676	**0.8092**	0.6552
S	0.58606	0.58627	0.59304	0.62443	**0.67478**	0.6356
MG	2.328	2.3236	2.3039	2.2832	**3.4056**	2.2302

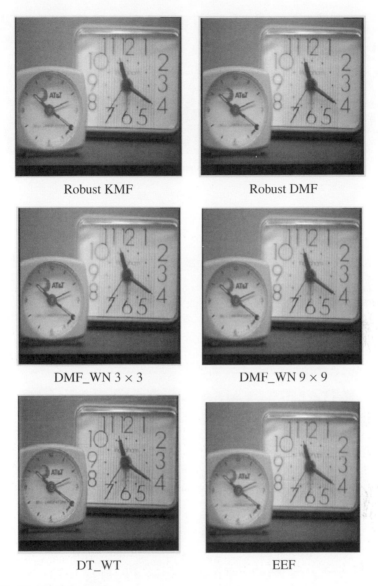

Robust KMF   Robust DMF

DMF_WN 3×3 DMF_WN 9×9

DT_WT   EEF

Figure 10.5 *The 'Clocks' example.*

the best performance achieved among the various methods in terms of the corresponding metric.

Example 2: Rice (see Figures 10.6 and 10.7; Table 10.2)
We applied a small amount of blur on the 'rice' image. While the DT_WT method works very well, the proposed methods are not visually very efficient. We see that the fused image remains blurred and the result is not very detailed. However, the metrics exhibit good values for the four proposed methods. This observation establishes the universally accepted rule that image fusion metrics do not always reflect the visual quality of an image.

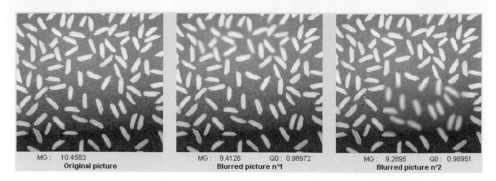

Figure 10.6

Table 10.2

	Robust KMF	Robust DMF	DMF_WN 3 × 3	DMF_WN 15 × 15	DT_WT	EEF
Q_0	0.995	0.99491	0.99616	0.99562	**0.99901**	0.99643
Q	0.9440	0.9433	0.9522	**0.9524**	0.9520	0.9498
Q_W	0.9757	0.9754	0.9786	**0.9792**	0.9734	**0.9792**
Q_e	0.9110	0.9102	0.9246	0.9198	**0.9338**	0.9262
S	0.8768	0.87583	0.88626	**0.88988**	0.86172	0.88272
MG	9.294	9.3048	9.278	9.1277	**10.5312**	9.1099

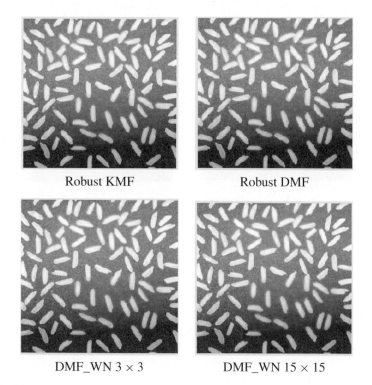

Figure 10.7   *The 'Rice' example.*

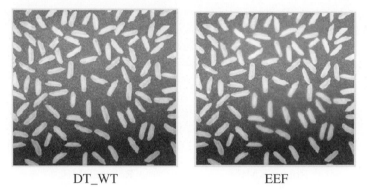

DT_WT   EEF

Figure 10.7 *(continued)*

Figure 10.8

Table 10.3

	Robust KMF	Robust DMF	DMF_WN 3 × 3	DMF_WN 15 × 15	DT_WT	EEF
Q_0	0.99265	0.99261	0.99334	**0.99605**	0.99455	0.98919
Q	0.8948	0.8948	0.8963	**0.9062**	0.8729	0.9040
Q_W	0.9509	0.9506	0.9542	0.9672	**0.9681**	0.9491
Q_e	0.8901	0.8892	0.9031	0.9302	**0.9487**	0.9022
S	0.87097	0.87019	0.8734	**0.89113**	0.88752	0.86997
MG	8.8364	8.8375	9.0073	9.0431	**10.1157**	8.8082

10.5.2 Case two: Multi-focus images, severe distortion

Example 1: Cameraman (see Figures 10.8 and 10.9; Table 10.3)
In this example we applied severe distortion on the 'cameraman' image. The proposed methods exhibit now a distinctively improved performance. The Robust DMF and Robust KMF methods are generally superior visually compared to the spatial domain EEF method in terms of the image fusion metrics used. Regarding the DMF_WN method with large sizes of local neighbourhood, it possibly exhibits comparable performance in terms of metrics with the DT_WT. It appears from a large number of experiments that the proposed methods are very efficient in the case of severe distortion. The DT_WT tends to

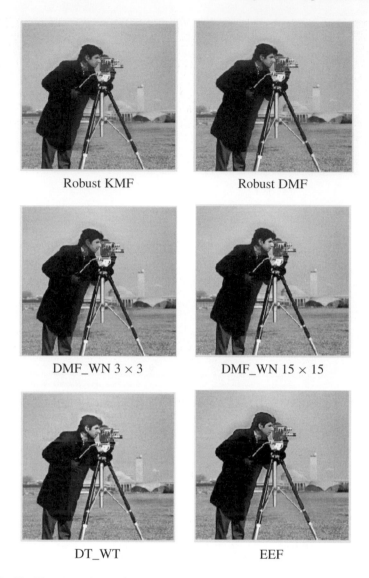

Robust KMF   Robust DMF

DMF_WN 3 × 3   DMF_WN 15 × 15

DT_WT   EEF

Figure 10.9 *The 'Cameraman' example.*

create some discontinuities in the image while the Robust DMF and Robust KMF provide really good visual results.

Example 2: Peppers (see Figures 10.10 and 10.11; Table 10.4)
Severe blur is also applied to various areas of the 'peppers' image. In the DT_WT method the distorted areas are still visible. However, in our methods it is harder to visualise where the original distortion was. Therefore, we can claim that in this example our methods give better results.

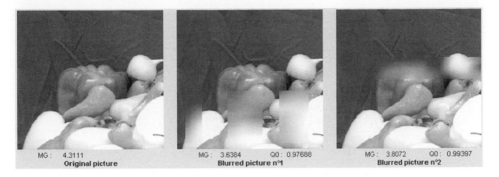

Figure 10.10

Table 10.4

	Robust KMF	Robust DMF	DMF_WN 3×3	DMF_WN 15×15	DT_WT	EEF
Q_0	0.99421	0.99435	0.99446	**0.99551**	0.99497	0.9941
Q	0.8964	0.8970	0.8971	0.9100	0.8549	**0.9140**
Q_W	0.8969	0.8981	0.8998	0.9210	**0.9462**	0.9292
Q_e	0.7495	0.7538	0.7651	0.8063	**0.9009**	0.8402
S	0.80689	0.80947	0.81066	0.82009	**0.82661**	0.82606
MG	3.7699	3.7244	3.7378	3.6913	**4.5186**	3.7229

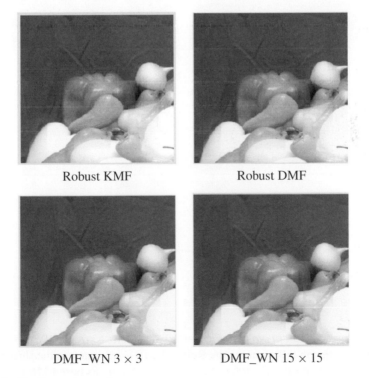

Robust KMF   Robust DMF

DMF_WN 3×3 DMF_WN 15×15

Figure 10.11 *The 'Peppers' example.*

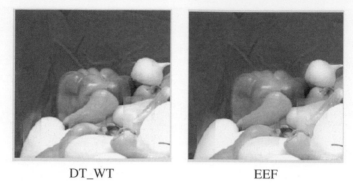

| DT_WT | EEF |

Figure 10.11   *(continued)*

MG : 4.6423
Blurred picture n°1

MG : 3.7451
Blurred picture n°2

Figure 10.12

Table 10.5

	Robust KMF	Robust DMF	DMF_WN 3 × 3	DMF_WN 15 × 15	DT_WT	EEF
Q	0.5537	0.3650	0.3761	0.4631	**0.6809**	0.5718
Q_W	0.6673	0.7055	0.7204	0.7921	**0.8402**	0.7680
Q_e	0.4184	0.4824	0.5375	0.6167	**0.7796**	0.6459
S	0.39238	0.41474	0.44032	0.49446	**0.60084**	0.48381
MG	3.6275	**7.6672**	7.2204	5.9965	6.7304	4.4191

10.5.3 Case three: Multi-sensor images

Example 1: Infrared/dark photo (see Figures 10.12 and 10.13; Table 10.5)
One can notice that in this example the Robust DMF fused image is the sharpest, even if some 'salt and pepper' noise artefacts seem to appear. The Robust KMF is nevertheless the clearest. Depending on what one expects from the fusion these two methods give very

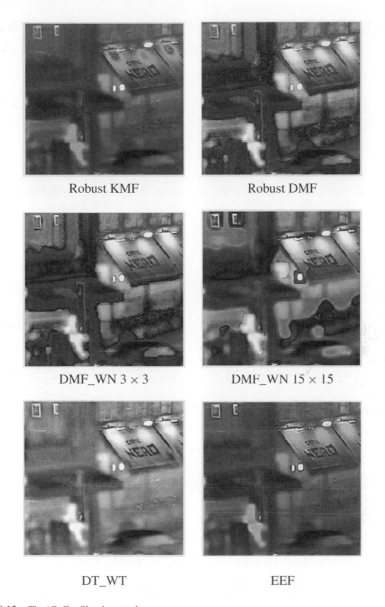

Figure 10.13 *The 'Coffee Shop' example.*

good results. It is interesting to observe that the DMF_WN method which gave us good results previously fails here. However, in terms of metric the DT_WT seems to perform better.

Example 2: Medical photos (see Figures 10.14 and 10.15; Table 10.6)
The Robust DMF and KMF provide good visual results although the corresponding metrics are not again the best.

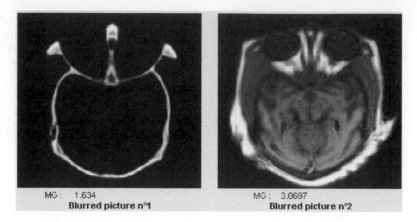

Figure 10.14

Table 10.6

	Robust KMF	Robust DMF	DMF_WN 3×3	DMF_WN 15×15	DT_WT	EEF
Q	0.6451	0.7114	**0.8138**	0.8085	0.7939	0.6499
Q_W	0.6417	0.7997	0.8231	0.7394	**0.8301**	0.7042
Q_e	0.3618	0.5632	0.6271	0.5086	**0.6605**	0.4818
S	0.51682	0.63921	**0.7067**	0.6442	0.70103	0.55065
MG	2.5105	4.1956	3.8514	3.5105	**4.2258**	2.4108

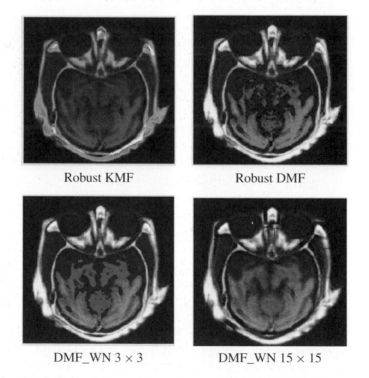

Figure 10.15 *The 'Medical' example.*

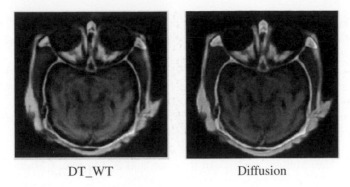

DT_WT	Diffusion

Figure 10.15 *(continued)*

10.6 Conclusions

Throughout this chapter we have described new spatial domain methods for multi-focus and multi-sensor image fusion. The mathematical background relevant to the proposed techniques is based on the iterative solution of two novel optimisation formulations related to the statistical properties of the original unknown image. In the experimental results presented in this chapter, it is highlighted that the proposed methods provide good results in almost every situation in terms of the widely used image fusion performance evaluation metrics. The only scenario where the proposed methods seem to be weak is the multi-focus scenario where the source images exhibit light distortion. Moreover, the visual assessment of the proposed methods is encouraging, although it is important to stress out the fact that the evaluation of image fusion results depends on the perception of the individual viewer. The introduction of the robust version of the proposed methods does not enhance visually the fusion results but enables us to obtain good results with less number of iterations. The local neighbourhood method yields better results although the optimal size of the neighbourhood is still a parameter under investigation and depends on the particular image set scenario. Among the proposed methods, for multi-focus image scenarios one will rather choose the DMF_WN or the Robust KMF method, while for multi-sensor image scenarios the Robust DMF or the Robust KMF would be more appropriate.

References

[1] I.W. Selesnick, R.G. Baraniuk and N.G. Kingsbury, 'The dual-tree complex wavelet transform', *IEEE Signal Processing Magazine*, Vol. 22, No. 6, 2005, pp. 123–151.

[2] N. Mitianoudis and T. Stathaki, 'Pixel-based and region-based image fusion schemes using ICA bases', *Elsevier Journal of Information Fusion*, Vol. 8, No. 2, 2007, pp. 131–142.

[3] J.R. Treichler and B.G. Agee, 'A new approach to multipath correction of constant modulus signals', *IEEE Transactions on Acoustics, Speech and Signal Processing*, Vol. 31, No. 2, 1983, pp. 459–472.

[4] Q. Li and T. Stathaki, 'Image fusion using dispersion minimisation', in *Proceedings of IEEE International Conference on Acoustic, Sound and Signal Processing*, Toulouse, France, 2006.

[5] D.L. Li, R.M. Mersereau and S. Simske, 'Blur identification based on kurtosis minimization', in *Proceedings of the IEEE International Conference on Image Processing*, Vol. 1, September 2005, pp. 905–908.

[6] A. Papoulis and S. Unnikrishna Pillai, *Probability, Random Variables and Stochastic Processes*, McGraw-Hill, 2002.

[7] J. Yang and R.S. Blum, 'A statistical signal processing approach to image fusion using hidden Markov models', in R. Blum and Z. Liu (eds.), *Multi-Sensor Image Fusion and Its Applications*, Marcel Dekker/CRC, 2005.

[8] N. Mitianoudis and T. Stathaki, 'Joint fusion and blind restoration for multiple image scenarios with missing data', *The Computer Journal*, in press.

[9] Z. Wang and A.C. Bovik, 'A Universal Image Quality Index', *IEEE Signal Processing Letters*, Vol. 9, No. 3, 2002, pp. 81–84.

[10] L. Wald, T. Ranchin and M. Mangolini, 'Fusion of satellite images of different spatial resolution: Assessing the quality of resulting images', *Photogrammetric Engineering and Remote Sensing*, Vol. 63, No. 6, 1997, pp. 691–699.

[11] G. Piella and H. Heijmans, 'A new quality metric for image fusion', in *Proceedings of the IEEE International Conference on Image Processing*, Vol. 3, September 2003, pp. 173–176.

11

Fusion of edge maps using statistical approaches

Stamatia Giannarou and Tania Stathaki

Communications and Signal Processing Group, Imperial College London, London, UK

This work aims at describing a new framework which allows for the quantitative fusion of edge maps that arise from both different preselected edge detectors and multiple image realisations. This work is inspired from the problem that despite the enormous amount of literature on edge detection techniques, there is no single one that performs well in every possible image context. Two approaches are proposed for this purpose. The first one is the so-called Receiver Operating Characteristics (ROC) analysis which is introduced for a sound performance evaluation of the edge maps that emerge from different parameter specifications. In the second one, the Kappa Statistics are employed in a novel fashion to estimate the accuracy of the above edge maps in order to form the optimum final edge image. This method is unique in the sense that the balance between the false detections (False Positives and False Negatives) is explicitly assessed in advance and incorporated in the estimation of the optimum threshold. The results of applying the above two techniques are demonstrated and compared.

11.1 Introduction

Let us have a set of source images describing different realisations of the same true scene. The available images have been acquired from different sensors (multi-sensor scenario) or they are of the same type but exhibit different types of distortion, as for example blurring (multi-focus scenario). Our aim is to integrate the signal information present in the source images in order to produce a single edge map. The composite edge map should contain a more useful description of the edges of the true scene than the one provided by any of the individual sources, and therefore, should be more useful for human visual or machine perception. The problem described in this work can be classified as a joint image fusion and edge detection problem. It is important to stress out that it is very common in many Image Processing, Computer Vision and Pattern Recognition applications to work with the edge map of an image than the image itself. This is because most of the information regarding an object related to its contour and not the characteristics of the inner region of the object.

Edge detection is by far the most common and direct approach for detecting discontinuities that could highlight object-boundary information in a digital image. Edge detection must be efficient and reliable since it is crucial in determining how successful subsequent processing stages will be. In order to fulfil the reliability requirement of edge detection, a great diversity of operators have been devised with differences in their mathematical and algorithmic properties.

Some of the earliest methods such as the Sobel [1] and Roberts [2], are based on the so-called 'Enhancement and Thresholding' approach [3]. According to that method, the image is convolved with small kernels (low-order high pass filters) and the result is thresholded to identify the edge points. Since then, more sophisticated operators have been developed. Marr and Hildreth [4] were the first to introduce the Gaussian smoothing as a pre-processing step in edge feature extraction. Their method detects edges by locating the zero-crossings of the Laplacian (second derivative) of Gaussian of an image. Canny [5] developed an alternative Gaussian edge detector based on optimising three criteria. He employed Gaussian smoothing to reduce noise and the first derivative of the Gaussian to detect edges. Deriche [6] extended Canny's work to derive a recursively implemented edge detector. Rothwell [7] designed a spatially adaptive operator which is able to recover reliable topological information.

An alternative approach to edge detection is the multiresolution one. In such a representation framework, the image is convolved with Gaussian filters of different sizes to produce a set of images at different resolutions. These images are integrated to produce a complete final edge map. Typical algorithms which follow this approach have been produced by Bergholm [8], Lacroix [9] and Schunck [10]. Parametric fitting is another approach used in edge detection. This involves fitting the image with a parametric edge model and then finding the parameters that minimise the fitting error. A detector that belongs to the above category is proposed by Nalwa and Binford [11]. Furthermore, the idea of replicating the human vision performance using mathematical models gave space to the development of feature detection algorithms based on the human visual system. A typical example is the edge detector developed by Peli [12]. Another interesting category of edge detectors is the Logical/Linear operators [13] which combine aspects of linear operators' theory and Boolean algebra.

Intuitively, the question that arises is which edge detector and detector-parameter settings can produce better results. In spite of the aforementioned work, an ideal scheme able to detect and localise edges with precision in many different contexts, has not yet been produced. This is getting even more difficult because of the absence of an evident 'correct edge map' (ground truth), on which the performance of an edge detector could be evaluated. While an edge detector may be robust to noise, it may fail to mark corners and junctions properly. Another common issue with edge detection is the incomplete contour representation. Problems like the above, strongly motivate the development of a general method for combining different edge detection schemes in order to take advantage of their strengths while overcoming their weaknesses.

Let us assume n original detectors, where a detector refers to a mathematical method that attempts to identify the presence (or absence) of an event. In our work we are interested in edge detectors which investigate the presence of edges in a digital image signal. These original detectors are transformed to a new set of detectors where each new detector is a function of all of the original detectors. This function is solely controlled by a parameter named Correspondence Threshold (CT) which will be explained in the main body of the chapter. Each one of the new detectors is associated with a specific value of the CT parameter; this value identifies uniquely the detector. The new detectors vary with respect to their strength, starting from weak detectors that highlight only the strong edges and are basically noise free, to strong detectors that also highlight weak edges and fine detains but exhibit significant amount of noise. In this work we are interested in selecting one of the new edge detectors as the final detection result. We present two novel contributions.

The first novel contribution is based on the use of the so-called Receiver Operating Characteristic (ROC) curve. The only related work was presented in [14]. However, in [14] the original edge maps are generated for different combinations of the parameter values of a singe edge detector and more specifically the Canny edge detector. In this work the original edge maps are different popular edge detectors which although follow similar mathematical techniques, they still produce diverse results.

The second novel contribution is based on the employment of a normalised and corrected edge detection performance statistical metric known as Kappa Statistic. The Kappa Statistic has been used solely in medicine [15]. We are seeking at optimising the Kappa Statistic which, in the specific framework, is a function of the available edge detectors and additionally a scalar parameter which controls the strength of the final detector and consequently the balance between false alarms and misdetections.

The later is the main novelty of this work. It is an important research contribution to the edge detection problem since it allows for the blind combination of multiple detectors and more importantly the pre-specified control of the type of preferred misclassifications.

The chapter is organised as follows. Section 11.2 concerns the brief analysis of a set of popular edge detectors that will be used in this work. Section 11.3 presents two novel approaches for the quantitative combination of multiple edge detectors. Section 11.4 contains experimental results yielded using our implementation of the automatic edge detection algorithms together with a comparative study of the methods' performance. Conclusions are given in Section 11.5.

11.2 Operators implemented for this work

Several approaches to edge detection focus their analysis on the identification of the best differential operator necessary to localise sharp changes of the image intensity. These approaches recognise the necessity of a preliminary filtering step, as a smoothing stage, since differentiation amplifies all high-frequency components of the signal, including those of the textured areas and noise. The most widely used smoothing filter is the Gaussian one which has been shown to play an important role in detecting edges.

Canny's approach [5] is a standard technique in edge detection. This scheme, in substance, identifies edges in the image as the local maxima of the convolution of the image with an 'optimal' operator. The operator's optimality is subject to three performance criteria defined by Canny and is a very close approximation to the first derivative of the Gaussian function $G(x, y)$. For example, the partial derivative with respect to x is defined as:

$$\frac{\theta}{\theta x} G(x, y) = \frac{\theta}{\theta x} \mathrm{e}^{-(x^2+y^2)/2\sigma^2}$$

where σ^2 denotes the variance of the Gaussian filter and controls the degree of smoothing. After this process, candidate edge pixels are identified as the pixels that survive an additional thinning process known as *non-maximal suppression* [16]. Then, the candidate edges are thresholded to keep only the significant ones. Canny suggests *hysteresis thresholding* to eliminate streaking of edge contours.

Using an approach similar to Canny's, Deriche [6] derived a different optimal operator. Contrary to Canny, whose operator is based on a finite antisymmetric filter, Deriche deals with an antisymmetric filter which has an infinite support region defined as:

$$f(x) = -c \cdot \mathrm{e}^{-a|x|} \cdot \sin \omega x$$

where a, c and ω are positive reals. This filter is sharper than the derivative of the Gaussian and is efficiently implemented in a recursive fashion. The procedure that follows in Deriche's method is the same as the one used in Canny's edge detection; non-maximal suppression and hysteresis thresholding is applied as described previously.

Although Canny's detector performs well in localising edges and suppressing noise, yet in several cases it fails to provide a complete boundary in objects. Rothwell's [7] operator is an improvement to earlier edge detectors, capable of recovering sound topological descriptions. It follows a line of work similar to Canny's. The uniqueness of this algorithm originates in the use of a dynamic threshold which varies across the image [17].

In general, it is very difficult to find a single scale of smoothing which is optimal for all the edges in an image. One smoothing scale may keep good localisation while giving detections sensitive to noise. Thus, multiscale edge detection is introduced as an alternative. In this approach, edge detectors with different filter sizes are applied to the image to extract edge maps at different smoothing scales. This information is then combined to result in a more complete final edge image.

Bergholm [8] introduced the coarse-to-fine tracking as an approach to multiscale edge detection. The initial steps of this method are based on Canny's approach. This algorithm relies on the fact that edge detection at a coarse resolution yields significant edges, while their accurate location is detected at a finer resolution. Therefore, the main idea is to initially detect the edges applying a strong Gaussian smoothing and then focus on these edges by tracking them over decreasing smoothing scale.

In [9], Lacroix introduces another algorithm for multiscale detection based on Canny's method. Contrary to Bergholm [8] who proposed the tracking of edges from coarse-to-fine resolution, in Lacroix's method the edge information is combined moving from fine-to-coarse resolution aiming at avoiding the problem of splitting edges. Schunck's work [10] is another study that advocates the use of derivatives of Gaussian filters with different variances to detect intensity changes at different resolution scales. The gradient magnitudes over the selected range of scales are multiplied to amplify significant edges while suppressing the weak ones. Hence, a composite edge image is formed.

In this work we use the six edge detectors mentioned in this section. The use of convolutional methods is justified by the fact that they are simple to implement while producing accurate detection results.

11.3 Automatic edge detection

In this chapter, we intend to throw light on the uncertainty associated with the parametric edge detection performance. The statistical approaches described here attempt to automatically form an optimum edge map, by combining edge images emerged from different detectors.

We begin with the assumption that N different edge detectors will be combined. The first step of the algorithm comprises the correspondence test of the edge images, E_i, for $i = 1, \ldots, N$. A correspondence value is assigned to each pixel and is then stored in a separate array, V, of the same size as the initial image. The correspondence value is the frequency of identifying a pixel as an edge by the set of detectors. Intuitively, the higher the correspondence associated with a pixel, the greater the possibility for that pixel to be a true edge. Hence, the above correspondence value can be used as a reliable measure to distinguish between true and false edges [14].

However, these data require specific statistical methods to assess accuracy of the resulted edge images- accuracy here being the extent to which detected edges agree with true edges. Correspondence values ranging from 0 to N produce $N + 1$ thresholds which correspond to edge detections with different combinations of true positive and false positive rates. The threshold that corresponds to correspondence value 0 is ignored. So, the main goal of the method is to estimate the correspondence threshold CT (from the set CT_i where $i = 1, \ldots, N$) which results in an accurate edge map that gives the finest fit to all edge images E_i. In this section we describe two different approaches for this purpose.

11.3.1 ROC analysis

In our case, the classification task is a binary one including the actual classes $\{e, ne\}$, which stand for the *edge* and *non-edge* event, respectively and the predictive classes, *predicted edge* and *predicted non-edge*, denoted by $\{E, NE\}$. Traditionally, the data obtained by an edge detector are displayed graphically in a 2×2 matrix, the *confusion matrix*, with the notation indicated in Table 11.1.

Table 11.1 *Confusion matrix.*

	e	ne
E	True Positives (TP)	False Positives (FP)
NE	False Negatives (FN)	True Negatives (TN)

In order to mathematically define the conditional probabilities in the confusion matrix we begin by considering an image of size $K \times L$. The probability of a pixel to be a true edge will be denoted as $p_{k,l}$, where $k = 1, \ldots, K$ and $l = 1, \ldots, L$. In a similar way, $q_{k,l}$ will represent the probability of a pixel to be detected as edge. The probability of a True Positive outcome over all the pixels (k, l) of an image is defined as:

$$TP = \mathrm{Mean}(p_{k,l} \cdot q_{k,l})$$

This leads to the following equation:

$$TP = P \cdot Q + \rho \cdot \sigma_p \cdot \sigma_q$$

where σ_p and σ_q stand for the standard deviation of the distribution of $p_{k,l}$ and $q_{k,l}$, respectively. The parameter P represents the *prevalence* of the detection while the parameter Q is the *level* of the detection [18].

The parameter ρ in the above equation denotes the correlation coefficient between $p_{k,l}$ and $q_{k,l}$. A positive correlation coefficient between two random variables indicates that these variables follow the same trend. In our case the random variables of interest are the true edge image and the detected edge image. Therefore, a positive correlation coefficient indicates that if the probability of a pixel $f(x_1, y_1)$ being a true edge is higher compared to the same probability for the pixel $f(x_2, y_2)$, then the probability of the pixel $f(x_1, y_1)$ detected as edge pixel is also higher compared to the same probability for the pixel $f(x_2, y_2)$. In this work we assume that for a legitimate edge detection the correlation coefficient between true and detected edges is positive. This is a realistic assumption since edge detection relies on mathematical methods that exploit the local edge intensity information. In the case of random edge detection where the edges are identified purely by chance, the correlation coefficient is equal to $\rho = 0$. All the probabilities, computed for legitimate and random edge detection, are presented in Table 11.2, where the $'$ symbol denotes the complement operator.

The term *prevalence* refers to the occurrence of true edge pixels in the image whereas the *level* of the diagnosis corresponds to the occurrence of pixels detected as edges. Clearly, the optimum edge detector is the one that identifies as edges all the true edge pixels and therefore satisfies the equality:

$$P = Q \tag{11.1}$$

The conditional probabilities presented in the confusion matrix and in particular certain combinations of them, are used to define basic measurements of detection accuracy.

In our case, the concept of accuracy refers to the quality of information provided by an edge map. Thus, the accuracy assessment is the place to start in the estimation of the

Table 11.2 *Probabilities for legitimate and random edge detection.*

	Legitimate edge detection	Random edge detection
TP	$P \cdot Q + \rho \cdot \sigma_p \cdot \sigma_q = P \cdot SE$	$P \cdot Q$
FP	$P' \cdot Q - \rho \cdot \sigma_p \cdot \sigma_q = P' \cdot SP'$	$P' \cdot Q$
FN	$P \cdot Q' - \rho \cdot \sigma_p \cdot \sigma_q = P \cdot SE'$	$P \cdot Q'$
TN	$P' \cdot Q' + \rho \cdot \sigma_p \cdot \sigma_q = P' \cdot SP$	$P' \cdot Q'$

optimum correspondence threshold. It is basically characterised using the metrics of sensitivity (SE) and specificity (SP) [19]. Both these measures describe the edge detector's ability to correctly identify true edges while it negates the false alarms. Sensitivity (*SE*) corresponds to the probability of identifying a true edge as edge pixel. It is also referred to as True Positive rate and is defined as follows:

$$SE = TP/(TP + FN)$$

or

$$TP_{\text{rate}} = TP/(TP + FN) \tag{11.2}$$

The term specificity (*SP*) expresses to probability of identifying an actual non-edge as non-edge pixel. The measure $1 - SP$ is known as False Positive rate. These measures are given by the equations:

$$SP = TN/(TN + FP)$$

or

$$FP_{\text{rate}} = 1 - TN/(TN + FP) \tag{11.3}$$

Relying on the value of only one of the above metrics for our accuracy estimation would be an oversimplification and will possibly lead to misleading inferences. Based on this idea, the Receiver Operating Characteristics (ROC) analysis [20,21] can be introduced to quantify detection accuracy. In fact, a ROC curve provides a view of all the True Positive/False Positive rate pairs emerged from varying the correspondence over the range of the observed data. In this work, the ROC curve is used to select the correspondence threshold *CT* that would provide an optimum trade-off between the True Positive and the False Positive rate of edge detectors.

In order to calculate the points on the ROC curve, we apply each correspondence threshold CT_i on the correspondence test outcome, i.e., the matrix *V* mentioned above. This means the pixels are classified as edges and non-edges according to whether their correspondence value exceeds a CT_i or not. Thus, we end up with a set of possible best edge maps M_j, for $j = 1, \ldots, N$, corresponding to each CT_i. Every M_j is compared to the set of the initial edge images, E_i, in order to calculate the True Positive, TP_{rate_j}, and the False Positive, FP_{rate_j}, rates associated with each of them. So, according to Equations (11.2) and (11.3), for the M_j map these rates are defined as:

$$TP_{\text{rate}_j} = \frac{\overline{TP}_j}{\overline{TP}_j + \overline{FN}_j} \tag{11.4}$$

$$FP_{\text{rate}_j} = 1 - \frac{\overline{TN}_j}{\overline{FP}_j + \overline{TN}_j} \tag{11.5}$$

where $\overline{TP}_j + \overline{FN}_j$ is the *prevalence*, denoted by P, which represents the average number of true edges in M_j.

Averaging in (11.4)–(11.5) refers to the joint use of multiple edge detectors as shown in the following equations:

$$\overline{TP}_j = \frac{1}{N} \sum_{i=1}^{N} \left(\frac{1}{K \cdot L} \sum_{k=1}^{K} \sum_{l=1}^{L} M_{jE} \cap E_{iE} \right) \tag{11.6}$$

$$\overline{FP}_j = \frac{1}{N} \sum_{i=1}^{N} \left(\frac{1}{K \cdot L} \sum_{k=1}^{K} \sum_{l=1}^{L} M_{jE} \cap E_{iNE} \right) \tag{11.7}$$

$$\overline{TN}_j = \frac{1}{N} \sum_{i=1}^{N} \left(\frac{1}{K \cdot L} \sum_{k=1}^{K} \sum_{l=1}^{L} M_{jNE} \cap E_{iNE} \right) \tag{11.8}$$

$$\overline{FN}_j = \frac{1}{N} \sum_{i=1}^{N} \left(\frac{1}{K \cdot L} \sum_{k=1}^{K} \sum_{l=1}^{L} M_{jNE} \cap E_{iE} \right) \tag{11.9}$$

where M_{jE} and M_{jNE} represent the pixels detected as edges and non-edges in the edge map M_j, respectively. The same notation is used in the case of the edge maps E_i. The variables K and L stand for the initial image dimensions. For instance, the probability measurement in Equation (11.6) indicates the average number of pixels detected as edges in M_j and match with edge pixels in all detections E_i.

Each edge map M_j generates a point $(FP_{\text{rate}_j}, TP_{\text{rate}_j})$ in the ROC plane, forming the ROC curve. The position of these points provides qualitative information about the detection accuracy of each edge map. As we mentioned in Equation (11.1), the optimum CT should correspond to a detection that gives *prevalence* value P equal to its *level Q*. By definition of True Positive and False Positive rate, the following is valid:

$$P' \cdot FP_{\text{rate}} + P \cdot TP_{\text{rate}} = Q$$

The above equation in conjunction with (11.1) leads to the following mathematical expression, that the optimum edge detection should satisfy:

$$P' \cdot FP_{\text{rate}} + P \cdot TP_{\text{rate}} = P \tag{11.10}$$

Equation (11.10) defines a line that connects the points $(0, 1)$ and (P, P) in the ROC plane, known as *diagnosis line*. Therefore, the optimum CT occurs at the intersection (or close to that) of the ROC curve and the diagnosis line. The value of the selected CT determines how detailed the final edge image, EGT, will be. In the case of a noisy environment there should be a trade-off between an increase in information and the decrease in noise.

11.3.2 Weighted Kappa Coefficient

In edge detection, it is prudent to consider the relative seriousness of each possible disagreement between true and detected edges when performing accuracy evaluation. This section is confined to the examination of an accuracy measure which is based on the acknowledgement that in detecting edges, depending on the specific application, the consequences of a False Positive may be quite different from the consequences of a False Negative. For this purpose, the *Weighted Kappa Coefficient* [15,22] is introduced for the estimation of the correspondence threshold that results in an optimum final edge map.

Consider a mathematical measure A_0 of agreement between the outcomes of two algorithms that both attempt to solve the problem of detection of the presence or absence of a condition. Let A_c be the value expected on the basis of agreement by chance alone and A_a the value expected on the basis of complete agreement, i.e., $A_a = \max\{A_0\}$. Based on the above definitions, the Kappa Coefficient defined below is introduced as a *corrected* and *normalised* measure of agreement [23]:

$$k = \frac{A_0 - A_c}{A_a - A_c} \tag{11.11}$$

In the problem of edge detection A_0 may be defined as a measure of agreement between true and detected edges. The definition of A_c and A_a is obvious.

A generalisation of the above coefficient can be made to incorporate the relative cost of False Positives and False Negatives into our accuracy measure. We assume that weights $w_{u,v}$, for $u = 1, 2$ and $v = 1, 2$, are assigned to the four possible outcomes of the edge detection process displayed in the confusion matrix. The observed weighted proportion of agreement is given as:

$$D_{0_w} = \sum_{u=1}^{2} \sum_{v=1}^{2} w_{u,v} d_{u,v} \tag{11.12}$$

where $d_{u,v}$ indicates the probabilities in the confusion matrix. Similarly, the chance-expected weighted proportion of agreement has the form:

$$D_{c_w} = \sum_{u=1}^{2} \sum_{v=1}^{2} w_{u,v} c_{u,v} \tag{11.13}$$

where $c_{u,v}$ refers to the above four probabilities but in the case of random edge detection, i.e., the edges are identified purely by chance. Both these proportions are calculated as shown in Table 11.3. Based on the definition of Kappa Coefficient described previously, Weighted Kappa is then given by:

$$k_w = \frac{D_{0_w} - D_{c_w}}{\max(D_{0_w} - D_{c_w})} \tag{11.14}$$

Table 11.3

	Legitimate edge detection	Random edge detection
TP	$d_{1,1} = P \cdot SE$	$c_{1,1} = P \cdot Q$
FP	$d_{1,2} = P' \cdot SP'$	$c_{1,2} = P' \cdot Q$
FN	$d_{2,1} = P \cdot SE'$	$c_{2,1} = P \cdot Q'$
TN	$d_{2,2} = P' \cdot SP$	$c_{2,2} = P' \cdot Q'$

Substituting (11.12) and (11.13) in (11.14) gives:

$$k_w = \frac{w_{1,1} \cdot P \cdot SE + w_{1,2} \cdot P' \cdot SP' + w_{2,1} \cdot P \cdot SE' + w_{2,2} \cdot P' \cdot SP}{\max(D_{0_w} - D_{c_w})}$$

$$- \frac{(w_{1,1} \cdot P \cdot Q + w_{1,2} \cdot P' \cdot Q + w_{2,1} \cdot P \cdot Q' + w_{2,2} \cdot P' \cdot Q')}{\max(D_{0_w} - D_{c_w})}$$

or

$$k_w = \frac{w_{1,1} \cdot P \cdot (SE - Q) + w_{1,2} \cdot P' \cdot (SP' - Q) + w_{2,1} \cdot P \cdot (SE' - Q')}{\max(D_{0_w} - D_{c_w})}$$

$$+ \frac{w_{2,2} \cdot P' \cdot (SP - Q')}{\max(D_{0_w} - D_{c_w})}$$

so

$$k_w = \frac{w_{1,1} \cdot P \cdot Q' \cdot k(1,0) + w_{1,2} \cdot P' \cdot (SP' - Q) + w_{2,1} \cdot P \cdot (SE' - Q')}{\max(D_{0_w} - D_{c_w})}$$

$$+ \frac{w_{2,2} \cdot P' \cdot Q \cdot k(0,0)}{\max(D_{0_w} - D_{c_w})} \tag{11.15}$$

where P', Q' are the complements of P and Q, respectively. $k(1,0)$ and $k(0,0)$ are the quality indices of sensitivity and specificity, respectively, defined as:

$$k(1,0) = \frac{SE - Q}{Q'} \quad \text{and} \quad k(0,0) = \frac{SP - Q'}{Q}$$

The major source of confusion in statistical methods related to the Weighted Kappa Coefficient is the assignment of weights. In the method analysed here the weights indicate gain or cost and they lie in the interval $0 \leqslant |w_{u,v}| \leqslant 1$. From (11.15) it can be deduced that the total cost W_1 for true edges being properly identified as edges or not, is equal to:

$$W_1 = |w_{1,1}| + |w_{2,1}|$$

Similarly, the total cost W_2 for the non-edge pixels is defined as:

$$W_2 = |w_{1,2}| + |w_{2,2}|$$

We propose that true detections should be assigned positive weights representing gain whereas, the weights for false detections should be negative, representing loss. It can be proven that no matter how the split of these total costs is made between true and false outcomes, the result of the method is not affected [18]. Hence, for the sake of convenience the total costs are split evenly. As a result, we end up with two different weights instead of four:

$$k_w = \frac{\frac{W_1}{2} \cdot P \cdot Q' \cdot k(1,0) + \left(-\frac{W_2}{2}\right) \cdot P' \cdot (SP' - Q)}{\max(D_{0_w} - D_{c_w})}$$

$$+ \frac{\left(-\frac{W_1}{2}\right) \cdot P \cdot (SE' - Q') + \frac{W_2}{2} \cdot P' \cdot Q \cdot k(0,0)}{\max(D_{0_w} - D_{c_w})}$$

A further simplification leads to:

$$k_w = \frac{W_1 \cdot P \cdot Q' \cdot k(1,0) + W_2 \cdot P' \cdot Q \cdot k(0,0)}{\max(D_{0_w} - D_{c_w})} \tag{11.16}$$

Taking into account the fact that the maximum value of the quality indices $k(1,0)$ and $k(0,0)$ is equal to 1, the denominator in (11.16) takes the form: $W_1 \cdot P \cdot Q' + W_2 \cdot P' \cdot Q$. Dividing both numerator and denominator by $W_1 + W_2$, the final expression of the Weighted Kappa Coefficient, in accordance with the quality indices of sensitivity and specificity, becomes:

$$k(r,0) = \frac{r \cdot P \cdot Q' \cdot k(1,0) + r' \cdot P' \cdot Q \cdot k(0,0)}{r \cdot P \cdot Q' + r' \cdot P' \cdot Q} \tag{11.17}$$

where

$$r = \frac{W_1}{W_1 + W_2} \tag{11.18}$$

and r' is the complement of r. The Weighted Kappa Coefficient $k(r,0)$ indicates the quality of the detection as a function of r. It is unique in the sense that the balance between the false detections is determined in advance and then is incorporated in the measure.

The index r is indicative of the relative importance of False Negatives to False Positives. Its value is dictated by which error carries the greatest importance and ranges from 0 to 1. If we focus on the elimination of False Positives in edge detection, W_2 will predominate in (11.16) and consequently r will be close to 0 as it can be seen from (11.18). On the other hand, a choice of r close to 1 signifies our interest in avoiding False Negatives since W_1 will predominate in (11.16). A value of $r = 1/2$ reflects the idea that both False Positives and False Negatives are equally unwanted. No standard choice of r can be regarded as optimum since the balance between the two errors shifts according to the application.

Thus, for a selected value of r, the Weighted Kappa Coefficient $k_j(r,0)$ is calculated for each edge map as it is given in (11.17). The optimum CT is the one that maximises the Weighted Kappa Coefficient.

11.3.3 Geometric approach for the Weighted Kappa Coefficient

The estimation of the Weighted Kappa Coefficient $k_j(r, 0)$ can also be done geometrically. Every edge map M_j, for $j = 1, \ldots, N$, can be represented as a point $(k_j(0, 0), k_j(1, 0))$ on a two-dimensional graph with coordinates $(k(0, 0), k(1, 0))$. The set of points $(k_j(0, 0), k_j(1, 0))$, $j = 1, \ldots, N$, consist the so-called Quality Receiver Operating Characteristic (QROC) curve. A great deal of information is available from visual examination of such a geometric presentation. Equation (11.17) for the jth edge map can be rewritten in the form:

$$\frac{k_j(r, 0) - k_j(1, 0)}{k_j(r, 0) - k_j(0, 0)} = -\frac{P' \cdot Q \cdot r'}{P \cdot Q' \cdot r} \tag{11.19}$$

Therefore, if we consider the straight line on the QROC-plane described by the equation:

$$k_j(r, 0) - k(1, 0) = -\frac{P' \cdot Q \cdot r'}{P \cdot Q' \cdot r}(k_j(r, 0) - k(0, 0)) \tag{11.20}$$

it is obvious from Equations (11.19) and (11.20) that the point $(k_j(0, 0), k_j(1, 0))$ lies on this line. This is called the *r-projection line* and its slope is:

$$s = -\frac{P' \cdot Q \cdot r'}{P \cdot Q' \cdot r} \tag{11.21}$$

It is obvious that the point $(k_j(r, 0), k_j(r, 0))$ lies also on the *r-projection line* and also on the main diagonal described by the equation $k(0, 0) = k(1, 0)$.

This means that the Weighted Kappa Coefficient $k_j(r, 0)$ can be calculated graphically by drawing a line, for any value r of interest, through the point $(k_j(0, 0), k_j(1, 0))$ with slope given by (11.21). The intersection point, $(k_j(r, 0), k_j(r, 0))$, of this line with the major diagonal in the QROC plane is clearly indicative of the $k_j(r, 0)$ value. Figure 11.1 presents an example for the calculation of the Weighted Kappa Coefficient for a test point for $r = 0.5$. The procedure is repeated for every CT_i to generate N different intersection points. The closer the intersection point to the upper right corner (ideal point), the higher the value of the Weighted Kappa Coefficient. Hence, the optimum correspondence threshold is the one that produces an intersection point closer to the point $(1, 1)$ in the QROC plane.

11.3.4 An alternative to the selection of the r parameter value

In the previous section, parameter r is evaluated according to Equation (11.18). By assigning more weight to the false detection we want to eliminate, the ratio in (11.18) yields the appropriate value of r. However, a more efficient analysis is necessary. An alternative analysis that justifies the previously described selection of r is presented in this section.

Our main concern is to examine the behaviour of the quality measure $k(r, 0)$ as a function of the level, Q, and the parameter r. Substituting in (11.17) the probabilities given in

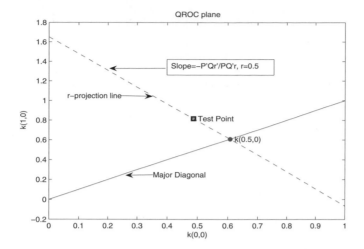

Figure 11.1 *Calculation of $k(0.5, 0)$ using a graphical approach on the QROC plane.*

Table 11.2, the Weighted Kappa Coefficient can be expressed as:

$$k(r, 0) = \frac{r \cdot P \cdot Q' \cdot k(1, 0) + r' \cdot P' \cdot Q \cdot k(0, 0)}{r \cdot P \cdot Q' + r' \cdot P' \cdot Q}$$

$$= \frac{r \cdot P \cdot (SE - Q) + r' \cdot P' \cdot (SP - Q')}{r \cdot P \cdot Q' + r' \cdot P' \cdot Q}$$

$$= \frac{r \cdot P \cdot (\rho \sigma_p \sigma_q / P) + r' \cdot P' \cdot (\rho \sigma_p \sigma_q / P')}{r \cdot P \cdot Q' + r' \cdot P' \cdot Q}$$

Thus, the quality measure, $k(r, 0)$, takes the form:

$$k(r, 0) = \frac{\rho \cdot \sigma_p \cdot \sigma_q}{r \cdot P \cdot Q' + r' \cdot P' \cdot Q}$$

The derivative of the Weighted Kappa Coefficient with respect to r is given by:

$$\frac{\mathrm{d}}{\mathrm{d}r} k(r, 0) = \rho \sigma_p \sigma_q \cdot \frac{Q - P}{(r \cdot P \cdot Q' + r' \cdot P' \cdot Q)^2} \tag{11.22}$$

The measures σ_p, σ_q are positive as they express standard deviations. The correlation coefficient, ρ, is positive, as well. Thus, it becomes obvious that the sign of the derivative, $(\mathrm{d}/\mathrm{d}r)k(r, 0)$, is determined by the value of Q relative to P.

A level, Q, greater than the prevalence, P corresponds to an edge detection that eliminates the misdetections by favouring the False Positives. In this case, according to (11.22), the derivative of the Weighted Kappa Coefficient is positive for any value of r and the quality measure $k(r, 0)$ is an increasing function of r. This means in applications where we are more interested in the elimination of False Negatives, a higher value of r in the interval [0, 1] will result in the selection of a more accurate edge map.

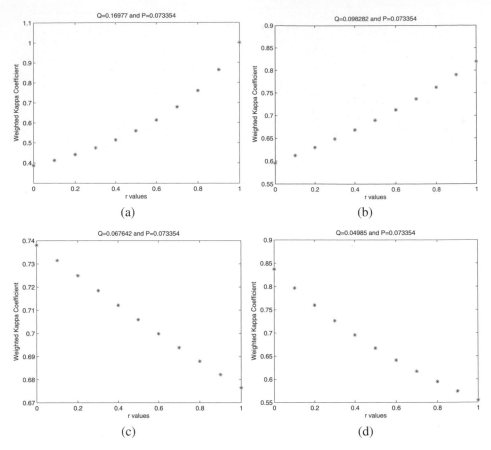

Figure 11.2 *Weighted Kappa Coefficient plots for edge maps that correspond to (a) CT = 1; (b) CT = 2; (c) CT = 3; (d) CT = 4.*

Equivalent conclusions are derived for the elimination of False Positives i.e. detections where the level is smaller than the prevalence. According to (11.22) the derivative, $(\mathrm{d}/\mathrm{d}r)k(r, 0)$, will be negative and the Weighted Kappa Coefficient will be a decreasing function of r. Therefore, small values of r in the interval $[0, 1]$ will yield CTs that correspond to more accurate edge maps.

The above conclusions are also verified experimentally. Figure 11.2 illustrates the values of the Weighted Kappa Coefficient as a function of r for four edge maps at different CTs. These plots are yielded from applying the Weighted Kappa Coefficient method on the Cameraman image to combine 6 edge detectors. Figures 11.2(a) and 11.2(b) correspond to CTs equal to 1 and 2, respectively, where the level values are greater than the prevalence. Observing these curves, it is clear that the Weighted Kappa Coefficient is an increasing function of r and high quality is achieved for values of r close to 1. On the contrary, as shown in Figures 11.2(c) and 11.2(d), the quality measure is a decreasing function of r for $CT = 3$ and $CT = 4$, where the level is smaller than the prevalence. In this case values of r close to 0 give edge maps with better quality.

(a)   (b)

Figure 11.3   *'Terrain' image: (a) Input 1; (b) Input 2.*

11.4 Experimental results and discussion

Using the framework developed, six edge detectors, proposed by Canny, Deriche, Bergholm, Lacroix, Schunck and Rothwell were combined to produce the optimum edge maps. Two sets of experiments are demonstrated in this section. The selection of the above mentioned edge detectors relies on the fact that they basically follow the same mathematical approach.

The performance of the proposed edge map fusion approach is demonstrated by present-ing two experimental results for fusing the edge maps of pairs of images which represent realisations of the same scene. The first image set is the 'Terrain' presented in Figure 11.3. It can be seen that although the two image realisations exhibit the same visual content, different areas are highlighted in each image and most importantly the human figure present in Figure 11.3(b) is missing in Figure 11.3(a). Therefore, the proposed edge map fusion approaches are appropriate in facilitating a blind object recognition task for the particular data set. The second image set is the 'Brain' presented in Figure 11.8. Again different areas and edges are highlighted in each image.

Specifying the value of the edge detection operators' input parameters was a crucial step. In fact, the parameter selection depends on the implementation and intends to maximise the quality of the detection. In our work, we were consistent with the parameters proposed by the authors of the selected detectors. The Bergholm algorithm was slightly modified by adding hysteresis thresholding to allow a more detailed result. In Lacroix technique we applied non-maximal suppression by keeping the size $k \times 1$ of the relative window fixed at 3×1 [9]. For simplicity, in the case of Schunck edge detection the non-maximal sup-pression method we used is the one proposed by Canny in [5] and hysteresis thresholding was applied for a more efficient thresholding.

For our experimental results, the standard deviation (*sigma*) of the Gaussian filter in Canny's algorithm [5] was set to *sigma* = 1, whereas, the *low* and *high thresholds* were automatically calculated by the image histogram. In Deriche's technique [6], the parameters' values were set to $a = 2$ and $w = 1.5$. The Bergholm [8] parameter set was a combination of *starting sigma*, *ending sigma* and *low* and *high threshold* and these where *starting sigma* $= 3.5$, *ending sigma* $= 0.7$ and the thresholds were automatically determined as previously. For the *Primary Rater* in Lacroix's method [9], the coarsest resolution was set to $\sigma_2 = 2$ and the finest one to $\sigma_0 = 0.7$. The intermediate scale σ_1 was computed according to the expression proposed in [9]. The gradient and homogeneity thresholds were estimated by the histogram of the gradient and homogeneity images, respectively. For the Schunck edge detector [10], the number of resolution scales was arbitrarily set to three as: $\sigma_1 = 0.7$, $\sigma_2 = 1.2$, $\sigma_3 = 1.7$. The difference between two consecutive scales was selected not to be greater than 0.5 in order to avoid edge pixel displacement in the resulted edge maps. The values for the *low* and *high thresholds* were calculated by the histogram of the gradient magnitude image. In the case of Rothwell method [7], the *alpha* parameter was set to 0.9, the *low threshold* was estimated by the image histogram again and the value of the smoothing parameter, *sigma*, was equal to 1. It is important to stress out that the selected values for all of the above parameters fall within the ranges proposed in the literature by the authors of the individual detectors.

In the approach to the estimation of the optimum correspondence threshold based on the maximisation of the '*Weighted Kappa Coefficient*,' the cost, r, is initially determined according to the particular quality of the detection (FP or FN) that is chosen to be optimised. For example as far as target object detection in military applications is concerned, missing existing targets in the image (misdetections) is less desirable than falsely detecting non-existing ones (false alarms). This is as well the scenario we assume in this piece of work, namely, we are primarily concerned with the elimination of FN at the expense of increasing the number of FP. Therefore, according to the analysis in the previous section, the cost value r should range from 0.5 to 1. Moreover, a trade-off between the increase in edge information and the decrease in noise in the final edge image is necessary when selecting the value of r.

In Figures 11.4–11.7 are shown the results of applying the '*ROC Analysis*' and the '*Weighted Kappa Coefficient*' approaches on the 'Terrain' image. The sample space, E_i $(i = 1, \ldots, 12)$, consisted of the edge detection outcomes produced by the six selected operators applied in the two image realisations of the set and is depicted in Figures 11.4 and 11.5. The probabilities given by Equations (11.4) and (11.5) were calculated for the statistical correspondence test of the edge detections E_i. The ROC curve implementation is illustrated in Figure 11.6(a). For the selected images, it can be observed that the intersection of the diagnosis line with the ROC curve occurs at a correspondence level close to 4. Thus, the optimum threshold is $CT = 4$. The final edge map, EGT, for this approach is presented in Figure 11.7(a). The experimental results when applying the '*Weighted Kappa Coefficient*' approach correspond to a value of r equal to 0.65. For this value of r, the calculation of the Weighted Kappa Coefficients yields $k_j(r, 0) = [0.262, 0.377, 0.386, 0.323, 0.236, 0.143, 0.061, 0.029, 0.0138, 0.007, 0.004, 0.001]$. Observing these results, it is clear that the Weighted Kappa Coefficient takes its maximum value at $k_3(r, 0)$. Thus, the optimum CT is equal to 3. The graphical estimation of $k_j(r, 0)$ for each CT is illustrated in Figure 11.6(b). The final edge map for this approach is presented in Figure 11.7(b).

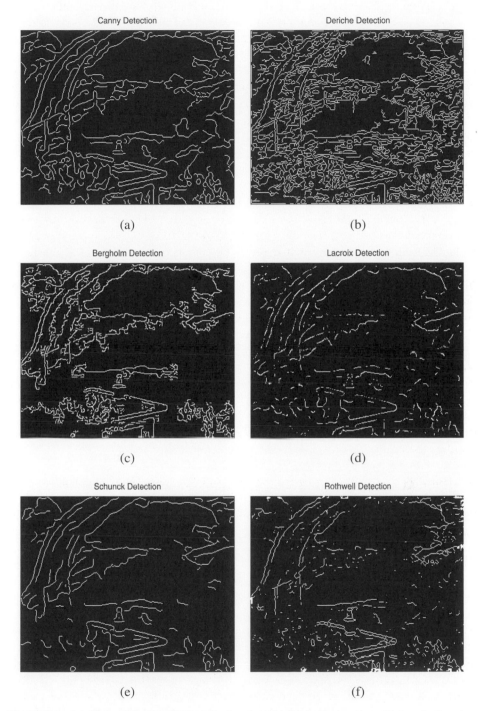

Figure 11.4 *(a) Canny detection; (b) Deriche detection; (c) Bergholm detection; (d) Lacroix detection; (e) Schunck detection; (f) Rothwell detection forming the sample set for the first 'Terrain' image.*

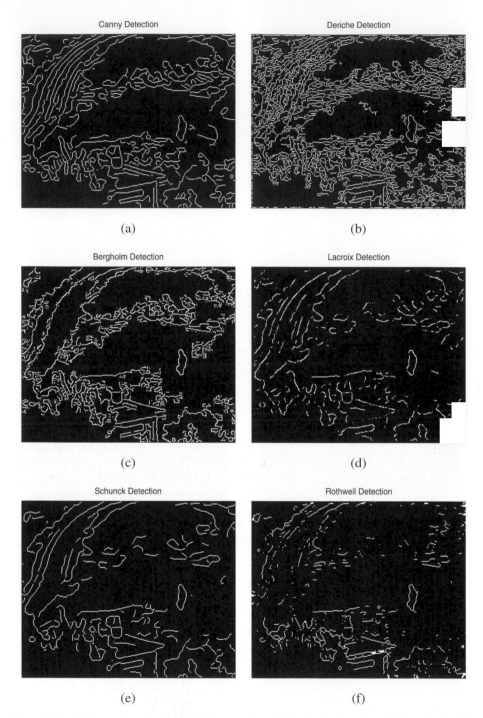

Figure 11.5 *(a) Canny detection; (b) Deriche detection; (c) Bergholm detection; (d) Lacroix detection; (e) Schunck detection; (f) Rothwell detection forming the sample set for the second 'Terrain' image.*

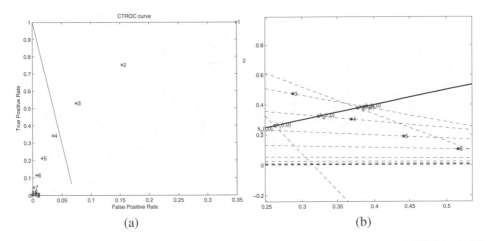

Figure 11.6 *Graphical estimation of the optimum CT for the 'Terrain' image applying (a) 'ROC Analysis' and (b) the 'Weighted Kappa Coefficient' approach with r = 0.65.*

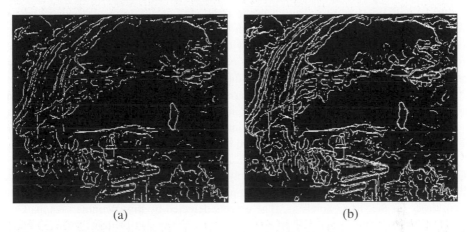

Figure 11.7 *Final edge map for the 'Terrain' images when applying (a) the 'ROC Analysis' and (b) the 'Weighted Kappa Coefficient' approach with r = 0.65.*

The performance of the proposed methods was also examined by fusing the two 'Brain' images shown in Figure 11.8. The 12 edge map results yielded from the six operators are illustrated in Figures 11.9 and 11.10. Applying the 'ROC' method, the correspondence threshold was estimated to be equal to 4. The final edge map for this approach is presented in Figure 11.12(a). The 'Weighted Kappa Coefficient' method was also applied for a value of r equal to 0.65 and the calculation of the Weighted Kappa Coefficients yields $k_j(r, 0) = [0.356, 0.447, 0.456, 0.388, 0.297, 0.189, 0.035, 0.019, 0.011, 0.005, 0.002, 0]$. We observe that the Weighted Kappa Coefficient takes its maximum value at $k_3(0.65, 0)$ and therefore, the optimum CT is equal to 3. The graphical estimation of the optimum CT for the 'ROC Analysis' and the 'Weighted Kappa Coefficient' method are illustrated in Figure 11.11. The final edge map is shown in Figure 11.12(b).

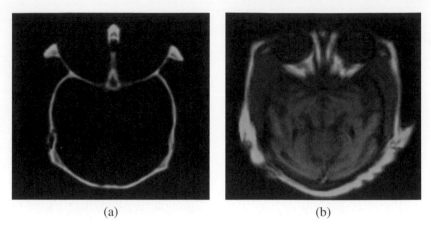

(a)   (b)

Figure 11.8   *'Brain' image: (a) Input 1; (b) Input 2.*

The above examples emphasise the ability of the proposed approaches to combine high accuracy with good noise reduction in the final edge detection result. Insignificant information is cleared, while the information preserved allows for easy, fast and accurate object recognition. This is particularly obvious in the areas of the path and the house's roof in the 'Terrain' image set and in the centre of the scalp in the 'Brain' image set, when comparing the final edge maps with the initial set of edge detections. Furthermore, it is interesting to note that objects which are not present in one of the source images are included in the final edge images, as for example the human figure in the 'Terrain' image set and the detailed sub-regions within the inner part of the brain in the 'Brain' image set. Finally, edges due to texture are suppressed in the final edge maps.

Comparing the edge maps produced by applying the above two approaches it is observed that the edge maps for the 'Weighted Kappa Coefficient' approach have better quality than those for the 'ROC Analysis.' The objects detected by the 'Weighted Kappa Coefficient' approach are better defined regarding their shape and contour; for example the background in the 'Terrain' image set. Furthermore, the number of detected edges in the same approach is greater. This is expected since the selected value of r is 0.65. Obviously, in Figure 11.12(b) the contour of the scalp and inner regions of the brain is clearly distinguishable whereas in Figure 11.12(a) the later is hardly detected. The same observation is made for the contour of the pathway in the case of the 'Terrain' image set. In addition, objects on the foreground of the 'Terrain' image set, such as the shape of the house, are more complete in Figure 11.7(b) rather than in Figure 11.7(a). Nevertheless, the performance of the 'Weighted Kappa Coefficient' approach for the particular choice of r, seems to be superior to 'ROC Analysis' since it is more sensitive to minor details. This is clearly noticeable on the background area of the 'Terrain' image set.

The computational cost of the proposed methods is obviously higher compared to that of applying each edge detector individually. However, this is acceptable since the goal is to form a more complete final edge image by fusing the pre-selected edge detectors in order to take advantage of their strengths while overcoming their weaknesses.

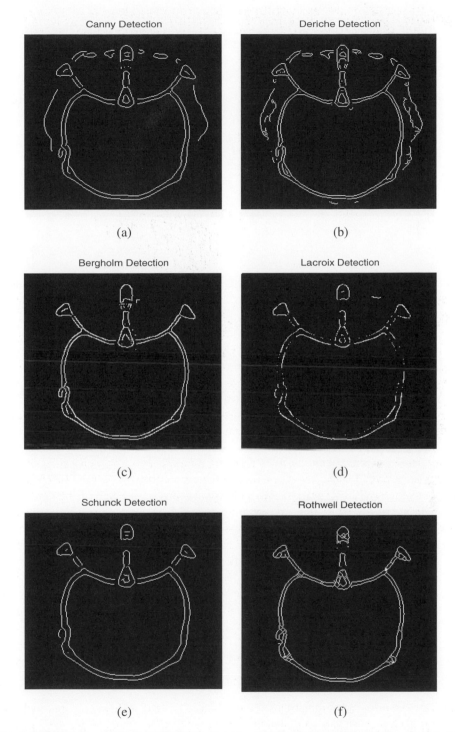

Figure 11.9 *(a) Canny detection; (b) Deriche detection; (c) Bergholm detection; (d) Lacroix detection; (e) Schunck detection; (f) Rothwell detection forming the sample set for the first 'Brain' image.*

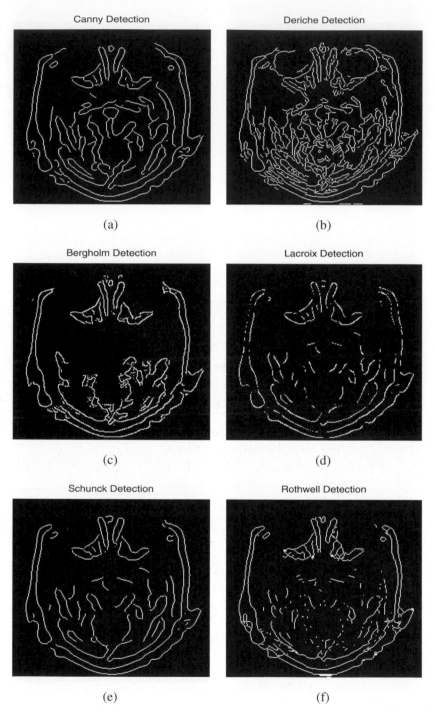

Figure 11.10   *(a) Canny detection; (b) Deriche detection; (c) Bergholm detection; (d) Lacroix detection; (e) Schunck detection; (f) Rothwell detection forming the sample set for the second 'Brain' image.*

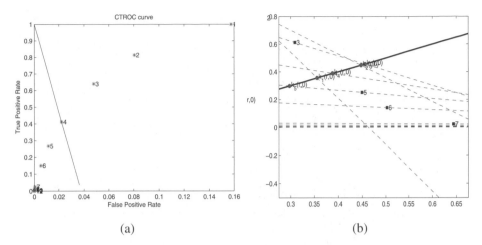

(a)   (b)

Figure 11.11 *Graphical estimation of the optimum CT for the 'Brain' image applying (a) the 'ROC Analysis' and (b) the 'Weighted Kappa Coefficient' approach with r = 0.65.*

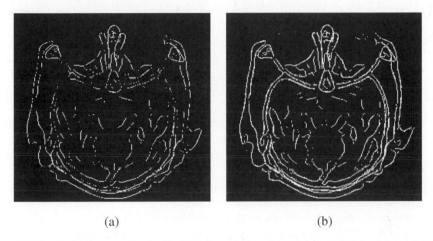

(a)   (b)

Figure 11.12 *Final edge map for the 'Brain' image when applying (a) the 'ROC Analysis' and (b) the 'Weighted Kappa Coefficient' approach with r = 0.65.*

The above conclusions arise after a large number of experimental results involving different types of images.

11.5 Conclusions

The selection of an edge detector operator is not a trivial problem, since different edge detectors often produce essentially varying edge maps, even if they follow similar mathematical approaches. In this chapter we propose two techniques for the automatic statistical analysis of the correspondence of edge images that have emerged from different operators and also different realisations of the same true scene; the ROC analysis and the Weighted Kappa Coefficient method. Both techniques integrate efficiently the preselected set of edge detectors in terms of both the quality of the highlighted features and the elimination of noise and texture. However, the Weighted Kappa Coefficient approach can be considered superior in the sense that the trade off between detection of minor edges and noise reduction can be quantified in advance as part of the problem specifications. The conclusions of this piece of work arise from a large number of experiments that involve different types of images.

References

[1] J. Matthews, 'An introduction to edge detection: The Sobel edge detector', available at http://www.generation5.org/content/2002/im01.asp, 2002.

[2] L.G. Roberts, *Machine Perception of 3-D Solids. Optical and Electro-Optical Information Processing*, MIT Press, 1965.

[3] I.E. Abdou and W.K. Pratt, 'Quantitative design and evaluation enhancement/thresholding edge detectors', *Proceedings of the IEEE*, Vol. 67, 1979, pp. 753–763.

[4] D. Marr and E.C. Hildreth, 'Theory of edge detection', *Proceedings of the Royal Society of London, Series B*, Vol. 207, 1980, pp. 187–217.

[5] J.F. Canny, 'A computational approach to edge detection', *IEEE Transactions on Pattern Analysis and Machine Intelligence*, Vol. 8, No. 6, 1986, pp. 679–698.

[6] R. Deriche, 'Using Canny's criteria to derive a recursive implemented optimal edge detector', *International Journal of Computer Vision*, Vol. 1, No. 2, 1987, pp. 167–187.

[7] C.A. Rothwell, J.L. Mundy, W. Hoffman and V.D. Nguyen, 'Driving vision by topology', in *Int. Symp. on Computer Vision*, Coral Gables, FL, 1995, pp. 395–400.

[8] F. Bergholm, 'Edge focusing', *IEEE Transactions on Pattern Analysis and Machine Intelligence*, Vol. 9, No. 6, 1995, pp. 726–741.

[9] V. Lacroix, 'The primary raster: A multiresolution image description', in *Proc. 10th International Conference on Pattern Recognition*, Vol. 1, 1990, pp. 903–907.

[10] B. Schunck, 'Edge detection with Gaussian filters at multiple scales', in *Proc. IEEE Computer Society Workshop on Computer Vision*, 1987, pp. 208–210.

[11] V.S. Nalwa and T.O. Binford, 'On detecting edges', *IEEE Transactions on Pattern Analysis and Machine Intelligence*, Vol. 8, No. 6, 1986, pp. 699–714.

[12] E. Peli, 'Feature detection algorithm based on visual system models', *Proceedings of the IEEE*, Vol. 90, 2002, pp. 78–93.

[13] L.A. Iverson and S.W. Zucker, 'Logical/linear operators for image curves', *IEEE Transactions on Pattern Analysis and Machine Intelligence*, Vol. 17, No. 10, 1995, pp. 982–996.

[14] Y. Yitzhaky and E. Peli, 'A method for objective edge detection evaluation and detector parameter selection', *IEEE Transactions on Image Processing*, Vol. 25, No. 8, 2003, pp. 1027–1033.

[15] H. Kraemer, V. Periyakoil and A. Noda, 'Tutorial in biostatistics: Kappa coefficients in medical research', *Statistics in Medicine*, Vol. 21, No. 14, 2002, pp. 2109–2129.

[16] J. Canny, 'Finding edges and lines in images', Master's thesis, MIT, 1983.

[17] C. Rothwell, J. Mundy, W. Hoffman and V. Nguyen, 'Driving vision by topology', Technical Report 2444, INRIA, 1994.

[18] H. Kraemer, *Evaluating Medical Test: Objective and Quantitative Guidelines*, Saga Publications, Newbury Park, CA, 1992.

[19] B. Kirkwood and J.A. Sterne, *Essential Medical Statistics*, Blackwell Science, Oxford, 2003.

[20] T. Fawcett, *ROC Graphs: Notes and Practical Considerations for Data Mining Researchers*, Knowledge Discovery and Data Mining, 2003.

[21] M. Zweig and G. Campbell, 'Receiver-operating characteristic (ROC) plots: A fundamental evaluation tool in clinical medicine', *American Association for Clinical Chemistry*, Vol. 39, No. 4, 1993, pp. 561–577.

[22] J. Fleiss, *Statistical Methods for Rates and Proportions*, Wiley Series in Probability and Mathematical Statistics, 1981.

[23] J. Sim and C. Wright, 'The kappa statistic in reliability studies: Use, interpretation, and sample size requirements', *Journal of the American Physical Therapy*, Vol. 85, No. 3, 2005, pp. 257–268.

12

Enhancement of multiple sensor images using joint image fusion and blind restoration

Nikolaos Mitianoudis and Tania Stathaki

Communications and Signal Processing Group, Imperial College London, London, UK

Image fusion systems aim at transferring 'interesting' information from the input sensor images to the fused image. The common assumption for most fusion approaches is the existence of a high-quality reference image signal for all image parts in all input sensor images. In the case that there are common degraded areas in at least one of the input images, the fusion algorithms cannot improve the information provided there, but simply convey a combination of this degraded information to the output. In this study, the authors propose a combined spatial-domain method of fusion and restoration in order to identify these common degraded areas in the fused image and use a regularised restoration approach to enhance the content in these areas. The proposed approach was tested on both multi-focus and multi-modal image sets and produced interesting results.

12.1 Introduction

Data fusion is defined as the process of combining data from sensors and related information from several databases, so that the performance of the system can be improved, while the accuracy of the results can be also increased. Essentially, fusion is a procedure of incorporating essential information from several sensors to a composite result that will be more comprehensive and thus more useful for a human operator or other computer vision tasks.

Image fusion can be similarly viewed as the process of combining information in the form of images, obtained from various sources in order to construct an artificial image that contains all 'useful' information that exists in the input images. Each image has been acquired using different sensor modalities or capture techniques, and therefore, it has different features, such as type of degradation, thermal and visual characteristics. The main concept behind all image fusion algorithms is to *detect strong salient features* in the input sensor images and *fuse* these details to the synthetic image. The resulting synthetic image is usually referred to as the *fused image*.

Let $x_1(\mathbf{r}), \ldots, x_T(\mathbf{r})$ represent T images of size $M_1 \times M_2$ capturing the same scene, where $\mathbf{r} = (i, j)$ refers to pixel coordinates (i, j) in the image. Each image has been acquired using different sensors that are placed relatively close and are observing the same scene. Ideally, the images acquired by these sensors should be similar. However, there might exist some miscorrespondence between several points of the observed scene, due to the different sensor viewpoints. *Image registration* is the process of establishing point-by-point correspondence between a number of images, describing the same scene. In this study, the input images are assumed to have negligible registration problems or the transformation matrix between the sensors' viewpoints is known. Thus, the objects in all images can be considered geometrically aligned.

As already mentioned, the process of combining the important features from the original T images to form a single enhanced image $y(\mathbf{r})$ is usually referred to as *image fusion*. Fusion techniques can be divided into *spatial domain* and *transform domain* techniques [1]. In spatial domain techniques, the input images are fused in the spatial domain, i.e. using localised spatial features. Assuming that $g(\cdot)$ represents the 'fusion rule,' i.e. the method that combines features from the input images, the spatial domain techniques can be summarised, as follows:

$$y(\mathbf{r}) = g\big(x_1(\mathbf{r}), \ldots, x_T(\mathbf{r})\big) \qquad (12.1)$$

Moving to a transform domain enables the use of a framework, where the image's salient features are more clearly depicted than in the spatial domain. Let $\mathcal{T}\{\cdot\}$ represent a transform operator and $g(\cdot)$ the applied fusion rule. Transform-domain fusion techniques can then be outlined, as follows:

$$y(\mathbf{r}) = \mathcal{T}^{-1}\big\{g\big(\mathcal{T}\{x_1(\mathbf{r})\}, \ldots, \mathcal{T}\{x_T(\mathbf{r})\}\big)\big\} \qquad (12.2)$$

Several transformations were proposed to be used for image fusion, including the *Dual-Tree Wavelet Transform* [1–3], *Pyramid Decomposition* [4] and image-trained Independent Component Analysis bases [5,6]. All these transformations project the input images onto localised bases, modelling sharp and abrupt transitions (edges) and therefore, describe the image using a more meaningful representation that can be used to detect and emphasise salient features, important for performing the task of image fusion. In essence, these transformations can discriminate between salient information (strong edges and texture) and constant or non-textured background and can also evaluate the quality of the provided salient information. Consequently, one can select the required information from the input images in the transform domain to construct the 'fused' image, following the criteria presented earlier on.

In the case of multi-focus image fusion scenarios, an alternative approach has been proposed in the spatial domain, exploiting current error estimation methods to identify high-quality edge information [7]. One can perform error minimisation between the fused and input images, using various proposed error norms in the spatial domain in order to perform fusion. The possible benefit of a spatial-domain approach is the reduction in computational complexity, which is present in a transform-domain method due to the forward and inverse transformation step.

In addition, following a spatial-domain fusion framework, one can also benefit from current available spatial-domain image enhancement techniques to incorporate a possible restoration step to enhance areas that exhibit distorted information in all input images. Current fusion approaches cannot enhance areas that appear degraded in any sense in all input images. There is a necessity for some pure information to exist for all parts of the image in the various input images, so that the fusion algorithm can produce a high quality output. In this work, we propose to reformulate and extend Jones and Vorontsov's [7] spatial-domain approach to fuse the non-degraded common parts of the sensor images. A novel approach is used to identify the areas of common degradation in all input sensor images. A double-regularised image restoration approach using robust functionals is applied on the estimated common degraded area to enhance the common degraded area in the 'fused' image. The overall fusion result is superior to any traditional fusion approach since the proposed approach goes beyond the concept of transferring useful information to a thorough fusion-enhancement approach.

12.2 Robust error estimation theory

Let the image $y(\mathbf{r})$ be a recovered version from a degraded observed image $x(\mathbf{r})$, where $\mathbf{r} = (i, j)$ are pixel coordinates (i, j). To estimate the recovered image $y(\mathbf{r})$, one can minimise an error functional $E(y)$ that expresses the difference between the original image and the estimated one, in terms of y. The error functional can be defined by

$$E(y) = \int_{\Omega} \rho\big(\mathbf{r}, y(\mathbf{r}), |\nabla y(\mathbf{r})|\big)\, d\mathbf{r} \tag{12.3}$$

where Ω is the image support, $\nabla y(\mathbf{r})$ is the image gradient. The function $\rho(\cdot)$ is termed the *error norm* and is defined according to the application, i.e. the type of degradation or the desired task. For example, a least square error norm can be appropriate to remove additive Gaussian noise from a degraded image. The extremum of the previous equation can be estimated, using the *Euler–Lagrange* equation. The Euler–Lagrange equation is an equation satisfied by a function f of a parameter t which extremises the functional:

$$E(f) = \int F\big(t, f(t), f'(t)\big)\, dt \tag{12.4}$$

where F is a given function with continuous first partial derivatives. The Euler–Lagrange equation is described by the following ordinary differential equation, i.e. a relation that contains functions of only one independent variable, and one or more of its derivatives with respect to that variable, the solution t of which extremises the above functional [8]:

$$\frac{\partial}{\partial f(t)} F\big(t, f(t), f'(t)\big) - \frac{d}{dt} \frac{\partial}{\partial f'(t)} F\big(t, f(t), f'(t)\big) = 0 \tag{12.5}$$

Applying the above rule to derive the extremum of (12.3), the following Euler–Lagrange equation is derived:

$$\frac{\partial \rho}{\partial y} - \nabla \left(\frac{\partial \rho}{\partial \nabla y} \right) = 0 \qquad (12.6)$$

Since $\rho(\cdot)$ is a function of $|\nabla y|$ and not ∇y, we perform the substitution

$$\partial \nabla y = \partial |\nabla y| / \mathrm{sgn}(\nabla y) = |\nabla y| \partial |\nabla y| / \nabla y \qquad (12.7)$$

where $\mathrm{sgn}(y) = y/|y|$. Consequently, the Euler–Lagrange equation is given by:

$$\frac{\partial \rho}{\partial y} - \nabla \left(\frac{1}{|\nabla y|} \frac{\partial \rho}{\partial |\nabla y|} \nabla y(\mathbf{r}) \right) = 0 \qquad (12.8)$$

To obtain a closed-form solution $y(\mathbf{r})$ from (12.8) is not straightforward. Hence, one can use numerical optimisation methods to estimate y. Gradient-descent optimisation can be applied to estimate $y(\mathbf{r})$ iteratively using the following update rule:

$$y(\mathbf{r}, t) \leftarrow y(\mathbf{r}, t-1) - \eta \frac{\partial y(\mathbf{r}, t)}{\partial t} \qquad (12.9)$$

where t is the time evolution parameter, η is the optimisation step size and

$$\frac{\partial y(\mathbf{r}, t)}{\partial t} = -\frac{\partial \rho}{\partial y} + \nabla \left(\frac{1}{|\nabla y|} \frac{\partial \rho}{\partial |\nabla y|} \nabla y(\mathbf{r}, t) \right) \qquad (12.10)$$

Starting with the initial condition $y(\mathbf{r}, 0) = x(\mathbf{r})$, the iteration of (12.10) continues until the minimisation criterion is satisfied, i.e. $|\partial y(\mathbf{r}, t)/\partial t| < \epsilon$, where ϵ is a small constant ($\epsilon \sim 0.0001$). In practice, only a finite number of iterations are performed to achieve visually satisfactory results [7]. The choice of the error norm $\rho(\cdot)$ in the Lagrange–Euler equation is the next topic of discussion.

12.2.1 Isotropic diffusion

As mentioned previously, one candidate error norm $\rho(\cdot)$ is the least-squares error norm. This norm is given by

$$\rho\left(\mathbf{r}, |\nabla y(\mathbf{r})|\right) = \frac{1}{2} |\nabla y(\mathbf{r})|^2 \qquad (12.11)$$

The above error norm smooths Gaussian noise and depends only on the image gradient $\nabla y(\mathbf{r})$, but not explicitly on the image $y(\mathbf{r})$ itself. If the least-squares error norm is substituted in the time evolution equation (12.10), we get the following update:

$$\frac{\partial y(\mathbf{r}, t)}{\partial t} = \nabla^2 y(\mathbf{r}, t) \qquad (12.12)$$

which is the *isotropic diffusion* equation having the following analytic solution [9]:

$$y(\mathbf{r}, t) = G(\mathbf{r}, t) * x(\mathbf{r}) \qquad (12.13)$$

where $*$ denotes the convolution of a Gaussian function $G(\mathbf{r}, t)$ of standard deviation t with $x(\mathbf{r})$, the initial data. The solution specifies that the time evolution in (12.12) is a convolution process performing Gaussian smoothing. However, as the time evolution iteration progresses, the function $y(\mathbf{r}, t)$ becomes the product of the convolution of the input image with a Gaussian of constantly increasing variance, which will finally produce a constant value. In addition, it has been shown that isotropic diffusion may not only smooth edges, but also causes drifts of the actual edges in the image edge, because of the Gaussian filtering (smoothing) [9,10]. These are two disadvantages that need to be seriously considered when using isotropic diffusion.

12.2.2 Isotropic diffusion with edge enhancement

Image fusion aims at transferring salient features to the fused image. In this work and in most fusion systems, saliency is interpreted as edge information and therefore, image fusion aims at highlighting edges in the fused image. An additional desired property can be to smooth out any possible Gaussian noise. In order to achieve the above tasks using an error estimation framework, the objective is to create an error norm that will enhance edges in an image and simultaneously smooth possible noise. The following error norm, combining *isotropic smoothing* with *edge enhancement*, was proposed in [7]:

$$\rho\big(\mathbf{r}, y(\mathbf{r}, t), |\nabla y(\mathbf{r}, t)|\big) = \frac{\alpha}{2}|\nabla y(\mathbf{r}, t)|^2 + \frac{\beta}{2}J_x(\mathbf{r})\big(y(\mathbf{r}, t) - x(\mathbf{r})\big)^2 \qquad (12.14)$$

where α, β are constants that define the level of smoothing and edge enhancement respectively that is performed by the cost function, t is the time evolution and J_x is commonly termed the *anisotropic gain function*, which is a Gaussian smoothed edge map. One possible choice for implementing a Gaussian smoothed edge map is the following:

$$J_x(\mathbf{r}) = \kappa \int |\nabla x(\mathbf{q})|^2 G(\mathbf{r} - \mathbf{q}, \sigma)\,\mathrm{d}^2\mathbf{q} \qquad (12.15)$$

where $G(\cdot)$ is a Gaussian function of zero-mean and standard deviation σ and κ is a constant. Another choice can be a smoothed Laplacian edge map. The anisotropic gain function has significantly higher values around edges or where sharp features are dominant compared to blurred or smooth regions.

Substituting the above error norm into the gradient descent update of (12.10) yields the following time evolution equation with anisotropic gain:

$$\frac{\partial y(\mathbf{r}, t)}{\partial t} = \alpha \nabla^2 y(\mathbf{r}, t) - \beta J_x(\mathbf{r})\big(y(\mathbf{r}, t) - x(\mathbf{r})\big) \qquad (12.16)$$

The above equation essentially smoothes noise while enhancing edges. The parameters α and β control the effects of each term. The parameter α controls the amount of noise smoothing in the image and β controls the anisotropic gain, i.e. the preservation and enhancement of the edges. For noiseless images, an evident choice is $\alpha = 0$ and $\beta = 1$. In this case, for short time intervals, the anisotropic gain function J_x induces significant changes dominantly around regions of sharp contrast, resulting in edge enhancement.

There is always a possibility that in some regions of interest, the anisotropic gain function is not high enough and therefore the above update rule can potentially degrade the quality of information that is already integrated into the input image and consequently in the enhanced image. To prevent such erasing effects, however small might be, John and Vorontsov [7] introduced the following modified anisotropic gain function:

$$J(\mathbf{r}, t) = J_x(\mathbf{r}) - J_y(\mathbf{r}, t) \tag{12.17}$$

The general update formula to estimate $f(\mathbf{r})$ becomes then

$$\frac{\partial y(\mathbf{r}, t)}{\partial t} = \alpha \nabla^2 y(\mathbf{r}, t) - \Theta\big(J(\mathbf{r}, t)\big) J(\mathbf{r}, t)\big(y(\mathbf{r}, t) - x(\mathbf{r})\big) \tag{12.18}$$

where

$$\Theta(J) = \begin{cases} 1, & J \geqslant 0, \\ 0, & J < 0 \end{cases} \tag{12.19}$$

The new term $\Theta(J)J$ allows only high quality information, interpreted in terms of edge presence, to transfer to the enhanced image. In the opposite case that $J_x(\mathbf{r}) < J_y(\mathbf{r})$, the information in the enhanced image has better edge representation than the original degraded image for several $\mathbf{r}$ and therefore, no processing is necessary. In the case of a single input image, the above concept might not seem practical. In the following section, the proposed concept is employed in a multiframe input scenario, where the aim is to transfer only high quality information to the enhanced image $y(\mathbf{r})$. In this case, this positive edge injection mechanism is absolutely vital to ensure information enhancement.

12.3 Fusion with error estimation theory

In this section, the authors propose a novel spatial-domain fusion algorithm, based on the basic formulation of John and Vorontsov. In [7], a sequential approach to image fusion based on error estimation theory was proposed. Assuming that we have a number of T input frames $x_n(\mathbf{r})$ to be fused, one can easily perform selective image fusion, by iterating the update rule (12.18) for the estimation of $y(\mathbf{r})$ using each of input images x_n consecutively for a number of K iterations. In a succession of intervals of K iterations, the synthetic frame finally integrates high-quality edge areas from the entire set of input frames.

The possibility of data fusion occurring in regions where the anisotropic gain function is not high enough, can potentially degrade quality information already integrated into the synthetic frame. To prevent such erasing effects, as mentioned in the previous section, a differential anisotropic gain function can be introduced to transfer only high quality information to the fused image $y(\mathbf{r})$. The proposed approach by John and Vorontsov can be applied mainly in the case of a video stream, where the quality of the observed image is enhanced, based on previous and forthcoming frames. However, this framework is not efficient in the case of fusion applications, where the input frames are simultaneously available for processing and fusion. In this case, a reformulation of the above procedure is needed and is described in full in the following section.

12.3.1 A novel fusion formulation based on error estimation theory

Assume there are T images $x_n(\mathbf{r})$ that capture the same observed scene. The input images are assumed to be registered and each image contains exactly the same scene. This assumption is valid, since in most real-life applications, the input sensors are arranged in a close-distance array and similar zoom level in order to minimise the need for registration or the viewpoint transformation matrix is known. Different parts of the images are blurred using different amounts and types of blur. The objective is to combine the useful parts of input information to form a composite ('fused') image.

The described setup can model a possible out-of-focus scenario of image capture. We have all witnessed the case, where we want to take a photograph of an object in a scene and the camera focuses on a background point/object by mistake. As a result, the foreground object appears blurred in the final image, whereas the background texture is properly captured. In a second attempt to photograph the object correctly, the foreground object appears properly and the background appears blurred. Ideally, we would like to combine the two images into a new one, where everything would appear in full detail. This is an example of a real-life application for the fusion of out-of-focus images. The same scenario can also appear in military surveillance and general surveillance applications, where one would like to enhance the surveillance output, by combining multiple camera inputs at different focal length.

The fused image $y(\mathbf{r}, t)$ can be constructed as a linear combination of the T input registered images $x_n(\mathbf{r})$. The fusion problem is usually solved by finding the weights $w_n(\mathbf{r}, t)$ that transfer all the useful information from the input images x_n to the fused image y [5,6]:

$$y(\mathbf{r}, t) = w_1(\mathbf{r}, t)x_1(\mathbf{r}) + \cdots + w_T(\mathbf{r}, t)x_T(\mathbf{r}) \tag{12.20}$$

where $w_n(\mathbf{r}, t)$ denotes the nth weight of the image x_n at position $\mathbf{r}$. To estimate these weights, we can perform error minimisation using the previously mentioned approach of Isotropic Diffusion with edge enhancement. The problem is now to estimate the weights w_n simultaneously, so as to achieve edge preservation. This cannot be accomplished directly by the scheme proposed by Jones and Vorontsov.

In other words, we need to estimate the derivative $\partial w_n/\partial t$ simultaneously, for all $n = 1, \ldots, T$. We can associate $\partial w_n/\partial t$ with $\partial y/\partial t$ that has already been derived before:

$$\frac{\partial y}{\partial t} = \frac{\partial y}{\partial w_n}\frac{\partial w_n}{\partial t} = x_n\frac{\partial w_n}{\partial t} \tag{12.21}$$

Therefore, we can use the previous update rule to estimate the contribution of each image to the fused one:

$$\frac{\partial w_n(\mathbf{r}, t)}{\partial t} = \frac{1}{x_n(\mathbf{r})}\frac{\partial y(\mathbf{r}, t)}{\partial t} \tag{12.22}$$

The fusion weight $w_n(\mathbf{r}, t)$ of each input image can then be estimated using sequential minimisation with the following update rule $\forall n = 1, \ldots, T$:

$$w_n(\mathbf{r}, t+1) \leftarrow w_n(\mathbf{r}, t) - \eta \frac{\partial w_n(\mathbf{r}, t)}{\partial t} \qquad (12.23)$$

where

$$\frac{\partial w_n(\mathbf{r}, t)}{\partial t} = -\frac{1}{x_n(\mathbf{r})} \Theta\big(J_n(\mathbf{r}, t)\big) J_n(\mathbf{r}, t)\big(y(\mathbf{r}, t) - x_n(\mathbf{r})\big) \qquad (12.24)$$

and $J_n(\mathbf{r}, t) = J_{x_n}(\mathbf{r}) - J_y(\mathbf{r}, t)$. To avoid possible numerical instabilities, for those $\mathbf{r}$ that $x_n(\mathbf{r}) = 0$, a small constant is added to these elements so as to become non-zero. All weights are initialised to $w_n(\mathbf{r}, t) = 1/T$, which represents the 'mean' fusion rule. As this scheme progresses over time, the weights are adapting and tend to emphasise more the useful details that exist in each image and suppress the information that is not very accurate. In addition, all the fusion weights are estimated simultaneously using this scheme. Therefore, after a couple of iterations the majority of the useful information is extracted from the input images and transferred to the composite image.

12.3.2 Fusion experiments of out-of-focus and multimodal image sets using error estimation theory

In this section, we perform several fusion experiments of both out-of-focus and multimodal images to evaluate the performance of the proposed approach. Most test images were taken from the Image Fusion server [11]. The numerical evaluation in most experiments was performed using the indexes proposed by Piella [12] and Petrovic [13].

In the first experiment, the system is tested with an out-of-focus example, the 'Disk' dataset. The ICA-based fusion algorithm, proposed in [5], was employed as a benchmark to the new proposed algorithm. We used 40 TopoICA 8×8 bases, trained from 10000 patches that were randomly selected from natural images. Then, the 'Weighted Combination' rule was selected to perform fusion of the input images. On the other hand, for the spatial-domain fusion scheme, the parameters were set to $\alpha = 0$ (no visible noise), $\beta = 0.8$ and the learning parameter was set to $\eta = 0.08$. The Gaussian smoothed edge map of (12.15) was calculated by extracting an edge map using the Sobel mask, which was subsequently smoothed by a Gaussian 5×5 kernel of standard deviation $\sigma = 1$. The fusion results of the two methods are depicted in Figure 12.1. We notice that the proposed approach produces sharper edges compared to the ICA-based method. The difference is more visible around the edges of the tilted books in the bookcase and the eye on the cover of the book that is in front of the bookcase. In Figure 12.2, the convergence rate of the estimation of one of the fusion weights is shown. The proposed algorithm demonstrates almost linear convergence, which is expected for a gradient algorithm.

In Table 12.1, the performance of the proposed method is compared with the ICA-based method, in terms of the Petrovic and Piella method. The metrics give slightly higher

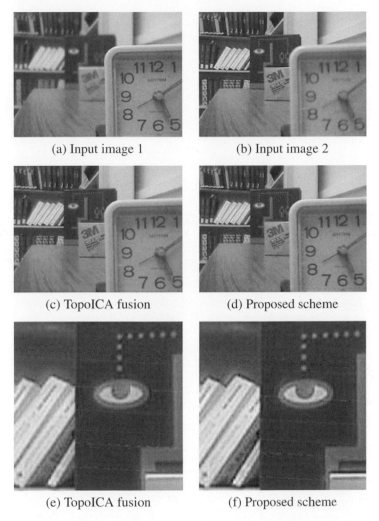

(a) Input image 1　　　　　　　　(b) Input image 2

(c) TopoICA fusion　　　　　　　(d) Proposed scheme

(e) TopoICA fusion　　　　　　　(f) Proposed scheme

Figure 12.1 *An out-of-focus fusion example using the 'Disk' dataset available by the Image Fusion server [11]. We compare the TopoICA-based fusion approach and the proposed Diffusion scheme.*

performance to the proposed methodology. However, we can observe an improvement in the visual representation of edges using the proposed method in the particular application of fusion of out-of-focus images.

The estimated fusion weights $w_1(\mathbf{r})$, $w_2(\mathbf{r})$ are depicted in Figure 12.3. It is clear that the weights w_1, w_2 highlight the position of high-quality information in the input images. The cost function that is optimised in this case aims at highlighting edges in the 'fused' image. This is essentially what is estimated by the weight maps $w_1(\mathbf{r})$, $w_2(\mathbf{r})$. This information can be used to identify common areas of inaccurate information in the input images. A restoration algorithm could be applied to these areas and enhance the final information that is conveyed to the 'fused' image.

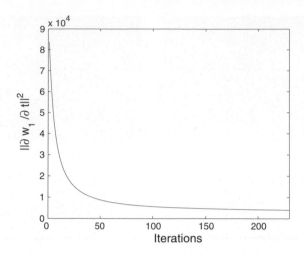

Figure 12.2 *Convergence of the estimated fusion weight w_1 using the proposed fusion algorithm in terms of* $\|\partial w_1/\partial t\|^2$.

Table 12.1 *Performance evaluation of the Diffusion approach and the TopoICA-based fusion approach using Petrovic [13] and Piella's [12] metrics.*

	Petrovic	Piella
TopoICA	0.6151	0.9130
Fusion with EE	0.6469	0.9167

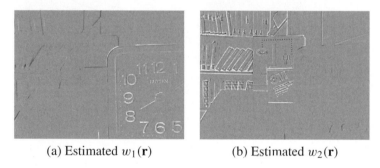

(a) Estimated $w_1(\mathbf{r})$ (b) Estimated $w_2(\mathbf{r})$

Figure 12.3 *The weights w_1, w_2 highlight the position of high quality information in the input images.*

The next step is to apply the proposed algorithm to a multimodal scenario. We will use an image pair from the 'UN camp' dataset of surveillance images from TNO Human Factors, provided by L. Toet [14] in the Image Fusion server [11]. We applied the TopoICA-based approach [5] using the 'max-abs' fusion rule and the proposed algorithm on the dataset, using the same settings as in the previous example. In Figure 12.4, we plot the fused results of the two methods and in Table 12.2, we plot their numerical evaluation using Petrovic and Piella's indexes.

According to the performance evaluation indexes, the ICA-based approach performs considerably better than the proposed approach. The same trend is also observed in the met-

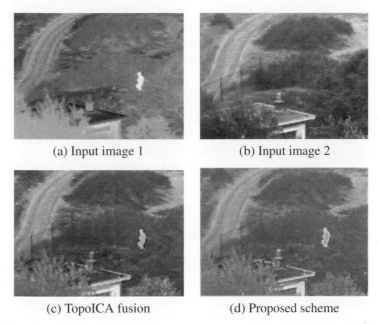

(a) Input image 1   (b) Input image 2

(c) TopoICA fusion   (d) Proposed scheme

Figure 12.4 *Comparison of a multimodal fusion example using the TopoICA method and the Diffusion approach. Even though the metrics demonstrate worse performance, the diffusion approach highlights edges giving a sharper fused image.*

Table 12.2 *Performance evaluation in the case of a multimodal example from the Toet database. The TopoICA-based approach is compared with the proposed fusion approach.*

	Petrovic	Piella
TopoICA	0.4921	0.7540
Fusion with EE	0.4842	0.6764

rics. However, the proposed approach performs differently to a common fusion approach. It aims at highlighting the edges of the input images to the fused image, due to the edge enhancement term in the cost function. This is can be observed directly in Figure 12.4(d). All edges and texture areas are highly enhanced in the fused image together with the outline of the important target, i.e. the hidden man in the middle of the picture. Consequently, one should also consult the human operators of modern fusion systems, apart from proposed fusion metrics [12,13], in order to evaluate efficiently the performance of these algorithms. Perhaps the outlined fusion result is more appealing to human operators and the human vision system in general and therefore may be also be examined as a preferred solution.

12.4 Joint image fusion and restoration

The basic Image Fusion concept assumes that there is some useful information for all parts of the observed scene at least in one of the input sensors. However, this assumption might not always be true. This means that there might be parts of the observed scene where there is only degraded information available. The current fusion algorithms will fuse all high quality information from the input sensors and for the common degraded areas will form a blurry mixture of the input images, as there is no high quality information available.

In the following section, the problem of identifying the areas of common degraded information in all input images is addressed. A mechanism is established for identifying common degraded areas in an image. Once this part is identified, an image restoration approach can be applied as a second step in order to enhance these parts for the final composite 'fused' image.

12.4.1 Identifying common degraded areas in the sensor images

The first task will be to identify the areas of degraded information in the input sensor images. An identification approach, based on local image statistics, will be pursued to trace the degraded areas.

The 'fused' image will be employed, as it emerges from the fusion algorithm. As mentioned earlier, the fusion algorithm will attempt to merge the areas of high detail to the fused image, whereas for the areas of degraded information, i.e. areas of weak edges or texture in all input images, will not impose any preference to any of the input images and therefore the estimated fusion weights will remain approximately equal to the initial weights $w_i = 1/T$. Consequently, the areas of out-of-focus distortion will be described by areas of low edge information in the fused image. Equivalently, some areas of very low texture or constant background also need to be excluded, since there is no benefit in restoring them. These areas can be traced, by evaluating the local standard deviation of an edge information metric in small local neighbourhoods around each pixel. The following algorithm for extracting common degraded areas is described in the following steps:

(1) Extract an edge map of the fused image f, using the Laplacian kernel, i.e. $\nabla^2 f(\mathbf{r}, t)$.
(2) Find the local standard deviations $V_L(\mathbf{r}, t)$ for each pixel of the Laplacian edge map $\nabla^2 f(\mathbf{r}, t)$, using 5×5 local neighbourhoods.
(3) Reduce the dynamic range by calculating $\ln(V_L(\mathbf{r}, t))$.
(4) Estimate $V_{sL}(\mathbf{r}, t)$, by smoothing $\ln(V_L(\mathbf{r}, t))$ using a 15×15 median filter.
(5) Create the common degraded area map $A(\mathbf{r})$ by thresholding $V_{sL}(\mathbf{r}, t)$. The mask $A(\mathbf{r})$ is set to 1, for those $\mathbf{r}$ that $q \min_r(V_{sL}(\mathbf{r}, t)) < V_{sL}(\mathbf{r}, t) < p \operatorname{mean}_r(V_{sL}(\mathbf{r}, t))$, otherwise is set to zero.

Essentially, we create an edge map, as described by the Laplacian kernel. The Laplacian kernel was chosen because it was already estimated during the fusion stage of the framework. The next step is to find the local activity in 5×5 neighbourhoods around each pixel in the edge map. A metric of local activity is given by the local standard deviation. A pixel of high local activity should be part of an 'interesting' detail in the image (edge, strong texture, etc.), whereas a point of low local activity might be a constant background or weak texture pixel. We can devise a heuristic thresholding scheme in order to iden-

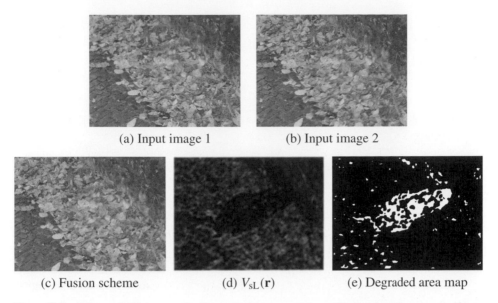

(a) Input image 1   (b) Input image 2

(c) Fusion scheme   (d) $V_{sL}(\mathbf{r})$ (e) Degraded area map

Figure 12.5   *If there exist blurry parts in all input images, common Image Fusion algorithms cannot enhance these parts, but will simply transfer the degraded information to the fused image. However, this area of degraded information is still identifiable.*

tify these areas of weak local activity, i.e. possible degraded areas in all input images for fusion. The next step is to reduce the dynamic range of these measurements, using a logarithmic nonlinear mapping, such as $\ln(\cdot)$. To smooth out isolated pixels and connect similar areas, we perform median filtering of the log-variance map. Consequently, the common degraded area map is created by thresholding the values of the log-variance map with a heuristic threshold set to $q \min_r (V_{sL}(\mathbf{r}, t)) < V_{sL}(\mathbf{r}, t) < p \operatorname{mean}_r (V_{sL}(\mathbf{r}, t))$, where p, q are constants. The aim is to avoid high quality edge/texture and constant background information. The level of detail along with the level of constant background differ for different images. In order to identify the common degraded area with accuracy, the parameters p, q need to be defined manually for each image. The parameter q defines the level of background information that needs to be removed. In a highly active image, q is usually set to 1, however, other values have to be considered for images with large constant background areas. The parameter p is the upper bound threshold to discriminate between strong edges and weak edges, possibly belonging to a common degraded area. Setting p around the mean edge activity, we can find a proper threshold for the proposed system. Values that were found to work well in experiments were $q \in [0.98, 1]$ and $p \in [1, 1.1]$. Some examples of common degraded area identification using the above technique are shown in Figures 12.5 and 12.6.

12.4.2 Image restoration

A number of different approaches for tackling the image restoration problem have been proposed in the literature, based on various principles. For an overview of image restoration methods, one can always possibly refer to Kundur and Hatzinakos [15] and Andrews

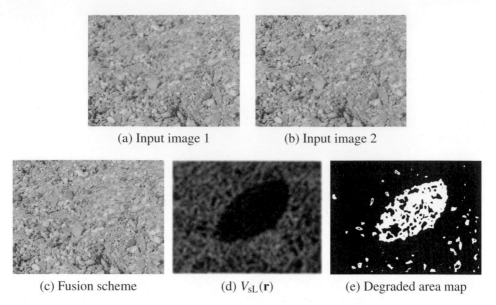

(a) Input image 1   (b) Input image 2

(c) Fusion scheme   (d) $V_{\text{sL}}(\mathbf{r})$ (e) Degraded area map

Figure 12.6 *Another example of degraded area identification in 'fused' images.*

and Hunt [16]. In this study, the double-weighted regularised image restoration approach in the spatial domain is pursued, that was initially proposed by You and Kaveh [17], with additional robust functionals to improve the performance in the case of outliers. The restoration problem is described by the following model:

$$y(\mathbf{r}) = h(\mathbf{r}) * f(\mathbf{r}) + d(\mathbf{r}) \tag{12.25}$$

where $*$ denotes 2D convolution, $h(\mathbf{r})$ the degradation kernel, $f(\mathbf{r})$ the estimated image and $d(\mathbf{r})$ possible additive noise.

12.4.2.1 Double weighted regularised image restoration

Conventional double weighted regularisation for blind image restoration [16] estimates the original image by minimising the cost function $Q(h(\mathbf{r}), f(\mathbf{r}))$ of the following quadratic form:

$$Q\big(h(\mathbf{r}), f(\mathbf{r})\big) = \underbrace{\frac{1}{2}\big\|A_1(\mathbf{r})\big(y(\mathbf{r}) - h(\mathbf{r}) * f(\mathbf{r})\big)\big\|^2}_{\text{residual}}$$

$$+ \underbrace{\frac{\lambda}{2}\big\|A_2(\mathbf{r})\big(C_f * f(\mathbf{r})\big)\big\|^2}_{\text{image regularisation}} + \underbrace{\frac{\gamma}{2}\big\|A_3(\mathbf{r})\big(C_h * h(\mathbf{r})\big)\big\|^2}_{\text{blur regularisation}} \tag{12.26}$$

where $\|\cdot\|$ represents the $\mathcal{L}_2$-norm. The above cost function has three distinct terms. The residual term, the first term on the right-hand side of (12.26), represents the accuracy of the restoration process. This term is similar to a second-order error-norm (least-squares estimation), as described in a previous paragraph. The second term, called the regularising term, imposes a smoothness constraint on the recovered image and the third term acts similarly to the estimated blur. Additional constraints must be imposed, including the *non-negativity* and *finite-support constraint* for both the blurring kernel and the image. Besides, the blurring kernel must always preserve the energy, i.e. all the coefficients should sum to 1. The regularisation operators C_f and C_h are high-pass Laplacian operators applied on the image and the PSF, respectively. The functions A_1, A_2 and A_3 represent spatial weights for each optimisation term. The parameters λ and γ control the trade-off between the residual term and the corresponding regularising terms for the image and the blurring kernel.

One can derive the same cost function through a Bayesian framework of estimating $f(\mathbf{r})$ and $h(\mathbf{r})$. To illustrate this connection, we assume that the blurring kernel $h(\mathbf{r})$ is known and the aim is to recover $f(\mathbf{r})$. A Maximum-A-Posteriori (MAP) estimate of $f(\mathbf{r})$ is given by performing $\max_f \log p(y, f|\mathbf{r}) = \max_f \log p(y|f, \mathbf{r})p(f|\mathbf{r})$, where $\mathbf{r}$ denotes the observed samples. Assuming Gaussian noise for $d(\mathbf{r})$, we have that $p(y|f, \mathbf{r}) \propto \exp(-0.5a\|y(\mathbf{r}) - h(\mathbf{r}) * f(\mathbf{r})\|^2)$. Assuming smoothness for the image profile, one can employ the image prior $p(f|\mathbf{r}) \propto \exp(-0.5b\|C_f * f(\mathbf{r})\|^2)$, which has been widely used by the engineering community [18] in setting constraints on first or second differences, i.e. restricting the rate of changes in an image (a, b are constants that can determine the shape of the prior). Using the proposed models, one can derive a MAP estimate by optimising a function that is the same as the first two terms of (12.26), illustrating the connection between the two approaches.

To estimate $f(\mathbf{r})$ and $h(\mathbf{r})$, the above cost function needs to be minimised. Since each term of the cost function is quadratic, it can simply be optimised by applying alternating Gradient Descent optimisation [16]. This implies that the estimates for the image and the PSF can be estimated alternatively, using the gradients of the cost function with respect to $f(\mathbf{r})$ and $h(\mathbf{r})$. More specifically, the double iterative scheme can be expressed as follows:

- At each iteration, update:

$$f(t+1) = f(t) - \eta_1 \frac{\partial Q(h(t), f(t))}{\partial f(t)} \tag{12.27}$$

$$h(t+1) = h(t) - \eta_2 \frac{\partial Q(h(t), f(t+1))}{\partial h(t)} \tag{12.28}$$

- Stop, if f and h converge.

The terms η_1 and η_2 are the step size parameters that control the convergence rates for the image and *Point Spread Function* (PSF) (blurring kernel) respectively. After setting the initial estimate of the image as the degraded image, and the PSF as a random mask, the cost function is differentiated with respect to the image first, while the PSF is kept constant, and vice versa. The required derivatives of the cost function are presented below:

$$\frac{\partial Q(h, f)}{\partial f} = -A_1(\mathbf{r})h(-\mathbf{r}) * \big(y(\mathbf{r}) - h(\mathbf{r}) * f(\mathbf{r})\big)$$

$$+ \lambda\big(A_2(\mathbf{r})C_f^T * \big(C_f * f(\mathbf{r})\big)\big) \tag{12.29}$$

$$\frac{\partial Q(h, f)}{\partial h} = -A_1(\mathbf{r})f(-\mathbf{r}) * \big(y(\mathbf{r}) - h(\mathbf{r}) * f(\mathbf{r})\big)$$

$$+ \gamma\big(A_3(\mathbf{r})C_h^T * \big(C_h * h(\mathbf{r})\big)\big) \tag{12.30}$$

where the superscript T denotes the transpose operation. Substituting (12.29) and (12.30) into (12.27) and (12.28) yields the final form of the algorithm (12.27) and (12.28), where the corresponding functions are iterated until convergence.

12.4.2.2 Robust functionals to the restoration cost function

There exist several criticisms regarding the conventional double regularisation restoration approach. One is the non-robustness of the least squares estimators employed in the traditional residual term, once the assumption of Gaussian noise does not hold [19]. Moreover, the quadratic regularising term penalises sharp grey-level transitions, due to the linearity of the derivative of the quadratic function. This implies that sudden changes in the image are filtered, and thus, the image edges are blurred. To alleviate this problem, we can introduce *robust functionals* in the cost function, in order to rectify some of the problems of this estimator. Therefore, the original cost function becomes

$$Q\big(h(\mathbf{r}), f(\mathbf{r})\big) = \frac{1}{2}\big\| A_1(\mathbf{r})\rho_n\big(y(\mathbf{r}) - h(\mathbf{r}) * f(\mathbf{r})\big)\big\|^2$$

$$+ \frac{\lambda}{2}\big\| A_2(\mathbf{r})\rho_f\big(C_f * f(\mathbf{r})\big)\big\|^2$$

$$+ \frac{\gamma}{2}\big\| A_3(\mathbf{r})\rho_d\big(C_h * h(\mathbf{r})\big)\big\|^2 \tag{12.31}$$

Three distinct robust kernels $\rho_n(\cdot)$, $\rho_f(\cdot)$ and $\rho_d(\cdot)$ are introduced in the new cost function and are referred to as the robust residual and regularising terms respectively. The partial derivatives of the cost function take the following form:

$$\frac{\partial Q(h, f)}{\partial f} = -A_1(\mathbf{r})h(-\mathbf{r}) * \rho_n'\big(y(\mathbf{r}) - h(\mathbf{r}) * f(\mathbf{r})\big)$$

$$+ \lambda\big(A_2(\mathbf{r})C_f^T * \rho_f'\big(C_f * f(\mathbf{r})\big)\big) \tag{12.32}$$

$$\frac{\partial Q(h, f)}{\partial h} = -A_1(\mathbf{r})f(-\mathbf{r}) * \rho_n'\big(y(\mathbf{r}) - h(\mathbf{r}) * f(\mathbf{r})\big)$$

$$+ \gamma\big(A_3(\mathbf{r})C_h^T * \rho_d'\big(C_h * h(\mathbf{r})\big)\big) \tag{12.33}$$

Robust estimation is usually presented in terms of the influence function $l(r) = \partial\rho/\partial r$. The influence function characterises the bias of a particular measurement on the solution. Traditional least squares kernels fail to eliminate the effect of outliers, with linearly increasing and non-bounded influence functions. On the other hand, they also tend to over-smooth the image's details, since such edge discontinuities lead to large values of smoothness error. Thus, two different kernel types are investigated, in order to increase the robustness and reject outliers in the context of the blind estimation.

To suppress the effect of extreme noisy samples ('outliers') that might be present in the observations, the derivative of an ideal robust residual term should increase less rapidly than a quadratic term in the case of outliers. One candidate function can be the following:

$$\rho'_n(x) = \frac{1}{1 + (x/\theta)^{2\upsilon}} \tag{12.34}$$

Obviously, the specific function associated with the residual term assists in suppressing the effect of large noise values in the estimation process, by setting the corresponding influence function to small values. Optimal values for the θ and υ parameters have been investigated in [20]. These parameters determine the 'shape' of the influence function and as a consequence the filtering of outliers.

In order to find a trade-off between noise elimination and preservation of high-frequency details, the influence functional for the image regularising term must approximate the quadratic structure at small to moderate values and alternatively deviate from the quadratic structure at high values, so that the sharp changes will not be greatly penalised. One possible formulation of the image regularising term is expressed by the absolute entropy function shown below, which reduces the relative penalty ratio between large and small signal deviations, compared with the quadratic function [19]. Hence, the absolute entropy function produces sharper boundaries than the quadratic one, and therefore can be employed for blind restoration.

$$\rho_f(x) = \left(|x| + e^{-1}\right) \ln\left(|x| + e^{-1}\right) \tag{12.35}$$

$$\rho'_f(x) = \frac{1}{2}\mathrm{sgn}(x)\left(\ln\left(|x| + e^{-1}\right) + 1\right) \tag{12.36}$$

For simplicity, the robust functional for the stabilising term of the Point Spread Function (PSF) is kept the same as the image regularising term ($\rho'_d(x) = \rho'_f(x)$). The actual PSF size can still be estimated at a satisfactory level. The PSF support is initially set to a large enough value. The boundaries of the assumed PSF support are trimmed at each iteration in a fashion which is described later, until it reduces to a PSF support that approximates the true support [17].

12.4.3 Combining image fusion and restoration

In this section, we propose an algorithm that can combine all the previous methodologies and essentially perform fusion of all the parts that contain valid information in at least one of the input images and restoration of those image parts that are found to be degraded in all input images.

The proposed methodology consists of splitting the procedure in several individual parts:

(1) The first step is to use the proposed fusion update algorithm of Section 12.3.1 to estimate the fused image $y(\mathbf{r})$. In this step, all useful information from the input images has been transferred to the fused image and the next step is to identify and restore the areas where only low quality information is available. In other words, this step ensures that all high quality information from the input images has been transferred to the fused image. The result of this step is the fused image $y(\mathbf{r})$.

(2) The second step is to estimate the common degraded area, using the previous methodology based on the Laplacian edge map of the fused image $y(\mathbf{r})$. More specifically, this step aims at identifying possible corrupted areas in all input images that need enhancement in order to highlight more image details that were not previously available. This will produce the common degraded area mask $A(\mathbf{r})$.

(3) The third step is to estimate the blur $h(\mathbf{r}, t)$ and the enhanced image $f(\mathbf{r}, t)$, using the estimated mask of the Common Degraded area as $A(\mathbf{r})$ and the produced fused image $y(\mathbf{r})$. This step is essentially enhancing only the common degraded area and not the parts of the image that have been identified to contain high quality information. The restoration is performed as described in the previous section, however, the updates for $f(\mathbf{r}, t)$ and $h(\mathbf{r}, t)$ are influenced only by the common degraded area. More specifically, the update for the enhanced image of (12.27) becomes

$$f(\mathbf{r}, t+1) = f(\mathbf{r}, t) - \eta_1 A(\mathbf{r}) \frac{\partial Q(h(\mathbf{r}, t), f(\mathbf{r}, t))}{\partial f(\mathbf{r}, t)} \tag{12.37}$$

In a similar manner the update for the Point Spread Function (PSF) needs to be influenced only by the common degraded area, i.e. in (12.33) $f(\mathbf{r})$ is always substituted by $A(\mathbf{r}) f(\mathbf{r})$.

12.4.4 Examples of joint image fusion and restoration

In this section, three synthetic examples are constructed to test the performance of the joint fusion and restoration approach. The proposed joint approach is compared to the performance of the Error-Estimation-based fusion and the previously proposed ICA-based image fusion approach. Three natural images are employed and two blurred sets were created from each of these images. These image sets are created so that: (i) a different type/amount of blur is used in the individual images, (ii) there is an area that is blurred in both input images, (iii) there is an area that is not blurred in any of the input images. We have to note that in this case, the ground truth image needs to be available, to evaluate these experiments efficiently. The enhanced images will be compared with the ground truth image, in terms of Peak Signal-to-Noise Ratio (PSNR) and Image Quality Index Q_0, as proposed by Wang and Bovik [21]. In these experiments, the fusion indexes proposed by Petrovic and Xydeas [13] and Piella [12], cannot be used since they measure the amount of information that has been transferred from the input images to the fused image. Since the proposed fusion–restoration approach aims at enhancing the areas that have low quality information in the input images, it makes no sense to use any evaluation approach that employs the input images as a comparison standard. The images used in this experimental section can be downloaded[1] or requested by email from the authors.

There were several parameters that were manually set in the proposed fusion–restoration approach. For the Fusion part, we set $\alpha = 0$ (noise-free examples 1–2) or $\alpha = 0.08$ (noisy example 3), $\beta = 0.8$, the learning parameter was set to $\eta = 0.08$. The Gaussian smoothed edge map of (12.15) was again calculated by extracting an edge map using the Sobel mask, which was subsequently smoothed by a Gaussian 5×5 kernel of standard deviation $\sigma = 1$. For the common degraded area identification step, a separate set of values for p, q will be given for each experiment. For the restoration step, we followed the basic guidelines proposed by You and Kaveh [17]. Hence, the regularisation matrices C_f, C_h were set as follows:

$$C_f = \begin{bmatrix} 0 & -0.25 & 0 \\ -0.25 & 1 & -0.25 \\ 0 & -0.25 & 0 \end{bmatrix}, \qquad C_h = \begin{bmatrix} 2 & -1 \\ -1 & 0 \end{bmatrix} \qquad (12.38)$$

Some parameters were fixed to $\lambda = 0.1$, $\gamma = 10$, $\eta_1 = 0.25$, $\eta_2 = 0.00001$. The functions $A_1(\mathbf{r})$ and $A_3(\mathbf{r})$ were fixed to 1, whereas $A_2(\mathbf{r})$ was adaptively estimated for each iteration step, to emphasise regularisation on low-detail areas according to local variance (as described in [17]). For the robust functionals, we set $v = 2$ and $\theta \in [1.5, 3]$ was set accordingly for each case. The estimate kernel $h(\mathbf{r})$ was always initialised to $1/L^2$, where $L \times L$ is its size. All elements of the kernel were forced to be positive along the adaptation and sum to 1, so that the kernel does not perform any energy change. This is achieved by performing the mapping $h(\mathbf{r}) \leftarrow |h(\mathbf{r})|/\sum_{\mathbf{r}} |h(\mathbf{r})|$. The size L was usually set in advance, according to the experiment. If we need to estimate the size of the kernel automatically, we can assume initially a 'large' size of kernel L. There is a mechanism to reduce the effective size of the kernel along the adaptation. The variance (energy) of a smaller $(L - 1) \times (L - 1)$ kernel is always compared to the variance (energy) of the $L \times L$ kernel. In the case that the smaller kernel captures more than 85 per cent of the total kernel variance, its size becomes the new estimated kernel size in the next step of the adaptation. For the ICA-based method, the settings described in Section 12.3.2 were used.

In Figure 12.7, the first example with the 'leaves' dataset is depicted. The two artificially created blurred input images are depicted in Figures 12.7(a) and 12.7(b). In Figure 12.7(a), Gaussian blur is applied on the upper left part of the image and in Figure 12.7(b) motion blur is applied on the bottom right part of the image. The amount of blur is randomly chosen. It is obvious that the two input images contain several areas of common degradation in the image centre and several areas that were not degraded at the bottom left and the top right of the image. In Figure 12.7(c), the result of the fusion approach using Isotropic Diffusion is depicted. As expected, the fusion algorithm manages to transfer all high quality information to the fused image, however, one area in the centre of the image still remains blurred since there is no high quality reference in any of the input images. Therefore, the output remains blurred in the fused image in the common

---

[1]http://www.commsp.ee.ic.ac.uk/~nikolao/Fusion_Restoration.zip.

(a) Input image 1   (b) Input image 2

(c) Fusion scheme   (d) Fusion + restoration scheme

(e) Fusion (affected area)   (f) Fusion + restoration (affected area)

Figure 12.7 *Overall fusion improvement using the proposed fusion approach enhanced with restoration. Experiments with the 'leaves' dataset.*

degraded area. The common degraded area can be identified by the algorithm as depicted in the previously illustrated Figure 12.5(e), using $p = 1.07$ and $q = 1$. In Figure 12.7(d), we can see the final enhanced image, after the restoration process has been applied on the common degraded area for $L = 5$. An overall enhancement to the whole image quality can be witnessed with a significant edge enhancement compared to the original fused image. In Figures 12.7(e) and 12.7(f), a focus on the common degraded area in the fused and the fused/restored image can verify the above conclusions. In Figure 12.8, we plot the convergence of the restoration part of the common degraded area, in terms of the update for the restored image $f(\mathbf{r})$ and the update for the estimated blurring kernel $h(\mathbf{r})$. In addition, the estimated kernel is also depicted in Figure 12.8. The estimated kernel follows our intuition of a motion blur kernel around $20°$, blurred by a Gaussian kernel. In Table 12.3, the performance of the TopoICA-based fusion scheme, the fusion scheme based on Error Estimation and the fusion + restoration scheme are evaluated in terms of Peak Signal-to-Noise Ratio (PSNR) and the Image Quality Index Q_0, proposed by Wang

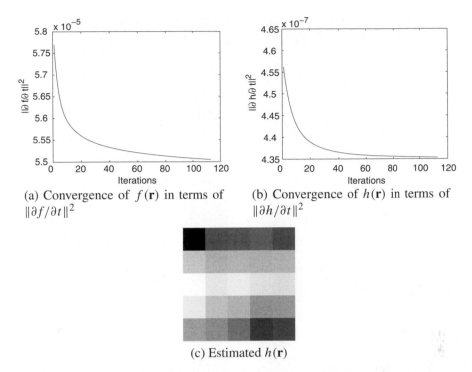

(a) Convergence of $f(\mathbf{r})$ in terms of $\|\partial f/\partial t\|^2$

(b) Convergence of $h(\mathbf{r})$ in terms of $\|\partial h/\partial t\|^2$

(c) Estimated $h(\mathbf{r})$

Figure 12.8 *Convergence of the restoration part and the final estimated $h(\mathbf{r})$ for the common degraded area in the 'leaves' example. The directivity of the estimated mask indicates the estimation of motion blur.*

Table 12.3 *Performance evaluation of the fused with isotropic diffusion and the combined fusion–restoration approach in terms of PSNR (dB) and Q_0.*

	Fused TopoICA		Fused error est.		Fused + restored	
	PSNR (dB)	Q_0	PSNR (dB)	Q_0	PSNR (dB)	Q_0
Leaves	17.65	0.9727	25.740	0.9853	25.77	0.9864
Pebbles	21.27	0.9697	25.35	0.9713	25.99	0.9755
Noisy BA747	17.35	0.9492	24.18	0.9757	24.41	0.9770
Porto	21.33	0.9860	22.94	0.9897	23.37	0.9907
Noisy Porto	19.71	0.9768	20.55	0.9818	20.62	0.9821

and Bovik [21]. The visible edge enhancement in the common degraded area, provided by the extra restoration step is also confirmed by the two metrics.

Similar conclusions follow the next example with the 'pebbles' dataset in Figure 12.9. The two artificially created blurred input images are depicted in Figures 12.9(a) and 12.9(b). In Figure 12.9(a), Gaussian blur is applied to the upper left part of the image and in Figure 12.9(b) Gaussian blur of different variance (randomly chosen) is applied to the bottom right part of the image. Again, the two input images contain an area of common degradation in the image centre and several areas that were not degraded in the bottom

(a) Input image 1   (b) Input image 2

(c) Fusion scheme   (d) Fusion + restoration scheme

(e) Fusion (affected area)   (f) Fusion + restoration (affected area)

Figure 12.9 *Overall fusion improvement using the proposed fusion approach enhanced with restoration. Experiments with the 'pebbles' dataset.*

left and the top right of the image. In Figure 12.9(c), the result of the fusion approach using Isotropic Diffusion is depicted. As expected, the fusion algorithm manages to transfer all high quality information to the fusion image except for the area in the centre of the image that still remains blurred. This common degraded area was properly identified by the proposed algorithm, using $p = 1.05$ and $q = 1$, as depicted in Figure 12.6(e). In Figure 12.9(d), the final enhanced image is depicted after the restoration process that has been applied on the common degraded area for $L = 3$. On the whole, the image quality has been enhanced compared to the original fused image. In Figures 12.9(e) and 12.9(f), a focus on the common degraded area in the fused and the fused/restored image can verify the above conclusions. The visible achieved enhancement of the new method is also supported by the PSNR and Q_0 measurements that are described in Table 12.3. The two

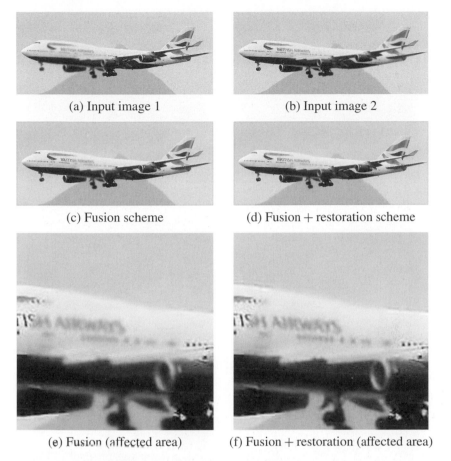

(a) Input image 1 (b) Input image 2

(c) Fusion scheme (d) Fusion + restoration scheme

(e) Fusion (affected area) (f) Fusion + restoration (affected area)

Figure 12.10 *Overall fusion improvement using the proposed fusion approach enhanced with restoration. Experiments with the 'British Airways (BA747)' dataset.*

methods based on error estimation also outperformed the ICA-based transform-domain method, as depicted in Table 12.3.

The third experiment demonstrates the capability of the proposed system to handle noisy cases as well. Two images were artificially created by blurring the upper left and down right respectively of an airplane image (British Airways – BA747) with randomly chosen Gaussian blur kernels. Additive white Gaussian noise of standard deviation 0.03 (input signals normalised to [0, 1]) was also added to both images, yielding an average SNR = 27 dB. As previously, there exists an area in the middle of the image, where the imposed degradations overlap, i.e. there is no ground truth information in any of the input images. The denoising term of the fusion step was activated by selecting $\alpha = 0.08$. In Figure 12.10(c), the result of the fusion approach using Isotropic Diffusion is depicted. As previously, the algorithm managed to perform fusion of the areas where valid information is available in the input images, and also suppress the additive Gaussian noise. The common degraded area was identified using $p = 1$ and $q = 0.99$. These images contain large areas of constant background, whereas the two previous images contained a lot

(a) Input image 1   (b) Input image 2

(c) Ground truth   (d) TopoICA fusion

(e) Fusion scheme   (f) Fusion + restoration scheme

Figure 12.11   *Overall fusion improvement using the proposed fusion approach enhanced with restoration and comparison with the TopoICA fusion scheme. Experiments with the 'Porto' dataset.*

of textural detail. In this case, it is essential to avoid these large areas of constant background to be estimated as part of the common degraded area, and therefore, we choose $q = 0.99$ instead of 1 as previously. The restoration step was applied with $L = 3$, offering an overall enhancement in the visual quality and the actual benchmarks, compared to the error-estimation fusion approach and the ICA-based fusion approach. The calculated metrics suggest that there is limited significant improvement, because the enhancement

(a) Input image 1   (b) Input image 2

(c) Ground truth   (d) TopoICA fusion

(e) Fusion scheme   (f) Fusion + restoration scheme

Figure 12.12 *Overall fusion improvement using the proposed fusion approach enhanced with restoration and comparison with the TopoICA fusion scheme. Experiments with the 'noisy-Porto' dataset.*

in the relatively small common degraded area is averaged with the rest of the image. However, one can observe that there is obvious visual enhancement in the final enhanced image, especially in the common degraded area.

Another final example demonstrates the capability of the proposed system to handle noisy cases and more complicated scenes. Two images were artificially created by blurring the foreground object (statue in Porto, Portugal) along with some adjacent area and another area surrounding the statue with randomly chosen Gaussian blur kernels. A noiseless and a noisy example were created with additive white Gaussian noise of standard deviation 0.04 and 0.05, respectively (input signals were again normalised to [0, 1]). As previously, there exists an area surrounding the statue, where the imposed degradations overlap. In Figure 12.11, the noiseless example is depicted along with the results of the fusion with error estimation approach, the combined fusion–restoration approach and the Topographic ICA with the 'max-abs' rule. The common degraded area was identified using $p = 0.92$ and $q = 1$. In the restoration step, the kernel size was chosen to be $L = 5$. In Figure 12.12, the corresponding results in the case of additive noise are depicted. The denoising term of the fusion step was activated by selecting $\alpha = 0.08$. As previously, the algorithm managed to perform fusion of the areas where valid information is available in the input images, and also suppress the additive Gaussian noise. The calculated performance indexes in Table 12.3 verify again the obvious visual enhancement in the final enhanced image, especially in the common degraded area.

12.5 Conclusions

The problem of image fusion, i.e. the problem of incorporating useful information from various modality input sensors into a composite image that enhances the visual comprehension and surveillance of the observed scene, was addressed in this study. More specifically, a spatial-domain method was proposed to perform fusion of both multi-focus and multi-modal input image sets. This method is based on error estimation methods that were introduced in the past for image enhancement and restoration and are solely performed in the spatial domain. In the case of multi-focus image sets scenarios the proposed spatial-domain framework seems to match the performance of several current popular transform-domain methods, as for example, the wavelet transform and the trained ICA technique. The proposed methodology exhibits also interesting results in the case of multi-modal image sets, producing outputs with distinctively outlined edges compared to transform-domain methods.

More specifically, a combined method of fusion and restoration was proposed as the next step from current fusion systems. By definition, fusion systems aim only at transferring the 'interesting' information from the input sensor images to the fused image, assuming there is proper reference image signal for all parts of the image in at least one of the input sensor images. In the case that there exist common degraded areas in all input images, the fusion algorithms cannot improve the information provided there, but simply convey this degraded information to the output. In this study, we proposed a mechanism of identifying these common degraded areas in the fused image and use a regularised restoration approach to enhance the content in this area. In the particular case of multi-focus images, the proposed approach managed to remove the blur and enhance the edges in the common degraded area, outperforming current transform-based fusion systems.

There are several potential applications of the proposed system. Military targeting or surveillance units can benefit from a combined fusion and restoration platform to improve their targeting and identification performance. Commercial surveillance appliances can also benefit from a multi-camera, multi-focus system that fuses all input information into a composite image with wide and detailed focus. In addition, there are several other applications such as increasing the resolution and quality of pictures taken by commercial digital cameras.

Acknowledgement

This work has been funded by the UK Data and Information Fusion Defence Technology Centre (DIF DTC) AMDF cluster project.

References

[1] P. Hill, N. Canagarajah and D. Bull, 'Image fusion using complex wavelets', in *Proc. 13th British Machine Vision Conference*, Cardiff, UK, 2002.

[2] N. Kingsbury, 'The dual-tree complex wavelet transform: A new technique for shift invariance and directional filters', in *Proc. IEEE Digital Signal Processing Workshop*, Bryce Canyon, UT, USA, 1998.

[3] S.G. Nikolov, D.R. Bull, C.N. Canagarajah, M. Halliwell and P.N.T. Wells, 'Image fusion using a 3-d wavelet transform', in *Proc. 7th International Conference on Image Processing and Its Applications*, 1999, pp. 235–239.

[4] G. Piella, 'A general framework for multiresolution image fusion: From pixels to regions', *Information Fusion*, Vol. 4, 2003, pp. 259–280.

[5] N. Mitianoudis and T. Stathaki, 'Pixel-based and region-based image fusion schemes using ICA bases', *Information Fusion*, Vol. 8, No. 2, 2007, pp. 131–142.

[6] N. Mitianoudis and T. Stathaki, 'Adaptive image fusion using ICA bases', in *Proceedings of the International Conference on Acoustics, Speech and Signal Processing*, Toulouse, France, May 2006.

[7] S. John and M.A. Vorontsov, 'Multiframe selective information fusion from robust error estimation theory', *IEEE Transactions on Image Processing*, Vol. 14, No. 5, 2005, pp. 577–584.

[8] D. Zwillinger, *Handbook of Differential Equations*, Academic Press, Boston, 1997.

[9] M.J. Black, G. Sapiro, D.H. Marimont and D. Heeger, 'Robust anisotropic diffusion', *IEEE Transactions on Image Processing*, Vol. 7, No. 3, 1998, pp. 421–432.

[10] P. Perona and J. Malik, 'Scale-space and edge detection using anisotropic diffusion', *IEEE Transactions on Pattern Analysis and Machine Intelligence*, Vol. 12, No. 7, 1990, pp. 629–639.

[11] The Image Fusion server, http://www.imagefusion.org/.

[12] G. Piella, 'New quality measures for image fusion', in *7th International Conference on Information Fusion*, Stockholm, Sweden, 2004.

[13] C. Xydeas and V. Petrovic, 'Objective pixel-level image fusion performance measure', in *Sensor Fusion IV: Architectures, Algorithms and Applications*, in *Proceedings of SPIE*, Vol. 4051, 2000, pp. 88–99.

[14] A. Toet, 'Detection of dim point targets in cluttered maritime backgrounds through multisensor image fusion', in *Targets and Backgrounds VIII: Characterization and Representation*, in *Proceedings of SPIE*, Vol. 4718, 2002, pp. 118–129.

[15] D. Kundur and D. Hatzinakos, 'Blind image deconvolution', *IEEE Signal Processing Magazine*, Vol. 13, No. 3, 1996, pp. 43–64.

[16] H.C. Andrews and B.R. Hunt, *Digital Image Restoration*, Prentice–Hall, 1997.

[17] Y.L. You and M. Kaveh, 'A regularisation approach to joint blur identification and image restoration', *IEEE Transactions on Image Processing*, Vol. 5, No. 3, 1996, pp. 416–428.

[18] R. Molina, A.K. Katsaggelos and J. Mateos, 'Bayesian and regularization methods for hyperparameter estimation in image restoration', *IEEE Transactions on Image Processing*, Vol. 8, No. 2, 1999, pp. 231–246.

[19] M.E. Zervakis and T.M. Kwon, 'On the application of robust functionals in regularized image restoration', in *Proceedings of the International Conference on Acoustics, Speech and Signal Processing*, April 1993, pp. V289–V292.

[20] D.B. Gennery, 'Determination of optical transfer function by inspection of the frequency-domain plot', *Journal of the Optical Society of America*, Vol. 63, 1973, pp. 1571–1577.

[21] Z. Wang and A.C. Bovik, 'A universal image quality index', *IEEE Signal Processing Letters*, Vol. 9, No. 3, 2002, pp. 81–84.

13

Empirical mode decomposition for simultaneous image enhancement and fusion

David Looney and Danilo P. Mandic

Imperial College London, UK

Image enhancement and restoration via information fusion is addressed using the inherent fission properties of Empirical Mode Decomposition (EMD). Fission via EMD describes the adaptive decomposition of an image signal into a set of oscillatory modes that act as a set of naturally derived basis functions. Embedded in the modes are the frequency scales of the image data. Given that a variety of humanly observable image features such as object texture or degradation effects such as noise can often be attributed to local variations in specific spatial frequencies, it follows that the behaviour of the extracted image modes reflect these features. Simultaneous restoration and enhancement can be achieved by fusion of the 'relevant' modes. The fusion process can be used to highlight specific image attributes to aid machine vision tasks such as object recognition. Examples on a variety of image processing problems from denoising through to the fusion of visual and thermal images support the analysis and also it is demonstrated how machine learning techniques such as adaptive filtering can be combined with EMD to enhance the fusion process.

13.1 Introduction

A variety of tasks within the field of image processing, such as segmentation, object recognition and tracking, are critically sensitive to external real world elements caused by environment (scene illumination) or poor quality data acquisition (noise). Furthermore, a given task may require that specific features of an image be emphasised for the purpose of efficient classification, as is true in the case of classifiers that seek object textures. This has highlighted the necessity for 'signal conditioning,' that is for optimal, fast and reliable pre-processing at the data level with the aim of simultaneous image restoration and enhancement. This is particularly important for images recorded in real world situations.

A number of different techniques for image restoration and enhancement have been proposed over the years [1]. Examples include: principal and independent component analysis [2]; maximum entropy and maximum likelihood methods; iterative methods, and the use of artificial neural networks [3]. Often it is the case with some of these approaches that unrealistic assumptions are made about the data. For example, decomposition methods such as principal component analysis (PCA) make assumptions of linearity and stationarity. Similar criticisms can be made of independent component analysis (ICA) which requires that the image components be statistically independent. It is possible to achieve superior image processing with high performance requirements (Bayesian and particle models [4]) but this comes at the cost of high computational complexity. It can also be argued that increasingly restrictive methods that limit the number of possible solutions can detract from the overall performance.

As image features (noise, texture, incident illumination effects) often correspond to variations in spatial frequencies, Fourier based methods are often employed for restoration (as, for example, denoising) or feature enhancement (image smoothing or sharpening). Despite the power of these techniques, they rely on a projection onto a linear set of predefined bases. This limits the abilities of Fourier methods when processing real world images that often display nonlinear and non-stationary behaviour. Additionally, the use of a fixed basis set restricts the analysis of high frequency content.

To that end, we consider image decomposition via Empirical Mode Decomposition (EMD) [5]. Uniquely, the approach can be seen as the result of a filtering process [6] as the decompositions are a set of narrow band components (called Intrinsic Mode Functions or IMFs) that inherently reflect the variations in the spatial frequencies of the image. Unlike Fourier methods, the approach is fully adaptive and makes no known prior assumptions of the data [5].

With the support of examples, this chapter will show how simultaneous image enhancement and restoration can be achieved through intelligent 'fusion via fission.' It is first demonstrated how EMD fits within the general framework of information fusion via fission and then the decomposition algorithm is rigorously explained. Image restoration and enhancement are illustrated on case studies of: (i) image denoising; (ii) shade removal; (iii) texture analysis; and (iv) fusion of multiple image modalities (thermal and visual).

13.2 EMD and information fusion

Data and information fusion is the approach whereby data from multiple sensors or components is combined to achieve improved accuracies and more specific inferences that could not be achieved by the use of only a single sensor [7]. Its principles have been employed in a number of research fields including information theory, signal processing and computing [7–10], and an overview can be found in [11].

Figure 13.1 *The 'waterfall model' of information fusion.*

Recent work [12] demonstrates that the decomposition nature of EMD provides a unifying framework for 'information fusion via fission,' where fission is the phenomenon by which observed information is decomposed into a set its components. More specifically, the stages of Signal Processing, Feature Extraction and Situation Assessment from the waterfall model (a well-established fusion model given in Figure 13.1) can all be achieved by EMD.

13.2.1 Empirical mode decomposition

Empirical mode decomposition [5] is a technique to adaptively decompose a given signal, by means of a process called the sifting algorithm, into a finite set of AM/FM modulated components. These components, called 'intrinsic mode functions' (IMFs), represent the oscillation modes embedded in the data. The IMFs act as a naturally derived set of basis functions for the signal; EMD can thus be seen as an exploratory data analysis technique. In fact, EMD and the Hilbert–Huang transform comprise the so-called 'Hilbert spectral analysis' [5]; a unique spectral analysis technique employing the concept of instantaneous frequency. In general, the EMD aims at representing an arbitrary signal via a number of IMFs and the residual. More precisely, for a real-valued signal $x[k]$, the EMD performs the mapping

$$x[k] = \sum_{i=1}^{N} c_i[k] + r[k] \tag{13.1}$$

where the $c_i[k]$, $i = 1, \ldots, N$ denote the set of IMFs and $r[k]$ is the trend within the data (also referred to as the last IMF or residual). By design, an IMF is a function which is characterised by the following two properties: the upper and lower envelope are symmetric; and the number of zero-crossings and the number of extrema are exactly equal or they differ at most by one.

In order to extract the IMFs from a real world signal, the *sifting* algorithm is employed, which is described in Table 13.1. Following the sifting process, the Hilbert transform can be applied to each IMF separately. This way, it is possible to generate analytic signals, having an IMF as the real part and its Hilbert transform as the imaginary part, that is $x + j\mathcal{H}(x)$ where $\mathcal{H}$ is the Hilbert transform operator. Equation (13.1) can therefore be augmented to its analytic form given by

$$X(t) = \sum_{i=1}^{n} a_i(t) \cdot e^{j\theta_i(t)} \tag{13.2}$$

where the trend $r(t)$ is purposely omitted, due to its overwhelming power and lack of oscillatory behaviour. Observe from (13.2), that now the time-dependent amplitude $a_i(t)$

Table 13.1 *The EMD algorithm.*

---

1. Connect the local maxima of the signal with a spline. Let U denote the spline that forms the upper envelope of the signal.
2. Connect the local minima of the signal with a spline. Let L denote the spline that forms the lower envelope of the signal.
3. Subtract the mean envelope $m = (U + L)/2$ from the signal to obtain a proto-IMF.
4. Repeat Steps 1, 2 and 3 above until the resulting signal is a proper IMF (as described above). The IMF requirements are checked indirectly by evaluating a stoppage criterion, originally proposed as

$$\sum_{k=0}^{T} \frac{|h_{n-1}[k] - h_n[k]|^2}{h_{n-1}^2[k]} \leqslant \text{SD}$$

   where $h_n[k]$ and $h_{n-1}[k]$ represent two successive sifting iterates. The SD value is usually set to 0.2–0.3.
5. After finding an IMF, this same IMF is subtracted from the signal. The residual is regarded as new data and fed back to Step 1 of the algorithm.
6. The algorithm is completed when the residual of Step 5 is a monotonic function. The last residual is considered to be the trend.

---

can be extracted directly and that we can also make use of the phase function $\theta_i(t)$. Furthermore, the quantity $f_i(t) = d\theta_i/dt$ represents the instantaneous frequency [13]; this way by plotting the amplitude $a_i(t)$ versus time t and frequency $f_i(t)$, we obtain a time–frequency–amplitude representation of the entire signal called the Hilbert spectrum. It is this combination of the concept of instantaneous frequency and EMD that makes the framework so powerful as a signal decomposition tool.

13.3 Image denoising

It is often the case, regardless of the quality of the camera used, that some level of unwanted noise contaminates the final image. Image denoising is therefore a vitally important aspect of image processing and has received much attention in recent years. Although there are many different algorithms designed to reduce noise levels, they often make assumptions about the data and can fail when the subject image does not conform to certain criteria. Projection based methods such as PCA rely on a correct choice of subspace for both the signal and noise and, as noted before, ICA assumes unrealistic independence conditions. Furthermore, it is desirable to use algorithms that are fast and that have low computational complexity. To this end, finite impulse response (FIR) adaptive filtering algorithms [14,15] are commonly used.

Adaptive filtering algorithms demonstrate fast and effective denoising capabilities, both supervised and blind. The image information is either processed as one-dimensional (1D) data by concatenating the rows and columns of the image matrix, or by direct two-dimensional (2D) processing. A crucial problem with either approach is the correct choice for the dimensions of the filter or the 'support region.' The choice is not straightforward [3], while a large support region can theoretically facilitate improved denoising performance, it may also introduce unwanted artifacts (such as scratching or blurring) which become stored in the memory and get propagated throughout the image. However, in the analysis of medical or multimedia recording and also in pattern recognition, the

Figure 13.2 *EMD and image denoising. The original image.*

Figure 13.3 *EMD and image denoising. The original image with added white Gaussian noise. SNR: 13 dB.*

performance of the high level algorithms is critically dependent on the 'perceptual quality' of the input images. It is therefore crucial that the desired denoising algorithm should not only be fast, but should maintain a balance between quantitative performance and retaining the quality of the original image.

As EMD fission decomposes a signal into components referring to its respective frequency scales, it is natural to assume that it has potential in image denoising. Because noise primarily corrupts the high frequency detail of the image, the high index IMFs are often dominated by the interference signal.

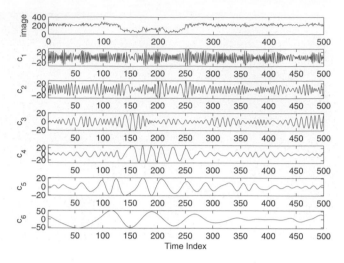

Figure 13.4 *EMD and image denoising. A segment of the image signal (top) and the first six of the extracted IMFs.*

Figure 13.5 *EMD and image denoising. Image restoration using the partial reconstruction approach. The 'best' IMF components are selected by visual inspection. SNR: 17.5 dB.*

Consider the original image in Figure 13.2. Zero mean white Gaussian noise is introduced so that image has a signal-to-noise-ratio (SNR) of 13 dB (Figure 13.3). Treating each column (or alternatively row) in the image matrix as a separable signal, we can convert the two-dimensional data into a single vector. In our example, this vector is decomposed into a set of 18 IMFs via EMD. Segments of the first 6 of these IMFs is shown in Figure 13.4.

Basic image restoration can be achieved by fusion of IMF components 4–18, as is shown in Figure 13.5. For this example, the selection was made empirically by noting visually which IMFs 'best' retained the image without noise. The SNR of the restored image is 17.5 dB. This partial reconstruction approach can also be accomplished in an automated

Figure 13.6 *EMD and image denoising. Image restoration using OEMD. SNR: 18.7 dB.*

fashion [16]. It follows the behaviour of EMD as a dyadic filter bank for white Gaussian noise [6,17] and the fact that the IMF log-variance follows a simple linear model controlled by the Hurst exponent. In [16], the use of confidence intervals which determine the level of noise energy in each of the IMF components was discussed for automatic noise removal by simply omitting IMFs dominated by noise energy. Although effective, the partial reconstruction approach is computationally complex (multiple simulations must be run to determine the noise-only confidence intervals) and is handicapped by its binary decision nature. Additionally, it is not based on any optimality criterion.

In [18], two linear algorithms were proposed which fuse the IMF components to produce the best estimate of the original signal in the least mean squares sense. The first (optimal EMD or OEMD), determines an optimal vector set of weights for estimating the original signal by linear weighting of the IMFs. The second (bidirectional optimal EMD or BOEMD) takes into account the temporal structure of a signal and determines an optimal two-dimensional weight matrix to approximate the original signal. It was shown that for several denoising applications, OEMD and BOEMD outperformed the standard partial reconstruction approach. Restoration of the corrupted image using OEMD is given in Figure 13.6, the computed SNR is 18.7 dB. Although more effective, the approach is still limited by its static block based nature, which does not facilitate local feature fusion.

It was suggested [19] that a more adaptive fusion process could be achieved by combining EMD with machine learning (adaptive filtering) so as to cater for the non-stationarity of the original image data and to facilitate local feature relevance estimation. The results of EMD combined with the Generalised Normalised Gradient Descent (GNGD) [20] algorithm are shown in Figure 13.7. It is noted that not only does the combined machine learning and EMD fusion approach give a high quantitative SNR performance (21 dB), but also the perceptual quality of the image is better retained compared to the OEMD results in Figure 13.6. Unlike standard filtering approaches, large support regions are not required [19], and scratching effects are dramatically reduced. The overall texture of the image reflects more accurately that of the original (compare the background of

Figure 13.7   *EMD and image denoising. Image restoration using EMD and machine learning. SNR: 21 dB.*

Figures 13.6 and 13.7) and the image edges are sharper (this is demonstrated by the level of detail restored to the patterns in the clothes).

13.4 Texture analysis

Texture analysis is extremely important in the field of image processing, allowing classifiers to identify objects of interest by locating their unique texture signatures. A variety of different schemes have been proposed over the years including Markovian analysis, fractal models and filtering methods [21]. A natural interpretation of textures is specific trends in high spatial frequencies. Given its nature, it has been demonstrated how image decomposition via EMD is capable of isolating texture detail [22].

A simple algorithm is proposed for the purposes of texture retrieval. An image matrix is decomposed into a set of row vectors and EMD is performed on each of the row vectors. Correspondingly, EMD is then performed on the set of column vectors. A threshold IMF index is arbitrarily chosen. Only IMF values corresponding to indices greater than the threshold are considered (high frequency IMFs). The average of these values, obtained from both the row and column decomposition stages, is then determined. The information contained within these averaged values, providing the threshold index is suitably high, represents the image texture, hereby referred to as the image 'texture layer.' The texture layer is removed from the original image exposing the underlying 'residue layer.'

Consider the image of a wood panel in Figure 13.8. Note that the characteristic wood texture is composed of high frequency texture detail superimposed onto vertical bands that alternate in intensity. Applying the algorithm proposed above and setting the threshold index to 3, texture and residue layers for the image are obtained (Figure 13.8). It is clear that the characteristic 'wood texture' is retained by the high frequency IMFs. Furthermore, the underlying intensity pattern of vertical bands is evident in the residue image. The algorithm is applied to an image of a carpet given in Figure 13.9. It is again evident

Figure 13.8 *EMD and texture analysis. Top: An original image of a wood panel. Bottom left: The texture layer. Bottom right: The residue layer.*

that the texture is captured in the texture layer composed of the high frequency IMFs (Figure 13.9), while the underlying chequered intensity pattern is visible in the residue image.

The simple algorithm described above demonstrates the inherent fission nature of EMD applied to images, facilitating the decomposition of an image signal into its respective texture layers. Although what was achieved was mere texture isolation, more advanced approaches utilising EMD for the purposes of texture classification are available. Work in [22] proposes a method whereby EMD is applied along several directions of the image for robust, rotation-invariant texture classification. Extensions of EMD to 2D for texture analysis are also available in the literature [23,24].

13.5 Shade removal

A key problem for a machine vision system is image changes that occur due to scene illumination. Incident light on a surface produces complex artifacts, making it difficult for the system to separate changes caused by local variations in illumination intensity and colour. This causes problems for a number of rudimentary vision operations such as object recognition and image segmentation.

Figure 13.9 *EMD and texture analysis. Top: An original image of a carpet texture. Bottom left: The texture layer. Bottom right: The residue layer.*

A number of solutions addressing the removal of illumination intensity or shade have been proposed. A recent solution [25], demonstrates that a 1D greyscale image free from incident illumination intensity may be constructed from a full colour 3D image. This greyscale data is then used to remove the shadows from the original 3D image. The procedure, although effective, is computationally complex requiring the image data to be projected into a space that minimises its entropy. Another algorithm [26] addresses the idea of creating comprehensive normalised images; so that a given object captured under a variety of different illumination conditions has the theoretically the same normalised appearance. It is effective and computationally light. Critically, however, it also requires a full colour 3D image to isolate illumination artifacts by examining changes across all colour channels.

The fission properties of EMD can alternatively be utilised to separate shade artifacts from an image. As observed in [27], it can be assumed that shade in images creates low valued regions with large extrema that change slowly. Therefore, it is likely that the effects of the shade will be isolated in the lower index IMFs. Intelligent fusion of the shade free image can be achieved by combining the relevant IMFs.

Shade removal was achieved on images in Figures 13.10 and 13.11 from their respective original images by removing the residue obtained from each and replacing it with its mean value. In both cases, low level Gaussian noise was added to the original images to aid the fission process. This idea of noise assisted EMD is described in [28]. It is shown

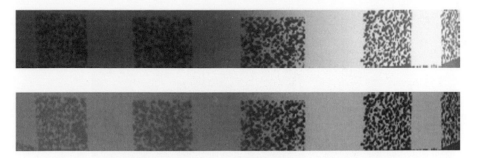

Figure 13.10 *EMD and shade removal. Top: The original image, note the increasing level of brightness from left to right across the image. Bottom: Shade removal. The shading is now uniform across the image surface.*

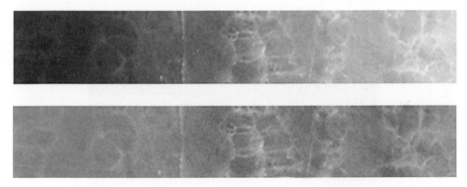

Figure 13.11 *EMD and shade removal. Top: The original image. Note the increasing level of brightness from left to right across the image. Bottom: Shade removal. The shading is now uniform across the image surface.*

that in certain conditions, where mode mixing between IMFs can occur, the true IMF components can be extracted by adding low level noise and ensemble averaging over several simulations. However, the averaging process (known as Ensemble EMD) mostly affects the high frequency IMF components. As this particular algorithm requires only the residue, no averaging is necessarily required. The solution is computationally efficient and crucially it can be performed on a greyscale image; it does not require information from separate colour channels.

13.6 Fusion of multiple image modalities

As different methods of image acquisition have become available, the principle of fusing images from several image sensors has become an interesting topic in the literature [29–31]. Ideally, the fused or hybrid image retains all 'relevant information' from the different sources while unwanted artifacts and noise are disregarded. Specifically, the increasing availability of thermal cameras has let to a development of fusion techniques to combine visual and thermal images. Examples include Wavelet [32] and PCA [33] based algorithms.

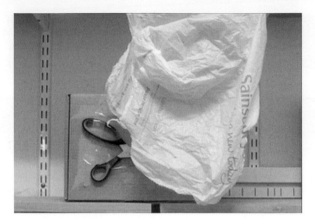

Figure 13.12 *EMD and fusion of multiple image modalities. The visual image. Note the level of detail and sharp definition.*

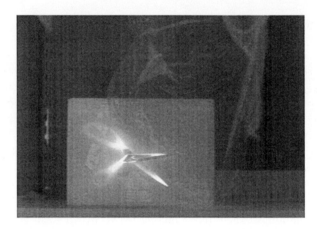

Figure 13.13 *EMD and fusion of multiple image modalities. The thermal image. The scissors is no longer obstructed by the plastic bag.*

In [34] the use of EMD was proposed, whereby the 1D IMF components extracted from both the visual and thermal images are fused via a linear weighting scheme. An example of a fused image is given in Figure 13.14, where the IMFs from the visual image shown in Figure 13.12 and the thermal image shown in Figure 13.13 are combined using an empirically selected set of weights. The thermal image was obtained using an Infratec VarioCAM hr head camera. On the one hand, the visual image clearly depicts a pair of scissors partially obstructed by a plastic bag. On the other hand, the scissors is fully visible in the thermal image but overall it lacks the definition of the visual image. Additionally it is more susceptible to noise. The fused image demonstrates how the relevant modalities from each image have been preserved. The optimal weighting scheme (proposed in [34]) was that which minimised the mutual information between the IMF components. Although the work did not provide an automated scheme for producing the weights, it was demonstrated that EMD fusion had the potential to outperform other existing algorithms.

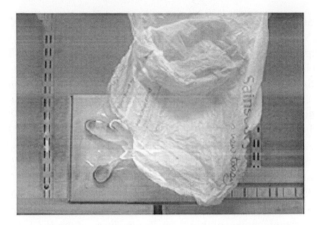

Figure 13.14 *EMD and fusion of multiple image modalities. The fused image which preserves the relevant modalities.*

13.7 Conclusion

It has been demonstrated how decomposition via EMD and the subsequent fusion of the extracted components is a powerful tool for the purposes of image enhancement and restoration under the general framework of fusion via fission. EMD is non-parametric and self-adaptive which is advantageous when processing real world images which display non stationary and nonlinear behaviour. Furthermore, the approach is computationally light and instantly intuitive compared to other existing algorithms. This has been supported by several simulations in applications such as image denoising and shade removal. Although 2D extensions of the algorithm are available in the literature, it has been illustrated how standard 1D EMD is suitably effective. This is important for fast and on-line processing. Fusion of the IMF decompositions can be achieved in an ad hoc or an automated fashion. It has also been shown how the fusion process can be combined with machine learning techniques for improved performance.

References

[1] M. Petrou and P. Bosdogianni, *Image Processing: The Fundamentals*, Wiley, 1999.

[2] A. Cichocki and S.I. Amari, *Blind Signal and Image Processing*, Wiley, 2002.

[3] A. Palmer, M. Razaz and D.P. Mandic, 'A spatially-adaptive neural network approach to regularized image restoration (invited paper)', *Journal of Intelligent & Fuzzy Systems*, Vol. 13, 2003, pp. 177–185.

[4] D. Sivia and J. Skilling, *Data Analysis: A Bayesian Tutorial*, Oxford University Press, 2006.

[5] N.E. Huang, Z. Shen, S.R. Long, M.L. Wu, H.H. Shih, Z. Quanan, N.C. Yen, C.C. Tung and H.H. Liu, 'The empirical mode decomposition and the Hilbert spectrum for nonlinear and non-stationary time series analysis', *Proceedings of the Royal Society, Series A*, Vol. 454, 1998, pp. 903–995.

[6] P. Flandrin, G. Rilling and P. Goncalves, 'Empirical mode decomposition as a filter bank', *IEEE Signal Processing Letters*, Vol. 11, No. 2, 2004, pp. 112–114.

[7] D.L. Hall and J. Llinas, 'An introduction to multisensor data fusion', *Proceedings of the IEEE*, Vol. 85, No. 1, 1997, pp. 6–23.

[8] D.F. Group, 'Functional description of the data fusion process', Technical report, Office of Naval Technology, 1992.

[9] L. Wald, 'Some terms of reference in data fusion', *IEEE Transactions on Geoscience and Remote Sensing*, Vol. 37, No. 3, 1999, pp. 1190–1193.

[10] E. Waltz and J. Llinas, *Multisensor Data Fusion*, Artech House, 1990.

[11] D.P. Mandic, D. Obradovic, A. Kuh, T. Adali, U. Trutschell, M. Golz, P.D. Wilde, J. Barria, A. Constantinides and J. Chambers, 'Data fusion for modern engineering applications: An overview', in *Proceedings of the IEEE International Conference on Artificial Neural Networks (ICANN'05)*, 2005, pp. 715–721.

[12] D.P. Mandic, M. Golz, A. Kuh, D. Obradovic and T. Tanaka, *Signal Processing Techniques for Knowledge Extraction and Information Fusion*, Springer, 2007.

[13] L. Cohen, 'Instantaneous anything', in *Proceedings of the IEEE International Conference on Acoustics, Speech and Signal Processing (ICASSP)*, Vol. 5, 1993, pp. 105–108.

[14] B. Widrow and S.D. Stearns, *Adaptive Signal Processing*, Prentice–Hall, Englewood Cliffs, NJ, 1985.

[15] W.B. Mikhael and S.M. Ghosh, 'Two-dimensional variable-step-size sequential adaptive gradient algorithms with applications', *IEEE Transactions on Circuits and Systems*, Vol. 38, 1991, pp. 1577–1580.

[16] P. Flandrin, P. Goncalves and G. Rilling, 'Detrending and denoising with empirical mode decompositions', in *Proceedings of the 12th European Signal Processing Conference (EUSIPCO'04)*, Vol. 2, 2004, pp. 1581–1584.

[17] Z. Wu and N.E. Huang, 'A study of the characteristics of white noise using the empirical mode decomposition method', *Royal Society of London Proceedings, Series A*, Vol. 460, 2004, pp. 1597–1611.

[18] B. Weng and K.E. Barner, 'Optimal and bidirectional optimal empirical mode decomposition', in *Proceedings of the IEEE International Conference on Acoustics, Speech, and Signal Processing (ICASSP'07)*, Vol. III, 2007, pp. 1501–1504.

[19] D. Looney, D. Mandic and T. Rutkowski, 'An adaptively regularised method for denoising of real world images', in *Proceedings of the 4th Institution of Engineering and Technology International Conference on Visual Information Engineering (VIE'07)*, 2007.

[20] D.P. Mandic, 'A generalized normalized gradient descent algorithm', *IEEE Transactions on Signal Processing*, Vol. 11, 2004, pp. 115–118.

[21] A.W. Smeulders, M. Worring, S. Santini, A. Gupta and R. Jain, 'Content-based image retrieval at the end of the early years', *IEEE Transactions on Pattern Analysis and Machine Intelligence*, Vol. 22, No. 12, 2000, pp. 1349–11379.

[22] Z.X. Liu, H.J. Wang and S.L. Peng, 'Texture classification through directional empirical mode decomposition', in *Proceedings of the 17th International Conference on Pattern Recognition (ICPR'04)*, 2004, pp. 803–806.

[23] J.C. Nunes, S. Guyot and E. Deléchelle, 'Texture analysis based on local analysis of the bidimensional empirical mode decomposition', *Machine Vision and Applications*, Vol. 16, No. 3, 2005, pp. 177–188.

[24] C.Z. Xiong, J.Y. Xu, J.C. Zou and D.X. Qi, 'Texture classification based on EMD and FFT', *Journal of Zhejiang University (Science)*, Vol. 7, No. 9, 2006, pp. 1516–1521.

[25] G.D. Finlayson, S.D. Hordley, C. Lu and M.S. Drew, 'On the removal of shadows from images', *IEEE Transactions on Pattern Analysis and Machine Intelligence*, Vol. 28, No. 1, 2006, pp. 59–68.

[26] G.D. Finlayson, B. Schiele and J.L. Crowley, 'Comprehensive colour image normalization', in *Proceedings of the 5th European Conference on Computer Vision (ECCV'98)*, Vol. 1, 1998, pp. 475–490.

[27] R. Bhagavatula and M. Savvides, 'Analyzing facial images using empirical mode decomposition for illumination artifact removal and improved face recognition', in *Proceedings of the IEEE International Conference on Acoustics, Speech, and Signal Processing (ICASSP'07)*, Vol. I, 2007, pp. 505–508.

[28] Z. Wu and N.E. Huang, 'Ensemble empirical mode decomposition: A noise-assisted data analysis method', Technical report 193, Center for Ocean–Land–Atmosphere Studies, 2004.

[29] L.P. Yaroslavsky, B. Fishbain, A. Shteinman and S. Gepshtein, 'Processing and fusion of thermal and video sequences for terrestrial long range observation systems', in *Proceedings of the 7th Annual International Conference of Information Fusion*, 2004, pp. 848–855.

[30] D.A. Fay, A.M. Waxman, M. Aguilar, D.B. Ireland, J.P. Racamato, W.D. Ross, W.W. Streilein and M.I. Braun, 'Color visualization, target learning and search', in *Proceedings of the 3rd Annual International Conference of Information Fusion*, 2000, pp. 215–219.

[31] D.A. Socolinsky and L.B. Wolff, 'Multispectral image visualization through first-order fusion', *IEEE Transactions on Image Processing*, Vol. 11, No. 8, 2002, pp. 923–931.

[32] H. Li, B.S. Manjunathand and S.K. Mitra, 'Multisensor image fusion using the wavelet transform', *Graphic Models and Image Processing*, Vol. 57, No. 3, 1995, pp. 235–245.

[33] G. Bebis, A. Gyaourova, S. Singh and I. Pavlidis, 'Face recognition by fusing thermal infrared and visible imagery', *Image and Vision Computing*, Vol. 24, No. 7, 2006, pp. 727–742.

[34] H. Hariharan, A. Gribok, M.A. Abidi and A. Koschan, 'Image fusion and enhancement via empirical mode decomposition', *Journal of Pattern Recognition Research (JPRR)*, Vol. 1, No. 1, 2006, pp. 16–32.

14

Region-based multi-focus image fusion

Shutao Li and Bin Yang

College of Electrical and Information Engineering, Hunan University, Changsha, China

14.1 Introduction

Recently, region-based image fusion has attracted considerable attention because of its perceived advantages, which include: (1) The fusion rules are based on combining regions instead of pixels. Thus, more useful tests for choosing proper regions from the source images, based on various properties of a region, can be implemented prior to fusion. (2) Processing semantic regions rather than at individual pixels can help overcome some of the problems with pixel-fusion methods such as sensitivity to noise, blurring effects and misregistration [1].

A number of region-based image fusion methods have been proposed [1–6]. Most of the existing methods are implemented in a transform domain. A generic diagram illustrating a region-based fusion method that uses multiresolution (MR) analysis is shown in Figure 14.1. Firstly, pre-registered images are transformed using a multiresolution analysis method. Regions representing specific image features are then extracted by an image segmentation method that utilises the information obtained from the transform coefficients. The regions are then fused based on region characteristics. Experimental results of these methods are encouraging. However, an image fused in this way may still lose some information present in the source images because of the implementation of some type of inverse multiresolution transform. To solve this problem, region-based image fusion approaches which are implemented in spatial domain have been proposed [7–9].

In this chapter, the principles of region-based image fusion in spatial domain are first described in detail. Then two region-based fusion methods are introduced. Experimental results arising from the proposed methods are also presented.

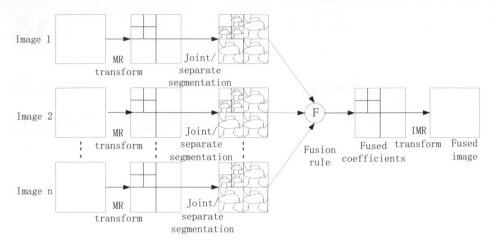

Figure 14.1 *Region-based image fusion scheme in transform domain.*

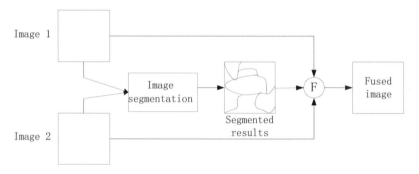

Figure 14.2 *Region-based image fusion scheme in spatial domain.*

14.2 Region-based multi-focus image fusion in spatial domain

A diagram that illustrates a generic fusion scheme in spatial domain is shown in Figure 14.2. It consists of three steps: image segmentation, region clarity calculation, and construction of the fused image. The source images are first segmented using a segmentation method. Then the clarity of every region in each source image is calculated according to specific clarity measures. In this context clarity refers to the quality of focus in an image area or the local saliency (sharpness, high frequency information, edge presence, etc.). Finally, the fused image is constructed by the fusion rules based on the clarity. Obviously, the result of image segmentation and the computation of a specific focus measure are of vital importance for the fusion outcome.

14.2.1 Image segmentation

The segmentation results are vital for the fusion outcome as the whole implementation of fusion is based on the segmented regions. For image fusion task, there are at least two source images available and the segmentation operation can be performed in two ways: a segmentation map may be obtained by considering either jointly all the input images or, otherwise, each of the input images to generate an independent segmentation map. The corresponding approaches are defined as joint and separate segmentation, respectively. A problem can occur for separately segmented images in areas where different images have different features or when similar features appear slightly different in size especially in image fusion scenarios where the source images come from different modalities. Where detected regions partially overlap, if the overlapped section is incorrectly dealt with, artifacts will be introduced in the fusion result. Moreover, dealing with those extra regions will increase the time of the fusion process and this is the main problem of separate segmentation. However, if the information from the segmentation process is going to be used to register the images or if input images are completely different, it can be more effective to separately segment the images.

14.2.2 Focus measures

The clarity measures are also of vital importance for the image fusion. There are three focus measures that can be used for fusion, namely, the L_2 norm of image gradient (L_2G), the absolute central moment (ACM), and the spatial frequency (SF) [10] defined as follows: the L_2 norm of image gradient (L_2G) is

$$L_2G = \sqrt{\sum_{m=0}^{M-1}\sum_{n=1}^{N-1}\left[\left(L_m * F(m,n)\right)^2 + \left(L_n * F(m,n-1)\right)^2\right]}$$

where the following are the edge detection Sobel masks

$$L_m = \begin{bmatrix} -1 & -2 & -1 \\ 0 & 0 & 0 \\ 1 & 2 & 1 \end{bmatrix}, \quad L_n = \begin{bmatrix} -1 & 0 & 1 \\ -2 & 0 & 2 \\ -1 & 0 & 1 \end{bmatrix} \tag{14.1}$$

Absolute central moment (ACM) is

$$ACM = \sum_{i=0}^{I-1} |i - \mu| p(i) \tag{14.2}$$

where μ is the mean intensity value of the image, and i is the grey level with maximum value I.

Spatial frequency (SF) is

$$SF = \sqrt{(RF)^2 + (CF)^2} \tag{14.3}$$

where RF and CF are the row frequency

$$RF = \sqrt{\frac{1}{MN} \sum_{m=0}^{M-1}\sum_{n=1}^{N-1} [F(m,n) - F(m,n-1)]^2}$$

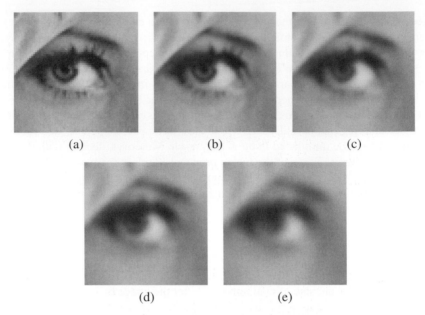

Figure 14.3 *Image region cropped from 'Lena' and its blurred versions: (a) Original image; (b) Gaussian blur with radius 0.5; (c) Gaussian blur with radius 0.8; (d) Gaussian blur with radius 1.0; (e) Gaussian blur with radius 1.5.*

and column frequency

$$CF = \sqrt{\frac{1}{MN} \sum_{m=1}^{M-1} \sum_{n=0}^{N-1} \left[F(m,n) - F(m-1,n)\right]^2}$$

respectively. From (14.3) we can easily deduct that *SF* is a single measurement of the amount of frequency content within the image of interest. All these measures can measure the clarity of a region effectively.

We use *SF* as the clarity measure in our experiments. Usually, the human visual system is too complex to be fully understood with present psychological means, but the use of spatial frequency has led to an effective objective quality assessment for image compression. An experiment which was designed in [7] verifies this argument. An image region of size 64×64 cropped from the standard 'Lena' image is shown in Figure 14.3(a). Figures 14.3(b) to 14.3(e) show the Gaussian blurred versions of Figure 14.3(a). The blurring radii are 0.5, 0.8, 1.0, and 1.5, respectively. The spatial frequencies of the cropped and blurred versions of 'Lena' are given in Table 14.1, from which it can be observed, as expected, that the spatial frequency content reduces, while the image gets more blurred.

A similar experiment is implemented to an image region selected from the standard 'Peppers' image. Its blurred versions are also generated by Gaussian blurring with radii of 0.5, 0.8, 1.0, and 1.5, respectively. The same conclusions as in the previous experiment are valid here.

Table 14.1 *Spatial frequency of the image regions in Figure 14.3.*

	Figure 14.3(a)	Figure 14.3(b)	Figure 14.3(c)	Figure 14.3(d)	Figure 14.3(e)
SF	16.10	12.09	9.67	8.04	6.49

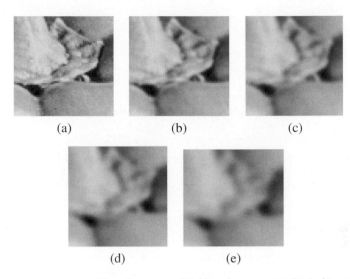

(a)   (b)   (c)

(d)   (e)

Figure 14.4 *Image region cropped from 'Peppers' and its blurred versions: (a) Original image; (b) Gaussian blur with radius 0.5; (c) Gaussian blur with radius 0.8; (d) Gaussian blur with radius 1.0; (e) Gaussian blur with radius 1.5.*

Table 14.2 *Spatial frequency of the image regions in Figure 14.4.*

	Figure 14.4(a)	Figure 14.4(b)	Figure 14.4(c)	Figure 14.4(d)	Figure 14.4(e)
SF	28.67	17.73	12.98	10.04	7.52

From the experimental results and the previous research, it can be concluded that the spatial frequency reflects the level of clarity (sharpness) of a still visual image. In the situation of combination of images which are out-of-focus at different areas, the objective is to obtain an image which is in-focus everywhere. Therefore, it is rational to use the spatial frequency as a sharpness evaluation criterion to fuse the multi-focus image set.

14.3 A spatial domain region-based fusion method using fixed-size blocks

The simplest region-based image fusion method is that in which all input images are decomposed into blocks with the same size. In this section a spatial domain region-based fusion method using fixed-size blocks is proposed. The fused image is obtained by combining corresponding blocks from the source images using some fusion rules.

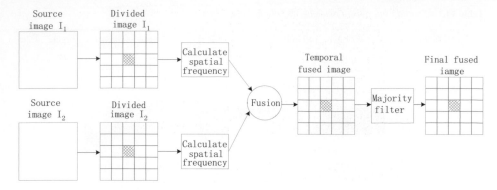

Figure 14.5 *Schematic diagram of the fix size block-based image fusion method.*

A simple case where there are only two source images with different focus was carried out.

14.3.1 The proposed method

Firstly, the two spatially registered source images are decomposed into fixed-size blocks. The spatial frequencies of the corresponding blocks between the two original images are compared to decide which block should be used to construct the fusion result. The proposed algorithm is computationally simple and can be applied in real time. Figure 14.5 shows the schematic diagram of the proposed multi-focus image fusion method.

The fusion process is accomplished by performing the following steps.

(1) Two registered source images I_1 and I_2 are decomposed into blocks of size $M \times N$. Let $BoI_1^{(i)}$ and $BoI_2^{(i)}$ denote the ith block of image I_1 and I_2, respectively.

(2) The spatial frequency of every image block is calculated. Let $SF_i^{I_1}$ and $SF_i^{I_2}$ be the spatial frequency of $BoI_1^{(i)}$ and $BoI_2^{(i)}$, respectively.

(3) Compare the spatial frequencies of corresponding blocks of the two source images to decide which should be used to construct the temporary fused image:

$$BoF_i = \begin{cases} BoI_1^{(i)} & SF_i^{I_1} > SF_i^{I_2} + TH, \\ BoI_2^{(i)} & SF_i^{I_1} < SF_i^{I_2} - TH, \\ (BoI_1^{(i)} + BoI_2^{(i)})/2 & \text{otherwise} \end{cases} \tag{14.4}$$

where BoF_i is the ith block of the fused image, TH is a threshold parameter, and $(BoI_1^{(i)} + BoI_2^{(i)})/2$ denotes a block formed by the pixel-by-pixel grey value average of $BoI_1^{(i)}$ and $BoI_2^{(i)}$.

(4) Consider a further processing of the so far obtained (temporary) fusion result. More specifically, each block of the fused image is separately examined. If the block of interest comes from image I_1 while the majority of the surrounding blocks come

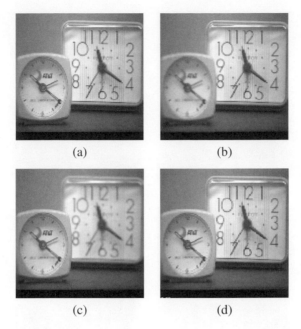

Figure 14.6 *Reference image, the blurred images, and the fusion result: (a) Reference image (in focus); (b) Image 1 (focus on right); (c) Image 2 (focus on left); (d) Fused image using the proposed method (block size 8 × 8, TH = 1.75).*

from image I_2, the block is switched to that of image I_2. This information can be easily deducted by formulating a matrix of size equal to the total size of blocks. We assume that if the block of interest comes from image I_1 the corresponding matrix element is 0, otherwise it is 1. A majority filter (which outputs 1 if the count of 1's outnumbers the count of 0's, and outputs 0 otherwise) and a 3 × 3 window are applied to the above mentioned matrix in order to obtain the final fusion result.

14.3.2 Experimental results

14.3.2.1 Generation of images with diverse focuses

To quantitatively evaluate the performance of the proposed method with different parameters, pairs of multi-focus image sets are generated by blurring the corresponding reference image with different out-of-focus types of distortion. A suitable candidate image could be the one which contains two objects with different distances from the camera. Firstly, the foreground object is blurred to generate one image. Then, the background object is blurred to generate the second image. From the test image of size 128 × 128, shown in Figure 14.6(a), two out-of-focus images shown in Figures 14.6(b) and 14.6(c) are created by Gaussian blurring with radius equal to 0.3 and 0.5, respectively. Additionally, in the second experimental example, from the image shown in Figure 14.7(a), whose size is 480 × 640, the pair of distorted source images shown in Figures 14.7(b) and 14.7(c) are generated by Adobe Photoshop using radial blurring with numbers equal to 6 and 2, respectively.

Figure 14.7 *Reference image, the blurred images, and the fusion result: (a) Reference image (in focus); (b) Image 3 (focus on right); (c) Image 4 (focus on left); (d) Fused image using the proposed method (block size 32 × 32, TH = 0.5).*

14.3.2.2 Comparison metric

The root mean square error (*RMSE*) is used as the evaluation criterion of the fusion method. The *RMSE* between the reference image R and the fused image F is

$$RMSE = \sqrt{\frac{\sum_{i=1}^{I} \sum_{j=1}^{J} [R(i,j) - F(i,j)]^2}{I \times J}} \tag{14.5}$$

where $R(i,j)$ and $F(i,j)$ are the pixel values at the (i,j) coordinates of the reference image and the fused image, respectively. The image size is $I \times J$.

14.3.2.3 Evaluation of the effect of block size and threshold

Firstly, the proposed technique is used to fuse Image 1 and Image 2, shown in Figures 14.6(b) and 14.6(c). Different block sizes and thresholds are employed to evaluate their effect on the fusion performance. The obtained *RMSE* measure is shown in Table 14.3. It can be observed from Table 14.3 that the optimal fusion result, shown in Figure 14.6(d), is obtained by block of size 8 × 8 and threshold of 1.75. Similar experiments are implemented for Image 3 and Image 4, shown in Figures 14.7(b) and 14.7(c),

Table 14.3 *The effects of different block size and TH on RMSE (Image 1 and Image 2).*

TH	Block size								
	4 × 4	4 × 8	8 × 8	8 × 16	16 × 16	16 × 32	32 × 32	32 × 64	64 × 64
0.00	0.8064	0.6339	0.5166	0.6902	0.5549	1.3832	1.8363	1.9682	1.9682
0.25	0.8038	0.6136	0.4940	0.6699	0.5550	1.3832	1.8363	1.9682	1.9682
0.50	0.7821	0.6104	0.4634	0.6710	0.5550	1.3856	1.3264	1.9682	1.9682
0.75	0.7467	0.6503	0.4822	0.6821	0.5550	1.3883	1.3352	1.9682	1.9682
1.00	0.7495	0.5916	0.4841	0.6367	0.5706	1.3883	1.3428	1.9793	1.9682
1.25	0.7462	0.5924	0.4718	0.6307	0.5903	1.3990	1.3428	1.9793	1.9682
1.50	0.7462	0.5924	0.4718	0.6307	0.5903	1.3990	1.3428	1.9793	1.9682
1.75	0.7394	0.6069	0.4493	0.6384	0.6295	1.4137	1.3428	1.9793	1.9682
2.00	0.7498	0.6234	0.4597	0.6352	0.6400	1.4267	1.3428	1.9793	1.9682
2.25	0.7556	0.6265	0.4764	0.6633	0.6626	1.4267	1.3428	1.9793	1.9682
2.50	0.7300	0.5838	0.4896	0.7014	0.6716	1.4267	1.3428	1.6003	1.9682
2.75	0.7588	0.5997	0.5316	0.7014	0.6716	1.4267	1.3428	1.6003	1.9682
3.00	0.7586	0.6545	0.5580	0.7014	0.7034	1.4267	1.3428	1.6003	1.6709

and the performance evaluation for different block sizes and thresholds is shown in Table 14.4, from which it can be seen that the optimal fusion result, shown in Figure 14.7(d), is obtained by block size of 32 × 32 and threshold of 0.5.

Through extensive experiments it can be concluded that the optimal block size depends on the size and the type of the source images. For specific types of source images, if the size of the decomposed blocks is too small, some blocking artefacts may appear in the fusion result, and if the block size is too large the fusion result would be deteriorated by uneven intensity distribution within smooth (low activity) areas. The general rule is: the bigger the source images, the bigger the divided blocks. Three different block sizes of 4 × 4, 32 × 32, and 120 × 128 are used to fuse Image 3 and Image 4 with threshold of 0.5. The corresponding image regions selected to demonstrate the fusion results are shown in Figures 14.8(a), 14.8(b), and 14.8(c), respectively. From these figures it can be seen that the fusion result with block size of 4 × 4 exhibits more intensively the problem of saw tooth type of edges, the result with block size of 120 × 128 demonstrates severe uneven grey distribution, and therefore, the fusion result with block size of 32 × 32 is the best.

From Tables 14.3 and 14.4, it can be seen that the optimal fusion results often occur with threshold ranging from 0.5 to 2.0. Furthermore, for certain sizes of the decomposed block (usually large), the effect of threshold on the fusion performance is minimal. We should stress out the fact that, as far as Equation (14.3) is concerned, as the threshold increases, the proposed algorithm converges to the simple average method.

14.3.2.4 Comparing with wavelet transform

The fusion of Image 3 and Image 4 using the wavelet transform was implemented to provide quantitative comparison with the proposed method. Five types of wavelet filter banks, Daubechies 'db4' and 'db10,' Coiflets 'coif5,' Symlets 'sym8,' and Biorthogonal 'bior3.5' were used [11]. Decomposition level ranges from 1 to 7. A region-based activity measurement is employed to reflect the active level of decomposed coefficients. The

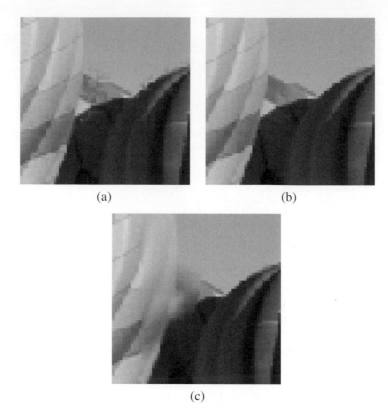

Figure 14.8 *Effects of different block size on fusion performance: (a) Block size* $= 4 \times 4$; *(b) Block size* $= 32 \times 32$; *(c) Block size* $= 120 \times 128$.

well known 'maximum selection rule' [2] is used as a method to combine the wavelet coefficients of the source images. A further window-based type of verification (as, for example, a majority filter) is applied to test for consistency [2]. These three algorithm specifications are optimal according to the experimental results of reference [1]. The performance measure *RMSE* of the fused results is presented in Table 14.5, from which it can be verified that the optimal fusion result is obtained with the wavelet filter bank 'coif5' and decomposition level of 7, which is, nevertheless, still worse than most of situations of Table 14.4. Another evaluation criterion, i.e. the mutual information between the original image and the fused image, has been tested by the authors to compare the wavelet-transform-based method with the proposed method. Similar conclusions are obtained, but due to limitations regarding the length of the chapter the detailed experimental results are not provided.

Three multi-focus image sets shown in Figures 14.9–14.11 are used as benchmarks to subjectively compare the proposed method and the technique based on wavelet transform. In Figures 14.9(a) and 14.9(b) the different focuses are on the Pepsi can and the testing card, respectively. Furthermore, Figures 14.10(a) and 14.10(b) show a pair of images containing two objects: one clock and one person with different distances from the camera,

Table 14.4 *The effects of different block size and TH on RMSE (Image 3 and Image 4).*

TH	Block size									
	4×4	4×8	8×8	8×16	16×16	16×32	32×32	32×40	40×40	40×64
0.0	3.1640	1.9681	1.2564	0.7050	0.5864	0.2107	0.2212	0.4901	0.4313	0.8010
0.25	3.1546	1.9186	1.2524	0.6516	0.5883	0.2099	0.2178	0.4212	0.4354	0.6836
0.50	3.1228	1.8898	1.2688	0.6682	0.5978	0.2403	0.1862	0.4239	0.4418	0.6882
0.75	3.0728	1.8772	1.2569	0.7033	0.5538	0.2759	0.2388	0.4440	0.4597	0.6975
1.0	3.0711	1.8356	1.2484	0.6796	0.5868	0.3141	0.2813	0.4846	0.4789	0.7121
1.25	3.0529	1.8403	1.2563	0.6849	0.6058	0.3526	0.3206	0.5311	0.5292	0.7243
1.50	3.0478	1.8488	1.2765	0.7187	0.6386	0.4175	0.3776	0.5587	0.5571	0.8525
1.75	3.0048	1.7744	1.2638	0.7393	0.6575	0.4412	0.3977	0.5587	0.5722	0.8525
2.00	2.9863	1.7737	1.2375	0.7528	0.6763	0.4412	0.4023	0.5787	0.5950	0.8837
2.25	2.9544	1.7611	1.2441	0.7949	0.6927	0.4644	0.4199	0.7129	0.6128	0.9114
2.50	2.9498	1.7643	1.2717	0.8071	0.7120	0.5012	0.4625	0.7289	0.7831	0.9114
2.75	2.9503	1.7718	1.2833	0.8179	0.7417	0.5170	0.6449	0.7611	0.8276	0.9186
3.00	2.9368	1.7539	1.3050	0.8523	0.7676	0.5558	0.6898	0.8044	0.8276	0.9525

Table 14.5 *Performance of various wavelet-based image fusion methods for Image 3 and Image 4 (RMSE).*

Wavelet	Decomposition level						
	1	2	3	4	5	6	7
Db4	7.3326	5.0334	2.9996	2.1537	2.0655	2.0652	2.0598
Db10	7.4248	5.0764	2.8370	2.0179	1.9483	1.9408	1.9400
Coif5	7.5434	5.0950	2.7625	1.8964	1.7911	1.7887	1.7821
Sym8	7.4810	5.0769	2.9147	1.9943	1.9044	1.9036	1.9016
Bior3.5	9.1313	8.0289	5.3517	3.3348	2.9505	2.8909	2.8909

and only one object in each image is in focus. Finally, Figures 14.11(a) and 14.11(b) are also two images with different focuses that represent the same original image.

The fusion results using the proposed algorithm and the wavelet-based approach are shown in Figures 14.9–14.11 (c) and (d), respectively. Comparing Figures 14.9(c) and 14.9(d), the small area above the string 'Re' in Figure 14.9(c) is vague, whereas that area in Figure 14.9(d) is clear. The edge of the table in Figure 14.9(d) is smooth as it appears in the source images, whereas the edge of the table in Figure 14.9(c) has some wrinkles. In Figures 14.10 and 14.11, similar conclusion can be drawn, i.e. that the proposed method outperforms or at least performs similarly to the wavelet transform approach.

14.3.2.5 Discussion

Objective and subjective comparisons between the proposed method and a number of wavelet-transform-based methods have been implemented. The experimental results have shown that the proposed method exhibits better performance for multi-focus image fusion scenarios. Adaptive and automatic methods for choosing the two parameters, i.e. the block size and threshold, should be further researched. Furthermore, the use of the proposed method in other image fusion applications, as for example fusion of visual and thermal images or millimetre-wave images, also requires further investigation.

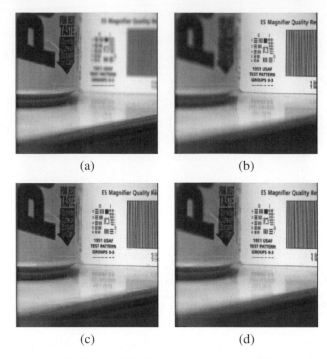

(a)   (b)

(c)   (d)

Figure 14.9   *Source images (size* 512×512*) and fusion results: (a) Image 5 (focus on the Pepsi can); (b) Image 6 (focus on the testing card); (c) Fused image using wavelet transform ('coif5' with decomposition level of 6); (d) Fused image using the proposed algorithm (block size* 32×32*, TH* $= 1.0$*).*

14.4 Fusion using segmented regions

For multi-focus image fusion scenarios, segmentation can be performed in two ways, i.e. the traditional segmentation based on image intensity and segmentation based on clarity. In both cases, the image is segmented into regions which represent specific features of the image. The fusion process consists of selecting each clearer region of the segmentation map from the source images.

14.4.1 Segmentation based on image intensity

14.4.1.1 Method

In this section, a new region-based multi-focus image fusion method is proposed. The intuitive idea behind this method is that images are perceived by humans in the region or object level instead of pixel level. The first step of the proposed method is to fuse the two source images by simple average. Then, the intermediate fused image is segmented by the so called normalised cuts segmentation technique [12]. Using the result of the segmentation, the two source images are partitioned. Finally, the corresponding regions of the two source images are fused using spatial frequency information. Figure 14.12 shows the schematic diagram of the proposed multi-focus image fusion method.

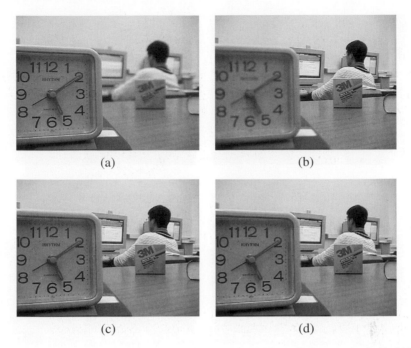

Figure 14.10 *Source images (size 480 × 640) and fusion results: (a) Image 7 (focus on the clock); (b) Image 8 (focus on the person); (c) Fused image using wavelet transform ('db8' with decomposition level of 5); (d) Fused image using the proposed algorithm (block size 40 × 40, TH = 0.75).*

The fusion process can be summarised in the following steps.

(1) An intermediate (temporary) fused image is obtained by averaging two registered source images I_1 and I_2.
(2) The intermediate fused image is segmented into several regions using normalised cuts algorithm [12], which is described in Section 14.4.1.2 below.
(3) The images I_1 and I_2 are partitioned using the results of step (2).
(4) The spatial frequencies of each region of the segmented versions of images I_1 and I_2 are computed.
(5) The spatial frequencies of corresponding regions of the two source images are compared to decide which should be used to construct the fused image:

$$RoF_i = \begin{cases} Ro_i^{(I_1)} & SF_i^{(I_1)} \geqslant SF_i^{(I_2)}, \\ Ro_i^{(I_2)} & \text{otherwise} \end{cases} \qquad (14.6)$$

where RoF_i is the ith region of the fused image, $SF_i^{(I_1)}$ and $SF_i^{(I_2)}$ are the spatial frequencies of the ith regions of image I_1 and I_2, respectively.

14.4.1.2 Region segmentation using normalised cuts

The idea of graph-based image segmentation is that the set of image points is represented as a weighted undirected graph $G = (V, E)$, where the nodes of the graph are the points

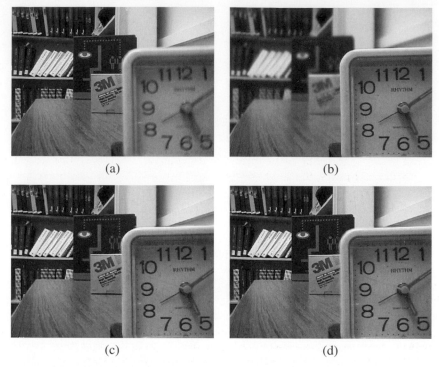

Figure 14.11   *Source images (size 480 × 640) and fusion results: (a) Image 9 (focus on the left); (b) Image 10 (focus on the right); (c) Fused image using wavelet transform ('db8' with decomposition level of 5); (d) Fused image using the proposed algorithm (block size 32 × 32, TH = 0.75).*

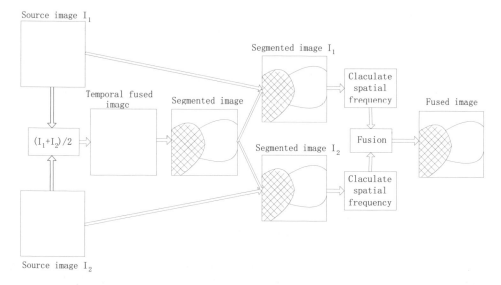

Figure 14.12   *Schematic diagram of the proposed fusion method using segmentation based on image intensity.*

in the image [12]. Every pair of nodes (i, j) is connected by an edge, and the weight on each edge $w(i, j)$ is a function of the similarity between nodes i and j. The graph $G = (V, E)$ is segmented into two disjoint complementary parts I_1 and I_2, $I_2 = V - I_1$, by removing the edges connecting two parts. The degree of dissimilarity between these two parts can be computed as the total weight of the edges that have been removed, denoted as $cut(I_1, I_2) = \sum_{u \in I_1, t \in I_2} w(u, t)$. The optimal bipartitioning of a graph is the one that minimises this *cut* value.

In order to deal with various drawbacks associated with the existing techniques which attempt to solve the above problem, instead of using the value of total edge weight connecting the two partitions, Shi and Malik proposed a disassociation measure to compute the *cut* cost as a fraction of the total edge connections to all the nodes in the graph [12]. It is called the normalised cut (*Ncut*) defined as follows:

$$Ncut(I_1, I_2) = \frac{cut(I_1, I_2)}{asso(I_1, V)} + \frac{cut(I_1, I_2)}{asso(I_2, V)} \tag{14.7}$$

where $asso(I_1, V) = \sum_{u \in I_1, t \in V} w(u, t)$ is the total connection from nodes in I_1 to all nodes in the graph and $cut(I_1, I_2) = \sum_{u \in I_1, t \in I_2} w(u, t)$ is defined above. The *Ncut* value has advantage over the existing criteria that favour cutting small sets of isolated nodes in the graph. Because the *cut* value will be a large percentage of the total connection from the small set to all other nodes, the *Ncut* value will not be small for the *cut* that partitions out small sets of isolating points. The algorithm is summarised as follows, assuming the image I is to be segmented [12]:

(1) Define the feature description matrix for a given image and a weighting function.
(2) Set up a weighted graph $G = (V, E)$, compute the edge weights, and summarise information into the matrices W and D. The weight $w(i, j)$ on the edge connecting two nodes (pixels in the context of image segmentation) (i, j) is defined a measure of the similarity between the two nodes. W is an $N \times N$ symmetric matrix which contains the weights $w(i, j)$. D is an $N \times N$ diagonal matrix with entries $d(i, i) = \sum_j w(i, j)$.
    The weight on an edge should reflect the likelihood that the two pixels connected by the edge belong to one object. Using just the brightness value of the pixels and their spatial location, we can define the graph edge weight connecting the two nodes i and j as [12]:

$$w(i, j) = e^{-\|F(i) - F(j)\|_2^2 / \sigma_I^2}$$
$$* \begin{cases} e^{-\|X(i) - X(j)\|_2^2 / \sigma_X^2} & \text{if } \|X(i) - X(j)\|_2 < r, \\ 0 & \text{otherwise} \end{cases} \tag{14.8}$$

where $X(i)$ is the spatial location of node i, and $F(i) = I(i)$, the intensity value. If nodes i and j are more than r pixels apart, the $w(i, j)$ will be zero. Generally, the value of σ is set to 10 to 20 per cent of the total range of the feature distance function.
(3) Solve $(D - W)x = \lambda Dx$ for eigenvectors with the smallest eigenvalues.
(4) Use the eigenvector with the second smallest eigenvalue to bipartition the graph by finding the splitting points such that *Ncut* is minimised.

(5) Decide if the current partition is stable and check the value of the resulting *Ncut*.
(6) Recursively repartition the segmented parts (go to step 2).
(7) Exit if *Ncut* for every segment is over some specified value – maximum allowed *Ncut*.

14.4.1.3 Experimental results

The two pairs of multi-focus image sets of Figures 14.9(a), 14.9(b) and Figures 14.10(a), 14.10(b) are fused using the proposed fusion method. The fused results shown in Figures 14.13 and 14.14 are used as benchmarks to compare between the proposed method and the technique based on wavelet transform [2].

For the wavelet-based fusion method, Daubechies 'db4' and decomposition level of 2 were used. A region-based activity measurement is employed to reflect the active level of decomposed coefficients. The well known 'maximum selection rule' [2] is used as a method to combine the wavelet coefficients of the source images. A further window-based type of verification (as, for example, a majority filter) is applied to test for consistency [2]. Window-based verification is applied to consistency verification. These three selections are optimal according to the experimental results in reference [2].

Figure 14.13(a) is the intermediate fused result by using simple average method. The segmented image using normalised cuts algorithm is shown in Figure 14.13(b). The parameter setting is $\sigma_I = 0.06$, $\sigma_X = 0.04$, $r = 8$. From the segmented results we can see that the intermediate fused image is segmented into several regions of different focus measures with the two source images. Figures 14.13(c) and 14.13(d) are the fused results by using the wavelet approach and the proposed method, respectively. Comparing Figures 14.13(c) and 14.13(d), the small area above the string 'ES Magnifier Quality Re' in Figure 14.13(c) is blurred, whereas that in Figure 14.13(d) is clear. To make better comparisons, the difference images between the fused image and the source images are given in Figures 14.13(e) to 14.13(h), too. The difference between a focused region in a source image and the fused image should ideally be zero. For example, in Figure 14.9(a) the Pepsi can is sharp (in focus), and in Figure 14.13(f) which is the difference between Figures 14.13(d) and 14.9(a) the Pepsi can region is indeed nearly zero, an observation which demonstrates that the entire focused area is contained in the fused image successfully. However, the difference between Figures 14.13(c) and 14.9(a) in the same region shown in Figure 14.13(e) is not zero, which shows that the fused result using wavelet transform is inferior compared to the proposed method.

In Figures 14.10(a) and 14.10(b), the person exhibits a slight motion and, therefore, the two original images cannot be registered strictly. This misregistration resulted in that the fused image by wavelet-based method is blurred in the person's head region. However, the fused image by the proposed method is superior compared to the wavelet-based result. From the difference images between the source images and the fused images the same conclusion as previously can be drawn, that the proposed method outperforms the wavelet transform approach.

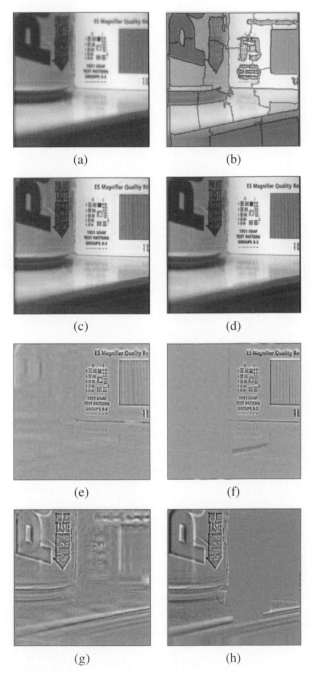

Figure 14.13 *Source images, fusion results and their differences: (a) Intermediate fused image using average method; (b) The segmented image; (c) Fused image using wavelet transform; (d) Fused image using the proposed algorithm; (e) Difference between Figures 14.13(c) and 14.9(a); (f) Difference between Figures 14.13(d) and 14.9(a); (g) Difference between Figures 14.13(c) and 14.9(b); (h) Difference between in Figures 14.13(d) and 14.9(b).*

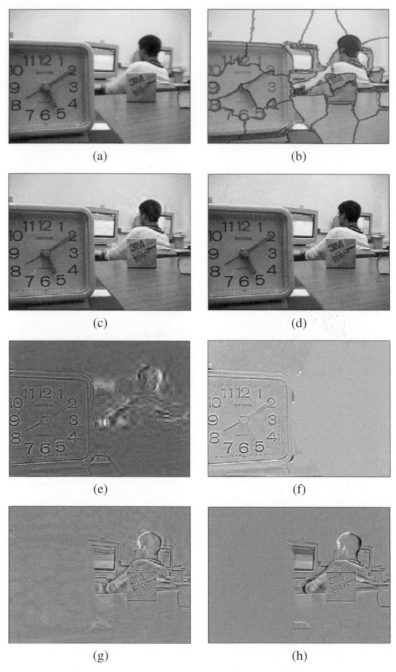

Figure 14.14 *Source images, fusion results and their differences: (a) Intermediate fused image using average method; (b) The segmented image; (c) Fused image using wavelet transform; (d) Fused image using the proposed algorithm; (e) Difference between Figures 14.14(c) and 14.10(a); (f) Difference between Figures 14.14(d) and 14.10(a); (g) Difference between Figures 14.14(c) and 14.10(b); (h) Difference between in Figures 14.14(d) and 14.10(b).*

14.4.2 Segmentation based on image clarity

14.4.2.1 Method

An effective algorithm which is suitable for combining multi-focus image sets of a scene is proposed in this section. The spatial frequency of a pixel's neighbour block is used to judge its sharpness and morphological (*MOR*) opening and closing are used for post-processing. Finally, the fused image is obtained by combining the focused regions. We assume two registered images I_1 and I_2 with different focuses to be fused. The algorithm consists of the following steps:

(1) Calculate the spatial frequency of each pixel within a 5×5 window in I_1 and I_2 using Equation (14.9), denoted by SF_1 and SF_2, respectively.
(2) Compare the values SF_1 and SF_2 to determine which pixel is in focus. The logical matrix Z (essentially a binary image) is constructed as

$$Z(m,n) = \begin{cases} 1 & SF_1(m,n) \geqslant SF_2(m,n), \\ 0 & \text{otherwise} \end{cases} \tag{14.9}$$

'1' in Z indicates that the pixel at position (m,n) in image I_1 is in focus, otherwise the pixel in I_2 is in focus.
(3) However, determination by SF alone is insufficient to discern all the focused pixels. There are thin protrusions, narrow breaks, thin gulfs, small holes, etc. in Z. To correct for these defects, morphological opening and closing constructed by combining dilation and erosion are employed [13]. Opening, denoted as $Z \circ B$, is simply erosion of Z by the structure element B, followed by dilation of the result by B. It removes thin connections and thin protrusions. Closing, denoted as $Z \bullet B$, is dilation followed by erosion. It joins narrow breaks and fills long thin gulfs. Holes larger than B cannot be removed simply using opening and closing operators. In practice, small holes are always judged incorrectly; therefore, a threshold, *TH*, is set to remove the holes smaller than the threshold. Opening and closing are again performed to smooth object contours.
(4) The fusion image is then constructed as

$$F(m,n) = \begin{cases} I_1(m,n) & Z(m,n) = 1, \\ I_2(m,n) & \text{otherwise} \end{cases} \tag{14.10}$$

14.4.2.2 Experimental results

We have compared our results with those obtained using a wavelet-based method, which is implemented in the following way. Firstly, the scaled images and detail images are obtained by using the wavelet transform up to 3 levels. Daubechies 'db1' is selected as the wavelet basis. Scaled images and detail images are then combined by choosing the pixel with the maximum absolute value. Consistency verification is implemented in this step. Specifically, if the centre pixel value comes from image I_1 while the majority of the surrounding pixel values come from image I_2, the centre pixel value is switched to that of image I_2. Finally, the inverse wavelet transform is implemented to recover the fused image. For the proposed algorithm, the structure element B is a 5×5 matrix

Figure 14.15 *Fusion example 1: (a) Focus on the right book; (b) Focus on the left book; (c) Z matrix of step 2; (d) Modified Z matrix of step 3; (e) Fusion result using the proposed algorithm; (f) Fusion result using a wavelet-based method.*

with logical 1's and the threshold is set to 1000. Two pairs of multi-focus image sets are used to test our algorithm against the wavelet-based method. The results of the first two experiments are shown in Figures 14.15 and 14.16. In Figure 14.16, we only show the fused results since the source images are illustrated previously in Figures 14.11(a) and 14.11(b). Carefully comparing the results we can see that the wavelet method loses sharpness and exhibits prominent blocking artefacts (the left book in Figures 14.15(e)

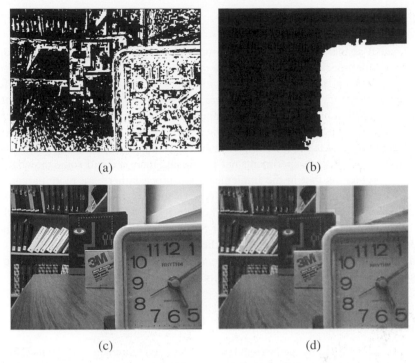

Figure 14.16 *Fusion example 2: (a) Z matrix of step 2; (b) Modified Z matrix of step 3; (c) Fusion result using the proposed algorithm; (d) Fusion result using a wavelet-based method.*

Table 14.6 *Performance of different fusion methods (EN: entropy; STD: standard deviation; SIM: similarity; DWT: wavelet-based method; SF-MOR: the proposed method.)*

Test images		Figure 14.13	Figure 14.14	Figure 14.15	Figure 14.16
EN	*DWT*	7.3436	7.1876	7.4536	7.4450
	SF-MOR	7.3494	7.2769	7.4583	7.6248
STD	*DWT*	59.9199	44.1591	47.3949	69.6140
	SF-MOR	61.4038	46.8276	48.3748	71.2814
SIM	*DWT*	0.8810	0.8661	0.8801	0.8347
	SF-MOR	0.9196	0.9128	0.9542	0.8986

and 14.15(f), and the left bookshelf in Figures 14.16(e) and 14.16(f)). To evaluate the performance of the proposed method quantitatively, three criteria, namely: entropy (*EN*), standard deviation (*STD*), and similarity (*SIM*), are used [6]. For these criteria, larger values indicate better fusion results. From the values shown in Table 14.6 we observe that the proposed algorithm outperforms the wavelet-based method.

It is important to stress out that the proposed algorithm resembles the manual cut-and-paste method, which is often used to obtain a standard fused image.

14.5 Discussion

Experimental results obtained from the proposed region-based fusion methods are encouraging. More specifically, in spite of the crudeness of the segmentation methods used, the results obtained from the proposed fusion processes which consider specific feature information regarding the source images are excellent in terms of visual perception. The presented algorithm in Section 14.3 is computationally simple and can be applied in real time. It is also valuable in practical applications. More sophisticated approaches proposed in Section 14.4 use some complex segmentation methods. Although the results we obtain from a number of experiments are promising, there are still some drawbacks. An indicative problem is that in the implementation process there are more parameters to be considered compared to an MR-based type of method, such as the wavelet method. For example, the block size and the threshold in the fixed size block-based method of Section 14.3, the parameters σ_I, σ_X, r of Section 14.4.1 and the structure element B and threshold of Section 14.4.2, all affect the fusion result. Adaptive methods for choosing those parameters should be further researched. In addition, further investigations are necessary for selecting more effective clarity measures.

Acknowledgements

This work is supported by the National Natural Science Foundation of China (No. 6040-2024) and Program for New Century Excellent Talents in University (NCET-2005). The multifocus 'Bookshelf,' 'Student,' and 'Pepsi' images are kindly supplied by Prof. Rick Blum of Lehigh University. These images are available online at www.imagefusion.org.

References

[1] J.J. Lewis, R.J. O'Callaghan, S.G. Nikolov, D.R. Bull and C.N. Canagarajah, 'Region-based image fusion using complex wavelets', in *Proceedings of the 7th International Conference on Information Fusion*, Stockholm, Sweden, June 2004, pp. 555–562.

[2] Z. Zhang and R. Blum, 'A categorization and study of multiscale-decomposition-based image fusion schemes', *Proceedings of the IEEE*, Vol. 87, No. 8, 1999, pp. 1315–1328.

[3] Z. Zhang and R. Blum, 'Region-based image fusion scheme for concealed weapon detection', in *Proceedings of the 31th Annual Conference on Information Sciences and Systems*, Baltimore, USA, March 1997, pp. 168–173.

[4] G. Piella, 'A region-based multiresolution image fusion algorithm', in *Proceedings of the 5th International Conference on Information Fusion*, Annapolis, USA, July 2002, pp. 1557–1564.

[5] G. Piella and H. Heijmans, 'Multiresolution image fusion guided by a multimodal segmentation', in *Proceedings of Advanced Concepts of Intelligent Systems*, Ghent, Belgium, September 2002, pp. 175–182.

[6] G. Piella, 'A general framework for multiresolution image fusion: from pixels to regions', *Information Fusion*, Vol. 4, No. 4, 2003, pp. 259–280.

[7] S. Li, J.T. Kwok and Y. Wang, 'Combination of images with diverse focus using spatial frequency', *Information Fusion*, Vol. 2, No. 3, 2001, pp. 169–176.

[8] B. Yang and S. Li, 'Multi-focus image fusion based on spatial frequency and morphological operators', *Chinese Optics Letters*, Vol. 5, No. 8, 2007, pp. 452–453.

[9] S. Li and B. Yang, 'Multifocus image fusion using region segmentation and spatial frequency', *Image and Vision Computing*, in press.

[10] S. Li and G. Chen, 'Clarity ranking for digital images', in *Proceedings of the 2nd International Conference on Fuzzy Systems and Knowledge Discovery*, Changsha, China, August 2005, pp. 610–613.

[11] I. Daubechies, *Ten Lectures of Wavelets*, Society for Industrial and Applied Mathematics, Philadelphia, 1992.

[12] J. Shi and J. Malik, 'Normalized cuts and image segmentation', *IEEE Transactions on Pattern Analysis and Machine Intelligence*, Vol. 22, No. 8, 2000, pp. 888–905.

[13] R. Gonzalez and R. Woods, *Digital Image Processing*, Prentice Hall, 2002.

15

Image fusion techniques for non-destructive testing and remote sensing applications

F.C. Morabito, G. Simone and M. Cacciola

University Mediterranea of Reggio Calabria, Faculty of Engineering, Dimet, Italy

There is a widespread recognition that the technique of image fusion can help in reaching processing and interpretation goals not achievable by single-sensor acquisition and processing. In this chapter, we present some algorithms of fusion based on multi-scale Kalman filtering and computational intelligence methodologies. The proposed algorithms are applied to two kinds of problems: a remote sensing segmentation, classification, and object detection application carried out on real data available from experimental campaigns and a non-destructive testing/evaluation problem of flaw detection using electromagnetic and ultrasound recordings. In both problems the fusion techniques are shown to achieve some superior performance with respect to the single-sensor image modality.

15.1 Introduction

Data fusion [1–3] can be defined as the synergistic use of knowledge from different sources to assist in the overall understanding of a phenomenon: data fusion algorithms can be broadly classified as either phenomenological or non-phenomenological. Phenomenological algorithms utilise knowledge of the underlying physical processes as a basis for deriving the procedure for fusing data. Several investigators are pursuing such approaches. However, such methods are likely to be difficult to derive and cumbersome to implement. Non-phenomenological approaches, in contrast, tend to ignore the physical process and attempt to fuse information using the statistics associated with individual segments of data. Within this framework, it is mandatory to develop effective data fusion techniques able to take advantage of such multi-sensor characteristics, aiming to a proper exploitation of the data measured by different sensors, or by the same sensor in differ-

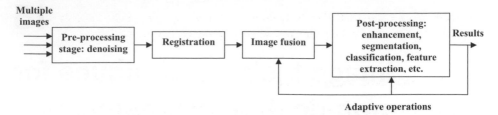

Figure 15.1   *Block schema of a general image fusion procedure.*

ent measuring context [4]. Research on the subject of data fusion has been motivated by a desire to obtain more comprehensive information about the system being analysed by combining sub-information from multiple sensors. As far as the imaging techniques are concerned, the recent advances in this field make it possible to combine information across the electromagnetic spectrum by the fusion of multi-modal images, i.e. by the so called image fusion.

The general procedure for image fusion is depicted in Figure 15.1 [5]. Here, the data gathered from multiple sources of acquisition are filtered through a denoising stage; a successive block must ensure the registration of multiple images/acquisition records. This stage consists of associating the corresponding pixels to the same physical points on the object. In this way, input images can be compared pixel by pixel. Subsequently, the registered data go through the image fusion process. Depending on the fusion algorithm, the fusion result could be a composite image, a thematic map, or a three-dimensional graph. Usually, when the fusion is implemented at the pixel level, a post-processing step may be applied to the fused image, such as classification, segmentation, or enhancement. In this way, it is possible to achieve quantitative results, which can be evaluated on the basis of predefined metrics according to the application requirements. The quantitative results from the evaluation may be used as feedback to guide the fusion process where applicable. An adaptive fusion process can be achieved, but this still remains a challenge for most applications.

A lot of different algorithms exploiting well-known image fusion techniques are available in the scientific literature (see [1–3] and references therein). The most known image fusion algorithms are based on Optimal Filtering (OF), Multi-Resolution Analysis (MRA), Bayesian inference, Dempster–Shafer Theory (DST), or even heuristic methods such as Artificial Neural Networks (ANN) and learning machines. Applications of image fusion are now ubiquitous and concerns many different research fields. Here, we shall discuss some of them, directed to Non-Destructive Testing and evaluation (NDT/NDE, see [6] and references therein) and Remote Sensing (see [7] and references therein). In this chapter, we first propose several data fusion algorithms, mainly referring to image fusion. The proposed algorithms will be briefly introduced through a theoretical perspective in Section 15.2. Then, a few applications will be proposed in order to show the usefulness of the above mentioned image fusion techniques. We first discuss the fusion of SAR and SIR remotely sensed images and, thus, we will comment the fusion of eddy currents and ultrasonic images for a NDT/NDE practical study case; the approached problems as well as our proposed techniques based on an image fusion approach will be described in Sections 15.3 and 15.4. Finally, Section 15.5 draws up our conclusions, showing the advantages of image fusion techniques in solving such kinds of problems.

15.2 The proposed image fusion techniques

In this section, a brief introduction of different image fusion techniques will be given. We shall propose the Multiple Kalman Filtering (MKF) [3,8,9], the Pixel Level (PL) [10], the Feature Level (FL) and the Symbol Level (SL) [11,12] data fusion techniques. As far as the PL, FL, and SL techniques are concerned, they are mainly exploiting ANN to carry out the data fusion procedure. It goes beyond the scope of this chapter to compare the proposed algorithms with different approaches proposed in the literature; however, some guidelines are given underlying the advantages of our approaches.

15.2.1 The MKF algorithm: how to merge multiple images at different scales

The MKF technique belongs to the realm of multiresolution stochastic processes [4,8], where one-dimensional time series or multidimensional images have a time-like variable, commonly referred to as scale. These processes have been modelled through windowed Fourier transform, sub-band coding, Laplacian pyramids, and wavelet theory. The performance evaluation of these techniques can be carried out by developing a stochastic process theory at multiple resolution scales. Whereas the usual signal representation is well localised in time – or in the case of images, in space – the Fourier representation of the signal is well localised in frequency. For signals with time varying spectra, i.e. signals where the frequency content is not constant over the time, a representation of the signal localised both in time and in frequency would be very useful for many purposes. The wavelet theory represents a compromise between good time resolution and good frequency resolution; it can be considered as a first attempt to introduce the multiscale representation of signals. In the sensor fusion cases, entire sets of signals or images from a suite of sensors can be considered collectively. The main need is not to decompose the signals at different resolutions, but to form a knowledge stream, upward from the finest scale to the coarsest scale, downward from the coarsest resolution to the finest resolution. Therefore, the MKF algorithm merges data at different resolution scales. Loosely speaking, the key of this multiple scale filtering is to consider the scale as an independent variable as the time, such that the description at a particular scale captures the features of the process up to scales that are relevant for the prediction of finer scale features. If the dataset consists of images at different resolutions, the MKF can be applied to add knowledge from finer resolution data to coarser resolution data and this information propagation is carried out also from the coarsest scale to the finest scale.

The multiscale models, in fact, provide accurate descriptions of a variety of stochastic processes and also lead to extremely efficient algorithms for optimal estimation and for the fusion of multiresolution measurements using multiscale and scale-recursive generalisations of classical Kalman filtering and smoothing. An image (2D signal) can be decomposed from a coarse to a fine resolution. At the coarsest resolution, the signal will consist of a single value (i.e. single image). At the next resolution, there are $q = 4$ values,

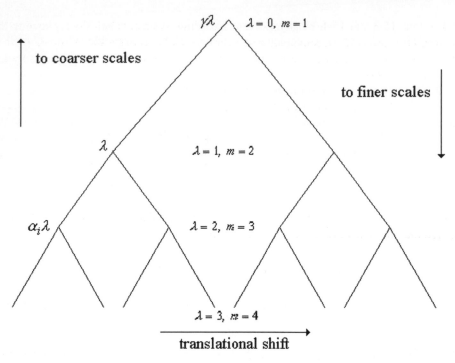

Figure 15.2 *Multiscale signal representation by dyadic tree, where each level of the tree corresponds to a single scale.*

and in general, at the mth resolution, we obtain q^m values. The values of the multiscale representation can be described on the index set (m, i, j), where m represents the resolution and (i, j) the location index. The scale-to-scale decomposition can be schematically depicted as a tree structure (see Figure 15.2 for details).

To describe the model, an abstract index λ is used to specify the nodes on the tree; $\gamma\lambda$ specifies the parent node of λ.

The aim of Kalman filtering [13] is to estimate a state vector $\mathbf{x}(t)$ of a dynamic process, when measurements $\mathbf{y}(t)$ about the process evolution are available. In the case of the MKF, the independent variable is not the time t, but the scale λ; when signals are represented by images (i.e. 2D signals), the aim is to estimate a state vector $\mathbf{X}(\lambda)$, on the basis of observations $\mathbf{Y}(\lambda)$ of the process at different resolutions. This scheme proceeds in two steps: downward and upward.

The multiple scale downward (coarse-to-fine resolution) model is given by [4,8]

$$\mathbf{X}(\lambda) = \mathbf{A}(\lambda) \cdot \mathbf{X}(\gamma\lambda) + \mathbf{B}(\lambda) \cdot \mathbf{W}(\lambda) \tag{15.1}$$

$$\mathbf{Y}(\lambda) = \mathbf{C}(\lambda) \cdot \mathbf{X}(\lambda) + \mathbf{V}(\lambda) \tag{15.2}$$

Since $\mathbf{X}(\gamma\lambda)$ represents the state at a resolution coarser than $\mathbf{X}(\lambda)$, $\mathbf{A}(\lambda) \cdot \mathbf{X}(\gamma\lambda)$ can be considered as a prediction term for the finer level; $\mathbf{B}(\lambda) \cdot \mathbf{W}(\lambda)$ is the new knowledge that we add from one scale to the next. The noisy measurements $\mathbf{Y}(\lambda)$ of the state $\mathbf{X}(\lambda)$, shown in Equation (15.2), combined with Equation (15.1), form the state estimation problem. The covariance matrices of the state and of the measurements are computed by the following equations:

$$\mathbf{P}_X(\lambda) \equiv E[\mathbf{X}(\lambda) \cdot \mathbf{X}^T(\lambda)] \tag{15.3}$$

$$\mathbf{R}(\lambda) \equiv E[\mathbf{V}(\lambda) \cdot \mathbf{V}^T(\lambda)] = \begin{bmatrix} \sigma_1^2 & 0 & \dots & 0 \\ 0 & \sigma_2^2 & \dots & 0 \\ \dots & \dots & \dots & \dots \\ 0 & \dots & 0 & \sigma_n^2 \end{bmatrix} \tag{15.4}$$

where σ_i^2, $i = 1, 2, \dots, n$, are the variances of the n available measurements. It can be proved that the state covariance can be computed by

$$\mathbf{P}_X(\lambda) = \mathbf{A}(\lambda) \cdot \mathbf{P}_X(\gamma\lambda) \cdot \mathbf{A}^T(\lambda) + \mathbf{B}(\lambda) \cdot \mathbf{B}^T(\lambda) \tag{15.5}$$

Equation (15.4) indicates that the measurements are independent and have different variances.

Corresponding to the above downward model, the upward (fine-to-coarse resolution) model is [4]

$$\mathbf{X}(\gamma\lambda) = \mathbf{F}(\lambda) \cdot \mathbf{X}(\lambda) + \overline{\mathbf{W}}(\lambda) \tag{15.6}$$

where

$$\mathbf{F}(\lambda) = \mathbf{P}_X(\gamma\lambda) \cdot \mathbf{A}^T(\lambda) \cdot \mathbf{P}_X^{-1}(\lambda) \tag{15.7}$$

and $\overline{\mathbf{W}}(\lambda)$ is an uncorrelated sequence with variance:

$$\mathbf{Q}(\lambda) \equiv E[\overline{\mathbf{W}}(\lambda) \cdot \overline{\mathbf{W}}^T(\lambda)]$$
$$= \mathbf{P}_X(\gamma\lambda) \cdot [\mathbf{I} - \mathbf{A}^T(\lambda) \cdot \mathbf{P}_X(\lambda)^{-1} \cdot \mathbf{A}(\lambda) \cdot \mathbf{P}_X(\gamma\lambda)] \tag{15.8}$$

We assume that each node has q children. We denote by $\alpha_i\lambda$, the ith child node of λ for $i = 1, \dots, q$. Also define:

- $\hat{\mathbf{X}}(\lambda|\alpha_i\lambda)$: predicted value of $\mathbf{X}(\lambda)$ using the estimate of child node α_i ($i = 1, \dots, q$) of λ;
- $\hat{\mathbf{X}}(\lambda|\lambda-)$: predicted value of $\mathbf{X}(\lambda)$ after merging the predictions of the q child nodes of λ;
- $\hat{\mathbf{X}}(\lambda|\lambda+)$: updated value of $\mathbf{X}(\lambda)$ using $\hat{\mathbf{X}}(\lambda|\lambda-)$ and the measurement $\mathbf{Y}(\lambda)$;
- $\hat{\mathbf{X}}(\lambda)$: estimated value of $\mathbf{X}(\lambda)$ after smoothing during the downward step.

The error co-variance matrices $\hat{\mathbf{P}}(\lambda|\alpha_i\lambda)$, $\hat{\mathbf{P}}(\lambda|\lambda-)$, $\hat{\mathbf{P}}(\lambda|\lambda+)$ and $\hat{\mathbf{P}}(\lambda)$ are defined similarly. Thus, the estimation by the MKF proceeds along the following steps:

(1) *Initialisation*: assign the following prior values at λ corresponding to the finest scale node

$$\hat{\mathbf{X}}(\lambda|\lambda-) = 0 \tag{15.9}$$

$$\hat{\mathbf{P}}(\lambda|\lambda-) = \mathbf{P}_X(\lambda) \tag{15.10}$$

where $\mathbf{P}_X(\lambda)$ is the prior error variance, i.e. the solution of (15.5), at the node λ.

(2) *Upward step*: to estimate the state and the error co-variance matrices, we can combine the available measurements and the predicted values $\hat{\mathbf{X}}(\lambda|\lambda-)$, by the following equations:

$$\hat{\mathbf{X}}(\lambda|\lambda+) = \hat{\mathbf{X}}(\lambda|\lambda-) + \mathbf{K}(\lambda) \cdot \left[\mathbf{Y}(\lambda) - \mathbf{C}(\lambda) \cdot \hat{\mathbf{X}}(\lambda|\lambda-) \right] \tag{15.11}$$

$$\hat{\mathbf{P}}(\lambda|\lambda+) = \left[\mathbf{I} - \mathbf{K}(\lambda) \cdot \mathbf{C}(\lambda) \right] \cdot \hat{\mathbf{P}}(\lambda|\lambda-) \tag{15.12}$$

where the Kalman gain matrix $\mathbf{K}(\lambda)$ is given by

$$\mathbf{K}(\lambda) = \hat{\mathbf{P}}(\lambda|\lambda-) \cdot \mathbf{C}^{\mathrm{T}}(\lambda) \cdot \left[\mathbf{C}(\lambda) \cdot \hat{\mathbf{P}}(\lambda|\lambda-) \cdot \mathbf{C}^{\mathrm{T}}(\lambda) + \mathbf{R}(\lambda) \right]^{-1} \tag{15.13}$$

Moving up to the parent node, we apply the Kalman filter prediction to get predictions from each child node by using

$$\hat{\mathbf{X}}(\lambda|\alpha_i\lambda) = \mathbf{F}(\alpha_i\lambda) \cdot \hat{\mathbf{X}}(\alpha_i\lambda|\alpha_i\lambda+) \tag{15.14}$$

where $\hat{\mathbf{X}}(\alpha_i\lambda|\alpha_i\lambda+)$ has been computed at the previous step. $\mathbf{F}(\alpha_i\lambda)$ is provided by the fine-to-coarse equation (15.7). We are supposing to know $\mathbf{P}_X(\lambda)$ at each resolution by Equation (15.5)

$$\hat{\mathbf{P}}(\lambda|\alpha_i\lambda) = \mathbf{F}(\alpha_i\lambda) \cdot \hat{\mathbf{P}}(\alpha_i\lambda|\alpha_i\lambda+) \cdot \mathbf{F}^{\mathrm{T}}(\alpha_i\lambda) + \hat{\mathbf{Q}}(\alpha_i\lambda) \tag{15.15}$$

where $\hat{\mathbf{P}}(\alpha_i\lambda|\alpha_i\lambda+)$ is related to the previous step and

$$\hat{\mathbf{Q}}(\alpha_i\lambda) = \mathbf{A}^{-1}(\alpha_i\lambda) \cdot \mathbf{B}(\alpha_i\lambda) \cdot \mathbf{Q}(\alpha_i\lambda) \cdot \mathbf{B}^{\mathrm{T}}(\alpha_i\lambda) \cdot \mathbf{A}^{-\mathrm{T}}(\alpha_i\lambda) \tag{15.16}$$

For each node we obtain q predictions from each of the q child nodes. They are merged to obtain a single prediction using:

$$\hat{\mathbf{X}}(\lambda|\lambda-) = \hat{\mathbf{P}}(\lambda|\lambda-) \cdot \sum_{i=1}^{q} \hat{\mathbf{P}}^{-1}(\lambda|\alpha_i\lambda) \cdot \hat{\mathbf{X}}(\lambda|\alpha_i\lambda) \tag{15.17}$$

$$\hat{\mathbf{P}}(\lambda|\lambda-) = \left[(1-q) \cdot \mathbf{P}_X^{-1}(\lambda) + \sum_{i=1}^{q} \hat{\mathbf{P}}^{-1}(\lambda|\alpha_i\lambda) \right]^{-1} \tag{15.18}$$

The upward step terminates when the recursion reaches the root node and we obtain the estimate $\hat{\mathbf{X}}(0) = \hat{\mathbf{X}}(0|0+)$.

(3) *Downward step*: the information is propagated downward after the upward step is completed. The estimators are

$$\hat{\mathbf{X}}(\lambda) = \hat{\mathbf{X}}(\lambda|\lambda+) + \mathbf{J}(\lambda) \cdot \left[\hat{\mathbf{X}}(\gamma\lambda) - \hat{\mathbf{X}}(\gamma\lambda|\lambda+) \right] \tag{15.19}$$

$$\hat{\mathbf{P}}(\lambda) = \hat{\mathbf{P}}(\lambda|\lambda+) + \mathbf{J}(\lambda) \cdot \left[\hat{\mathbf{P}}(\gamma\lambda) - \hat{\mathbf{P}}(\gamma\lambda|\lambda+) \right] \cdot \mathbf{J}^{\mathrm{T}}(\lambda) \tag{15.20}$$

where

$$\mathbf{J}(\lambda) = \mathbf{P}(\lambda|\lambda+) \cdot \mathbf{F}^{\mathrm{T}}(\lambda) \cdot \hat{\mathbf{P}}^{-1}(\gamma\lambda|\lambda+) \tag{15.21}$$

The estimate at a particular node in the downward step (Equation (15.19)) is equal to the sum of its estimates in the upward step and the difference in the estimates of the parent node in the downward and upward step weighted by a suitable coefficient.

15.2.2 PL, FL, and SL data fusion techniques

These three data fusion techniques belong to the macro-class of non-phenomenological methods. PL method can be applied when sensors are used to generate data in the form of images. The statistical characteristics of the images combined with information concerning the relationship between the sensors are used to develop the fusion strategy. FL technique implies the fusion of a reduced set of data representing the signal, called features. Features are an abstraction of the raw data intended to provide a reduced set that accurately and concisely represents the original data. SL fusion represents the highest level of fusion: such techniques call for extracting abstract elements of information called symbols. The symbols are manipulated using reasoning as a basis to generate better information. Potential benefits of data fusion include more accurate characterisation and often an ability to observe features that are otherwise difficult to perceive with a single sensor. The benefits are closely connected to the notion of redundant and complementary information. We witness redundancy in information when sensors observe the same features from the test specimen. In contrast, the fusion of complementary information allows features in the specimen to be observed that would otherwise not be seen.

In the scientific literature PL as well as FL techniques have been commonly used in the context of the ANN approach (see, for instance, [10,11]), which were deputed to fuse multiple measurements coming from different kinds of sensors with the aim of estimation the Bayes posterior probability that a sample of the measured signal has a certain information, according to the study case application. The ANN, in fact, can be considered as a black box model which is able to approximate nonlinear functions within a predefined accuracy by means of a learning procedure that exploits the availability of a dataset of examples. In the case of multilayer ANN, the nodes of the network are arranged in three layers (input, hidden, and output layer, respectively), whose size depends on the in-study case. This choice is due to the so called Kolmogorov's theorem: any continuous function can be computed by an opportune three-layer totally connected recurrent network having n input, $2n + 1$ hidden, and m output nodes (i.e. neurons) [14]. Layers are connected through links with associate weights that measure the strength of the connection, which are the objectives of the learning: they can indeed be updated during the training stage in order to minimise the error of the network in estimating the target values of the output quantities. Therefore, the ANN basically works on learning the numerical relationship between the inputs and the outputs of the system. Thus, in the case of image fusion techniques, ANN can be exploited to associate the measurements coming from multiple sensors with the information useful to inspectors and technicians, according to the inspected problem. In the PL based approach, the ANN processes just the intensity levels of each input image, by considering each image pixel by pixel. The PL data fusion approach cannot depict particular kind of information, such as a crack within a metallic plate, if it is not well depicted by imagery measurements (i.e. if it is not so much spatially extended), or if the used sensor has not a high resolution. This happens, since the PL based data fusion system does not acquire the information contextual to the pixel to be classified: in this sense, if we consider a window centred on the incoming pixel, we can extract useful information about the considered pixel, e.g. by computing the statistical moments of its neighbouring pixels. Therefore, this latter approach is based on the processing of particular features extracted from the available measurements (i.e. the images) and given as inputs to the ANN; in other words, it is the FL based data fusion approach. It allows us to gain information not only about the considered pixel but also about the context where it is immersed.

Both the previous systems are based on the intensity of the pixels to be classified: the former exploits just the intensity of the pixel, while the latter takes into account the intensity by computing, for instance, the statistical moments. Nevertheless, the disadvantage due to the sensor's resolution can still be present. Furthermore, another disadvantage can be introduced if the used sensors have different behaviour with respect to the Signal-to-Noise Ratio (SNR). Usually, high resolution corresponds to a low SNR and vice versa. Thus, data showing a low SNR can be considered to evaluate the macro-information, whereas the signals having a high SNR can be exploited to obtain the micro-information. In this way, the joint use of these sensors in a sort of SL approach allows us simultaneously to reduce the negative action of the noise (improving the quantitative information), thanks to the high SNR of the latter signals, and to correctly detect the macro-information (i.e. qualitative information), thanks to the high resolution of the former signals. In short, the SL based technique extracts symbols from different sensors as complementary information and evaluates the data by using these symbols.

15.3 Radar image fusion by MKF

This section reports the application of the MKF to the fusion of images of the same scene, acquired by different radars operating with different resolutions. In the case of remotely sensed data [13], since images acquired by multiple sources or by the same source in multiple acquisition contexts are available, and since very often the data provided by these different sources are complementary to each other, the merging operations can bring to a more consistent interpretation of the scene. The dataset processed here has been kindly provided by the Jet Propulsion Laboratory (JPL), CA, USA. It includes some images acquired by an airborne Synthetic Aperture Radar (SAR) [9] and by the SIR-C (Spaceborne Imaging Radar-C) [10]; the dataset refers to various parts of the San Francisco bay area. The images of the dataset are obviously affected by the topographic changes, and, therefore, a radiometric correction is needed to reduce these negative effects. The images have been spatially registered by identifying the corresponding image points, and by generating geometric relationships between the images. The input data have been referred to a common regular grid, and each pixel of each registered image corresponds to homologous pixels of the other images. To construct this grid, an image from the dataset has been chosen as a reference, and the others have been registered to it. The registered data are now the input to the fusion algorithm: the MKF merges data at different resolution scales.

15.3.1 *Dataset description*

We have applied the MKF to a dataset that consists of images acquired by two kinds of radar: a Synthetic Aperture airborne Radar (SAR) and a Spaceborne Imaging Radar-C (SIR-C). The SAR dataset consists of two images (Figures 15.3(a) and 15.3(b)) acquired in April 1995 during a JPL-AIRSAR mission (CM5599 and CM5598). They refer to the San Francisco Bay area, in particular to the Golden Gate site at latitude 37.6133° and at longitude $-121.7333°$. Two different bandwidths are adapted: 40.00 MHz for CM5599 and 20.00 MHz for CM5598. The spatial resolutions of the AIRSAR images are: 6.662 m along range and 18.518 m along azimuth for the CM5599 image, and 13.324 m along range and 18.518 m along azimuth for the CM5598 image.

The SIR-C dataset consists of an image depicting the same above described geographic area (Figure 15.4). The image that has been acquired in October 1994 has a lower resolution than the AIRSAR images. The spatial resolutions of the SIR-C image, in fact, are 25 m along range and 25 m along azimuth.

Before fusing the data, the main need is to refer the data to a common regular grid [17]; in this way, each pixel of each image will correspond to the homologous pixels of the remaining images. Our dataset includes three images and we decided to choose the CM5599 image as reference image, and we have matched the other images (CM5598 and SIR-C) to the reference one. The results of the data registration are depicted in Figure 15.5. Comparing the size of the SIR-C image depicted in Figure 15.4 to the size of the registered SIR-C image in Figure 15.5(b), it can be seen that it has been oversampled to obtain the same size of the reference CM5599 image. Now, the dataset, that we would fuse, consists of the registered data (Figures 15.2(b), 15.5(a) and 15.5(b)). We have included another image in this dataset, simulated by averaging a window of 4×4 neighbourhood pixels in the CM5599 image. In this way, we have obtained another image with a resolution higher than the SIR-C image, but lower than the full resolution CM5599 image, and than the registered CM5598 image (see Figure 15.5(c)).

15.3.2 *Image fusion MKF model*

The model explained in Section 15.2.1 is applied to fuse the image dataset as depicted in Figure 15.6. The model has to be specified for our case, and particularly we have to specify the values of the model parameters $\mathbf{B}(\lambda)$ and $\mathbf{C}(\lambda)$. Since $\mathbf{Y}(\lambda) \in \Re_{N,M}$, and

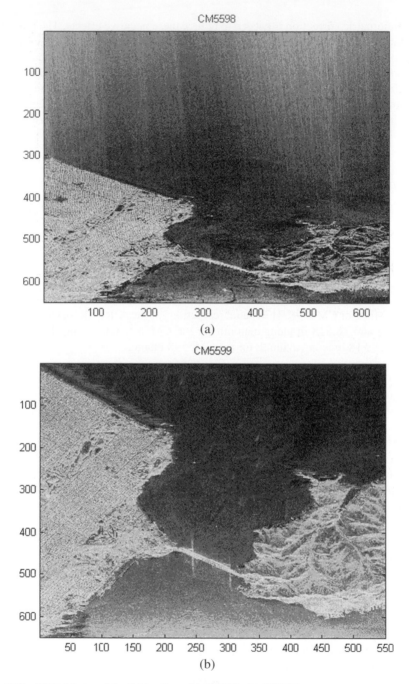

Figure 15.3 *AIRSAR image of the Golden Gate: (a) CM5598; (b) CM5599.*

we have four different measurements, the aim is to estimate the state $\mathbf{X}(\lambda) \in \Re_{N,M}$ on the basis of the measurements at four different resolution levels. We have chosen the following model:

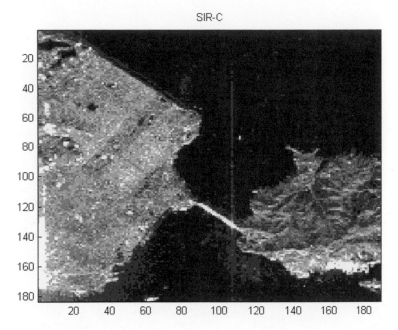

Figure 15.4 *SIR-C image of the Golden Gate.*

$$\mathbf{X}(\lambda) = \mathbf{A}(\lambda) \cdot \mathbf{X}(\gamma\lambda) + \mathbf{W}(\lambda) \tag{15.22}$$

$$\mathbf{Y}(\lambda) = \mathbf{X}(\lambda) + \mathbf{V}(\lambda) \tag{15.23}$$

This model has been chosen as in the case of the classical Kalman filter applied to estimate a scalar random constant, when measurements of the constant are available: $\mathbf{B}(\lambda) = \mathbf{C}(\lambda) \equiv \mathbf{I}_{N,M}$ [18]. $\mathbf{A}(\lambda)$ has been estimated by the minimum mean square error (MMSE) technique, and it takes into account the different radar reflectivity of the images of the dataset.

The matrices have the following sizes: $\mathbf{P}_X, \mathbf{Q}, \mathbf{R} \in \mathfrak{R}_{N,N}$ and $\hat{\mathbf{P}}, \mathbf{K}, \mathbf{J} \in \mathfrak{R}_{N,N}$; the matrix $\mathbf{P}_X(0) = \sigma^2(\mathbf{Y}(m=1)) \cdot \mathbf{I}_{N,N}$. Since we have just 4 measurements, the dyadic tree consists of 4 different levels, and at the finest resolution we have the CM5599 image, the CM5598 image is the measurement at the level $m = 3$, the low resolution CM5599 is that at $m = 2$, and the SIR-C is that at $m = 1$.

By comparing the registered SIR-C image (Figure 15.5(b)) with the fused image at resolution $m = 1$ (Figure 15.7(a)), it can be noted that the knowledge from the finer scales has been propagated and transferred to the coarsest resolution image. In fact, by observing the estimated state at resolution $m = 1$, one can see that more details have been added, and these details have been provided by the CM5598 and CM5599 measurements. The same observation applies to the estimated state at resolutions $m = 2, 3, 4$; as a reference image we use the full resolution image, that is, the fused image at $m = 4$ resolution level. In this image we can distinguish three different areas: urban (left), mountain (right–centre), and sea areas. A comparison between the input data and the fused data can be carried out

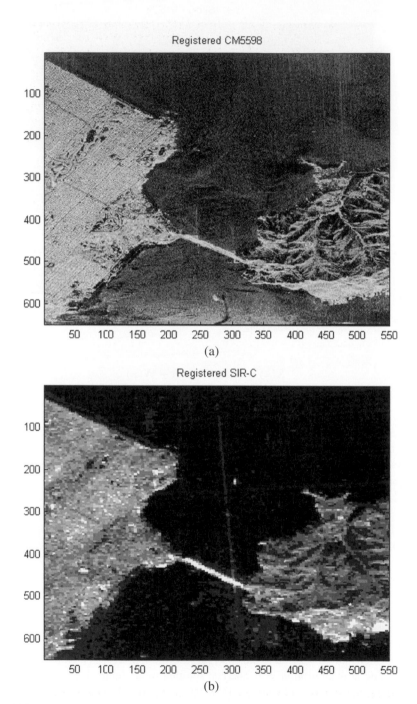

Figure 15.5 *(a) CM5598 is registered to CM5599; (b) SIR-C is registered to CM5599; (c) simulated low resolution CM5599 image.*

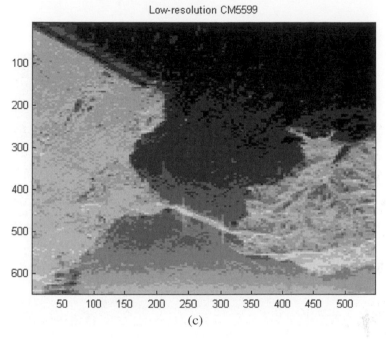

Low-resolution CM5599

(c)

Figure 15.5   *(continued)*

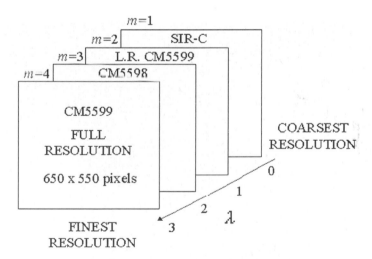

Figure 15.6   *The dataset consists of four different images: SIR-C, low resolution CM5599, CM5598, full resolution CM5599.*

by computing the standard deviations of different samples related to these areas. Three samples for each area have been considered:

- urban area: sample #1 (101:164, 1:64); sample #2 (401:464, 1:64); sample #3 (351:414, 1:64);

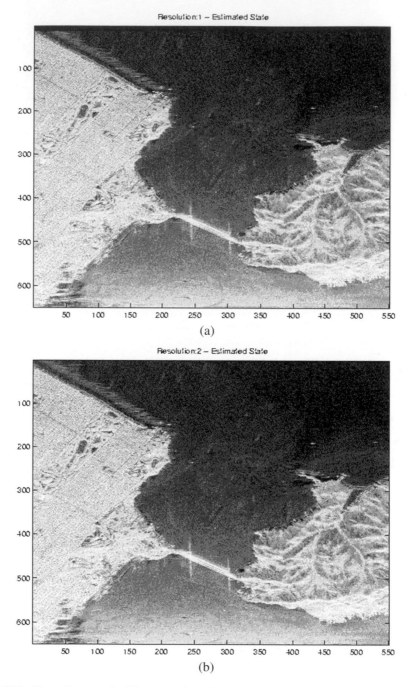

Figure 15.7   *Merged image at the following resolution level: (a) m = 1; (b) m = 2; (c) m = 3; (d) m = 4.*

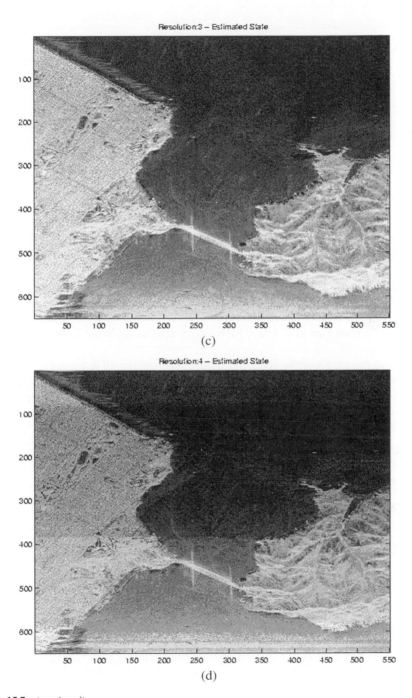

(c)

(d)

Figure 15.7   *(continued)*

Table 15.1 *Standard deviations for samples from CM5599.*

CM5599	Sample #1	Sample #2	Sample #3
Urban area	46.96	41.19	41.90
Mountain area	43.89	44.08	32.71
Sea area	16.83	63.94	22.73

Table 15.2 *Standard deviations for samples from merged image at resolution m = 4.*

Merged $m = 4$	Sample #1	Sample #2	Sample #3
Urban area	31.54	31.69	30.59
Mountain area	33.78	33.71	26.00
Sea area	14.25	44.22	17.67

- mountain area: sample #1 (401:464, 401:464); sample #2 (401:464, 351:414); sample #3 (465:528, 465:528);
- sea area: sample #1 (1:64, 401:464); sample #2 (301:364, 401:464); sample #3 (501:564, 151:214).

The values of the standard deviation for the full resolution CM5599 image and for the fused image at resolution $m = 4$ have been reported in the following Tables 15.1 and 15.2.

We can note that each sample in the fused image has a standard deviation lower than the corresponding sample of the input image. Therefore, in the fused image at full resolution, neighbourhood pixels corresponding to the same area have an intensity level more similar, and this fact means that the fusion process has been able to reduce the negative effects of the speckle noise typically affecting remote sensed images.

15.3.3 A detection test on straight lines

In order to demonstrate that the fused images have information content higher than the single measurements, we propose a detection test on straight lines. This test is carried out to the input CM5599 full resolution image and to the fused image at the resolution $m = 4$. The experiments are based on the use of the Hough Transform (HT) [19], i.e. a 2D non-coherent operator which maps an image to a parameter domain.

Since the image is unavoidably affected by the speckle noise, we need a system able to distinguish between spurious peaks related to noise or background effects and peaks related to straight lines really present in the input image. A Constant False Alarm Rate (CFAR) detection algorithm has been applied in the Hough plane to detect the correct peaks. Figure 15.8(a) depicts the HT computed on the portion (401:528, 201:328) of the full resolution CM5599 (Figure 15.3(b)); Figure 15.8(b) depicts the HT computed on the same portion of the merged image at $m = 4$ resolution level. This portion contains the Golden Gate Bridge. The aim is to detect the bridge which can be considered in the context of the image as a straight line. It can be noted that the bridge is detected in both

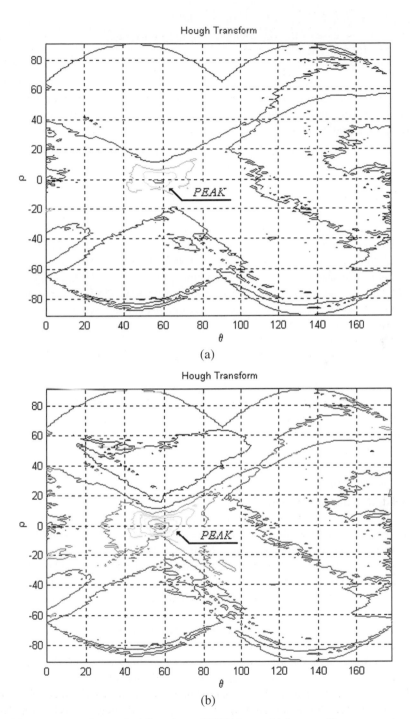

Figure 15.8 *HT computed on (a) the full resolution CM5599 image; (b) the merged image at m = 4 resolution level.*

images (as indicated in Figures 15.8(a) and 15.8(b)) by comparing the probability of false alarm (P_{fa}) in the case of the HT computed on the fused image and the P_{fa} of the HT of the CM5599 image. The bridge in the first case (merged data) has been detected with a lower P_{fa} than in the second case (CM5599 input image); the P_{fa} in the case of the fused image is 10 per cent lower than the P_{fa} in the case of the CM5599 image. Therefore, the knowledge added from the coarsest resolution to the finest resolution image can be measured by taking into account the fact that the lines in the fused image are more evident and detectable, and furthermore, the fused image exhibits a lower probability of false alarm.

15.4 An NDT/NDE application of FL, PL, and SL

In this section, a data fusion approach to the classification of eddy current and ultrasonic measurements [15] is proposed in an applicative example from the framework of defect detection/recognition methods for non-destructive testing/evaluation systems. The experimental data processed here have been kindly provided by the Material Assessment Research Group, Iowa State University, Ames, USA. The purpose is to demonstrate that a multi-sensor approach that combines the advantages carried by each sensor is able to locate potential cracks on the inspected specimen, maximising the probability of detection and minimising the probability of false alarm with respect to the cases where just one sensor is used. Within this framework, PL, FL, and SL data fusion approaches have been compared. The experimental results carried out on an aluminium plate pointed out the ability of the symbol level proposed approach to classify the input images within a prescribed accuracy level, by taking into account both the probability of detection of a defect and the related probability of false alarm, i.e. to decide in favour of defect presence where it is absent. These quantities often have an economic value in the context of testing.

The layout of the inspected specimen is described by Figure 15.9. Here, a sort of 'butterfly' shaped defect is present. Both eddy current and ultrasonic testing retrieved an imaging map of the inspected plate. Each technology offers its own set of advantages and disadvantages. Ultrasonic imaging techniques offer excellent resolution; however, the method is sensitive to a wide variety of measurement conditions, including surface roughness and coupling which affects the signal-to-noise ratio by reducing it relevantly. In contrast, eddy current techniques do not require coupling and are relatively insensitive to surface roughness conditions. The disadvantages associated with the eddy current method lie in its poor resolution capabilities. Although eddy current methods offer excellent flaw detection capabilities, they are not an effective method for characterising small flaws due to their poor resolution characteristics. Therefore, it is required to jointly process the ultrasonic and eddy current measurements in order to obtain good results in terms of classification performances. In the proposed case study, Figure 15.10(a) shows the eddy current image obtained at 8 kHz excitation frequency using a Zetec® E-144-P pancake probe which has inner diameter of 0.11 inch with 0.05 inch ferrite core (the grey levels represent the 8 bit scaled intensity of the magnitude signal). It can be noted that the resolution of the sensor is low, since the magnitude of the signal increases when the sensor approaches the defect; nevertheless, the measurement has a very high signal-to-

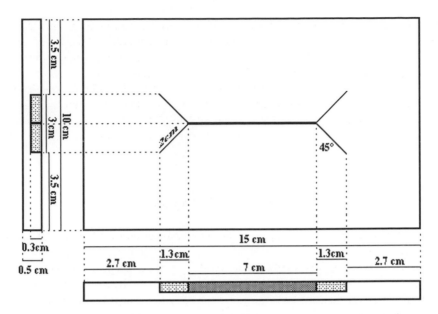

Figure 15.9 *The inspected specimen for the NDT related problem: it is possible to note a 'butterfly' shaped defect at the centre of the plate.*

noise ratio. On the other hand, Figure 15.10(b) depicts the ultrasonic measurement carried out on the inspected aluminium plate, obtained by scanning acoustic microscope system with a 60 MHz focused transducer. In the case of ultrasonic measurements, the sensors are mostly sensitive to the specimen roughness, that implies a low signal-to-noise ratio; nevertheless, the measurement resolution is very high; indeed, evident modifications of the measured signal occur just in the location where the defect is present. Problems can occur when the defect is too thin when compared with the spatial resolution used during the scanning of the sensor over the inspected plate. In our case, the wings of the butterfly are too thin and their internal portions are not detected by the ultrasonic signal.

To process the data available from the experimental campaigns, as far PL and FL data fusion techniques are concerned, a multilayered feed-forward ANN has been used. Its first layer (a linear layer) acts as the input layer by acquiring the incoming samples; the input of the data was preceded by a suitable normalisation of the data. The second (hidden) layer is made of nonlinear (sigmoidal) nodes of varying number whose optimal size is decided through standard pruning and growing procedures. The third layer outputs the probability that the considered pixel belongs to the 'defect' class. After the training phase, the testing data are processed: pixels are assigned to the 'defect' class on the basis of a thresholding operation on the estimated probability. In the PL based approach, the ANN processes just the intensity levels of each image. The result of the procedure, in terms of classified image, is depicted in Figure 15.11(a) (the 'defect' pixels are depicted black). In order to estimate the performance of the algorithm, a truth map of the specimen has been computed on the basis of some a priori knowledge on the inspected specimen; thus, the probability of defect detection, P_d, and the probability of false alarm, P_{fa}, have been evaluated on the classified testing data. The probability of false alarm expresses the event

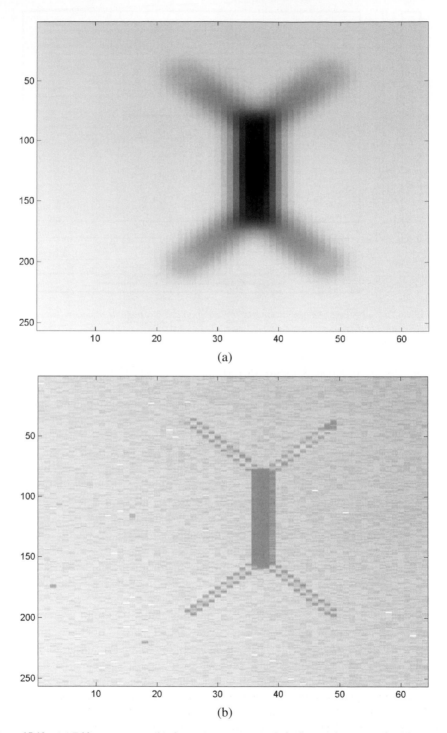

Figure 15.10   *(a) Eddy current map; (b) ultrasonic measurement. In both maps the magnitude of the measured signal is plotted.*

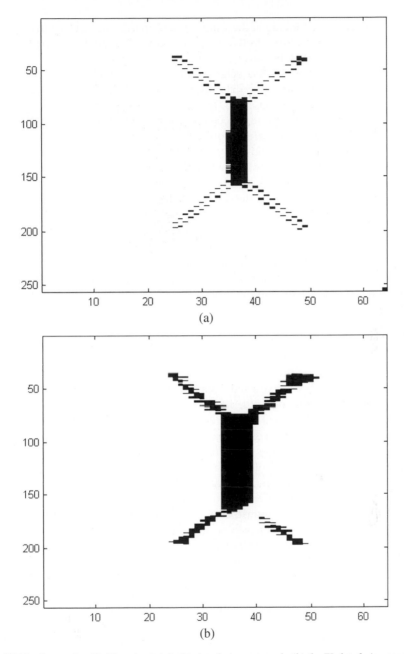

Figure 15.11 *Image classified by using (a) the PL data fusion approach; (b) the FL data fusion approach.*

that a 'no-defect' pixel can be classified as a 'defect' pixel. In the PL case, we obtained the following results: $P_{\mathrm{d}} = 65\%$ and $P_{\mathrm{fa}} = 1\%$. The achieved probability of false alarm is very low; however, in NDT applications the probability of detection has a greater penalty. It is thus needed to improve the performance of the system by maximising the P_{d}. In fact, the PL data fusion approach is not able to detect particular areas of the specimen where the defect is present. By looking at the wings of the butterfly, we can say that the system is not able to detect these parts. In the ultrasonic measurement, due to the fact that the wings are very thin and that the flaw is not very deep, the inner parts of the wings do not appear, and therefore, the pixel level data fusion classify the related pixels as background pixels. This fact happens, since the data fusion system does not acquire the contextual information of the pixel to be classified. In this sense, if we consider a window centred on the incoming pixel, by computing specific statistical moments of its neighbouring pixels, we will be able to extract useful information about the considered pixel. Therefore, for each pixel of the ultrasonic and eddy current images, we compute the statistical moments (mean, standard deviation, skewness, and kurtosis) of the pixels in a window that surrounds the pixel of interest: the ANN now processes as inputs these features and classifies the whole input image, as in the case of the PL approach. This FL based approach allows us to obtain information not only about the considered pixel but also about the context where it is located. By computing the probability of detection and of false alarm, we obtained the following results: $P_d = 84\%$ and $P_{fa} = 5\%$. Figure 15.11(b) depicts the image classified by using this FL data fusion scheme. This experimental result confirms the usefulness of the FL with respect to the PL approach, by substantially increasing the probability of detection, although the probability of false alarm is slightly increased as well.

Both previous systems exploit the information related to the pixels to be classified: the first one exploits just the intensity of the pixel, while the second one investigates the statistical moments of the intensity within a neighbourhood of the pixel. The relevant information about the flaw is extracted from the two different acquiring technologies in a different fashion. By looking at the input readings (Figure 15.9), we can note that in the eddy current image the signal is more and more evident as the sensor approaches the defect, while in the ultrasonic image we have a clean change in the intensity of the signal when the ultrasonic beam reaches the defect. Therefore, the ultrasonic sensor produces measurements with very high resolution; however, in our experiments, the ultrasonic data suffer of two main drawbacks: first, the wings are too thin to be detected effectively; then, the ultrasonic measurements suffer from the roughness of the specimen, thus generating a low signal-to-noise ratio. In contrast, even if the eddy current readings imply a low resolution, they have a very high signal-to-noise ratio. This kind of information can be used to implement a defect location system that could maximise the probability of detection while fixing the acceptable level for the probability of false alarm. The high resolution offered by the ultrasonic measurements when the sensor reaches the border of the defect can be exploited to improve the detection of the borders of the flaw by using the variations of the signal intensity. In practice, by using a Sobel mask [16], we can extract the edges from the ultrasonic measurements and we can use these edges to locate the defect borders (see Figure 15.12(a)). The borders extracted from the ultrasonic measurements are used as a guide to correct the eddy current classified image. The defect areas that are not detected by the ultrasonic sensor are detected by the eddy current measurement. This allows us to reduce the negative action of the noise because of the high signal-to-noise ratio of the eddy current signal. At the same time, we are able to detect correctly the shape of the defect, thanks to the high resolution of the ultrasonic signal. Figure 15.12(b) shows the classified image by using the so defined SL data fusion approach. In terms of the probability of detection and of false alarm, we obtained the following results: $P_d = 98\%$ and $P_{fa} = 3\%$.

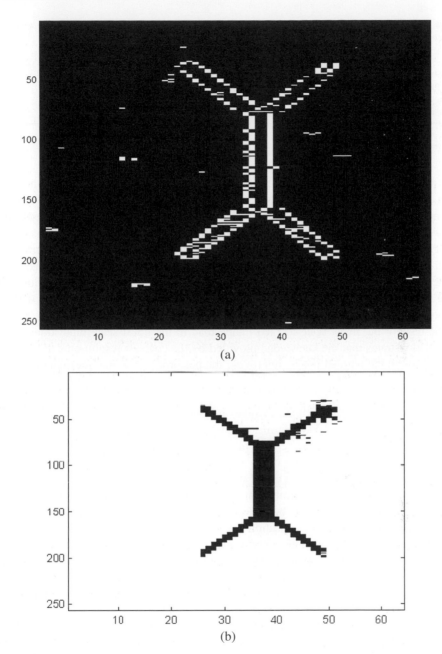

Figure 15.12 *(a) Edge extracted by the ultrasonic measurement, by using a Sobel mask; (b) image classified by using the SL data fusion approach.*

15.5 Conclusions

In this chapter, we proposed two relevant applications of several image fusion techniques in industrially significant fields, namely the NDT/NDE and the segmentation classifica-

Table 15.3 *Comparison of the performance of the data fusion techniques by means of the probabilities of detection and false alarm.*

	Probability of detection	Probability of false alarm
PL approach	65%	1%
FL approach	84%	5%
SL approach	98%	3%

tion and detection of defence targets in remote sensing imagery. In particular, an image fusion algorithm based on MKF has been applied to combine remotely sensed data acquired by radars having different resolutions. The considered images have been acquired during the AIRSAR Mission and SIR-C/X-SAR Mission. The data have been co-registered to refer each pixel of each image to a common regular grid. The image fusion algorithm has been tested, and the merged images have been presented at different resolutions. A lineament detection algorithm based on the HT has been applied to the full resolution input data and to the full resolution merged data. The Golden Gate Bridge has been detected in both images, and it is shown that the computed probability of false alarm is lower in the case of the finest scale merged image than in the finest scale input image. This fact demonstrates that the knowledge provided by the coarsest resolution data has been transferred to the merged image, improving the performance of the lineament detection algorithm.

The second presented application involves non-phenomenological image fusion methods, such as PL, FL, and SL data fusion. They have been applied in the framework of defect detection/recognition system, to fuse eddy current and ultrasonic measurements: the techniques classify each pixel of the acquired measurements in two different 'defect' and 'no-defect' classes. The joint use of the eddy current and ultrasonic measurements is suggested because of the poor results that are obtained by processing each single recorded type of signals alone. Therefore, both measurements are jointly processed, and the information used to perform the classification has been extracted at three different levels: pixel, feature, and symbol. The numerical performance of these techniques has been compared by using the probability of detection and the probability of false alarm. Experiments carried out on real data confirmed the effectiveness of the proposed SL based approach, by maximising the probability of detection and by achieving an acceptable probability of false alarm with respect to the PL and FL fusion techniques.

The results limited to the presented applications confirmed the usefulness of image fusion techniques.

Acknowledgements

Authors are very grateful to Dr. Ellen O'Leary, Science Coordinator of the AIRSAR Mission (Jet Propulsion Laboratory, CA, USA), and to the Material Assessment Research Group, Iowa State University, Ames, USA, for providing the datasets used in this work.

The authors also explicitly thank other co-workers and co-authors of the papers cited in the references.

References

[1] X.E. Gros, *NDT Data Fusion*, Arnold, 1997.

[2] X.E. Gros, *Applications of NDT Data Fusion*, Kluwer Academic Publishers, 2000.

[3] R.S. Blum and Z. Liu (eds.), *Multi-Sensor Image Fusion and Its Applications. Signal Processing and Communications*, CRC Press/Taylor & Francis, 2005.

[4] K.C. Chou, A.S. Willsky and A. Benveniste, 'Multiscale recursive estimation, data fusion, and regularization', *IEEE Transactions on Automatic Control*, Vol. 39, No. 3, 1994, pp. 464–477.

[5] Z. Liu, D.S. Forsyth, P. Ramuhalli and A. Fhar, 'A data fusion framework for multiple nondestructive inspection images', in: C.H. Chen (ed.), *Ultrasonic and Advanced Methods for Nondestructive Testing and Material Characterization*, World Scientific Publishing, 2008, pp. 1–20, in press.

[6] G. Simone and F.C. Morabito, 'NDT image fusion using eddy current and ultrasonic data', *COMPEL: International Journal for Computation and Mathematics in Electrical and Electronic Engineering*, Vol. 20, No. 3, 2001, pp. 857–868.

[7] G. Simone, A. Farina, F.C. Morabito, S.B. Serpico and L. Buzzone, 'Image fusion techniques for remote sensing applications', *Information Fusion*, Vol. 3, No. 1, 2002, pp. 3–15.

[8] M. Basseville, A. Benveniste, K.C. Chou, S.A. Golden, R. Nikoukhah and A.S. Willsky, 'Modeling and estimation of multiresolution stochastic processes', *IEEE Transactions on Information Theory*, Vol. 38, No. 2, 1992, pp. 766–784.

[9] R. Wilson, 'Multiresolution image modelling', *Electronics & Communication Engineering Journal*, 1997, pp. 90–96.

[10] X.E. Gros, Z. Liu, K. Tsukada and K. Hanasaki, 'Experimenting with pixel level NDT data fusion techniques', *IEEE Transactions on Instrumentation and Measurement*, Vol. 49, No. 5, 2000, pp. 1083–1090.

[11] L. Bruzzone, D.F. Prieto and S.B. Serpico, 'A neural-statistical approach to multitemporal and multisource remote-sensing image classification', *IEEE Transactions on Geoscience and Remote Sensing*, Vol. 37, No. 3, 1999, pp. 1350–1359.

[12] L. Bruzzone, 'An approach to feature selection and classification of remote sensing images based on the Bayes rule for minimum cost', *IEEE Transactions on Geoscience and Remote Sensing*, Vol. 38, No. 1, 2000, pp. 429–438.

[13] A. Farina and F.A. Studer, *Radar Data Processing*, Research Studies Press/John Wiley & Sons, 1984.

[14] V. Kurková, 'Kolmogorov's theorem and multilayer neural networks', *Neural Networks*, Vol. 5, 1992, pp. 501–506.

[15] M. Buonsanti, M. Cacciola, S. Calcagno, F.C. Morabito and M. Versaci, 'Ultrasonic pulse–echoes and eddy current testing for detection, recognition and characterisation of flaws detected in metallic plates', in *9th European Conference on Non-Destructive Testing, ECNDT 2006*, Berlin, Germany, September 2006, CD-ROM.

[16] J.C. Russ, *The Image Processing Handbook*, second ed., CRC Press, 1995.

[17] F.W. Leberl, *Radargrammetric Image Processing*, Artech House, 1990.

[18] A. Signorini, A. Farina and G. Zappa, 'Application of multiscale estimation algorithm to SAR images fusion', in *Proc. of International Symposium on Radar, IRS98*, Munich, September 1998, pp. 1341–1352.

[19] F.C. Morabito, G. Simone and A. Farina, 'Automated lineament detection in SAR images based on the joint use of wavelet and hough transforms', in *Proc. of 5th International Conference on Radar Systems*, Brest, France, May 1999, pp. 1.2.13–18.

16

Concepts of image fusion in remote sensing applications

Pushkar Pradham, Nicolas H. Younan and Roger L. King

Department of Electrical and Computer Engineering, Mississippi State University, USA

16.1 Image fusion

16.1.1 Introduction

Image fusion refers to the process of combining two or more images into one composite image, which integrates the information contained within the individual images [1]. The result is an image that has a higher information content compared to any of the input images. The goal of the fusion process is to evaluate the information at each pixel location in the input images and retain the information from that image which best represents the true scene content or enhances the utility of the fused image for a particular application. Image fusion is a vast discipline in itself, and refers to the fusion of various types of imagery that provide complementary information. For example, thermal and visible images are combined to aid in aircraft landing [2]. Multispectral images are combined with radar imagery because of the ability of the radar to 'see' through cloud cover.

This chapter focuses on image fusion techniques for remote sensing applications to fuse multispectral (MS) and panchromatic (PAN) images. The objectives of this chapter are to explain to the reader pan sharpening algorithms and some of the applications of pan sharpening. The chapter focuses heavily on pan sharpening algorithms based on multiresolution analysis methods. A detailed mathematical explanation of multiresolution theory is also given for those who are not familiar with this area. There are many variations within the multiresolution-based pan sharpening algorithms; some of the main ones are explained here. Given that there are many multiresolution-based pan sharpening algorithms, it can be confusing for someone who wants to implement or use a pan sharpening technique as to which one to choose. The later part of the chapter presents some recommendations on the effectiveness and the performance of these methods. These recommendations are based on a comprehensive study that evaluates various pan sharpening algorithms by applying them to various satellite images from different sensors and based on meaningful quantitative metrics.

16.1.2 Characteristics of remote sensing imagery

Remote sensing images are characterised by their spectral, spatial, radiometric, and temporal resolutions. Spectral resolution refers to the bandwidth and the sampling rate over which the sensor gathers information about the scene. High spectral resolution is characterised by a narrow bandwidth (e.g., 10 nm). Spatial resolution refers to the smallest features in the scene that can be separated (resolved). The radiometric resolution refers to the dynamic range or the total number of discrete signals of particular strengths that the sensor can record. A larger dynamic range for a sensor results in more details being discernible in the image. The Landsat 7 sensor records 8-bit images; thus it can measure 256 unique grey values of the reflected energy while Ikonos-2 has an 11-bit radiometric resolution (2048 grey values). In other words, a higher radiometric resolution allows for simultaneous observation of high and low contrast objects in the scene. The temporal resolution refers to the time elapsed between consecutive images of the same ground location taken by the sensor. Satellite-based sensors, based on their orbit, may dwell continuously on an area or revisit the same area every few days. The temporal characteristic is helpful in monitoring land use changes [3].

Due to system tradeoffs related to data volume and signal-to-noise ratio (SNR) limitations, remote sensing images tend to have either a high spatial resolution and low spectral resolution or vice versa [4]. The following section explains the relationship between the spatial resolution and spectral resolution.

16.1.3 Resolution tradeoffs

All sensors have a fixed signal-to-noise ratio that is a function of the hardware design. The energy reflected by the target must have a signal level large enough for the target to be detected by the sensor. The signal level of the reflected energy increases if the signal is collected over a larger instantaneous field of view (IFOV) or if it is collected over a broader spectral bandwidth. Collecting energy over a larger IFOV reduces the spatial resolution while collecting it over a larger bandwidth reduces its spectral resolution. Thus, there is a tradeoff between the spatial and spectral resolutions of the sensor. As noted above, a high spatial resolution can accurately discern small or narrow features like roads, automobiles, etc. A high spectral resolution allows the detection of minor spectral changes, like those due to vegetation stress or molecular absorption [4].

Most optical remote sensing satellites carry two types of sensors – the panchromatic and the multispectral sensors. The multispectral sensor records signals in narrow bands over a wide IFOV while the panchromatic sensor records signals over a narrower IFOV and over a broad range of the spectrum. Thus, the multispectral (MS) bands have a higher spectral resolution, but a lower spatial resolution compared to the associated panchromatic (PAN) band, which has a higher spatial resolution and a lower spectral resolution. Table 16.1

Table 16.1 *Spectral and spatial resolutions of Landsat 7 bands.*

Band number	Spectral range (nm)	Spatial resolution (m)
1	450–515	30
2	525–605	30
3	630–690	30
4	750–900	30
5	1550–1750	30
6	1040–1250	60
7	2090–2350	30
Panchromatic	520–900	15

Table 16.2 *Spectral and spatial resolutions of SPOT 5 bands.*

Band number	Spectral range (nm)	Spatial resolution (m)
1	500–590	10
2	610–680	10
3	780–890	10
4	1580–1750	20
Panchromatic	480–710	2.5

lists the spectral and spatial resolutions of the Landsat 7 ETM+ sensor [5]. Table 16.2 gives the same information for the SPOT 5 sensor [6]. The Landsat 7 MS bands 1–5 have a spatial resolution of 30 m while the PAN band has a 15 m resolution. Similarly, the MS bands 1–4 of SPOT 5 have a spatial resolution (10 m), which is four times worse than that of the PAN band (2.5 m). Quickbird, Ikonos, and the Indian Remote Sensing Satellite (IRS) are other satellite systems that carry high spatial resolution and high spectral resolution sensors.

Figures 16.1 and 16.2 represent the MS and PAN images taken from a Quickbird satellite of the same scene. The MS and PAN images have a spatial resolution of 4 and 1 m, respectively. The MS image is stretched to the same size as the PAN image to illustrate the missing details. The scene details are much clearer in the PAN image but on the other hand it has no colour information.

16.1.4 Pan sharpening

Researchers and customers who buy satellite imagery desire both high spatial and spectral resolutions simultaneously in order to extract the maximum information content from the imagery. Thus, the information from both the PAN and MS channels needs to be integrated into one channel. Many different image fusion methods are found in the literature to combine MS and PAN images. These techniques combine the spatial details from a high spatial resolution-low spectral resolution (PAN) image with the low spatial resolution-high spectral resolution (MS) image to create a high spatial, high spectral resolution image. In the remote sensing literature, this image fusion process is popularly referred to as 'pan sharpening' since the details of the PAN image are used to 'sharpen'

Figure 16.1   *Quickbird multispectral image.*

the MS imagery. A more specific and accurate definition of the pan sharpening would be the enhancement of the spatial resolution of a low spatial resolution image by the integration of higher resolution details from an available higher spatial resolution image. This general terminology must be used since the higher spatial resolution image does not necessarily have to be that of a panchromatic band. Thus, a Landsat 7 MS image of 30 m spatial resolution could be fused with any MS band of SPOT 5 10 m spatial resolution to increase the Landsat 7 MS image's resolution by a factor of three. One could also use a higher resolution aerial photograph of the area if it was available.

16.1.5 Applications of pan sharpening

Many applications such as mapping of land use, vegetation and urban areas benefit from pan sharpening. The different objects or classes observed in the scene can be better distinguished or classified due to the high spectral resolution of multispectral images. However, the maps created will have a coarse appearance due to the low spatial resolution of the MS image. On the other hand, the different classes cannot be separated in the panchromatic imagery as they have almost identical grey values [7], but the higher spatial resolution,

Figure 16.2   *Quickbird PAN image.*

of the panchromatic band leads to a more accurate delineation of the structures and the boundaries between them [8]. Since the pan sharpened image has both a high spectral resolution and a high spatial resolution, the objects can be classified efficiently as well as delineated with higher accuracy. The maps created from the pan sharpened images leads to enhanced visual interpretation. Zhang et al. [7] used a pan sharpened image obtained from the fusion of Quickbird MS and PAN images to extract road networks in Fredericton, New Brunswick, Canada. Similarly pan sharpened images can be effectively used to extract other urban features like buildings and is an effective tool for urban mapping.

16.2 Pan sharpening methods

This section describes some pan sharpening algorithms implemented in practice along with their pros and cons. The popular methods for pan sharpening are based on Intensity Hue Saturation (IHS), the Principal Components Analysis (PCA), and Multiresolution Analysis (MRA) transformation. To explain the pan sharpening algorithms in this chapter the following notations will be used: B represents the set of MS bands, A – the PAN image and F – the pan sharpened bands. The notations B and F will be used to represent all the multispectral bands collectively. If a particular band is to be addressed a subscript shall be used, e.g. B_1 refers to the first band in the MS set. Most of the techniques require that the MS images first be resampled to the pixel size of the PAN image, thus the resampled MS images will be noted by the superscript *, i.e. B^*.

16.2.1 The intensity hue saturation method

16.2.1.1 The IHS transform

Digital images are generally displayed by an additive colour composite using the three primary colours – red (R), green (G), and blue (B) [3]. However, colours can be described by an alternate representation: Intensity, Hue, and Saturation (IHS). Intensity represents the total luminance of the image, hue represents the dominant wavelength contributing to the colour, and saturation describes the purity of the colour relative to grey. The IHS transformation separates the spatial and spectral information in a RGB image. The intensity component represents the spatial information while the hue and saturation describe the spectral information [3]. The spatial information can then be manipulated by performing some mathematical operation on the 'intensity' component to enhance the image without altering its spectral representation. This principle is used in the IHS pan sharpening scheme. The intensity at each pixel is defined as the maximum value of the R, G, and B values at that pixel position. The details to calculate IHS from RGB and the reverse RGB to IHS are given below [9]:

RGB TO IHS

1) $I := \max(R,G,B)$
2) Let $X := \min(R,G,B)$
3) $S := (I - X)/I$, if $S = 0$ return
4) Let $r := (I - R)/(I - X)$, $g := (I - G)/(I - X)$, $b := (I - B)/(I - X)$
5) If $R = I$ then $H := (\text{if } G = X \text{ then } 5 + b \text{ else } 1 - g)$;
   If $G = I$ then $H := (\text{if } B = X \text{ then } 1 + r \text{ else } 3 - b)$,
   else $H := (\text{if } R = X \text{ then } 3 + g \text{ else } 5 - r)$
6) $H := H/6$.

IHS TO RGB

1) $H := H^*6$
2) Let $J := \text{floor}(H)$, $F := H - J$
3) Let $M := I^*(1 - S)$,
   $N := I^*(1 - (S^*F))$,
   $K := I^*(1 - S^*(1 - F))$
4) Switch(J):
   Case 0: $(R,G,B) := (I,K,M)$,
   Case 1: $(R,G,B) := (N,I,M)$,
   Case 2: $(R,G,B) := (M,I,K)$,
   Case 3: $(R,G,B) := (M,N,I)$,
   Case 4: $(R,G,B) := (K,M,I)$,

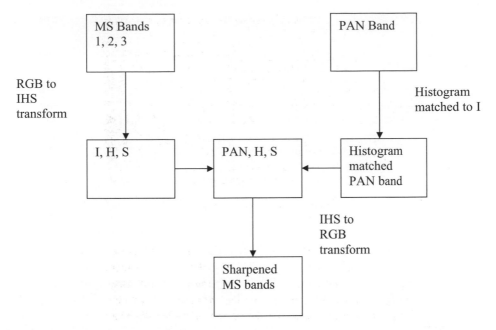

Figure 16.3 *IHS-based pan sharpening scheme.*

Case 5: (R,G,B) := (I,M,N).

In the above transforms, the RGB space is first normalised to the range [0, 1] and thus all RGB and HSV values fall in this range. A point to note is that if the saturation (S) is zero it means all R, G, B are equal at this point, and the hue is not defined at this point. The hue is then replaced by the immediate previous value in the image.

16.2.1.2 *The IHS-based pan sharpening*

The IHS pan sharpening technique is the oldest known data fusion method and one of the simplest. Figure 16.3 illustrates this technique for convenience. In this technique the following steps are performed:

1. The low resolution MS imagery is co-registered to the same area as the high resolution PAN imagery and resampled to the same resolution as the PAN imagery.
2. The three resampled bands of the MS imagery B^*, which represent the RGB space are transformed into IHS components.
3. The PAN imagery is histogram matched to the 'I' component. This is done in order to compensate for the spectral differences between the two images, which occurred due to different sensors or different acquisition dates and angles.
4. The intensity component of MS imagery is replaced by the histogram matched PAN imagery. The RGB of the new merged MS imagery is obtained by computing a reverse IHS to RGB transform.

In the above algorithm, replacing the spatial component of the MS imagery with the PAN imagery allows the details of the PAN imagery to be incorporated in to the MS imagery.

Figure 16.4   *IHS pan sharpened image.*

16.2.1.3 Comments on the IHS method

The IHS technique is fairly easy to understand and implement. Moreover, it requires very little computation time compared to the more sophisticated MRA techniques. However, it severely distorts the spectral values of the original colour of the MS image. This is seen from Figure 16.4, which shows the IHS pan sharpened Quickbird MS and PAN images given in Figures 16.1 and 16.2, respectively. The colours of the buildings in the lower right region of the image (below the diagonal road) have changed from white or light blue to a strong blue. The vegetation above the road also appears lighter in colour. Thus the IHS technique is good only for visual analysis, and not machine classification based on the spectral signatures of the original MS image. Moreover, it is also limited to three bands at a time.

The Principal Component Analysis (PCA) method is quite similar to IHS except that, instead of the IHS transformation, the Principal Components (PC) of the MS bands are calculated and the first PC is replaced by the PAN band. The advantage of PCA over IHS is that it does not have the three band limitation and can be applied to any number of bands at a time. However, this technique also introduces spectral distortion in the pan sharpened image like the IHS method.

Figure 16.5 *Quickbird image pan sharpened using a substitutive-based multiresolution method.*

16.2.2 Multiresolution analysis-based pan sharpening

Section 16.1.5 showed that there are various applications of pan sharpening ranging from land use mapping to road extraction. All these applications involve classification of the imagery. In order that the pan sharpened imagery is classified correctly, the spectral information or the radiometry of the MS imagery must be preserved. One target application is the supervised classification of the pan sharpened imagery by using the spectral signatures derived from the original MS imagery [10].

Thus, preserving the spectral information of the original MS images in the pan sharpened images is important. This means that there should be ideally zero or minimal change in the radiometry or digital number (DN) values of the image. The change or loss of the original radiometry is also termed as 'spectral distortion' in literature. It is found that multiresolution analysis (MRA)-based methods seem to be very effective at producing pan sharpened images with least spectral distortion or high spectral fidelity with respect to the original MS images. Figure 16.5 shows the Quickbird MS image given in Figure 16.1 but pan sharpened using a multiresolution-based scheme. There are many variations in the multiresolution-based pan sharpening methods. The specific multiresolution scheme used to pan sharpen this image was a substitutive method using the Redundant Wavelet Transform. The Redundant Wavelet Transform and the substitutive method are explained in Sections 16.2.3.2 and 16.2.4.3. It can be seen that the colours in this pan sharpened image are much closer to the original MS image compared to the IHS pan sharpened image.

On the other hand, if spectral distortion is not of much concern and the goal is to produce pan sharpened images to serve as maps for better visual interpretation, any previously defined technique like the IHS or PCA should be sufficient.

The remainder of this chapter focuses on MRA methods for pan sharpening. The following couple of sections give an introduction to multiresolution theory and how it can be applied to produce pan sharpened images.

16.2.3 Multiresolution theory

Pattern recognition in images encompasses the automatic identification of variable sized objects in the image. For example, if there is a requirement to detect all the edges in an image, these edges could be large or small depending on how rapidly the image intensity changes. The small edges can be easily identified by applying small neighbourhood detection operators while the large edges can be identified by large neighbourhood operators. However, it is recognised that using a large operator to detect a larger feature is computationally very inefficient [11]. Rather than scaling the operator to the scale of the object, it is more efficient to vary the scale of the image (or the object). Thus, the image analysis (pattern recognition) is more efficient if the image is analysed at different resolutions [12].

An image at a given resolution can be divided into coarser approximations at a lower resolution. Suppose the original image has a resolution r_j and its lower resolution approximation image has resolution r_{j-1}. Then, the details missing in the lower resolution representation are given by the difference between the original image and the approximation image [12]. At coarser resolutions only the large objects are visible and the viewer gets only a rough idea of the image context. The original image can be reconstructed as successive details are added to the approximations and the finer details of the image become visible. A mathematical definition and an explanation of multiresolution are given below.

Multiresolution: A continuous function $f(t)$ can be decomposed into many subspaces, where each subspace contains a part of the whole function. Each of these different parts contains a projection of the function at different resolutions. This decomposition of the function onto subspaces at different scales or different resolutions can be defined as multiresolution [13].

In order to explain the multiresolution theory, two subspaces V_j and W_j are used and j denotes the scale of the subspace. The projection of $f(t)$ on the subspace V_j is denoted by $f_j(t)$. As the scale $j \to \infty$ the projection $f_j(t)$ better approximates the function $f(t)$. It should also be noted that each subspace V_j is contained within the next higher subspace

V_{j+1}, which can be generalised by the following equation:

$$V_0 \subset V_1 \subset \cdots \subset V_j \subset V_{j+1} \subset \cdots \qquad (16.1)$$

The details missing in V_j to construct V_{j+1} are contained in the subspace W_j, thus the subspace V_{j+1} can also be written as

$$V_{j+1} = V_j \oplus W_j \qquad (16.2)$$

The above equation can be generalised for any two consecutive subspaces (i.e., $V_1 = V_0 \oplus W_0$) which leads to the following summation:

$$V_{j+1} = V_0 \oplus W_0 \oplus W_1 \cdots W_j \qquad (16.3)$$

In addition to the above properties, the subspaces must also satisfy the dilation and translation requirements. The dilation requirement states that all the rescaled functions in V_j will be in V_{j+1}. If a scaling factor of two is chosen, then it means that if the subspace V_j contains frequencies up to f, V_{j+1} must contain the frequencies up to $2f$. This is given by the following equation:

$$f(t) \in V_j \quad \Leftrightarrow \quad f(2t) \in V_{j+1} \qquad (16.4)$$

While the translation requirement states that if $f_j(t)$ is in V_j then a shifted version of $f_j(t) - f_j(t-k)$ must also be in V_j:

$$f_j(t) \in V_j \quad \Leftrightarrow \quad f_j(t-k) \in V_j \qquad (16.5)$$

Finally, a scaling function $\phi(t)$ must be defined that generates the approximations of the function $f_j(t)$ on each subspace V_j. The translations of the $\phi(t) - \phi(t-k)$ must span the whole space V_0 and be orthonormal. The scaling function $\phi_{j,k}(t)$ for each subspace V_j is generated by the dilations and translations of $\phi(t)$:

$$\phi_{j,k}(t) = 2^{j/2} \phi(2^j t - k) \qquad (16.6)$$

The details that are present in the subspace W_j can be taken by the simple difference between two successive approximations $f_j(t)$ and $f_{j+1}(t)$ or by decomposing the function $f(t)$ on the wavelet function $-\psi(t)$. The wavelet functions for each subspace W_j are obtained by dilations and translations of this basic wavelet function:

$$\psi_{j,k}(t) = 2^{j/2} \psi(2^j t - k) \qquad (16.7)$$

This multiresolution analysis is not restricted to the continuous functions $f(t)$, but can be extended to discrete functions, which can be a one-dimensional signal or a two-dimensional image. The Laplacian Pyramid (LP), A Trous Wavelet Transform (AWT), Discrete Wavelet Transform (DWT), and the Redundant Wavelet Transform (RWT) are a few examples of transforms that perform multiresolution analysis of signals and images using different approaches. Each of these transforms is explained in detail in the following sections.

16.2.3.1 The Discrete Wavelet Transform (DWT)

According to the multiresolution theory, a function can be decomposed into its approximations and details. Thus, a given signal at original resolution J can be decomposed into details at successively lower resolutions $J - 1, J - 2, \ldots, J_0$ and the approximation at the lowest resolution J_0. This is given by the following equation [14]:

$$f(t) = \sum_k a_{J_0}(k)\phi_{J_0 k}(t) + \sum_k \sum_{j=J_0}^{J} d_j(k)\psi_{j,k}(t) \tag{16.8}$$

The coefficients $a_{J_0}(k)$ and $d_j(k)$ are the Discrete Wavelet Transform (DWT) coefficients of the function and they are calculated by the inner products of the function with the scaling and wavelet functions respectively. In practice, the DWT coefficients are not calculated in this way. Mallat described a filter bank implementation of the DWT [12] where a set of low pass and high pass filters obtained from the scaling and wavelet functions are convolved with the signal and the output of both the filters are downsampled. The outputs of the low pass and high pass filters correspond to the approximation (low frequency) and detail (high frequency) components of the signal, respectively. This procedure can be recursively applied to the output of the low pass filter to obtain successively lower approximations, while the output of the high pass filter at each iteration level is retained. The DWT algorithm can be applied for up to $\log_2 N$ iterations for a signal of length N until we are left with only one sample of the signal.

An image is a two-dimensional signal. Thus, if we want to compute the DWT coefficients of an image, we can apply the DWT algorithm on the rows first and then on the columns. This is possible because the DWT is a separable transform. Figure 16.6 illustrates the process of obtaining the DWT of an image. The symbols h_φ and h_ψ are, respectively, the low pass and high pass wavelet filters used for decomposition. In the first step, the rows of the image are convolved with the low pass and high pass filters and the result is downsampled by a factor of two along the columns. The high pass or detailed coefficients characterise the image's high frequency information with vertical orientation while the low pass component contains its low frequency vertical information. Both subimages are again filtered column-wise with the same low pass and high pass filters and downsampled along rows. This DWT produces four subimages at a lower scale – A, H, V, and D. Each subimage is one-fourth the size of the original image. These four components can be interpreted as follows [11]:

1) Approximation coefficients (A) – the intensity or grey-level variations of the image.
2) Horizontal coefficients (H) – variations along the columns.
3) Vertical coefficients (V) – variations along the rows.
4) Diagonal coefficients (D) – variations along the diagonals.

Sequential DWTs can be performed on the approximation image A to yield images at successively lower resolutions.

The Inverse Discrete Wavelet Transform (IDWT) allows us to reconstruct the image at increasingly higher resolutions by using the four subimages, A, H, V, and D, obtained from

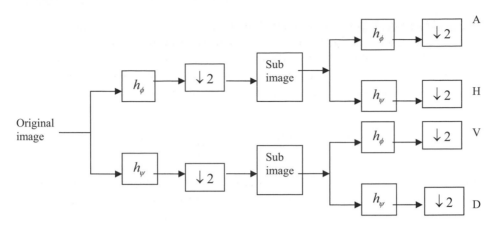

Figure 16.6 *Two-dimensional discrete wavelet transform.*

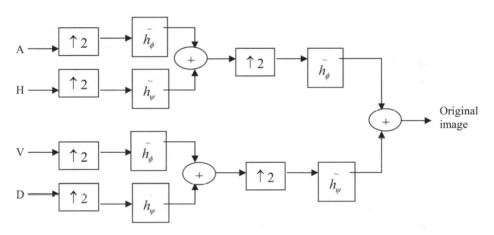

Figure 16.7 *Two-dimensional inverse discrete wavelet transform.*

the DWT of the original image. The symbols $\tilde{h}_\varphi$ and $\tilde{h}_\psi$ are the low pass and high pass reconstruction filters respectively. Figure 16.7 illustrates the process of reconstructing the original image from its DWT components. Each step is the inverse of a corresponding step in the DWT. The four subimages are upsampled along rows first, since the last step in the DWT was downsampling along rows. The approximation and the vertical subimages are convolved with the low pass filter while the horizontal and detailed subimages are convolved with the high pass filter column-wise (in the DWT we filtered along the columns second). The upsampled and convolved versions of the vertical (V) and diagonal (D) coefficients are summed together, and the upsampled and convolved versions of the approximation (A) and horizontal (H) coefficients are summed together. Both these intermediate outputs are upsampled along columns and then filtered row wise with the high pass and low pass reconstruction filters, respectively, and summed to yield the original image.

It is noted that some authors use a reverse notation to denote the resolution, i.e. an increasing subscript is used to denote lower resolutions. Thus, in the above discussion on DWT if the original image has a resolution J, successive DWTs will yield images at resolutions $J + 1$, $J + 2$, and so on. This reverse notation is used to explain the RWT, LP, and AWT in the following sections to maintain consistency with the original texts.

16.2.3.2 The redundant wavelet transform

If the wavelet transform is applied to a shifted copy of the signal, the wavelet coefficients should merely be a shifted version of the coefficients that were obtained by applying the wavelet transform on the original signal. This property of the wavelet transform is called shift invariance [15].

The DWT described above is not shift invariant, since the wavelet coefficients of the DWT change when the signal is shifted. Shift variance results from the application of subsampling in the wavelet transform. The main step in all the wavelet transforms is convolving the signal (or image) with a filter bank to obtain the approximation and the detail images. In the DWT algorithm the output of the filtering is critically subsampled, i.e. the outputs of the filter banks are decimated by a factor of two (the most common case). This subsampling causes the coefficients to change when the input is shifted. The shift variance of the DWT can sometimes be a problem in applications like pattern recognition or image fusion. For example, the use of the DWT in image fusion is known to cause artifacts in the fused images.

The problem of shift variance is overcome by oversampling or removing the subsampling step at each scale in the DWT and instead upsampling the filters at each scale. Since the subsampling is eliminated there is redundant data in the approximation and detail images at each scale and the transform is no longer orthogonal. Thus the oversampled DWT is known as the Redundant Wavelet Transform (RWT). Suppose the original image is at resolution $j = 0$, then the lowpass and highpass filters at each scale are upsampled by inserting $2^j - 1$ zeros between its non-zero coefficients [15]. Thus, the filter coefficients at each successive resolution j are given as follows:

$$h_\phi^j = h_\phi \uparrow 2^j \tag{16.9}$$

$$h_\psi^j = h_\psi \uparrow 2^j \tag{16.10}$$

Figure 16.8 shows the diagram of the RWT. Since the downsampling is eliminated, the A, H, V, and D images are all the same size as the original image. Thus, the RWT is both computation and data intensive compared to the DWT.

16.2.3.3 The Laplacian pyramid

The Laplacian Pyramid (LP) was first proposed by Burt et al. [1] for compact image representation. The basic steps of the LP are as follows:

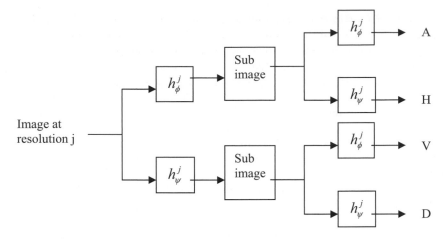

Figure 16.8   *Redundant wavelet transform.*

1. Convolve the original image g_0 with a lowpass filter w (e.g., the Gaussian filter) and subsample it by two to create a reduced lowpass version of the image – g_1.
2. This image is then upsampled by inserting zeros in between each row and column and interpolating the missing values by convolving it with the same filter w to create the expanded lowpass image g_1', which is subtracted pixel by pixel from the original to give the detail image – L_0 given by

$$L_0 - g_0 - g_1' \tag{16.11}$$

In order to achieve compression, rather than encoding g_0, the images L_0 and g_1 are encoded. Since g_1 is the lowpass version of the image it can be encoded at a reduced sampling rate and since L_0 is largely decorrelated it can be represented by far fewer bits than required to encode g_0. The above steps can be performed recursively on the lowpass and subsampled image g_1 a maximum of N number of times if the image size is $2^N \times 2^N$ to achieve further compression. Thus the end result is a number of detail images $L_0, L_1, \ldots, L_N$ and the lowpass image g_N. Each recursively obtained image in the series is smaller in size by a factor of four compared to the previous image and its centre frequency reduced by an octave.

The inverse transform to obtain the original image g_0 from the N detail images L_0, $L_1, \ldots, L_N$ and the lowpass image g_N is as follows:

1. g_N is upsampled by inserting zeros between the sample values and interpolating the missing values by convolving it with the filter w to obtain the image g_N'.
2. The image g_N' is added to the lowest level detail image L_N to obtain the approximation image at the next upper level:

$$g_{N-1} = L_N + g_N' \tag{16.12}$$

3. Steps 1 and 2 are repeated on the detail images $L_0, L_1, \ldots, L_{N-1}$ to obtain the original image.

16.2.3.4 The A Trous wavelet transform

The A Trous Wavelet Transform (AWT) is very similar to the LP except that the lowpass images are never subsampled. Thus, the approximation and detail images at consecutive scales are the same size as the original image [16]. Since the approximation and detail images of the AWT are not subsampled, in an effort to avoid confusion a different notation will be used to denote them a_j and d_j, respectively. Prior to filtering the image at each scale j, the filter w is upsampled by inserting $2^j - 1$ zeros between its sample values and consequently, it is denoted by w_j to indicate this. The following two equations describe how the approximation and detail images for each scale are obtained. The detail image at each scale is simply a difference of the approximation image at that scale and at the immediate higher scale.

$$a_j = a_{j-1} * w_j \qquad (16.13)$$

$$d_j = a_{j-1} - a_j \qquad (16.14)$$

The inverse transform to obtain the original image a_0 is very simple. The detail images at each scale are simply added one by one to the approximation image a_N at the lowest level:

$$a_0 = a_N + d_0 + d_1 + \cdots + d_N \qquad (16.15)$$

16.2.3.5 Summary of multiresolution transforms

Any MRA transform used generates two types of subband images – the approximation or low frequency image at the lowest scale requested and one or more detail or high frequency images at each scale depending on the transform. For example, the RWT and the DWT generate three detail images (H, V, D) at each scale corresponding to the details of the image in three directions – horizontal, vertical, and diagonal, respectively. While the LP and the AWT generate only a single detail image at each scale without any spatial orientation. The DWT and RWT compute the details by decomposing the signal over the wavelet function while the LP and AWT obtain the details by simply taking the difference between the approximation images at successive scales. Another point to note is that the DWT and the LP are subsampled transforms (i.e. the images are decimated by a factor of two at each consecutive scale) while the RWT and the AWT are oversampled transforms because there is no decimation at consecutive scales.

It is found that the shift variance of the DWT causes artifacts in the pan sharpened images. These artifacts can be minimised by using linear phase or bi-orthogonal filters [17] but they cannot be entirely eliminated. Thus it is recommended that DWT not be used for pan sharpening images.

16.2.4 Multiresolution-based pan sharpening

The basic principle of multiresolution-based pan sharpening is as follows. The MRA transform is applied on the PAN image to separate the low and high frequency information in the image. The low frequency PAN image is discarded while the high frequency

PAN image is substituted into the MS image to construct the pan sharpened image. There are many variants of the MRA pan sharpening method. These variations arise from the freedom to choose the MRA transform (e.g., RWT, AWT, etc.) applied and whether the MRA transform is applied on only the PAN or both the PAN and MS images. In this chapter some of the popular MRA-based techniques are explained.

16.2.4.1 Coefficient synthesis method

Once the MS and PAN images have been decomposed into their detail and approximation coefficients there are different techniques to combine them and synthesise the pan sharpened images. Primarily, there are two variants of the MRA-based method – additive and substitutive methods, explained in the following subsections. Out of the two, the substitutive method and its variants are more popular. Since each of them can be implemented using almost any MRA transform, the explanations will be given with a generic notation.

In the following discussion, the detail coefficients (or images) obtained by applying the MRA transform will be denoted by d while the approximation image will be denoted by a. The detail coefficients of the MS image will be subscripted by B and thus denoted by d_B and those of the PAN image by d_A. Similarly, the approximation images from the MS and PAN images will also be subscripted by B and A, respectively. The combined or selected coefficients, used to reconstruct the pan sharpened image, are subscripted with $F - d_F$, a_F. The approximation or detail coefficients will be indexed by a vector $\vec{p} = (i, j, k, l)$ as suggested by [1], where the (i, j) term represents the pixel location in the transformed images, k denotes the scale of the wavelet transform and l the orientation of the detail image. For transforms like the AWT or LP l is one since there is only one detail image at each level in the pyramid, while for the RDWT l can be one, two or three corresponding to the three detail images – H, V, D. Since any MRA transform can be used to implement a particular scheme, the acronym MT will be used to convey the meaning that a forward MRA transform is applied and IMT to mean that an inverse MRA transform is applied.

16.2.4.2 Additive method

In the additive method, the MT is applied only to the PAN image to decompose it into the approximation and detail coefficients and then the detail or high frequency coefficients are injected into the MS bands to enhance it. Figure 16.9 shows the flowchart of the additive pan sharpening method. The main steps in the additive method are as follows:

1. Both the images are coregistered and the MS imagery is resampled to the same spatial resolution as the PAN imagery.
2. The PAN image is histogram matched to each MS band to be sharpened (this step is optional). Hence, if this step is performed there will be one histogram matched PAN band per MS band to be sharpened.
3. The multiresolution transform is applied and the PAN image is decomposed into its approximation and detail images:

$$\{a_A, d_A\} = \text{MT}(A) \tag{16.16}$$

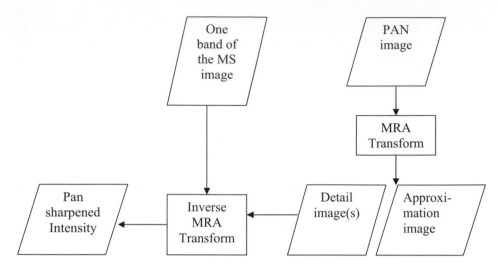

Figure 16.9   *Additive MRA-based pan sharpening.*

4. The approximation image of the PAN band, a_A, is discarded and the resampled MS image B_i^* substituted in its place. The inverse wavelet transform is computed on them to obtain the sharpened MS band:

$$F_i = \mathrm{IMT}(B_i^*, d_A) \tag{16.17}$$

5. Steps 3–4 are repeated for each MS band.

16.2.4.3 Substitutive method
In this technique, the MRA is performed on both the PAN and the MS images. Since the aim is to incorporate the details from the PAN image into the MS imagery, the detail coefficients from the MS image and the approximation image of the PAN image are discarded. The approximation image of the MS image is used along with the detail image of the PAN image for synthesising the pan sharpened image. The method is explained step by step below for sharpening each MS band separately. Figure 16.10 gives the flowchart of the method for convenience.

1. Both the images are coregistered and the MS imagery is resampled to the same spatial resolution as the PAN imagery.
2. The PAN image is histogram matched to each MS band (as described above this step is optional). If the histogram matching is performed there are three copies of the PAN image corresponding to each MS image – A_i.
3. The MS and PAN images are decomposed into their approximation and detail images:

$$\{a_{B_i}, d_{B_i}\} = \mathrm{MT}(B_i^*) \tag{16.18}$$

$$\{a_{A_i}, d_{A_i}\} = \mathrm{MT}(A_i) \tag{16.19}$$

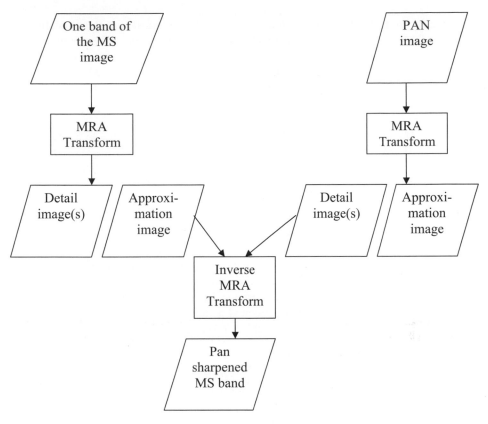

Figure 16.10 *Substitutive MRA-based pan sharpening.*

4. The detail images of the MS image are replaced by those of the PAN image and the inverse MRA transform is computed to obtain the pan sharpened imagery:

$$a_{F_i} = a_{B_i} \tag{16.20}$$

$$d_{F_i} = d_{A_i} \tag{16.21}$$

$$F_i = \text{IMT}(a_{F_i}, d_{F_i}) \tag{16.22}$$

Some substitutive methods based on conditional injection of the detail coefficients from the PAN image have been proposed. The aim of these techniques is to minimise the spectral distortion in the pan sharpened images with respect to the original MS images. The basic principle behind these techniques is to not discard the detail images of the MS image completely but rather choose the detail coefficients from either the MS or the PAN image at each pixel location based on a selection rule. These selection rules aim to choose the coefficient from either image which has a higher energy. The detail coefficients with higher magnitude or energy correspond to sharp or drastic changes in the image intensity and thus represent the 'salient' features (e.g., edges) in the image [18]. Since the goal of the pan sharpening is to enhance the edge information in the pan sharpened image, this logic makes sense. The logic behind these rules is that if the MS detail coefficients have the intensity to define the spatial quality they should be retained. The detail coefficients from the PAN image should be injected into the fused image only when required, thus because of this conditional injection of details the change in the radiometry of the MS image is minimised. The various selection rules differ in the way in which 'energy' is defined. A few coefficient selection rules are described below.

16.2.4.3.1 Maximum amplitude selection

In the Maximum Amplitude Selection rule (MAS), the detail coefficients at each pixel are retained from either the PAN or the MS image depending on which one has a higher absolute magnitude. The mathematical formula for this rule is given by [19]

$$d_F(\vec{p}) = \max\{\text{abs}(d_A(\vec{p})), \text{abs}(d_B(\vec{p}))\} \tag{16.23}$$

16.2.4.3.2 Window-based salience

This Window-Based Saliency rule (WBS) was proposed in [1], Burt et al. reason that since the salient features in the image are generally larger than one pixel, the selection rule must also be applied on more than one coefficient at a time. The salience at a pixel is computed as the variance or energy over a $m \times m$ window centred over it (where m is usually 3 or 5) as given below:

$$S_A(i, j, k, l) = \sum_{x=-m}^{+m} \sum_{y=-m}^{+m} d_A(i+x, j+y, k, l)^2 \tag{16.24}$$

$$S_B(i, j, k, l) = \sum_{x=-m}^{+m} \sum_{y=-m}^{+m} d_B(i+x, j+y, k, l)^2 \tag{16.25}$$

After measuring the salience of each coefficient in the image, the source coefficients are combined using either 'selection' or 'averaging,' based on a match measure between them. The match measure M_{AB} is given by the normalised correlation computed between the two coefficients from A and B around the same neighbourhood as used to compute the salience:

$$M_{AB}(i, j, k, l) = \frac{2 \sum \sum d_A d_B}{S_A + S_B} \tag{16.26}$$

This quantity is a measure of how similar the two images are at a particular pixel location. If the match measure is near $+1$ it means the images are very similar locally while values near -1 indicate that they are quite different. If the match measure M_{AB} exceeds a threshold α the source coefficients are combined through a weighted average:

$$d_F(\vec{p}) = w_A(\vec{p})d_A(\vec{p}) + w_B(\vec{p})d_B(\vec{p}) \tag{16.27}$$

For the weighted average situation the weights are computed as follows based on the match measure and the threshold:

$$w_{\min} = \frac{1}{2} - \frac{1}{2}\left(\frac{1 - M_{AB}}{1 - \alpha}\right) \tag{16.28}$$

$$w_{\max} = 1 - w_{\min} \tag{16.29}$$

The higher weight $w_{\max}$ is given to the coefficient with the higher energy and the lower weight $w_{\min}$ is assigned to the other coefficient:

$$S_A > S_B \quad \Rightarrow \quad \{w_A = w_{\max}, \ w_B = w_{\min}\} \tag{16.30}$$

$$S_B > S_A \quad \Rightarrow \quad \{w_B = w_{\max}, \ w_A = w_{\min}\} \tag{16.31}$$

If M_{AB} is below the threshold, the source coefficient which has a higher energy is selected out of the two similar to the MAS rule.

$$S_A > S_B \quad \Rightarrow \quad d_F(\vec{p}) = d_A(\vec{p}) \tag{16.32}$$

$$S_B > S_A \quad \Rightarrow \quad d_F(\vec{p}) = d_B(\vec{p}) \tag{16.33}$$

The logic behind selecting the coefficient with higher salience when the match measure is below a certain threshold is to retain a higher contrast in the fused image. Generally, regions with opposite contrast are associated with a low match measure. For example, if the MS image has a high contrast while the PAN image has a low contrast, in case their average is taken, the overall contrast or salience of the fused coefficient is reduced, resulting in blurry looking images.

The equations for calculating the weights show that when the match measure exceeds the threshold, $w_{\min}$ is extremely low while $w_{\max}$ is high. As M_{AB} increases towards one the weights become nearly equal. Burt et al. suggested using a quite high threshold ($\alpha = 0.85$); however, they also noted that change in the neighbourhood size or threshold did not affect the results greatly.

In reference to the selection rules that were described above, the basic substitutive method which throws away the detail coefficients of the MS image, will be referred to as the NULL rule since it is equivalent to not applying any selection rule. The NULL and the MAS rules are the most popular in pan sharpening while the window-based rules like WBS are more popular in image fusion applications such as fusion of visible and thermal images, images with different regions out of focus, etc.

16.3 Evaluation metrics

16.3.1 Significance

From the previous section, we can see that there are many variations of the MRA-based pan sharpening methods. Thus, if someone wants to implement a MRA-based pan sharpening method to regularly pan sharpen imagery they would be overwhelmed by the number of choices available to them. Thus a quantitative set of metrics must be used to evaluate the various pan sharpening algorithms and narrow down the choice to a few that perform better than others. Since the goal of pan sharpening is to enhance the spatial quality of the MS image and also preserve its spectral properties, two sets of metrics must be used to – spectral and spatial. A few spectral and spatial quality metrics are described below.

16.3.2 Spectral quality metrics

The goal of pan sharpening is to preserve the radiometry of the original MS images as much as possible, thus any metric used must measure the amount of change in DN values in the pan sharpened image compared to the original image. The following subsections define a few spectral quality metrics.

16.3.2.1 Correlation coefficient

The correlation coefficient measures the closeness or similarity between two images [20]. It can vary between -1 to $+1$. A value close to $+1$ indicates that the two images are very similar, while a value close to -1 indicates that they are highly dissimilar. The formula to compute the correlation between two images A and B, both of size $N \times N$ pixels is given by

$$\text{Corr}(A|B) = \frac{\sum_{i=1}^{N} \sum_{j=1}^{N} (A_{i,j} - \bar{A})(B_{i,j} - \bar{B})}{\sqrt{\sum_{i=1}^{N} \sum_{j=1}^{N} (A_{i,j} - \bar{A})^2 \sum_{i=1}^{N} \sum_{j=1}^{N} (B_{i,j} - \bar{B})^2}} \tag{16.34}$$

Various correlation coefficients are computed to evaluate the spectral quality of the pan sharpened images [20]. The inter-correlation between each pair of the unsharpened bands and the sharpened bands can be computed and compared. For example, $\text{Corr}(B_1, B_2)$ is the interband correlation between bands 1 and 2 before fusion, and $\text{Corr}(F_1, F_2)$ is the interband-correlation after fusion. Ideally a zero change in the correlation values would be desirable, i.e. if $\text{Corr}(B_1, B_2)$ was 0.94, the ideal value for $\text{Corr}(F_1, F_2)$ would be 0.94. Thus, if there are three bands being fused the inter-correlation between each pair of the three unsharpened bands and the three sharpened bands was computed and compared:

$$\text{Corr}(B_1|B_2), \ \text{Corr}(F_1|F_2) \tag{16.35}$$

$$\text{Corr}(B_2|B_3), \ \text{Corr}(F_2|F_3) \tag{16.36}$$

$$\text{Corr}(B_1|B_3), \ \text{Corr}(F_1|F_3) \tag{16.37}$$

Then, the correlation between each sharpened and unsharpened band $\text{Corr}(B_i^*, F_i)$ (for $i = 1, 2, \ldots, N$ if there are N bands) is computed. The ideal value for this is 1. Since the pan sharpened images are larger (more pixels) than the original MS image 'B' it is not possible to compute the correlation directly between them. Thus, the resampled MS image B^* is used for this comparison. This is acceptable since the resampling produces little change in the radiometry of the original images. Similarly, the correlation between the resampled MS image and the PAN image $\text{Corr}(B_i^*|A)$ is computed and it is compared with the correlation between the fused and PAN images $\text{Corr}(F_i|A)$. Ideally, the correlation between the fused and PAN image should be the same as that between the original MS and PAN image.

16.3.2.2 Root mean square error

The Root Mean Square Error (RMSE) between each unsharpened MS band and the corresponding sharpened band can also be computed as a measure of spectral fidelity [21]. It measures the amount of change per pixel due to the processing (e.g., pan sharpening) and is described by

$$\text{RMSE}_k = \frac{\sum_{i=1}^{N} \sum_{j=1}^{N} \sqrt{(B_k^*(i, j) - F_k(i, j))^2}}{N^2} \qquad (16.38)$$

During our research, it was found that the RMSE has a higher resolution compared to the correlation coefficient. This statement means that if the performance of the two algorithms is almost identical to each other, then the RMSE can better distinguish which one is better. For example, if the pan sharpened images produced by algorithms 1 and 2 have a correlation coefficient of 0.99 with respect to the MS image, it means the spectral quality of both algorithms is identical. On the other hand, if the RMSE values for the two corresponding images are 2.34 and 2.12, respectively, clearly algorithm 2 results in a higher spectral quality compared to algorithm 1, and only the RMSE can clarify this distinction.

In addition to the correlation coefficient or RMSE, the histograms of the original MS and the pan sharpened bands can also be compared [20]. If the spectral information has been preserved in the pan sharpened image, its histogram will closely resemble the histogram of the original image.

16.3.3 Spatial quality metrics

The evaluation of the spatial quality of the pan sharpened images is equally important since the goal is to retain the high spatial resolution of the PAN image. A survey of the pan sharpening literature revealed that very few articles evaluated the spatial quality of the pan sharpened imagery. A couple of spatial quality metrics proposed by researchers are explained below.

16.3.3.1 Mean gradient

Image gradients have been used as a measure of image sharpness [22]. The gradient at any pixel is the derivative of the DN values of neighbouring pixels. Generally sharper images have higher gradient values. Thus, any image fusion method should result in increased gradient values because this process makes the images sharper compared to the low resolution image. The mean gradient defines the contrast between the details variation of pattern on the image and the clarity of the image [23]. The mean gradient $\bar{G}$ of an image X is given by

$$\bar{G} = \frac{1}{N^2} \sum_{i=1}^{N} \sum_{j=1}^{N} \sqrt{\frac{\Delta I_x^2 + \Delta I_y^2}{2}} \qquad (16.39)$$

$$\Delta I_x = X(i+1, j) - X(i, j) \qquad (16.40)$$

$$\Delta I_y = X(i, j+1) - X(i, j) \tag{16.41}$$

In the above equations ΔI_x and ΔI_y are the horizontal and vertical gradients per pixel.

16.3.3.2 High pass correlation coefficient

This approach was first proposed by Zhou et al. [24] to measure the amount of edge information from the PAN image is transferred into the fused images. The high spatial resolution information missing in the MS image is present in the high frequencies of the PAN image. The pan sharpening process injects the higher frequencies from the PAN image into the MS image. Thus, they propose that the correlation coefficient between the highpass filtered PAN and the pan sharpened images would indicate how much spatial information from the PAN image has been incorporated into the MS image. A higher correlation between the two highpass filtered images implies that the spatial information has been retained. This correlation coefficient is called the High Pass Correlation Coefficient (HPCC).

16.3.3.3 Limitations of the spatial quality metrics

The goal of the MRA-based pan sharpening is to produce pan sharpened images with the highest spectral fidelity, while the spatial quality is sufficient such that all the structures observed in the PAN image can be observed easily in the pan sharpened image. Generally the algorithm is tweaked to give as high a spectral fidelity as possible assuming that even the slightest improvement will be beneficial. The same cannot be said about the spatial quality. It is sufficient that the user should be able to observe all the details of the PAN image in the sharpened image.

Thus, the task is to define the term 'sufficient spatial quality.' This is a very difficult task and cannot be measured by any metric (i.e. how can it be said if a HPCC value of 0.94 is sufficient or not?). Perhaps a HPCC of 0.95 is sufficient, but a value of 0.94 is not sufficient, i.e. the sharpened image do not contain all the details observed in the PAN image. Moreover, this value could be different for different images. Whether the pan sharpened image has all the details in the PAN image can only be determined through visual analysis. Thus, visual analysis plays an important role in determining the spatial quality of the images. Perhaps this is the reason for so few studies on the spatial quality of the pan sharpened images.

The above mentioned metrics were collected and researched further. Based on our study, few anomalies were found with both of them. For example, in one of the datasets, the HPCC values of the images pan sharpened with the Brovey method were lower than those of images sharpened using the wavelet-based method, implying the Brovey method sharpened images contain less high frequency information than those obtained from the wavelet based method. However, visual analysis clearly contradicted this result. Thus, the spatial quality of the pan sharpened images is analysed in this work using visual analysis.

16.4 Observations on the MRA-based methods

This section presents a comprehensive analysis of the performance of various MRA-based pan sharpening methods. One of the goals of this chapter is to make recommendations regarding which methods must be preferred and report whether there are any advantages or drawbacks of certain methods. The observations are based on an exhaustive research on MRA-based pan sharpening done by [25] and using the quantitative metrics described above. In order to avoid dataset specific observations, images from two different satellites – LANDSAT 7 and Quickbird – were used for analysis. The LANDSAT MS, PAN images have a resolution ratio of two while the Quickbird MS, PAN images have a resolution ratio of four. Thus, it also helps to make generalised conclusions for different resolution ratios.

16.4.1 Comparison of additive and substitutive methods

Three MRA transforms – RWT, AWT and LP – were used to pan sharpen imagery using the additive method and compared with the same image pan sharpened using the RWT-based substitutive method. It was found that applying the additive method with RWT as the MRA transform leads to poor spectral fidelity and excessively strong colours. However, closer inspection also reveals that the pan sharpened image obtained from the RWT-based additive method appears grainy or freckled compared to those obtained from the AWT or LP. Thus the RWT-based additive method leads to poor spectral and spatial quality and is not suitable for pan sharpening. Figure 16.11 illustrates this using a LANDSAT TM image which is pan sharpened using the RWT- and AWT-based additive methods and also the RWT-based substitutive method. To preserve space the LP-based pan sharpened image is not shown.

The AWT- and LP-based additively merged pan sharpened images appear to have good spatial quality and also the spectral distortion is smaller than the RWT-based substitutive method. Table 16.3 shows the RMSE between the pan sharpened images and the resampled MS images for each of the methods. The three values are for the three MS bands. The acronyms RWT-SUB, RWT-ADD and AWT-ADD stand for RWT-based substitutive method, RWT-based additive method and AWT-based additive method, respectively.

The same methods were used to pan sharpen another Quickbird scene whose PAN (1 m) and MS (4 m) images have a resolution ratio of four. A detailed visual analysis of these pan sharpened images shows that the spatial quality of the pan sharpened image created using the AWT-based additive method is not as good as the pan sharpened image created using one of the substitutive method. The LP-based additive merger is deemed the most suitable of the three additive methods studied. It did not give the blotchy appearance like the RWT and retained higher spatial quality compared to the AWT method when it was used to pan sharpen images with a resolution ratio of four.

Thus it is found that the LP-based additive method results in high spectral and high spatial quality pan sharpened images that are comparable to those produced by the substitutive method. One of the benefits of the additive methods in general is that only the PAN

(a) Original MS image resampled to PAN size

(b) Pan sharpened image obtained using RWT-based substitutive method

(c) Pan sharpened image obtained from RWT-based additive method

(d) Pan sharpened image obtained from AWT-based additive method

Figure 16.11 *Comparison of additive and substitutive pan sharpening methods.*

Table 16.3 *Spectral distortion of additive- and substitutive-based pan sharpened images.*

Error	RWT-SUB	RWT-ADD	AWT-ADD
$\text{RMSE}(B_1^*, F_1)$	2.438	28.10	2.065
$\text{RMSE}(B_2^*, F_2)$	4.465	35.03	3.756
$\text{RMSE}(B_3^*, F_3)$	4.834	39.83	4.082

image has to be decomposed. Thus, it will be computationally cheaper compared to the substitutive method in which both the PAN and MS images must be decomposed.

A disadvantage of the additive method is the inconvenience caused when the images to be merged do not have a resolution ratio that is a power of two. For example, suppose a LANDSAT 7 MS image at 30 m resolution is to be merged with a SPOT PAN image at 10 m resolution of the same scene. The resolution ratio of the two images is three, which is not a power of two. A first level decomposition of the PAN band would result in an approximation image at 20 m while the MS band has a resolution of 30 m. This would mean substituting an image at resolution 30 m in place of a 20 m image. Even if more levels of decomposition are performed the scales of the lowest approximation image of the PAN and the MS image will always be different. For example, after two decomposition levels the approximation image of the PAN will have a resolution of 40 m.

A workaround this problem is to use an M-scale wavelet transform or a pyramid scheme that decomposes images by a scale other than two, where M is the resolution ratio of the images to be merged. This approach was taken by Shi et al. [23] and Blanc et al. [26] who used a 3-band wavelet transform to merge the TM MS and SPOT P images. Thus applying a wavelet transform that decomposes by a scale of three, the resolution of the approximation image of the PAN after performing one decomposition level will be 30 m and then the 30 m MS image can be substituted in its place. However, very few wavelet software packages come with wavelet transforms that decompose images at a scale other than two, and there is also very little wavelet literature dealing with general M-scale transforms. This is a big inconvenience for remote sensing scientists who are mostly experts in earth sciences rather than electrical engineers. Moreover, even if a general M-scale transform is designed it would have to be provided for each resolution ratio that is not a multiple of another resolution ratio (e.g., 3, 5, 7). Similarly it is also possible to implement a generalised LP [27] that can decompose images by any scale ratio.

In comparison, the substitutive method is quite flexible and works for any resolution ratio as the MS image is first resampled to the same resolution as the PAN and then both images can be decomposed to the same scale. For example, in the above LANDSAT MS, SPOT PAN merging situation the LANDSAT MS image at 30 m would be first resampled to 10 m and then both images would be decomposed to the same scale – 20 m.

16.4.2 *Effect of the wavelet basis or filter kernel*

After one has chosen a MRA transform (e.g., RWT or AWT) to perform the pan sharpening one has the option to choose from a large set of filter or wavelet coefficients to decompose and reconstruct the images. Researchers have designed various filters and wavelets satisfying certain mathematical properties that are suitable for solving problems in their specific domains. For example, the bi-orthogonal family of wavelets is suitable for DWT-based image compression tasks. The effect of the wavelet bases or the filter kernels on the pan sharpened image was investigated in detail using quantitative metrics.

16.4.2.1 *RWT-based pan sharpening*
If the RWT is chosen as the basis for MRA transform, there are two main sets of wavelet bases available – the Daubechies and the bi-orthogonal wavelets. The Daubechies wavelets are nonlinear phase while the bi-orthogonal wavelets are linear phase [14]. Daubechies wavelets with different number of vanishing moments were used to observe

the effect of the number of vanishing moments on the results. We will use the notation 'db$N$' to denote a Daubechies wavelet having N vanishing moments (e.g., 'db2' denotes a Daubechies wavelet with two vanishing moments). The experiments were limited to the Daubechies wavelets – 'db1,' 'db2,' 'db5,' 'db10,' and 'db20.' These wavelets were studied to see the effect of very short, medium and long wavelets on quality of the pan sharpened images.

The Daubechies wavelets belong to a class of filters called the nonlinear phase filters. The choice of these filters in image compression leads to higher artifacts and less coding gain compared to another class of filters called the 'linear phase filters.' There were two purposes for performing these set of experiments – one is to find the wavelet that results in the highest spectral fidelity and the second was to observe if the linear phase filters had any advantage over the nonlinear filters. All the bi-orthogonal filters that were available in the MATLAB toolbox were investigated.

Pan sharpening many datasets using the different Daubechies wavelets, it was found that as wavelets with higher number of vanishing moments are applied, the spectral distortion in the pan sharpened images increases. One exception to this was the 'db1' wavelet also known as the Haar wavelet which resulted in maximum spectral distortion, even higher than the 'db20' wavelet. Thus using a 'db2' wavelet is recommended.

The pan sharpened images obtained by applying different bi-orthogonal wavelets were also studied. It was found that the bi-orthogonal wavelets do not give any better results than the Daubechies wavelets. The bi-orthogonal wavelet with 2 vanishing moments in the analysis function and 2 vanishing moments in the scaling function resulted in pan sharpened images which had a spectral fidelity close to the 'db2' generated image. Hence, it is seen that there is no particular advantage of using the bi-orthogonal wavelets for pan sharpening.

16.4.2.2 LP- and AWT-based pan sharpening

Various filters used for the LP- and AWT-based pan sharpening were also evaluated in order to determine their effect on spectral distortion in the pan sharpened images. There are many filters that can be applied to the LP – the binomial filters, Quadrature Mirror Filters (QMF), and the Gaussian filters. The interested reader is referred to [13] for an explanation of the properties of each of these filters. These filters were used to find an empirically best filter for LP- and AWT-based pan sharpening. Since the AWT is just an oversampled version of the LP, the same filters were used for the AWT also to find the one which gives the highest spectral fidelity.

Binomial filters of order one to twelve, i.e. 2- to 24-tap were applied to LP-based pan sharpening. It was found that the first-order (2-tap) binomial filter caused artifacts in the pan sharpened image. The images appeared pixilated and had the stair step effect almost all over the image. Actually, it was observed that all the even length binomial filters cause heavy spectral distortion in the pan sharpened images. The 3-tap (second-order) binomial filter resulted in the highest spectral fidelity among all the binomial filters. It was found that as the filter order was increased the spectral distortion also increased. The spectral

Table 16.4   *Gaussian filter coefficients obtained by varying the central weight.*

Filter coefficients	$a = 0.3$	$a = 0.375$	$a = 0.4$	$a = 0.5$	$a = 0.6$
$w(-2)$	0.10	0.0625	0.05	0	−0.05
$w(-1)$	0.25	0.25	0.25	0.25	0.25
$w(0)$	0.30	0.375	0.40	0.5	0.6
$w(1)$	0.25	0.25	0.25	0.25	0.25
$w(2)$	0.10	0.0625	0.05	0	−0.05

distortion caused by even length filters is much higher than odd length filters. The high spectral distortion caused by even length filters could be explained by the asymmetry of even length filters.

The same binomial filters were applied to AWT-based pan sharpening and similar observations were made. Lower-order odd binomial length filters give the least spectral distortion while even length filters cause heavy spectral distortion. One exception noted is that the 2-tap binomial filter does not produce artifacts in the pan sharpened image. This can be attributed to the fact that the AWT is an oversampled transform like the RWT.

Next, the LP-based pan sharpening is performed using QMFs. Again it is seen that the odd length QMF filters give good results but the even length QMF filters give artifacts. The artifacts are not very severe but they do exist in some regions of the image. These do not appear as pixilated or stair stepped edges in the image but rather seem to be occurring due to the smearing of spectral signatures of sharp features like buildings or roads outside the feature and thus also result in poor spectral quality. The QMF filters used were 5, 8, 9, 12, 13 and 16 tap, respectively. It was also seen that as the length of the odd length QMF filters is increased the spectral quality improves, this is opposite of what was observed for binomial filters. Comparison between the best binomial filter (3-tap) and the best QMF filter (13-tap) showed that the QMF filter gave higher spectral fidelity or lesser spectral distortion. Even the smallest 5-tap QMF filter gave higher spectral quality pan sharpened images than the 3-tap binomial filter. Thus it can be concluded that QMF filters must be preferred over the binomial filters. However, the same observation cannot be made for AWT-based pan sharpening. For some datasets the QMF filters gave higher spectral fidelity compared to the binomial filters while for some the binomial filters gave higher spectral fidelity.

Finally, a 5-tap Gaussian filter with varying central coefficients was applied to LP- and AWT-based pan sharpening. This filter was also chosen because many research articles investigating LP- or AWT-based pan sharpening methods use this filter. The different filter coefficients obtained by varying the central coefficient are given in Table 16.4. In the table each column corresponds to the filter coefficients obtained by changing the weight of the central coefficient, a, from 0.3 to 0.6. When a is increased beyond 0.5 it is seen that the border coefficients become negative and the filter becomes trimodal.

The effect of varying a from 0.3 to 0.6 is that the spectral fidelity increases as a increases. However, it was seen that the pan sharpened images corresponding to $a = 0.6$ filter have artifacts. These artifacts are quite minor and only a closer inspection identifies them.

These artifacts seem to be caused by the trimodal nature of the Gaussian function for $a = 0.6$. Thus the Gaussian filter corresponding to $a = 0.6$ will not be considered further for LP-based fusion. Thus the filter corresponding to $a = 0.5$ seems the most suitable among the Gaussian filters. Similar observations were made for the AWT-based pan sharpening except no artifacts are observed for images produced using a Gaussian filter with $a = 0.6$. This can be again attributed to the fact that the AWT is an oversampled transform. The spectral quality of LP-based pan sharpened images produced with Gaussian filters was compared with those from binomial filters and QMFs. The spectral fidelity of the Gaussian filter pan sharpened images is poorer than those of the binomial or QMFs. However, for the AWT-based pan sharpening the Gaussian filter corresponding to $a = 0.5$ consistently produced higher spectral fidelity images compared to the binomial or QMFs.

In summary, even length binomial and QMF filters are not recommended for pan sharpening images. For LP-based pan sharpening, QMF filters give the least spectral distortion compared to the binomial or Gaussian filters. While for AWT-based pan sharpening the Gaussian filter with a central coefficient of 0.5 is recommended.

16.4.3 Choice of the selection rule

A few selection rules in the context of the substitutive method to combine the detail coefficients of the MS and PAN images were described in Sections 16.2.4.3.1 and 16.2.4.3.2. These rules increase the spectral fidelity of the pan sharpened images. The MAS and WBS rules were evaluated with the NULL rule as a reference by pan sharpening different datasets. The RWT was used to do the pan sharpening although the observations made about these rules are applicable for the other transforms also.

In order to apply the MAS and the WBS rules, the MS images must be resampled to the PAN image's size using a resampling technique that results in a smooth image. A bilinear or bicubic interpolation technique is suitable for this purpose but not the nearest neighbour technique. If nearest neighbour resampling is done it results in a blocky and very pixilated looking resampled image. In this case if the coefficients from the MS detail images have a higher energy than the PAN detail coefficients they are selected by the rule and the output pan sharpened image appears blocky and pixilated in those regions. The WBS scheme can be applied on a sliding window of 3×3, 5×5 or in general $m \times m$ pixels. It was found for the WBS rule that as the window size is increased the spectral fidelity starts to decrease, thus the optimal window size is found to be 3×3.

The MAS and WBS rules show significant improvement in the spectral fidelity compared to the NULL rule. The MAS rule results in a higher spectral fidelity compared to the WBS rule. Quantitatively the MAS rule resulted in a 10–20 per cent improvement over the NULL rule in the spectral quality metrics depending on the dataset and the band.

The spatial quality of the pan sharpened images produced by using all the rules seems similar and thus the MAS rule must be favoured since it always results in the highest spectral fidelity. However, on close inspection of some of the pan sharpened images it

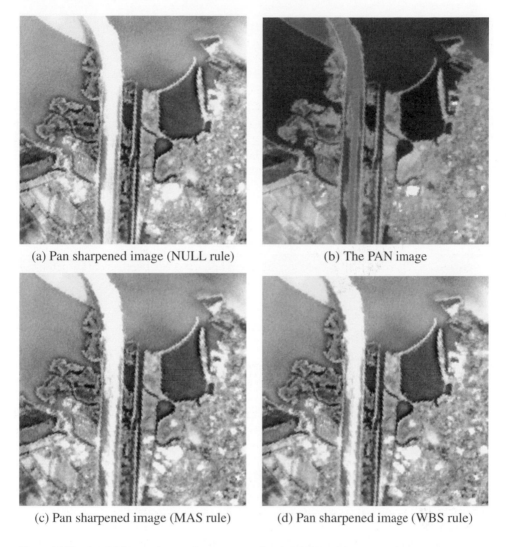

(a) Pan sharpened image (NULL rule)   (b) The PAN image

(c) Pan sharpened image (MAS rule)   (d) Pan sharpened image (WBS rule)

Figure 16.12   *LANDSAT pan sharpened images using the MAS, WBS and the NULL selection rules.*

was found that the MAS and WBS rules have some problems. In one of the LANDSAT TM scene of Vorarlberg region (Austria) it was found that using the MAS and WBS scheme resulted in pan sharpened images that had pixilated edges, i.e. it appeared unsatisfactorily sharpened in some regions. Figure 16.12 is used to illustrate this phenomenon by displaying a part of this scene, in the centre of this image is the Rhine River. The pan sharpened images corresponding to the NULL, MAS and the WBS methods are given. The original PAN image is also shown as a reference for the spatial quality of the pan sharpened images. This pixilated effect is observed along the entire length of the river from North to South in the MAS and WBS pan sharpened images but not in the image obtained by using the NULL rule.

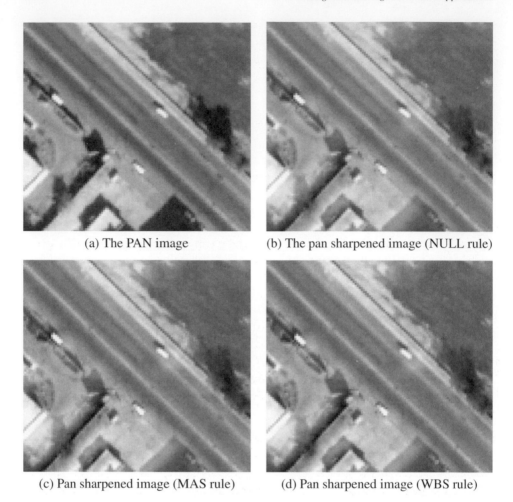

(a) The PAN image   (b) The pan sharpened image (NULL rule)

(c) Pan sharpened image (MAS rule)   (d) Pan sharpened image (WBS rule)

Figure 16.13   *Quickbird pan sharpened images obtained by using MAS, WBS, and the NULL rules.*

In another Quickbird dataset shown in Figure 16.13, some of the automobiles on the road in the bottom right hand corner have a smeared effect in the pan sharpened image created with MAS or WBS rules. This smearing effect is not present in the original PAN image. The pan sharpened image created with the NULL rule is also free of this effect. Although this problem is not very serious because it can hardly be observed at the original scale of the image but quite obvious when the images are observed at a smaller scale or only to a keen observer. The original MS images when resampled have a slightly smeared effect around sharp structures that have small spectral signatures. This smearing seems to be retained in the pan sharpened images created using the MAS or WBS rules because its energy exceeds the energy of the coefficients coming from the PAN image.

Thus, the MAS and WBS selection rules can be used to improve the spectral quality of the pan sharpened images. However, these selection schemes can cause artifacts sometimes. These artifacts are few and only observable at a higher scale upon careful visual analysis, but nevertheless they are present.

Table 16.5   *Relative improvement of LP and RWT over the AWT.*

Dataset	LP (%)			RWT (%)		
LANDSAT TM 1	11	12	7	13	15	10
LANDSAT TM 2	22	20	21	22	20	23
Quickbird 1	12	10	9	14	12	11
Quickbird 2	11	8	8	13	11	11

16.4.4 Comparison of various MRA transforms

The three main transforms used to perform pan sharpening were compared with each other – the RWT, AWT and the LP. It was seen from Section 16.4.2 that the choice of the wavelet or filter coefficients affects the spectral quality of the pan sharpened images. Thus, before comparing these transforms against each other the optimal wavelet or filter must be chosen for each of them. Thus in the experiments making the comparisons the second-order Daubechies wavelet ('db2') was chosen to pan sharpen images using the RWT. For the LP-based pan sharpening the 13-tap QMF was chosen and for the AWT the Gaussian filter corresponding to $a = 0.5$ was chosen.

It was seen that the spectral fidelity of the RWT is the highest, the LP is the second best. However, quantitatively the spectral fidelity of the pan sharpened images created by using the LP is quite close to the RWT. The AWT provides the least spectral fidelity compared to the LP and the RWT.

Table 16.5 gives the relative improvement in the spectral fidelity for the RWT and LP over the AWT method. The metrics are given for four datasets – two are LANDSAT TM images and the other two are Quickbird images. The three values in each cell are for the three MS bands merged. Each value in the columns for RWT and LP is the RMSE for the pan sharpened band (compared to the MS band) for the RWT or LP image minus the RMSE of the pan sharpened image for the AWT divided by the RMSE of the AWT pan sharpened image. The improvement for the second LANDSAT dataset is more than 20 per cent while for the other three datasets it is somewhat less around 10–15 per cent. It can also be inferred that the quality gain from using RWT over LP is not much.

Quite often complete LANDSAT or Quickbird scenes have to be pan sharpened, which are typically thousands of pixels in width and height and at least three to four bands have to be pan sharpened. Thus, there is a lot of data to process. The RWT is computationally a very expensive operation as it does not decimate any of the coefficient images at each scale. Moreover, if a selection rule, like the MAS, is applied, three detail images have to be processed at each level. In comparison, the LP decimates the approximation image by a factor of four at each scale and thus the data to process at each scale decreases, moreover there is only one detail image to process. Since the LP is computationally much faster than a RWT and marginally inferior in performance, if a practical solution is desired the LP-based substitutive method is recommended.

Table 16.6   *Recommended decomposition levels.*

Resolution ratio (*n*)	Urban feature extraction	Classification for land use mapping
2	4	2
3	4	2
4	5	3
5	5	3

16.4.5 Number of decompositions in the MRA transform

In the substitutive MRA method, both the MS and the PAN images are decomposed a number of times (generally one to four) and the approximation image of the MS image at the lowest scale is merged with the detail images from the PAN at each scale during the inverse transform. The spectral and spatial qualities of the pan sharpened images are affected by the number of decompositions applied. If fewer levels of decomposition are applied, the spatial quality of the pan sharpened images is inferior compared to the PAN image. In other words, the spatial information from the PAN image is not sufficiently merged with the MS image. The end result will be a pan sharpened image that will be somewhat blurry compared to the PAN image especially in regions with sharp features in the images like roads, automobiles, and buildings. On the other hand, the spectral similarity between the original MS and pan sharpened images decreases if excessive levels are applied and there is no gain in the spatial quality. It was found through experiments that if good spatial quality pan sharpened images are desired the number of decomposition levels must be chosen based on the resolution ratio of the PAN-MS pair.

In general, as the resolution ratio between the MS-PAN images increases, the number of decompositions required to extract the spatial information from the PAN image increases. In Table 16.6 we make two recommendations for choosing the optimal decomposition level for different resolution ratios based on the application. These recommendations are based on the experiments and empirical evaluations by [28] based on spectral quality metrics and visual analysis of the pan sharpened images to determine their spatial quality. The recommendations are made for resolution ratios between two to five as these are the most commonly encountered cases in pan sharpening. The decision should be driven by the application of the pan sharpened image. For example, if the pan sharpened images are to be used for automatic or supervised classification in order to create land use maps the lower number of decomposition levels given in the table must be chosen, since they better preserve the spectral similarity with respect to the MS images. For tasks like automated extraction of urban objects (e.g., roads, buildings) which require rich spatial information as well as preservation of the original spectral information, the higher number of decomposition levels is justified.

16.5 Summary

It is seen that multiresolution-based pan sharpening methods are effective for pan sharpening multispectral images while maintaining their spectral integrity. There are mainly two types of multiresolution-based pan sharpening methods – the additive and the substitutive method. It is seen that the additive method works well only the Laplacian Pyramid is used as the multiresolution transform. However, the drawback is that it is difficult to implement them when the images to be merged do not have a resolution ratio that is not a power of two. On the other hand, the substitutive method is quite flexible and works without modification for any resolution ratio images.

The effect of various wavelet bases or filters on the spectral distortion introduced in the pan sharpened images was also studied and recommendations made as to which wavelet or filter basis to choose depending on the transform. It is seen that out of the three multiresolution transforms evaluated for the substitutive method (RWT, LP, and AWT) the RWT and the LP pan sharpened images result in least spectral distortion. Since in terms of computation the LP is much efficient compared to the RWT and its performance is nearly as good as the RWT it is recommended to use the LP over the RWT if the highest spectral integrity is not desired. It was also discovered that some selection rules that minimise spectral distortion by retaining detail coefficients from the multispectral images can introduce artifacts in the pan sharpened image. Finally, recommendations are made for choosing the number of decompositions to be applied by the multiresolution transform during the pan sharpening process based on the resolution ratio of the images to be merged.

References

[1] P.J. Burt and E.H. Adelson, 'The Laplacian pyramid as a compact image code', *IEEE Transactions on Communications*, Vol. 31, No. 4, 1993, pp. 532–540.
[2] R. Sharma, 'Probabilistic model-based multisensor image fusion', Ph.D. dissertation, Oregon Graduate Institute of Science and Technology, 1999.
[3] T. Lillesand and R. Kiefer, *Remote Sensing and Image Interpretation*, third ed., John Wiley and Sons, 1994.
[4] Remote Sensing Technologies, http://chesapeake.towson.edu/data/tech.asp.
[5] R. King and J. Wang, 'A wavelet based algorithm for pan sharpening Landsat 7 imagery', in *Proceedings of the International Geoscience and Remote Sensing Symposium*, Vol. 2, 2001, pp. 849–851.
[6] The SPOT website, http://www.spotimage.fr.
[7] Y. Zhang and W. Wang, 'Multi-resolution and multi-spectral image fusion for urban object extraction', in *Proceedings of the 20th ISPRS Commission*, Vol. 3, 2004, pp. 960–966.
[8] M. Gonzalez-Audicana, L.S. Jose, R.G. Catalan and R. Garcia, 'Fusion of multispectral and panchromatic images using improved IHS and PCA mergers based on wavelet decomposition', *IEEE Transactions on Geoscience and Remote Sensing*, Vol. 42, No. 6, 2004, pp. 1291–1299.
[9] A.R. Smith, 'Color gamut transform pairs', in *Proceedings of the 5th Annual Conference on Computer Graphics and Interactive Techniques*, SIGGRAPH, 1978, pp. 12–19.
[10] ERDAS, *The ERDAS Field Guide*, seventh version, GIS & Mapping, LLC, Atlanta, GA, 2003.

[11] K.R. Castleman, *Digital Image Processing*, Prentice Hall, 1996.

[12] S.G. Mallat, 'A theory for multiresolution signal decomposition: The wavelet representation', *IEEE Transactions on Pattern Analysis and Machine Intelligence*, Vol. 11, No. 7, 1989, pp. 674–693.

[13] G. Strang and T. Nguyen, *Wavelets and Filter Banks*, third ed., Wellesley–Cambridge Press, Wellesley, MA, 1996.

[14] C.S. Burrus, R.A. Gopinath and H. Guo, *Introduction to Wavelets and Wavelet Transforms – A Primer*, Prentice Hall, 1998.

[15] J.E. Fowler, 'The redundant discrete wavelet transform and additive noise', *IEEE Signal Processing Letters*, Vol. 12, No. 9, 2005.

[16] J. Núñez, X. Otazu, O. Fors, A. Prades, V. Palà and R. Arbiol, 'Multiresolution-based image fusion with additive wavelet decomposition', *IEEE Transactions of Geoscience and Remote Sensing*, Vol. 37, No. 3, 1999, pp. 1204–1211.

[17] P. Hill, N. Canagarajah and D. Bull, 'Image fusion using complex wavelets', in *Proceedings of the Thirteenth British Machine Vision Conference*, 2002.

[18] H. Li, B.S. Manjunath and S.K. Mitra, 'Multisensor image fusion using the wavelet transform', *Graphical Models and Image Processing*, Vol. 57, No. 3, 1995, pp. 235–245.

[19] G.P. Lemeshewsky, 'Multispectral multisensor image fusion using wavelet transforms', *Proceedings of the SPIE*, Vol. 3716, 1999, pp. 214–222.

[20] V. Vijayaraj, N.H. Younan and C.G. O'Hara, 'Quantitative analysis of pansharpened images', *Optical Engineering*, Vol. 45, No. 4, 2006.

[21] S. Li, J.T. Kwok and Y. Wang, 'Using the discrete wavelet frame transform to merge Landsat TM and SPOT panchromatic images', *Information Fusion*, Vol. 3, 2002, pp. 17–23.

[22] R. Ryan, B. Baldridge, R.A. Schowengerdt, T. Choi, D.L. Helder and B. Slawomir, 'IKONOS spatial resolution and image interpretability characterization', *Remote Sensing of Environment*, Vol. 88, No. 1, 2003, pp. 37–52.

[23] R. Shi, C. Zhu, C. Zhu and X. Yang, 'Multi-band wavelet for fusing spot panchromatic and multispectral images', *Photogrammetric Engineering and Remote Sensing*, Vol. 69, No. 5, 2003, pp. 513–520.

[24] J. Zhou, D.L. Civco and J.A. Silander, 'A wavelet transform method to merge Landsat TM and SPOT panchromatic data', *International Journal of Remote Sensing*, Vol. 19, No. 4, 1998, pp. 743–757.

[25] P.S. Pradhan, 'Multiresolution based, multisensor, multispectral image fusion', Ph.D. dissertation, 2005.

[26] P. Blanc, T. Blu, T. Ranchin, T. Wald and T. Aloisi, 'Using iterated rational filter banks within the ARSIS concept for producing 10 m LANDSAT multispectral images', *International Journal of Remote Sensing*, Vol. 19, No. 12, 1998, pp. 2331–2343.

[27] B. Aiazzi, L. Alparone, A. Barducci, S. Baronti and I. Pippi, 'Multispectral fusion of multisensor image data by the generalized Laplacian pyramid', in *Proceedings of the IEEE International Geoscience and Remote Sensing Symposium*, 1999, pp. 1183–1185.

[28] P.S. Pradhan, R.L. King, N.H. Younan and D.W. Holcomb, 'Estimation of the number of decomposition levels for a wavelet-based multiresolution multisensor image fusion', *IEEE Transactions on Geoscience and Remote Sensing*, Vol. 44, No. 12, 2006, pp. 3674–3686.

17

Pixel-level image fusion metrics

Costas Xydeas[a] and Vladimir Petrović[b]

[a] Head of DSP research, Department of Communication Systems, Infolab21, Lancaster University, Lancaster, UK

[b] Imaging Science and Biomedical Engineering, University of Manchester, Oxford Road, Manchester, UK

The topic of objective metrics that are designed to quantify the performance of pixel-level image fusion systems is discussed in this chapter. The key principle that drives both the development and the subjective relevance of these metrics is the preservation of input image related visual information. Of course, image visual information can be defined in many ways and this can lead to the formulation of several possible pixel-level image fusion metrics. Initially, important visual information is considered that relates to 'edge'-based information and associated image regions. This concept is then expanded with the involvement of Human Visual System modelling within the metric formulation process. A number of such image fusion measures are therefore developed and their characteristics are examined for accurately predicting fusion system performance as compared to that obtained from subjective tests. Extensive experimentation allows for the detailed analysis, optimisation, and understanding of the comparative behaviour of these image fusion performance measures.

17.1 Introduction

Imaging arrays have become reliable sources of information in a growing range of applications. However, in order to fully exploit all image information that is obtained from sensing a scene at different spectral ranges, considerable processing effort is required. Furthermore, when the intended user is a human operator, displaying simultaneously multiple image modalities leads to confusion and overload, while integrating information across a group of users is almost impossible [1]. Signal-level image fusion systems deal with this problem by effectively 'compressing' several input images into a single output fused image and as a result they have attracted considerable research attention [2–8,30]. Whereas it is relatively straightforward to obtain a fused image, e.g. a simple but primitive method is to average input image signals, assessing the performance of signal-level fusion algorithms is a much harder proposition.

So far, the most reliable and direct method for evaluating image fusion performance is based on subjective tests that employ samples of representative users [9–13]. As recent

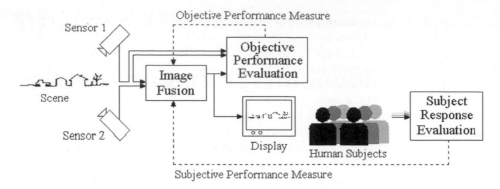

Figure 17.1 *Subjective and objective image fusion process evaluation.*

studies have shown, however, this approach is expensive in terms of time, effort, and equipment required [9,12,13]. Figure 17.1 illustrates this process where subjects evaluate, under tightly controlled conditions, fused images either by comparing them to each other or by performing specific visually oriented tasks. Preferences or task performance data are logged and processed further to yield meaningful performance evaluations.

An alternative and highly desirable approach will be one that requires no subjective tests but instead employs a subjectively meaningful, objective fusion evaluation metric. The comparative advantage of objective fusion assessment is obvious from the point of view of effort and time expended on the evaluation process. The implementation of such objective image fusion assessment algorithms takes the form of computer code and reduces assessment time from days or even weeks to a matter of a few minutes. A significant by-product of this process is the opportunity to use objective fusion performance metrics in the development of image fusion systems. Thus, an accurate objective evaluation metric could be used to provide immediate system fusion performance change information and, thus, to quickly and effectively guide the algorithmic design process of image fusion system parameter optimisation.

So far, only a limited number of relatively application-dependent image fusion performance metrics has been published in the literature [3,5,6,14,15]. In general, these methods are based on application specific knowledge and do not consider the generic goals and issues which underpin the image fusion process. Thus, target signature consistency, which includes both qualitative (subjective) and quantitative (objective) considerations, is used in [14] as an evaluation criterion for image fusion systems in detection/recognition applications. Furthermore, the idea of comparing the output fused image to an 'ideal' fused image in order to form a fusion performance estimate from corresponding differences is introduced in [5,6]. In both cases images with different focus points are fused manually (cut and paste) to produce the ideal fusion reference. In general, however, these methods are not applicable to other fusion applications where the ideal fused image is ill defined and cannot be obtained manually [5,6]. Rockinger and Fechner [3] and Qu et al. [29] meanwhile proposed mutual information as the general approach for both sequence and still image fusion performance assessment. A useful treatment of image quality metrics is given in [16].

The topic of objective metrics that are designed to quantify the performance of pixel-level image fusion systems is discussed in this chapter in a way that brings effectively together material published previously by the authors in [17,18]. The key principle that drives both the development and the subjective evaluation of these metrics is the preservation of image related visual information. Of course, image visual information or image content can be defined in many ways and this gives rise to several possible pixel-level image fusion metrics. A number of such image fusion measures are therefore developed and their characteristics are examined for accurately predicting fusion performance as compared to that obtained from subjective tests. Important visual information relates initially to 'edge'-based information and associated image regions. This concept is then expanded with the incorporation of Human Visual System modelling in the metric formulation process. Extensive experimentation allows for the detailed analysis, understanding, and thus, presentation of the 'behaviour' of these fusion performance measures.

17.2 Signal-level image fusion performance evaluation

The most obvious and natural approach that can be used to evaluate a fusion process would be to seek an *ideal* fused image against which all others could be compared. Such 'ground truth' information is unfortunately only available in very specific fusion applications [5,6]. Alternatively, one can start by defining certain goals to be achieved in signal-level image fusion and then proceed to measure the extent to which these goals are fulfilled. The second approach is adopted bellow as part of a generic framework for objectively measuring the success or otherwise of an image fusion process.

17.2.1 Signal-level image fusion

In theory, the goal underpinning signal-level image fusion can be simply defined as: *to represent within a single fused output image and without the introduction of distortion or loss of information, the visual information that is present in a number of input images.*

In practice, however, displaying all the visual information from several input images into a single output image is only possible when the inputs contain spatially and spectrally independent information. Since in real fusion applications this is never the case, the more practical objective of 'faithfully representing at the output fused image only the most visually important input information' is generally accepted. Furthermore, when considering that the most common application of fused imagery is for display [1–13] purposes, the preservation of *perceptually* important visual information becomes an important issue. The ideal fused output image should therefore contain all the important visual information, as perceived by an observer viewing the input images. Additionally, a signal-level fusion system must ensure that no distortions or other 'false' information are introduced in the fused image.

Thus, an image fusion performance measure must be able to: (i) identify and localise visual information in the input and fused images, (ii) evaluate its perceptual importance, and (iii) measure (quantify) the accuracy with which input information is represented in

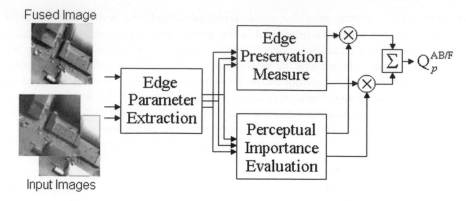

Figure 17.2   *Basic structure of the edge-based image fusion performance measure.*

the fused image. Finally, such a metric must also be able to distinguish between true scene (input) information and fusion artefacts which may appear in the output fused image.

17.2.2 'Edge'-based image fusion performance evaluation framework

Objective image fusion performance metrics can be founded upon the association of visual information to the 'edge' or gradient information that is present in an image. It is an established fact that the human visual system (HVS) processes visual stimuli (images) with the aim of resolving uncertainty and focuses on the extraction of information by inspecting areas where uncertainty is greatest. Thus, images convey information through 'uncertainties' that exist in the input signal and which manifest themselves in the changes (edges) rather than in individual signal (pixel) values [19,20]. The term 'edge' in this context refers to all changes in image signal value, also known as the image gradient [21]. This association is particularly important in the context of image fusion where signal values change independently of the physical effects which create them (e.g. illumination in visible range and thermal radiation in IR sensors).

When applying this broad definition of visual information to image fusion evaluation, an ideally fused output becomes an image that contains all the 'edge' information that is present in the input images. Often, however, fusion algorithms will partially transfer 'edge' information from the inputs to the fused output image and fusion system performance can be therefore evaluated by measuring the degree of completeness of edge information transfer. Furthermore, in order to correctly quantify the effect of different edges on fusion performance one must take into account their value of perceptual importance.

The general structure of this fusion performance metric development framework is shown in Figure 17.2. Here edge strength and orientation parameters are extracted from the input and fused output images at each pixel location (requirement (i) in Section 17.2.1). An edge preservation measure then is formed that estimates how well the edge information found in the fused image represents that of the input images. This results in two 'map' type signals that quantify similarity between edge information in the fused and each of the input signals at every pixel location (i.e. requirement (iii) in Section 17.2.1). Using this information, the relative perceptual importance of individual input image pixels is then determined and quantified as a numerical coefficient (i.e. requirement (ii) in Section 17.2.1). Finally, relative importance coefficients are used as perceptual weighting factors in a normalised summation of the edge preservation coefficients over the whole input data set. The result is a numerical score that directly represents the success or failure of the signal-level fusion process.

17.2.2.1 Edge related information estimates

The first step in this fusion performance assessment approach is the extraction of edge information from input and fused images. In real images, edges are distributed according to image content and their spatial location forms part of the uncertainty and hence information that an observer attempts to resolve and comprehend [19,20]. Furthermore, information is not only conveyed by strongly detectable edges, since relatively homogeneous image areas also contain a small amount of information. This fact is taken into consideration and gives a motivation for using all pixels in the formulation of the objective metric. Our discussion here assumes for simplicity that only two input images A and B are fused to yield an output fused image F.

Edge parameter information is extracted from the images using the Sobel operator [21] defined by two 3×3 templates that measure horizontal and vertical edge components. Filtering the input images A and B and the output fused image F with Sobel templates provides two images S_I^x and S_I^y, $I \in \{A, B, F\}$, which highlight edge information in the x- and y-directions, respectively.

The edge strength

$$g_I(m, n) = \left[S_I^x(m, n)^2 + S_I^y(m, n)^2\right]^{1/2}$$

and orientation

$$\alpha_I(m, n) = \arctan\left[S_I^y(m, n)/S_I^x(m, n)\right]$$

parameters then evaluated for each location (m, n), $1 \leqslant m \leqslant M$ and $1 \leqslant n \leqslant N$, where M, N are the dimensions of the image. Both parameters have values within a finite range. More specifically, $0 \leqslant g \leqslant g_{\max}$ (i.e. from no change to maximum contrast) for strength, and $-\pi/2 \leqslant \alpha \leqslant \pi/2$ (i.e. from vertical $(-\pi/2)$, through horizontal (0), to the inverted vertical $(\pi/2)$) for orientation. An example of this type of information is shown for a pair of input images, see Figures 17.3(a) and 17.3(d). Their edge strength parameter maps in Figures 17.3(b) and 17.3(e) easily identify those regions in white where the gradient exhibits large values; areas with insignificant signal change activity appear as dark. The complementary nature of the visual information contained in these images is also evident. Note that the orientation parameter maps shown in Figures 17.3(c) and 17.3(f) (horizontal – black, vertical – white) appear to be more random in nature, especially where edges are rather weak.

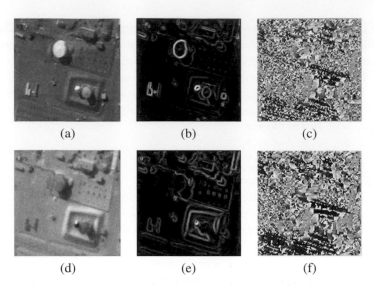

Figure 17.3 *Fusion input images (a, d) and corresponding edge strengths (b, e) and edge orientations (c, f).*

17.2.2.2 Perceptual edge information preservation estimates

Edge Information Preservation (EIP) values are derived for a given pair of input and corresponding fused images. EIP quantifies how well edge information in the output fused image represents the edge information found in the input image. EIP is based on a pixel-by-pixel comparison of fused image edge parameters to the corresponding parameters of an input image and by doing so it models the perceived information loss incurred by fusing (i.e. transferring) edges from the input into the fused image. This part of the system also discriminates between distortions and true scene information.

More specifically, an input edge is perfectly represented in the fused image only if both its strength and orientation are preserved. Any change in one or both of these parameters indicates loss of information and initially Relative Change Coefficients are evaluated for both properties. Thus, when a loss of contrast exists (i.e. the fused edge is weaker than the corresponding edge in the input signal) the change of edge strength from say A into F, i.e. G^{AF}, is evaluated as the ratio of the fused to the input edge strengths:

$$G^{AF}(m, n) = \begin{cases} \frac{g_F(m,n)}{g_A(m,n)} & \text{when } g_A(m, n) > g_F(m, n), \\ \frac{g_A(m,n)}{g_F(m,n)} & \text{otherwise} \end{cases} \tag{17.1}$$

In the opposite case, contrast enhancement in the fused image is treated as distortion, as this is not the objective of the fusion process and the ratio is inverted. G^{AF} is unity when the fused edge strength $g_F(m, n)$ is a perfect representation, i.e. is equal to input strength $g_A(m, n)$.

The edge orientation quantity is cyclic in nature, and it is more complicated to define what constitutes loss of information in terms of change of orientation. The situation is simplified, however, by assuming that inverted edges (analogous to photographic negative) represent each other perfectly and are therefore of the same orientation. This means

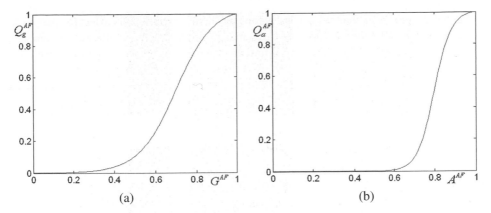

(a) (b)

Figure 17.4 *Perceptual strength and orientation preservation parameters, which model the non-linear nature of the HVS: (a) strength and (b) orientation.*

that edge orientation values lie in the region between horizontal, $\alpha_A(m, n) = 0$, and vertical, $|\alpha_A(m, n)| = \pi/2$. Change of orientation in F with respect say to A, i.e. A^{AF} is then evaluated as a normalised relative distance between the input and fused edge orientation:

$$A^{AF}(m, n) = \frac{||\alpha_A(m, n) - \alpha_F(m, n)| - \pi/2|}{\pi/2} \qquad (17.2)$$

In a way that is analogous to G^{AF}, A^{AF} is unity when the input and fused image orientation values are equal and decreases linearly as these values diverge.

Parameters G^{AF} and A^{AF} describe in a linear fashion the change in these two basic parameters of visual edge information. However, to model the perceived information loss detected by an observer, the non-linear nature of HVS response to various types of stimuli must be taken into account [20,22]. Thus, the loss of edge information with respect to the loss of edge strength and orientation is modelled by sigmoid functions. These non-linearities produce the Perceptual Strength and Orientation Preservation estimates Q_g^{AF} and Q_α^{AF}:

$$Q_g^{AF}(m, n) = \frac{\Gamma_g}{1 + e^{-\kappa_g(G^{AF}(m,n) - \sigma_g)}} \qquad (17.3)$$

$$Q_\alpha^{AF}(m, n) = \frac{\Gamma_\alpha}{1 + e^{-\kappa_\alpha(A^{AF}(m,n) - \sigma_\alpha)}} \qquad (17.4)$$

Note that sigmoid types of functions are extensively used for modelling the behaviour of biological systems and in this context they capture well the response of observers to the loss of visual information, see Figure 17.4. The parameter set $(\kappa_g, \sigma_g, \kappa_\alpha, \sigma_\alpha)$ which determines the exact shape of the sigmoid curves of Figure 17.4 is itself determined via an optimisation process that maximises a 'correspondence' measure between objective and subjective image fusion assessment results. Furthermore, Γ_g and Γ_α are scaling constants selected such that for optimal values $\kappa_g, \sigma_g, \kappa_\alpha, \sigma_\alpha$ and A^{AF}, G^{AF} equal to 1, the parameters Q_g^{AF}, Q_α^{AF} will also be equal to 1.

In general, very small changes in edge parameters do not significantly affect the perceived similarity between edges in input and fused images. In fact, as these changes increase and once visual differences are perceived, further increases in the difference between edge parameters have little perceptual effect. Relatively speaking, HVS is less sensitive to the perceived loss of contrast (strength) as compared to orientation information. Thus, even relatively small changes in edge orientation values can have a significant perceptual effect. Therefore, Q_α^{AF} decreases very rapidly, and for a 35° change ($A^{AF} = 0.61$), almost all orientation information is lost ($Q_\alpha^{AF} = 0.006$), see Figure 17.4(b). Note that the values of both preservation parameters Q_g^{AF} and Q_α^{AF} lie between 0 and 1.

Next, the above strength and orientation preservation parameters, Q_g^{AF} and Q_α^{AF} can be combined into a single edge preservation map denoted by the coefficient $Q^{AF}(m, n)$ that models how truthfully the fused image F represents the input image A at each location (m, n):

$$Q^{AF}(m, n) = \sqrt{Q_g^{AF}(m, n) Q_\alpha^{AF}(m, n)} \qquad (17.5)$$

Like Q_g^{AF} and Q_α^{AF}, Q^{AF} values fall within the $[0, 1]$ range, where $Q^{AF} = 0$ corresponds to complete loss of information and $Q^{AF} = 1$ indicates 'ideal fusion' and complete transfer of information from A into F.

This procedure works well when modelling information loss of input edges which can be detected by an observer and have strength larger than a threshold T_d. Input image edges of strength less than T_d signify that 'there is no change in illumination at that location' and need to be considered differently. Specifically, whether this information is preserved in the fused image determines the value of Q^{AF}. If both the input and the fused 'edges' cannot be detected, the information is preserved perfectly and Q^{AF} values are set to 1. However, if an input edge is undetectable (i.e. strength $< T_d$) but the corresponding fused edge can be perceived, distortion is present and the process described above (Equations (17.1) and (17.3)) is applied which results in very small Q^{AF} values.

17.2.3 Edge-based image fusion metric

Given the edge preservation maps Q^{IF}, $I \in \{A, B\}$, a single measure $Q^{AB/F}$ of success of a signal-level image fusion process operating on input images A and B to produce an output fused image F, can be obtained as the normalised sum of local edge preservation Q^{AF} and Q^{BF} maps weighted by respective perceptual importance coefficients w_A and w_B (see Figure 17.2):

$$Q^{AB/F} = \frac{\sum_{m=1}^{M} \sum_{n=1}^{N} [w_A(m, n) Q^{AF}(m, n) + w_B(m, n) Q^{BF}(m, n)]}{\sum_{m=1}^{M} \sum_{n=1}^{N} [w_A(m, n) + w_B(m, n)]} \qquad (17.6)$$

In this way, perceptually important information influences more the final performance measure value, and vice versa.

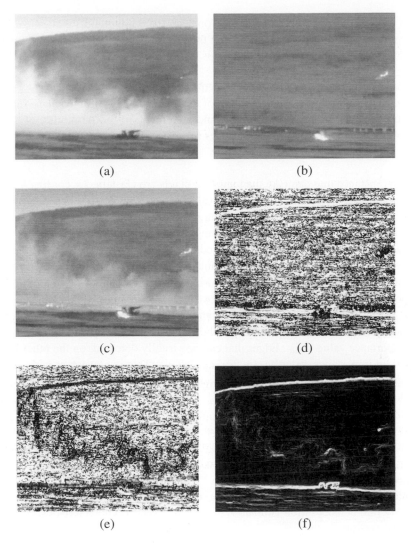

Figure 17.5 *Fusion performance measure: (a, b) input images A, B and (c) fused image F, (d, e) edge preservation measure Q^{AF} and Q^{BF}, and (f) feature importance map w_A.*

$Q^{AB/F}$ values also lie between 0 and 1, where zero indicates the total loss of input information and one implies ideal fusion. However, these extreme values are rarely achieved. Thus, image fusion algorithms that produce higher $Q^{AB/F}$ values than those obtained from other systems are also expected to perform 'subjectively' better.

A graphic example that relates to the formulation of this fusion performance metric is shown in Figure 17.5. Input images A and B, shown in Figures 17.5(a) and 17.5(b), are fused into F as shown in Figure 17.5(c), using Laplacian pyramid fusion [23]. The process gives rise to edge preservation maps Q^{AF} and Q^{BF}, shown in Figures 17.5(d) and 17.5(e) (with black $\equiv 0$ and white $\equiv 1$). The fused image is a good representation of the scene, hence the predominantly 'light' appearance of the edge preservation maps in

Figures 17.5(d) and 17.5(e). In particular, one can trace important input image features that have been transferred successfully into the fused image as white structures and lines in the corresponding edge preservation maps.

Equation (17.6) is indicative of (i) the dependence of the proposed image fusion performance metric framework on the definition of the perceptual importance maps formulated by the coefficients $w_A(m, n)$, $w_B(m, n)$, and therefore, (ii) the possibility of obtaining different image fusion performance metrics for different $w_A(m, n)$, $w_B(m, n)$ formulations.

In the first instance one can take the view that pixels are classified according to whether they contain detectable changes in signal value or not. This in turn depends on a number of factors [24] with the most significant being contrast (edge strength) information. Stronger edges are more powerful in attracting attention and are, therefore, more likely to be noticed by an observer. Thus, a simple edge strength threshold can be defined which corresponds to the weakest change in illumination level that is detectable by an observer. Under normal viewing conditions and ignoring effects such as visual masking, humans are reported to be able to discriminate between 64 illumination levels (shades of grey) [20,24], which corresponds to an edge strength threshold level of $g_d = 2$. In practice, the majority of pixels fall bellow this threshold and they are simply assigned a small but finite importance value w_{min}. Edges that can be visually detected, on the other hand, are assigned an importance value $w_A(m, n) = g_A(m, n)$. This is consistent with the generally accepted view that edges that attract more attention hold more value to the observer [20,24].

The perceptual importance coefficients map shown in Figure 17.5(f) is produced using this simple but effective $w_A(m, n) = g_A(m, n)$ methodology. The objective fusion performance score in this pixel-level image fusion example is calculated as $Q_{\text{Laplace}}^{AB/F} = 0.625$.

An alternative way for defining the $w_A(m, n)$, $w_B(m, n)$ information, by taking more rigorously into consideration Human Visual System (HVS) characteristics, is discussed in the next section.

17.2.4 Visible differences and associated image fusion metrics

The Visual Difference Predictor scheme that has been developed by Daly [25] models the Human Visual System and estimates the visibility, near the visual threshold, of the differences between two versions A and B of an image. The output of the Visual Difference Predictor is a two-dimensional map P, whose elements $0 \leqslant P(m, n) \leqslant 1$ indicate the probability of detecting visual differences between the two images at every pixel location (m, n). $P(m, n) = 1$ indicates that differences are *suprathreshold* and completely detectable, whereas $P(m, n) = 0$ indicates that the difference between the two images at this location cannot be detected. Notice that this Visual Difference Predictor output map does not discriminate between different suprathreshold visual distortions, which can be 'fully' detected.

Figure 17.6	*Visible differences between two input images and the fused image.*

Thus, Visual Difference (VD) relationships can be defined between each of the input images and the output fused image. Figure 17.6 illustrates these relationships on a real image fusion example where two multisensor images (A and B) are fused using Laplacian multiresolution type of fusion [23] to produce image F. In this figure the VD probability defined as the map that illustrates the probability of visible differences between A or B and F denoted by P^{AF} and P^{BF}, respectively, assumes 'white' values when the probability of a difference being detected is high. These maps clearly highlight the effects of the information fusion process. That is, areas taken from A are dark in P^{AF} and light in P^{BF}, and vice versa.

VD maps $P^{AF}(m, n)$ and $P^{BF}(m, n)$ can be used directly to evaluate fusion system performance. A simple measure can be the average probability of noticing a difference between the inputs and the fused image. If the fusion process is successful, input image information will be faithfully represented in the fused image resulting in a small probability that observers will notice differences.

Fusion systems that produce low $P^{AF}(m, n)$ and $P^{BF}(m, n)$ values therefore perform well. Thus, a VD-based metric can be defined as

$$\text{VDP} = 1 - \frac{1}{2MN} \sum_{m} \sum_{n} \left(P^{AF}(m, n) + P^{BF}(m, n) \right) \tag{17.7}$$

where $M \times N$ is the image dimension.

Now if one is to consider that many fusion algorithms rely on some form of spatial feature selection by effectively choosing to faithfully represent at every (m, n) pixel location only one of the input images, then at least one of $P^{AF}(m, n)$, $P^{BF}(m, n)$ values should be small. Accordingly, another VD type of metric denoted by $\text{VDP}_{\min}$ can be formulated as

$$\text{VDP}_{\min} = 1 - \frac{1}{MN} \sum_m \sum_n \min\left(P^{AF}(m, n), P^{BF}(m, n)\right) \qquad (17.8)$$

Finally, if one is to consider the mechanism of visual attention distribution and the fact that during subjective fusion evaluation trials the observers' attention is captured by only the most significant differences between the input and fused images, there is a rationale to restrict the measurement to such areas only. Applying a visual detection threshold T_d to $P^{AF}(m, n)$ and $P^{BF}(m, n)$ simulates such behaviour and another Visible Differences Area (VDA) metric can be defined which employs only those input image locations that exhibit significant changes between the input and fused images:

$$\text{VDA} = 1 - \frac{1}{2MN} \sum_m \sum_n \left(P^{AF}(m, n) + P^{BF}(m, n)\right)\big|_{\forall m,n, \, P^{AF,BF}(m,n) > T_d} \qquad (17.9)$$

T_d effectively determines the probability level at which image differences are likely to be noticed and, thus, effect perceived fusion performance.

It is also worth noting here that all the above metrics, i.e. Equations (17.7)–(17.9) assume values in the range [0, 1] where 0 signifies the 'worst' and 1 the 'best' possible fusion performance.

In addition to the above metrics, the concept of visible differences can be applied to enrich the previously discussed $Q^{AB/F}$ metric [5,6]. The rationale for combining the VD and $Q^{AB/F}$ evaluation approaches lies in the disparate nature of image properties these metrics address and, therefore, the resulting 'Hybrid' scheme promises a more 'rounded' assessment process. Visual differences (VD) information is therefore combined with $Q^{AB/F}$ in two useful ways: (i) use the VD maps within the $Q^{AB/F}$ framework in a hybrid Q_{VD} metric and (ii) evaluate both types of metrics independently and combine them *a posteriori* into a new class $f(Q^{AB/F}, M_{\text{VD}})$, $M_{\text{VD}} \in \{\text{VDP}, \text{VDP}_{\min}, \text{VDA}\}$ of metrics using simple arithmetic rules. Thus, VD information can be introduced into the $Q^{AB/F}$ framework and a preferred way is the following Q_{VD} metric:

$$Q_{\text{VD}} = \frac{\sum_{m=1}^{M} \sum_{n=1}^{N} [w_A(m, n) P^{AF}(m, n) Q^{AF}(m, n) + w_B(m, n) P^{BF}(m, n) Q^{BF}(m, n)]}{\sum_{m=1}^{M} \sum_{n=1}^{N} [w_A(m, n) P^{AF}(m, n) + w_B(m, n) P^{BF}(m, n)]}$$

$$(17.10)$$

In this manner the metric takes into account the influence of visible differences on attention distribution by assigning relatively more importance to locations that exhibit significant changes between the fused and input images. Alternatively, one could use only P^{AF} and P^{BF} concentrating only on the differences between the signals.

The second *ad hoc* hybrid alternative type of metric can be based on a simple linear combination of a VD type of metric and the $Q^{AB/F}$ formulations. The resulting

$f(M_{VD}, Q^{AB/F})$ metrics are of the form:

$$f(M_{VD}, Q^{AB/F}) = bQ^{AB/F} + (1 - b)M_{VD}, \quad M_{VD} \in \{VDP, VDP_{min}, VDA\} \quad (17.11)$$

The coefficient $b \in [0, 1]$ is defined experimentally in such a fashion so that the agreement between Q and corresponding subjective results is maximised.

17.3 Comparison of image fusion metrics

Objective fusion performance metrics should be compared and their performance judged with respect to their subjective relevance, i.e. the level of correspondence achieved between their objectively made assessment of fusion system performance and the fusion assessment results produced by subjective tests.

This requires some form of ground truth, i.e. a 'calibrated' set of subjective results against which the computer simulation results obtained from using a metric can be compared. Thus, results from subjective preference tests that are similar to those successfully used for image and video quality metric validation [26], can be employed. In particular, subjective tests involve displaying series of image sets, each consisting of two different fused images that are produced from a common input image pair. Subjects scrutinise these sets of input and fused images and are asked to decide which of the two output fused images better represents the visual information found in the input images.

Of course, the possibility also exists of the case whereby both fused images represent input information equally well. Votes from subjects are aggregated for each image set and after normalising with the number of subjects, preference scores S_1, S_2 are obtained for the two fused images and also a score S_0 for the case of equal preference. An example of input images and output fused image sets is shown in Figure 17.7.

Both fused images are then evaluated using the metric and an objective preference is recorded as $O_p = 1$, $p \in \{1, 2\}$, for the image with the higher metric score, or no preference ($O_0 = 1$) if the scores are within a 'similarity margin' (1.5 per cent found to be sufficient given the limited practical range of the metrics) [7].

Using these subjective and objective preference scores, two distinct measures of correspondence can be evaluated [17,18]. The first is the 'Correct Ranking' (CR) measure. This is the proportion of all image pairs in which the subjective and objective ranking of offered fused images correspond. A value close to 1 (or 100 per cent) is desirable since it means ideal agreement while the 'minimum' is the random guess rate of 33 per cent for the three options.

The second, so called 'relevance' measure (r) takes into account the relative certainty of the subjective scores. When the subjects are unanimous in their choice of a fusion scheme, the corresponding subjective preference is 1 and so is the 'certainty.' Alternatively, when each of the three preference options receives an equal number of votes, subjective preference and corresponding certainty are 33 per cent. Relevance r is thus the sum with respect

Figure 17.7   *Input images (a, b) and two different fused images (c, d).*

to all image sets of subjective preferences which correspond to the fused result for which
the objective metric score used is maximum. The sum is further normalised to a range
[0, 1] between the smallest and largest possible relevance scores given by the subjective
test results. Thus,

$$
r = \frac{\sum_{i=1}^{N} S_K^i \big|_{O_K^i = 1}}{\sum_{i=1}^{N} \max(S_0^i, S_1^i, S_2^i)} = \frac{\sum_{i=1}^{N} \sum_{K=0}^{2} S_K^i O_K^i}{\sum_{i=1}^{N} \max(S_0^i, S_1^i, S_2^i)}
\tag{17.12}
$$

An r value of 1 therefore means that the metric predicted the subject's preferred choice
in all image pairs. Globally, compared to CR, r places more emphasis on cases where
subjects are more unanimous.

17.3.1 Objective metric optimisation

Given the above framework for comparing different objective metrics, these metrics must
be individually optimised for maximum performance. Again metric performance optimisation relates to subjective tests and associated results.

Let us consider first the metric of Equation (17.6) with $w_A(m, n) = g_A(m, n)$.

Subjective results were obtained from eight separate subjective preference tests related
to the comparison of various pairs of fusion techniques. Results of the first six tests were

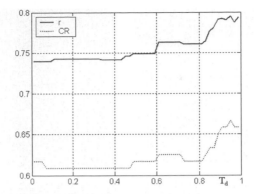

Figure 17.8 *VDA-based scheme. r and CR values as a function of T_d.*

used to optimise the edge preservation parameter set $(\kappa_g, \sigma_g, \kappa_\alpha, \sigma_\alpha)$, see Equations (17.3) and (17.4). These tests were based on 23, 10, 15, 15, 15, and 12 image sets and involved 13, 11, 28, 19, 19, and 9 subjects, respectively. Test 1 compared the DWT pyramid fusion technique to the image averaging fusion. The well-known 'maximum selection rule' was used as a method to combine the wavelet coefficients of the source images [6]. Test 2 [7] involved two further DWT schemes described in [5] and [7]. Tests 3 to 5 compared multiresolution fusion, i.e. a gradient-based method [27], DWT fusion [5], and RoLP (Ratio of Low-Pass) pyramid fusion [28] against each other over matching input image sets. Finally, Test 6 compared the efficient multiscale fusion described in [8] to that of [5]. Tests were performed individually under controlled conditions (darkened room, sufficiently large screen, no time limitations, etc.) [5,7].

The optimisation process was an exhaustive search within the perceptually meaningful limits of a four-dimensional parameter space. The 'optimal' parameter vector was found to be $(\kappa_g, \sigma_g, \kappa_\alpha, \sigma_\alpha) = (0.7, 11, 0.8, 24)$ and was used to obtain the sigmoid functions displayed in Figures 17.4(a) and 17.4(b). Using these parameters the $Q^{AB/F}$ metric predicted a subjective preference CR of 61 out of the 90 image sets used in the optimisation, that is, 68 per cent. Also a perceptual validation measure score $r = 0.825$ indicates (as compared to CR = 0.68) that the metric on average performed better in more relevant sets where subjects' agreement was high.

Using this $Q^{AB/F}$ optimised metric two further subjective tests were performed (Tests 7 and 8). Both tests employed 15 image sets and 15 subjects with the aim of comparing the Laplacian pyramid [23] and QMF (DWT) based on 'maximum selection rule' fusion schemes and the DWT multiscale and multiresolution fusion schemes [3,6]. Objective metric results for this data correlate even better to subject responses with correct classification in 26 out of the 30 input image sets; that is, CR = 0.87, while $r = 0.93$.

Let us now consider the VDA metric that relies on the visual detection threshold T_d and also the $f(Q^{AB/F}, M_{VD})$ metrics that rely on the linear combination factor b. Again the values of these parameters are optimised for maximum subjective relevance. 120 input image pairs were used in this optimisation process which, besides determining an optimal operating point for each metric, also provided a useful insight into the nature and

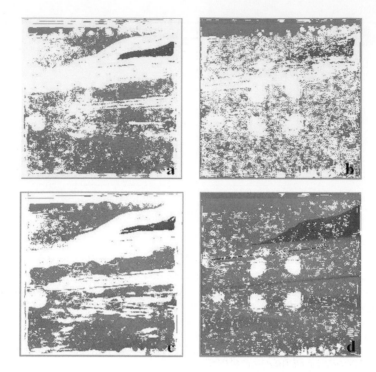

Figure 17.9   *Visible differences between input images (a, b) and resulting fused images (c, d) of Figure 17.7.*

robustness of the metrics. Thus VDA performance (in terms of r and CR) for various values of T_d is illustrated in Figure 17.8. There is a clear trend in that performance improves as the detection threshold value increases. For $T_d > 0.9$, fusion performance is relatively robust, while overall $r > 0.73$ and CR > 0.61. For higher values of T_d the metric considers progressively smaller areas of the image, and when $T_d = 0.95$ the areas affected by visible changes and used in the evaluation cover approximately 43 per cent of the image. This means that subjects form their decisions by considering relatively small proportions of the displayed images. A high level of relevance r also indicates that these areas correspond to the areas where the differences between the inputs and the fused output image are the greatest.

The effect of the visibility threshold ($T_d = 0.95$) is illustrated in Figure 17.9 where the visible difference maps P between the fused images shown in Figures 17.7(c) and 17.7(d) and their corresponding input images are provided. It is subjectively clear that the fused image in Figure 17.7(d) provides a much better representation of the scene and results in a much smaller number of visible differences (white pixels) compared to both inputs. This example also illustrates well the operation of the Visual Difference Predictor algorithm, since one can see that all the most probable differences are located around salient parts of the signal.

Finally, the results of the optimisation of the three linear combination metrics, i.e. VDP $+ Q$, VDP$_{\min} + Q$, and VDA $+ Q$, with respect to the coefficient b are shown in Figure 17.10. The behaviour of all three metrics is similar, reaching a performance

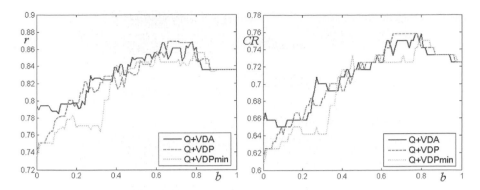

Figure 17.10 *Optimisation of the VDA + $Q^{AB/F}$ metric.*

peak at $b \approx 0.7$, which indicates a greater influence for the $Q^{AB/F}$ metric on the correct scores. Again, this indicates that the appearance of important 'edge' information in the fused image makes a 'stronger' impression on the subjects than general differences that they may be detected in the signals.

17.3.2 Performance of fusion metrics

A comparative experimental study of all the objective metrics, i.e. VDP, VDP$_{min}$, VDA, Hybrid Q_{VD}, the three VD + Q metrics, $Q^{AB/F}$, as well as that of the mutual information metric (QZY) of Qu et al. [29], was performed. Performance results are shown in Table 17.1. These are obtained using the same set of subjective test as in the previous section.

The VDP and VDP$_{min}$ metrics achieve a respectable level of success with r values of 0.74 and 0.736 and CR values of 61.6 and 60.8 per cent, respectively, on a par with the mutual information QZY metric. The VDA metric, on the other hand, performs better ($r = 0.795$ and CR = 66.7 per cent) enforcing the hypothesis that subjects have a tendency to selectively focus only on sections and not on the whole image area in order to make their decision.

However, it is also obvious from Table 17.1 that a purely VD-based evaluation approach presents no improvement over the gradient-based $Q^{AB/F}$ method. At best the VDA metric correctly ranks 80 out of the 120 fused image pairs (CR = 66.7 per cent) while $Q^{AB/F}$ achieves 87 out of 120. Furthermore performance is improved when the two approaches are combined. The hybrid Q_{VD} metric with $r = 0.829$ and CR = 72.5 per cent performs significantly better than the VD metrics and equally well as $Q^{AB/F}$. The best performance is, however, achieved when using the linear combination of these two main approaches. All three linearly combined metrics, i.e. VDP + Q, VDP$_{min}$ + Q, and VDA + Q, offer gains in terms of both r and CR when compared to $Q^{AB/F}$ and Q_{VD}. VDP + Q is slightly in front with correctly ranking 91 of the 120 image pairs (CR = 75.8 per cent) and achieving an r value of 0.869.

Table 17.1 *Subjective correspondence of different image fusion performance metrics.*

Metric	VDP	VDP_{min}	VDA	Q_{VD}	$VDP_{min} + Q$	$VDP + Q$	$VDA + Q$	$Q^{AB/F}$	QZY
r	0.74	0.736	0.795	0.829	0.855	0.869	0.868	0.833	0.742
CR	61.6%	60.8%	66.7%	72.5%	75.0%	75.8%	75.8%	72.5%	62.5%

Table 17.2 *Fusion performance scores of different metrics for the fused images in Figures 17.7 and 17.11.*

Fused image	VDP	VDP_{min}	VDA	Q_{VD}	$VDP_{min} + Q$	$VDP + Q$	$VDA + Q$	$Q^{AB/F}$	QZY
17.7(c)	0.16	0.27	0.40	0.53	0.42	0.39	0.49	0.49	1.03
17.7(d)	0.31	0.50	0.64	0.68	0.57	0.51	0.63	0.60	0.76
17.11(c)	0.43	0.82	0.57	0.77	0.78	0.66	0.69	0.76	1.79
17.11(d)	0.36	0.68	0.48	0.73	0.71	0.61	0.65	0.72	1.27

Figure 17.7 can be used in discussing further the performance of these objective metrics. Here the input images provide two different views of an outdoor scene, one via a visible and the other via an infrared range sensor. The output fused images in Figures 17.7(c) and 17.7(d) exhibit different characteristics. The image in Figure 17.7(c) suffers from reduced contrast and the introduction of significant fusion artefacts while Figure 17.7(d) provides a much clearer view of the scene. Individual metric scores for both of these images are given in Table 17.2. For this particular image pair, 27 out of the 28 subjects that took part in the test opted in favour of image 17.7(d) with no equal preferences. All the VD and combined metrics successfully predict this result by having a higher score for image 17.7(d), with varying difference margins. The exception is the QZY metric, which is based on a mutual information approach which employs 'histograms' rather than actual image features, and scores higher for image 17.7(c).

A different example where fused images do not differ so significantly is shown in Figure 17.11. Subjects opted here 9 to 4 in favour of the fused image 17.11(c), as opposed to 17.11(d), with 2 indicating equal preference. The reason is the appearance of fusion artefacts in the form of shadowing effects in image 17.11(d) and the generally lower contrast of this image compared to 17.11(c). While the Visual Difference Predictor maps between the fused images and the input 17.11(b), shown in Figures 17.11(g) and 17.11(h), provide no decisive information to differentiate between them, the effects of fusion artefacts are clearly visible on the Visual Difference Predictor map between the fused image 17.11(d) and the input 17.11(a) shown in Figure 17.11(f), in the form of white blotches indicating the areas where differences are visible. In comparison, the Visual Difference Predictor map between the fused image 17.11(c) and the input 17.11(a), shown in Figure 17.11(e), exhibits no such effects. In terms of the numerical fusion performance scores all the metrics agree with the subjects in this case, see Table 17.2.

Tables 17.1 and 17.2 indicate that the VD approach to image fusion assessment has validity. The good scores of the hybrid Q_{VD} metric prove the usefulness of the VD information in correctly emphasising areas of interest which are subjectively important. The success of this hybrid metric indicates that subject attention is guided by saliency in the displayed signal as well as by the perceived differences in visual features that should be identical

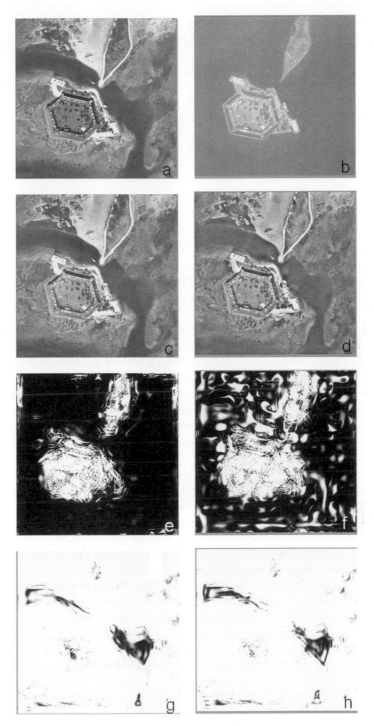

Figure 17.11 *Input images (a and b). Fused images (c and d). Visual Difference Predictor maps between images (c), (a) and (d), (a) (e and f). Visual Difference Predictor maps between images (c), (b) and (d), (b) (g and h).*

(in input and fused images). Equally, the improved performance of Q_{VD} with respect to the purely VD-based metrics indicates that absolute probabilities of detection of visible differences present a poorer measure of perceived information loss as compared to the gradient approach used in $Q^{AB/F}$ and Q_{VD}. Further justification for this argument is the optimal operating point of the linearly combined metrics at $b \approx 0.7$. This effectively indicates that the underlying $Q^{AB/F}$ framework still provides the majority of discriminatory information. At the same time, VD information provides an essential and important ingredient to overall fusion performance assessment. Finally, it can be argued that linear combinations of Hybrid measures are more robust than the individual VD or $Q^{AB/F}$ metrics in terms of sensitivity to system parameter values and input data types.

17.4 Conclusions

The formulation of objective but subjectively relevant pixel-level image fusion performance metrics is addressed in this chapter. Two different metric development avenues are explored which lead in to several such metrics, and which are then combined to produce additional 'Hybrid' forms of objective metrics.

The starting point in the development and subjective evaluation of these metrics is the preservation in the output fused image of the visual information that exists in the input images. In the first approach visual information is approximated with edge type of image information and, as a result, an edge-preservation-based image fusion metric framework is developed that attempts to take into account high level knowledge of HVS behaviour. However, not all edge preservation related information that is extracted from an image has the same perceptual significance and therefore the same contribution in the estimation of the objective metric. This is achieved by weighting edge preservation information with the corresponding edge strength values.

The second approach that takes fully into consideration HVS properties is based on measuring visual differences on a per pixel basis. In this case visual difference (VD) information is passed through a threshold and as a result only high VD image areas contribute to the value of the objective metric.

Another important aspect of this chapter is the development of the measures CR and r that are used to quantify the correspondence and, therefore, 'agreement' between objective and subjective image fusion performance results. These measures are used to optimise the parameters of the objective metrics and also to produce a comparative study of performance for all the previously defined objective metrics.

Thus, it was found that pure VD metrics achieve respectable performance but do not improve upon edge (gradient)-preservation-based algorithms. On the other hand, Hybrid metrics that combine both the VD and gradient preservation approaches outperform all other metrics. Such results clearly indicate that further development of this 'hybrid' evaluation technology is warranted not only in the field of image fusion evaluation but also in the wider field of image quality assessment. In the context of image fusion, the proven usefulness and dispersed nature of the VD information provides an exciting opportunity for exploration of the next incremental step in image fusion evaluation; a step designed to provide a much deeper insight into different aspects of the image information fusion process in general and a breakdown of fusion performance characterisation into important constituent parts, in particular.

References

[1] L. Klein, *Sensor and Data Fusion Concepts and Applications*, SPIE Press, 1999.

[2] Z. Zhang and R. Blum, 'A categorization of multiscale-decomposition-based image fusion schemes with a performance study for a digital camera application', *Proceedings of the IEEE*, Vol. 87, No. 8, 1999, pp. 1315–1326.

[3] O. Rockinger and T. Fechner, 'Pixel-level image fusion: The case of image sequences', *Proceedings of SPIE*, Vol. 3374, 1998, pp. 378–388.

[4] C. Pohl and J. van Genderen, 'Multisensor image fusion in remote sensing: Concepts, methods and applications', *International Journal of Remote Sensing*, Vol. 19, No. 5, 1998, pp. 823–854.

[5] H. Li, B. Munjanath and S. Mitra, 'Multisensor image fusion using the wavelet transform', *Graphical Models and Image Processing*, Vol. 57, No. 3, 1995, pp. 235–245.

[6] Y. Chibani and A. Houacine, 'Multiscale versus multiresolution analysis for multisensor image fusion', in *Proceedings of Eusipco 98*, Rhodes, 1998.

[7] V. Petrović and C. Xydeas, 'Multiresolution image fusion using cross band feature selection', *Proceedings of SPIE*, Vol. 3719, 1999, pp. 319–326.

[8] V. Petrović and C. Xydeas, 'Computationally efficient pixel-level image fusion', in *Proceedings of Eurofusion 99*, Stratford-upon-Avon, October 1999, pp. 177–184.

[9] A. Toet, J. Ijspeert, A. Waxman and M. Aguilar, 'Fusion of visible and thermal imagery improves situational awareness', *Proceedings of SPIE*, Vol. 3088, 1997, pp. 177–188.

[10] A. Toet, N. Schoumans and J. Ijspeert, 'Perceptual evaluation of different nighttime imaging modalities', in *Proceedings of Fusion 2000*, Paris, 2000, pp. TuD3-17–TuD3-23.

[11] A. Toet and E.M. Franken, 'Perceptual evaluation of different image fusion schemes', *Displays*, Vol. 24, No. 1, 2003, pp. 25–37.

[12] D. Ryan and R. Tinkler, 'Night pilotage assessment of image fusion', *Proceedings of SPIE*, Vol. 2465, 1995, pp. 50–67.

[13] P. Steele and P. Perconti, 'Part task investigation of multispectral image fusion using gray scale and synthetic color night vision sensor imagery for helicopter pilotage', *Proceedings of SPIE*, Vol. 3062, 1997, pp. 88–100.

[14] R. Sims and M. Phillips, 'Target signature consistency of image data fusion alternatives', *Optical Engineering*, Vol. 36, No. 3, 1997, pp. 743–754.

[15] M. Ulug and C. McCullough, 'A quantitative metric for comparison of night vision fusion algorithms', *Proceedings of SPIE*, Vol. 4051, 2000, pp. 80–88.

[16] S. Chandra, 'Texture image reconstruction from irregularly sampled data', Ph.D. thesis, Surrey University, UK, 2006.

[17] V. Petrović and C. Xydeas, 'Objective evaluation of signal level image fusion', *Optical Engineering*, Vol. 44, No. 8, 2005, pp. 087003-1–087003-8.

[18] V. Petrović and C. Xydeas, 'Evaluation of image fusion performance using visible differences', ECCV, May 2004, pp. 380–391.

[19] D. Marr, *Vision*, W.H. Freeman, San Francisco, 1982.

[20] W. Handee and P. Wells, *The Perception of Visual Information*, Springer, New York, 1997.

[21] M. Sonka, V. Hlavac and R. Boyle, *Image Processing Analysis and Machine Vision*, PWS Publishing, Pacific Grove, 1998.

[22] H. Schiffman, *Sensation and Perception*, John Wiley & Sons, New York, 1996.

[23] A. Akerman, 'Pyramid techniques for multisensor fusion', *Proceedings of SPIE*, Vol. 1828, 1992, pp. 124–131.

[24] P. Barten, *Contrast Sensitivity of the Human Eye and Its Effects on Image Quality*, SPIE Press, Bellingham, WA, 1999.

[25] S. Daly, 'The visible differences predictor: An algorithm for the assessment of image fidelity', in *Digital Images and Human Vision*, MIT Press, Cambridge, MA, 1993, pp. 179–206.

[26] VQEG, 'Final report from the Video Quality Experts Group on the validation of objective models of video quality assessment', 2000; available at http://www.vqeg.org.

[27] V. Petrović and C. Xydeas, 'Gradient based multiresolution image fusion', *IEEE Transactions on Image Processing*, Vol. 13, No. 2, 2004, pp. 228–237.

[28] A. Toet, 'Hierarchical image fusion', *Machine Vision and Applications*, Vol. 3, 1990, pp. 3–11.

[29] G. Qu, D. Zhang and P. Yan, 'Information measure for performance of image fusion', *Electronic Letters*, Vol. 38, No. 7, pp. 313–315.

[30] P. Burt and R. Kolczynski, 'Enhanced image capture through fusion', in *Proc. 4th International Conference on Computer Vision*, Berlin, 1993, pp. 173–182.

18

Objectively adaptive image fusion

Vladimir Petrović and Tim Cootes

Imaging Science and Biomedical Engineering, University of Manchester, Oxford Road, Manchester, UK

18.1 Introduction

Multiple sensor modalities offer additional information and potentially improved robustness to a range of imaging applications but come at a price of a huge increase in raw data that needs to be processed. This can often overwhelm human observers and machine vision systems entrusted with the task of extracting useful information from the imagery. Image fusion combines information from a number of multi-sensor images into a single fused one, making processing more efficient and display of information more reliable.

A wide variety of image fusion algorithms have been proposed [1–7]. Arithmetic fusion, a weighted combination of the inputs, is the simplest approach. Under the right conditions it is capable of producing good results but in general it is unreliable. Multi-scale approaches, which currently dominate the literature [1–7], are considerably less efficient. They add robustness to the process by decomposing information into separate signals according to scale (and orientation), collectively called *image pyramids*. This allows easy overlay of features across scales from different inputs by constructing a new, fused pyramid from input representations. The manner in which image details are represented allows salient features to be identified and preserved in the fused image using spatially selective fusion strategies [1–7].

Most fusion algorithms, despite using complex rules for integrating input information, still overwhelmingly rely on assuming fixed properties of their inputs and pre-determine their parameters accordingly. Collectively the algorithms can be tuned to achieve optimal performance for a variety of applications but this has to be done in an offline manner and requires a representative set of training images [4,6,8]. If real conditions differ from those used to select the parameters, performance is unpredictable and usually sub-optimal. This is of particular concern during fusion of live video streams over long periods as input conditions may change considerably, for instance, the variation in scenes observed out of an airplane coming in to land at a busy city airport. Robustness is therefore an issue of increasing importance as fusion systems see more widespread and varied application.

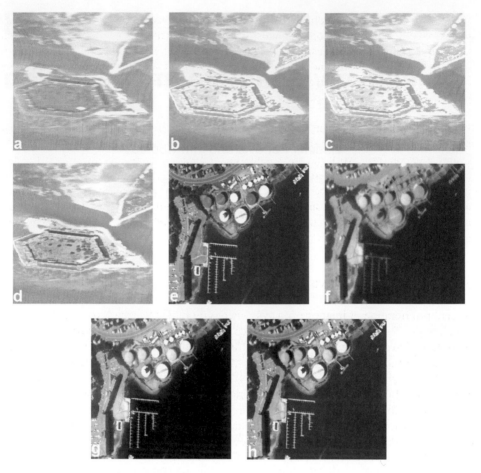

Figure 18.1   *Pre-set parameters provide sub-optimal fusion of (a) and (b) into (c), adapted parameters do better (d). A better match is achieved for fusion of (e) and (f), and only a small improvement is visible between (g) and (h).*

Figure 18.1 illustrates this on two pairs of multi-sensor images (a), (b) and (e), (f). Using parameters pre-tuned for a different multi-sensor image set (Figure 18.5) results in significant signal saturation and a loss of scene information for the first pair (c) while performance is close to optimal for the second pair (g) (both sets were fused using the BSF algorithm [7]).

This chapter presents techniques which aim to improve the robustness of general fusion algorithms by employing methods used in objective fusion evaluation to optimally adapt the fusion parameters to existing input conditions. Several unsupervised objective metrics (requiring no ideally fused image) have been demonstrated as relevant in evaluating fusion in a wide range of applications [9–16]. Using such independent metrics a fully automated, *objectively adaptive* framework for an online optimisation of the fusion process can be constructed. In particular, two distinct adaptive fusion mechanisms are considered: the 'forward adaptive' approach focusing on efficiency that merely guides the information fusion process and the 'feedback adaptive' approach which explicitly aims to optimise performance. The focus is on the general fusion architecture of the adaptive approach that may be applied to any fusion algorithm. We also demonstrate its potential in providing robustness in both still and sequence (video) fusion. Practical details of an objectively adaptive fusion system, such as the choice of objective metric and underlying fusion algorithm are left for a more specific consideration outside the scope of this chapter.

In the following, state-of-the-art objective fusion evaluation is considered as a crucial aspect of the proposed approach. This is followed by an introduction of the objectively adaptive fusion framework and an examination and evaluation of specific architectures to deal with still image and video fusion. Their advantages and weaknesses are demonstrated on representative, established fusion algorithms and realistic data sets from a variety of applications. Throughout, a specific focus will be on fusion for display which has broad appeal in a wide range of applications such as medical imaging, night vision, avionics, and remote sensing.

18.2 Objective fusion evaluation

Recent proliferation of image fusion systems has prompted the need for reliable ways of evaluating and comparing their performance. As a result, a plethora of fusion evaluation algorithms have emerged [8–17] clustered around a number of key ideas. The earliest attempts [3] focused on defining an *ideally* fused image against which the actual result could be compared using simple measures such as Root Mean Square Error (RMSE). Ideal fusion, however, is generally ill defined and only obtainable in very specific applications, such as fusion of multi-focus images. Another approach uses a more general, subjective fusion evaluation where representative audiences of observers are asked to perform some tasks with or simply view and evaluate fused imagery [8,17]. Although such trials produce reliable evaluation, they are highly impractical.

More recently, objective fusion metrics have emerged [9–16] to provide a computational alternative that requires no display equipment or complex organisation of an audience. Such metrics are fully automatic, performing evaluation based entirely on the inputs and the fused image. They do not need a ground truth, and produce a single numerical score reflecting fusion performance.

The earliest, an edge representation based metric $Q^{AB/F}$ [9,10] is based on the idea that a fusion algorithm that transfers input gradient information into the fused image more accurately performs better. For the fusion of input images A and B resulting in a fused image F, gradient strength $\mathbf{g}$ and orientation $\boldsymbol{\alpha}$ ($\in [0, \pi]$) are extracted at each location (n, m) from each image using the Sobel operator and used to define relative strength and orientation 'change' factors G and A, between each input and the fused image:

$$\left(G_{n,m}^{AF}, A_{n,m}^{AF}\right) = \left(\frac{g_{n,m}^{F}}{g_{n,m}^{A}}^{M}, 2\pi^{-1}\left||\alpha_{n,m}^{A} - \alpha_{n,m}^{F}| - \pi/2\right|\right) \tag{18.1}$$

where M is 1 for $g^{F} > g^{A}$ and -1 otherwise. An edge information preservation measure Q^{AF} models information loss between A and F with respect to the 'change' parameters with sigmoid functions defined by constants Γ, κ_g, σ_g, κ_α, and σ_α:

$$Q_{n,m}^{AF} = \frac{\Gamma}{\sqrt{(1 + e^{k_g(G_{n,m}^{AF} - \sigma_g)})(1 + e^{k_\alpha(A_{n,m}^{AF} - \sigma_\alpha)})}} \tag{18.2}$$

Total fusion performance $Q^{AB/F}$ is evaluated as a sum of local information preservation estimates between each of the inputs and fused, Q^{AF} and Q^{BF}, weighted by local perceptual importance factors w^A and w^B usually defined as local gradient strength:

$$Q^{AB/F} = \frac{\sum_{\forall n,m} Q_{n,m}^{AF} w_{n,m}^A + Q_{n,m}^{AF} w_{n,m}^B}{\sum_{\forall n,m} w_{n,m}^A + w_{n,m}^B} \tag{18.3}$$

$Q^{AB/F}$ is in the range $[0, 1]$ where 0 signifies complete loss of input information and $Q^{AB/F} = 1$ indicates 'ideal fusion' [9,10]. Augmentation of $Q^{AB/F}$ with estimates of visual differences was shown to slightly improve on its robustness [13]. A similar effect was observed when more abstract information obtained from robust image segmentation is introduced to evaluate information preservation estimates across important image regions [18].

Information theoretic measures such as entropy and mutual information have also been successfully used in fusion evaluation [12,14–16]. These metrics compare global image statistics between the inputs and the fused image and explicitly ignore local structure and spatial information. Despite this apparent shortcoming, when considering reasonable fusion algorithms (that aim to preserve spatial structure of the inputs), mutual information based on gradient strength statistics, for example, achieves very high levels of evaluation accuracy [16]. The basic approach defines the metric as the sum of mutual information (MI) estimates between the intensities in each input image and the resulting fused ($I_{F\{A,B\}}$):

$$M_F^{AB} = I_{FA}(F, A) + I_{FB}(F, B) \tag{18.4}$$

More recently, Tsallis entropy [15] and conditional entropy [14] have been suggested as more robust alternatives to Shannon entropy.

Another important fusion evaluation approach is based on the Universal Image Quality Index by Wang and Bovik [11]. Local image statistics are used to define a similarity Q_0 between images x and y,

$$Q_0 = \frac{4\sigma_{xy}\bar{x}\bar{y}}{(\bar{x}^2 + \bar{y}^2)(\sigma_x^2 + \sigma_y^2)} \tag{18.5}$$

for all 8×8 blocks (w) across the scene. These are then aggregated into three different metrics, Q_0, Q_w, and Q_e [11]. The most subjectively relevant Q_w [16] uses saliency weighting $\lambda(w)$ of local estimates Q_0 as defined in the equation

$$Q_w(a, b, f) = \sum_{w \in W} c(w)\big(\lambda(w)Q_0(a, f|w) + \big(1 - \lambda(w)Q_0(b, f|w)\big)\big) \tag{18.6}$$

where $Q_0(a, f|w)$ is defined according to (18.5) between input a and fused image f for local window w [11].

The Universal Image Quality Index approach considers local statistics and is somewhere between the highly localised $Q^{AB/F}$ and global statistical approaches such as mutual information. Q_w is thus considered alongside the original $Q^{AB/F}$ [10] and MI intensity metric [12] within the objectively adaptive fusion approach. All of the three metrics are representative of wider approaches in fusion evaluation and were shown to be subjectively meaningful against an extensive set of subjective trial results [16,17].

18.3 Objectively adaptive fusion

Two distinct approaches exist for adapting fusion parameters to the existing input conditions to optimise fusion performance using objective fusion evaluation. The approaches differ in the information used in adapting the parameters and the flow of the evaluation results. They can be summarised as follows:

- 'forward adaptive' or representation fusion based on an objective evaluation of input information only and
- 'feedback adaptive' or full optimisation fusion based on an objective evaluation of a full, provisional fused result.

Figure 18.2 illustrates these two approaches. In most fusion algorithms an estimate of image saliency such as local contrast or 'activity' is used to determine the relative importance of inputs across the scene. This information is then used to select the most relevant source for the fused image at any one location by adapting the appropriate fusion parameters, such as a selection map. Forward adaptive fusion shown in Figure 18.2(a) relies on information evaluation of objective fusion metrics as a more reliable and relevant manner of achieving this. Objective metrics incorporate robust mechanisms for evaluating and comparing information content proven against subjective fusion evaluation results [16,17]. A more accurate comparison of the inputs should result in better fused images and greater robustness of the entire fusion system. For this approach to function, however, an asymmetric evaluation metric M that satisfies condition

$$M(A, B) \neq M(B, A) \tag{18.7}$$

is required.

Multi-sensor imaging is founded on the idea that one sensor may show information that another is missing. When comparing fusion inputs directly symmetric metrics, such as mutual information, are unable to identify this and equality in (18.7) holds always making them unable to distinguish the usefulness of fusion inputs. Being able to identify which input is more relevant at which location in the scene would lead to more relevant feature selection maps and thus more robust fusion.

Conversely, the feedback adaptive approach (Figure 18.2(b)) explicitly optimises fusion parameters using a performance metric as an objective function. It starts with an initial

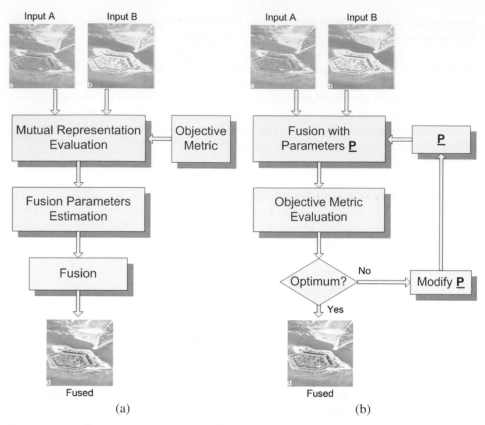

Figure 18.2 *Structure of the two objectively adaptive fusion approaches: (a) forward adaptive and (b) feedback adaptive.*

set of fusion parameters and produces a proxy fused image that is then evaluated using an objective metric. Fusion parameters are then modified according to the score using an optimisation algorithm and the process is repeated until performance converges. This should guarantee that an optimal solution will be found within the limits of the chosen objective metric and optimisation algorithm. High robustness is an explicit advantage of feedback adaptive fusion as it is by definition able to cope with any input conditions, while inefficiency is undoubtedly its main disadvantage. Whereas there are no theoretical limits to which fusion parameters can be optimised, practical considerations impose some restrictions. Table 18.1 summarises potential application scenarios for the proposed approaches within the contexts of still image and continuous (video) fusion.

18.3.1 Optimal still image fusion

Most of the advanced still image fusion systems today employ some form of multi-scale analysis [1–7,19], as fusion efficiency is usually forgone for the benefit of robust representation and optimal integration of features across scales. A crucial step in such a fusion process is *pyramid fusion* [5,7] that evaluates input features and determines the

Table 18.1 *Application fields of the two objectively adaptive fusion approaches.*

Application	Approach	
	Forward adaptive	Feedback adaptive
Still image fusion	Local evaluation of inputs for optimal feature selection and fusion parameters	Full fusion parameter optimisation with final fused image objective score as the objective function
Continuous (video) fusion	Local evaluation of inputs for optimal feature selection and fusion parameters	Running optimisation of key fusion parameters in time with current frame fusion score as objective function for next frame

mechanisms for their transfer from the input to a new fused multi-scale pyramid. The goal is usually to reproduce the most important, normally identified as the most salient, input features at all locations and scales across the scene. In the detail (high frequency) sub-bands of the pyramid, either a binary selection,

$$F(n, m, s) = \begin{cases} A(n, m, s) & \text{if } P(n, m, s) = 1, \\ B(n, m, s) & \text{if } P(n, m, s) = 0 \end{cases} \tag{18.8}$$

where (n, m) is the location and s the scale, or an arithmetic combination of the inputs,

$$F(n, m, s) = A(n, m, s)k_A + B(n, m, s)k_B \tag{18.9}$$

is used to construct the fused value. Low-pass residual signals containing coarse global features meanwhile are usually fused as arithmetic combinations with arbitrarily determined parameters, e.g. averaging (Equation (18.9)). Although more robust importance evaluation rules have emerged over the years [4,6] most still rely on simple rules such as (18.8) and (18.9) to transfer information into the fused image.

Both P and $\underline{k} = [k_A \; k_B]$ are parameters of the fusion process and can be optimised with respect to an objective metric in a full feedback adaptive fusion arrangement, see Figure 18.2. In practice these parameters are relatively independent of each other and can be optimised separately. In the case of residual signals arithmetic weights k_A and k_B can be optimised using any approach applicable to 2-dimensional variables. Figure 18.3 shows two of the most common optimisation surfaces for residual fusion using $OF = 1 - Q^{AB/F}$ as the objective function and F according to (18.9). They correspond to a case where inputs are significant but opposite (e.g., photographic negatives) – Figure 18.3(a) and a case where one input is dominant or they are similar – Figure 18.3(b). In both cases the surface is smooth and the optimum is easily reached within a couple of steps of the simple pattern search algorithm initialised from the middle point ($\underline{k} = [0.5 \; 0.5]$) (see the dots on the surface of plot 3(b)).

The same feedback approach can be applied to arithmetic or binary fusion of pixels in detail sub-bands whereby the objective metric is evaluated only for the small area around each location. This does not eliminate all dependence on neighbouring pixels and the optimisation should run over several iterations or until all coefficients converged (a potentially long process). The system can be bootstrapped by setting the initial positions for

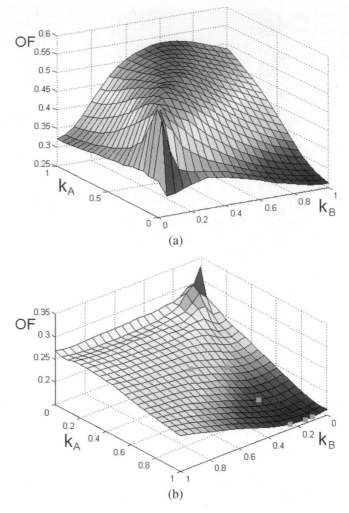

Figure 18.3 *Optimisation surfaces for arithmetic low-pass residual fusion, using* $1 - Q^{AB/F}$ *as the objective function.*

optimisation according to the result of the conventional selection approach likely to be relatively close to the optimum, e.g. select absolute max.

Forward adaptive fusion provides a more efficient alternative, particularly for large images. Mutual input evaluation, see Figure 18.2, is provided by $Q^{AB/F}$ [10] which satisfies the asymmetric evaluation requirement in (18.7) when used in the form $Q^{AA/B}$ and $Q^{BB/A}$, and with M set to 1 in Equation (18.1). Equation (18.2) then gives pixel-wise mutual representation estimates weighted using local importance given by local gradient amplitude to define how well each input represents the other: $Q^{AA/B} > Q^{BB/A}$ means input A represents the information in B and the actual scene better than the other way around. These estimates can be used directly to guide the feature selection process and ensure an optimal source is selected for the fused image at each location and scale. In

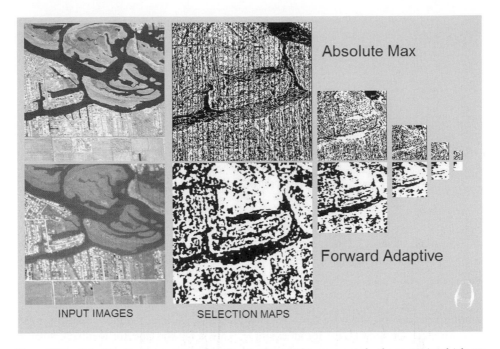

Figure 18.4 *Selection maps across a number of scales for a multi-sensor pair by the conventional 'choose absolute max' (top) and proposed forward adaptive framework (below).*

practice, $Q^{AA/B}$ and $Q^{BB/A}$ are only defined at the highest resolution level and for integration to different scale levels in multi-scale fusion are propagated by standard Gaussian filtering and decimation.

Figure 18.4 illustrates feature selection by the forward adaptive framework compared to the conventional 'choose absolute max' for the case of multi-resolution differential DWT fusion [6] of a pair of multi-sensor images. White indicates pixels drawn from top and black pixels drawn from the bottom input. Objective input evaluation achieves a much more stable selection pattern that better corresponds to meaningful image regions compared to the 'salt and pepper' selection pattern provided by absolute max selection.

Figure 18.5 shows the performance gains achieved using adaptive fusion for a number of established fusion algorithms: Bi-Scale fusion [7], Contrast Pyramid [1], conventional wavelet fusion DWT [2,3], Gradient Pyramid GP [20], and Laplacian Pyramid fusion. In each case the reference was conventional select absolute max fusion. All were evaluated on a data set illustrated in Figure 18.6 containing 171 multi-sensor image pairs containing a wide variety of representative scenes and contents as well as sensor combinations. In order to obtain a relevant assessment and avoid measuring the results with metrics explicitly optimised during fusion, fusion evaluation was performed using a further gradient MI metric based on mutual information between input and fused gradient strength images shown to be the most subjectively relevant [16].

Forward adaptive fusion improves performance for all of the systems considered with the biggest gains obtained for the simplest and perhaps least robust BSF system [7]. Improved

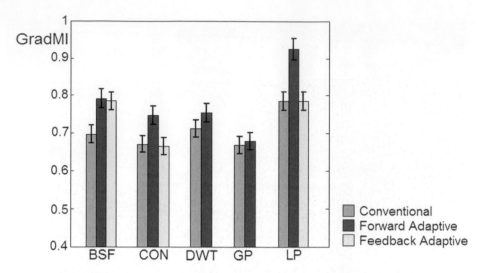

Figure 18.5 *Fusion performance of objectively adaptive fusion compared to the conventional approach.*

feature selection provided by this approach reconstructs input features faithfully in the fused image; see Figure 18.1 for an example. This is not the case for feedback adaptive fusion which generally fails to improve on the reference system apart from simple BSF fusion. The most likely cause is that the optimisation falls into a local minimum around the initialisation provided by the select max approach.

Figure 18.7 shows an example of fusion using the conventional select max [1] and the forward adaptive contrast pyramid fusion. The adaptive framework significantly improves the performance of the conventional algorithm, by reconstructing the roof structures on the houses as well as exhibiting clearer boundaries between meaningful segments in the scene.

18.3.2 Optimal video fusion

Image sequence or video fusion is an application of great practical importance. Efficiency is vital as up to 30 fused images are produced each second which precludes the use of some algorithms that work well on still images, most importantly many of the multi-scale approaches. We consider efficient Bi-Scale Fusion (BSF) scheme specifically designed for video fusion, Figure 18.8 [7], within the continuous adaptive fusion framework. BSF decomposes images into only two ranges of scale using sub-optimal but extremely efficient uniform templates with spectral separation directly determined by the size of the template. Residual low pass images are fused as a weighted sum, Equation (18.9), while detail images are combined in a select absolute max fashion [7].

The performance of this simple scheme is highly dependant on its parameters: the spectral separation (template size) and residual fusion weights. Parameter values that provide optimal performance depend on the inputs and can be adaptively optimised in a feedback adaptive fashion, Figure 18.8. This would ensure that parameters remain optimal

Figure 18.6 *Example images from the multi-sensor still image dataset used in the tests.*

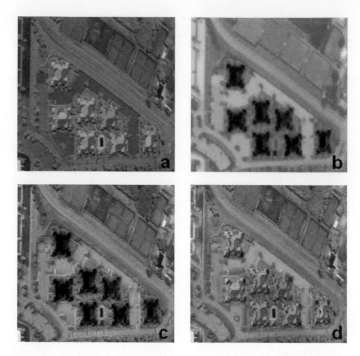

Figure 18.7   *Input images (a and b) and images fused using the Contrast Pyramid in a conventional arrangement (c) and proposed forward adaptive approach (d).*

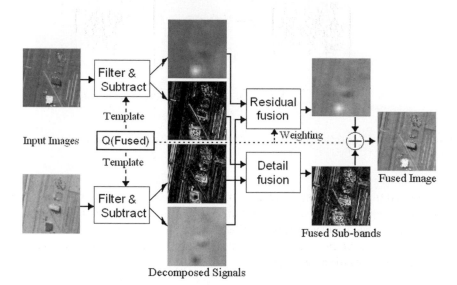

Figure 18.8   *Structure of the standard (BSF) (solid lines) and the adaptive bi-scale fusion (also dashed lines).*

throughout the sequence even when input conditions change. As inputs usually change slowly due to high imaging dynamics the parameters are optimised every single and every N frames. Residual fusion optimisation, as that described for still images, is performed

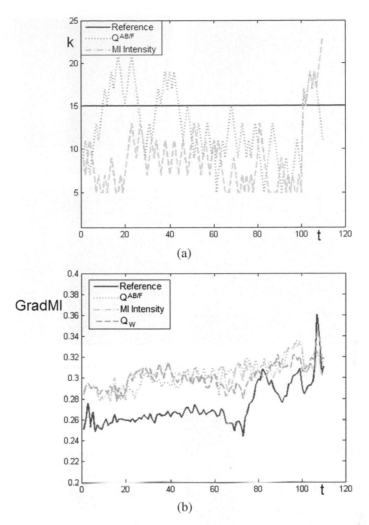

Figure 18.9 *Optimisation of template size against frame number (a) and resulting fusion performance (b) for sequence MS01.*

every $N = 20$ frames and $\underline{k}$ kept constant for the frames in between. Furthermore, a number of fixed template values are applied, evaluated using the objective metric and the best one chosen for the subsequent frames. This helps avoid local minima while a frame-to-frame optimisation of the size using step size 2 and the metric as the objective function drives the value towards a local optimum.

Figure 18.9(a) shows template size against frame number for an adaptively fused, well known MS01 sequence. Corresponding fusion performance achieved using the three objective metrics and measured by the gradient MI metric is shown in Figure 18.9(b). The reference produced with a fixed template and averaging between residuals is shown as the solid line on both plots. The adaptive framework makes a significant impact on fusion performance of BSF on sequence MS01. The parameters adapt to the changes in the

Figure 18.10   *Fusion of a multi-sensor frame (a and b) from the sequence MS01, by the reference (c) and adaptively optimised (d).*

Table 18.2   *Gradient MI results for adaptive fusion of several well-known multi-sensor sequences (Kayak sequences had two different sensors combinations).*

Sequence	Reference (non-adaptive)	Adaptive $Q^{AB/F}$	MI adaptive	Q_w adaptive
MS01	0.2730	0.3028	0.3041	0.3012
Uncamp	0.3189	0.3243	0.2950	0.3302
Kayak1 (1)	0.3460	0.4256	0.3996	0.4148
Kayak1 (2)	0.2130	0.2525	0.2488	0.2487
Kayak2 (1)	0.1279	0.1555	0.1480	0.1552
Kayak2 (2)	0.1298	0.1460	0.1436	0.1410
Dublin	0.7553	0.6930	0.7345	0.7693

inputs and fusion performance remains better than the reference system throughout the sequence. Figure 18.10 demonstrates fusion of a frame from this sequence. Adaptively fused image (Figure 18.10(d)) exhibits significantly better contrast and sharper details than the one obtained by the reference system (Figure 18.10(c)).

Numerical results of mean gradient MI measured performance for a number of multi-sensor sequences are provided in Table 18.2. In almost all cases the objectively adaptive approach improves the performance compared to the non-adaptive case. This demonstrates the advantage of introducing the adaptation of fusion parameters to deal with changing input conditions and how it adds robustness to the fusion process. If the forward adaptive framework, based on $Q^{AB/F}$, is also applied to feature selection the performance improves further, e.g. MS01 to around 0.305 for all three adaptive systems.

The exception is the AIC Thermal/Visible Night-Time Dataset sequence [21,22] where only one of the schemes improves performance. This demonstrates an important weak-

Figure 18.11 *Example from the 'Dublin' sequence: inputs (a and b); $Q^{AB/F}$ adaptive fused (c); and Q_w adaptive fused (d).*

ness of the objective metrics and derived objectively adaptive fusion. The IR input in this sequence suffers from a low SNR that confuses the metrics. They attempt to recreate the noise in the fused image causing fusion parameters to diverge from their optimal values. Figure 18.11 shows fusion of an example frame from this sequence. Noisy artefacts can be clearly seen on the $Q^{AB/F}$ adaptive fused (Figure 18.11(c)), while they are much less apparent when Q_w adapts the parameters (Figure 18.11(d)), although some important information (person in the foreground) is almost missing.

18.4 Discussion

This chapter has shown that an objectively optimised, adaptive image fusion approach can improve the performance and robustness of fusion schemes. By employing the methods used in objective fusion evaluation, parameters can be adapted to current input conditions in order to achieve optimal fusion performance.

On still images, of the two approaches presented, the 'forward adaptive' is the more robust despite no explicit guarantee that it will lead to a global optimum of fusion parameters. Feedback adaptive fusion, employing a full optimisation approach was shown to be too prone to local minima when attempted on spatially specific higher frequency details. Optimal results were achieved when forward adaptive approach was applied to fusion of detailed smaller scale information while residual combination parameters were optimised in a feedback adaptive loop. This indicates that an accurate evaluation of the detail information contained in the inputs is crucial for robust image fusion.

Feedback optimisation of fusion parameters was shown to be effective on fusion of image sequences where high dynamics were used to make the process more efficient by optimising parameters only every N frames or from one frame for the next. All metrics were shown to improve fusion performance and adapt the parameters to changing conditions in the inputs successfully. Here too, forward adaptive fusion presents an advantage although it is not clear whether a realistic real time application is possible using the $Q^{AB/F}$ metric to evaluate the inputs.

Finally, a weakness in the objective fusion evaluation of fusion was identified on the example of noisy input data. Noise is a problem of image fusion in general and robust mechanisms for avoiding the situation where noise dominates the fused image are certain to be a topic of research in the future.

Although this chapter provided only a proof of concept, the simple and efficient schemes shown were found to provide much improved fusion robustness. However, the demonstrated systems are merely one take on a broader approach of adaptive fusion. Other, potentially more robust, optimisation and adaptation strategies surely exist and can be applied in conjunction with other objective metrics and fusion algorithms.

Acknowledgements

The authors would like to acknowledge TNO in the Netherlands, University of Bristol, University of Dublin and Defence Research Authority Valcartier for the use of their imagery in this work.

References

[1] A. Toet, 'Hierarchical image fusion', *Journal of Machine Vision and Applications*, Vol. 3, 1990, pp. 3–11.
[2] S. Nikolov, P. Hill, D. Bull and C. Canagarajah, 'Wavelets for image fusion', in A. Petrosian and F. Meyer (eds.), *Wavelets in Signal and Image Analysis*, Kluwer Academic Publishers, Dordrecht, 2001, pp. 213–244.
[3] H. Li, B. Munjanath and S. Mitra, 'Multi-sensor image fusion using the wavelet transform', *Journal of Graphical Models and Image Processing*, Vol. 57, No. 3, 1995, pp. 235–245.
[4] Z. Zhang and R. Blum, 'A categorization of multi-scale-decomposition-based image fusion schemes with a performance study for a digital camera application', *Proceedings of the IEEE*, Vol. 87, No. 8, 1999, pp. 1315–1326.
[5] V. Petrović and C. Xydeas, 'Gradient based multiresolution image fusion', *IEEE Transactions on Image Processing*, Vol. 4, No. 2, 2004, pp. 163–183.
[6] G. Piella, 'A general framework for multiresolution image fusion: from pixels to regions', *International Journal of Information Fusion*, Vol. 4, 2003, pp. 258–280.

[7] V. Petrović and C. Xydeas, 'Computationally efficient pixel-level image fusion', in *Proceedings of Eurofusion 99*, Stratford-upon-Avon, October 1999, pp. 177–184.

[8] A. Toet and E. Franken, 'Perceptual evaluation of different image fusion schemes', *Displays*, Vol. 24, 2003, pp. 25–37.

[9] C. Xydeas and V. Petrović, 'Objective image fusion performance measure', *Electronics Letters*, Vol. 36, No. 4, 2000, pp. 308–309.

[10] V. Petrović and C. Xydeas, 'Objective evaluation of signal-level image fusion performance', *Optical Engineering*, Vol. 44, No. 8, 2005, 087003.

[11] G. Piella and H. Heijmans, 'A new quality metric for image fusion', in *Proceedings of the International Conference on Image Processing*, Vol. 3, 2003, pp. 173–176.

[12] G. Qu, D. Zhang and P. Yan, 'Information measure for performance of image fusion', *Electronics Letters*, Vol. 38, No. 7, 2002, pp. 313–315.

[13] V. Petrović and C. Xydeas, 'Evaluation of image fusion performance with visible differences', in *Proceedings of ECCV, Prague, 2004*, in *Lect. Notes Comput. Sci.*, Vol. 3023, 2004, pp. 380–389.

[14] V. Tsagaris and V. Anastassopoulos, 'Global measure for assessing image fusion methods', *Optical Engineering*, Vol. 45, No. 2, 2006, 026201.

[15] N. Cvejić, C. Canegarajah and D. Bull, 'Image fusion metric based on mutual information and Tsallis entropy', *Electronics Letters*, Vol. 42, No. 11, 2006, pp. 626–627.

[16] V. Petrović and T. Cootes, 'Information representation for image fusion evaluation', in *Proceedings of Fusion 2006*, Florence, ISIF, July 2006.

[17] V. Petrović, 'Subjective tests for image fusion evaluation and objective metric validation', *Information Fusion*, Vol. 8, No. 2, 2007, pp. 208–216.

[18] N. Cvejić, D. Bull and C. Canegarajah, 'A new metric for multimodal image sensor fusion', *Electronics Letters*, Vol. 43, No. 2, 2007, pp. 95–96.

[19] O. Rockinger and T. Fechner, 'Pixel level image fusion: The case of image sequences', *Proceedings of the SPIE*, Vol. 3374, 1998, pp. 378–388.

[20] P. Burt and R. Kolczynski, 'Enhanced image capture through fusion', in *Proceedings of the 4th International Conference on Computer Vision*, Berlin, 1993, pp. 173–182.

[21] 'The online resource for research in image fusion', http://www.imagefusion.org, 2005.

[22] C. O'Conaire, N. O'Connor, E. Cooke and A. Smeaton, 'Comparison of fusion methods for thermo-visual surveillance tracking', in *International Conference on Information Fusion*, ISIF, 2006.

19

Performance evaluation of image fusion techniques

Qiang Wang, Yi Shen and Jing Jin

Department of Control Science and Engineering, Harbin Institute of Technology, P.R. China

Image fusion performance evaluation methods form an essential part in the development of image fusion techniques. This chapter discusses various performance evaluation measures that have been proposed in the field of image fusion and also analyses the effects of fusion structures on the outcomes of fusion schemes. Indicative experiments on applying these measures to evaluate a couple of widely used image fusion techniques are also presented to demonstrate the usage of the measures, as well as to verify their correctness and effectiveness.

19.1 Introduction

Since image fusion techniques have been developing fast in various types of applications in recent years, methods that can assess or evaluate the performance of different fusion technologies objectively, systematically, and quantitatively have been recognised as an urgent requirement.

As far as image fusion is concerned, the acceptable quality for the fused image is set by the receiver of the image which is usually the human observer. Therefore, quality assessments of fused images are traditionally carried out by visual analysis. At an early stage in image fusion investigations, it was recognised that the quality of research in image fusion would be enhanced by an understanding of the human visual system so that an image quality measure could be found that would correspond to the way in which the human visual system assesses image fidelity. Many attempts have been made to derive suitable distortion measures based on the various models of the human visual system, producing a large array of candidate measures of image quality. However, despite these investigations, an accurate visual model and hence, an accurate distortion measure, has not been found, since visual perception is still not well understood.

Since the most appropriate test of image fusion is whether the quality of the fused image is adequate, a decision best made by the viewer, some image fusion systems are assessed through a subjective rating methodology.

However, in general, the receiver of the fused images will not be a human viewer but some form of automated image processing system. The major difficulty is that the fused image will exhibit some loss of information in comparison to the individual input (source) images. The severity of this loss is explicitly related to the image fusion application. In other words, the information lost in one particular application may be necessary for another. Therefore, a general performance measure is required that can be applied, even if the application of image fusion is unknown in advance.

So far, only a few objective and quantitative performance metrics have been developed.

Li et al. [1] defined the standard deviation of the difference between the ideal image (ground truth) and the fused image to be a performance measure. The basic idea of the method is to create simulation environments for which the ideal fusion result is known. Other statistical measurements from digital signal analysis such as Signal-to-Noise Ratio (SNR), Peak Signal-to-Noise Ratio (PSNR), and Mean Square Error (MSE) are also commonly used measures in assessing image fusion techniques in cases where the ground truth image is available [2–4].

However, in a practical situation, an ideal image is not available. In [5] two new parameters, the Fusion Factor (FF) and the Fusion Symmetry (FS), have been presented and provide useful guidelines in selecting the best fusion algorithm.

A Mutual Information (MI) criterion has been used as the measure for evaluating the performance of a fusion method in [6]. MI represents how much information has been transformed from the source images to the fused image.

Xydeas and Petrovic [7] proposed a framework for measuring objectively image fusion performance on a pixel-level basis, which is explicitly related to perceptual information. In [7] the visual information is associated with the 'edge' information, while the region information is ignored.

Wang and Shen [8] have proposed a Quantitative Correlation Analysis (QCA) method to evaluate the performances of hyperspectral image fusion techniques. Moreover, a fast version of the QCA method has been proposed, useful in cases where the number of source images increases and the size of the image expands [9].

The perceptual approach in [7] can only assess image fusion techniques that fuse two source images only into a single image. On the other hand, the QCA and fast QCA method can assess image fusion techniques which are applied on multiple source images and also produce potentially multiple output images. However, they are based on the Correlation Information Entropy (CIE) which is developed from the concept of the correlation coefficient which can only describe linear correlation between two variables.

In order to overcome the drawbacks of the QCA, the authors in [10,11] have presented a performance evaluation method for different image fusion techniques based on nonlinear correlation measures, which investigates a general type of relationship between the source images and the fused images and not only linear relationships.

Other schemes that concentrate on specific applications can be also found in the literature [12,13].

This chapter is organised as follows. Firstly, we will introduce some statistical measures such as the SNR, PSNR, and MSE, which require an ideal or reference image when applied. Secondly, in Section 19.3, nonlinear correlation measures such as the MI, FF, and FS will be discussed. Section 19.4 discusses an edge information based objective measure. Section 19.5 discusses the effects of fusion structures on the performance of image fusion algorithms. Two evaluation schemes, i.e. the Nonlinear Correlation Analysis and the Information Deviation Analysis, which can be applied to fusion techniques that handle multiple input and multiple output images, are introduced in Section 19.6. A discussion is given at the end.

19.2 Signal-to-Noise-Ratio (SNR), Peak Signal-to-Noise Ratio (PSNR) and Mean Square Error (MSE)

Signal-to-Noise Ratio (SNR), Peak Signal-to-Noise Ratio (PSNR), and Mean Square Error (MSE) are commonly used measures in assessing image fusion techniques, that consider an image as a special type of signal. The quality of a signal is often expressed quantitatively with the signal-to-noise ratio defined as [2,3]

$$SNR - 10\log_{10}\left(\frac{\text{Energy}_{\text{signal}}}{\text{Energy}_{\text{noise}}}\right) \qquad (19.1)$$

where $\text{Energy}_{\text{signal}}$ is the sum of the squares of the signal values and $\text{Energy}_{\text{noise}}$ is the sum of the squares of the noise samples. In the context of a signal estimation algorithm, the signal refers to the estimated signal and the noise to the difference (error) between the estimated and the original signal. SNR is unitless and therefore independent of the data units. As far as the image is concerned, the SNR can be written as

$$SNR = 10\log_{10}\frac{\sum_{m=1}^{S_1}\sum_{n=1}^{S_2}z(m,n)^2}{\sum_{m=1}^{S_1}\sum_{n=1}^{S_2}[z(m,n)-o(m,n)]^2} \qquad (19.2)$$

where $z(m,n)$ and $o(m,n)$ denote the intensity of the pixel of the estimated and original image, respectively, at location (m,n). The size of the images is $S_1 \times S_2$. High values of SNR show that the error of the estimation is small and, therefore, among various image fusion methods the ones that exhibit higher SNR's can be considered of better performance.

Figure 19.1 *(a) Original AVIRIS image; (b) test image with noise added to the upper-left corner; (c) test image with noise added to the lower-left corner.*

The PSNR and the MSE are measures similar to the SNR defined as [4]

$$\text{PSNR} = 10\log_{10}\frac{255^2}{\sum_{m=1}^{S_1}\sum_{n=1}^{S_2}[z(m,n)-o(m,n)]^2} \tag{19.3}$$

$$\text{MSE} = \frac{\sum_{m=1}^{S_1}\sum_{n=1}^{S_2}[z(m,n)-o(m,n)]^2}{255^2} \tag{19.4}$$

When assessing the performance of an image fusion technique using the above mentioned measurements, we require knowledge of the original image (ground truth). For that reason these measurements can be used only with synthetic (simulated) data.

The above measurements exhibit the drawback of providing a global idea regarding the quality of an image. In cases where the fused image exhibits artefacts concentrated within a small area, these measurement can still produce an acceptable value even if the image is visually unacceptable.

19.2.1 Experiment

In order to evaluate the capabilities of the fusion measures presented, two test images are generated from a remote sensing image with varying SNR's and types of backgrounds. The original image used is from an AVIRIS (Airborne Visible/Infrared Imaging Spectrometer) data set which is downloaded from LARS (Laboratory for Applications of Remote Sensing) at Purdue University. More specifically, the original image is a portion of an AVIRIS data taken in June 1992, which covers a mixture of agricultural/forestry land in the Indian Pine Test Site in Indiana. To create two test images to be fused, two noise realisations with uniform distribution within the interval (0.0, 1.0) are first scaled by different constant parameters, then convolved with different lowpass filters, and finally added into two different regions, i.e. the upper-left and lower-left corners of the original image. Figure 19.1(a) provides the original AVIRIS image, and Figures 19.1(b), 19.1(c) give the two distorted images denoted by S_A and S_B.

The test images are fused using two widely used algorithms known to exhibit generally good performance. In the first algorithm the fused image F_a is simply the average of the

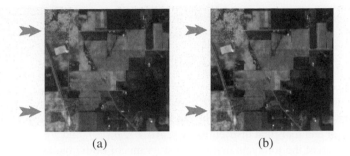

(a)   (b)

Figure 19.2 *Fusion results of test images using correlated noise added to the original image. (a) Average image; (b) WTF fused image.*

Table 19.1 *Measures for the fused test images containing correlated noise.*

Fusion method	Average image	WTF fused image
SNR	28.4023	29.0421
PSNR	165.0642	165.8180
MSE	6.7819e$-$008	6.2894e$-$008

two source images S_A and S_B, i.e. $F_a = (S_A + S_B)/2$. The second algorithm is a Wavelet Transform based Fusion (WTF) described in [8,14] which produces a fused image F_w. The resulting fused images using the two algorithms are displayed in Figure 19.2.

Values of the described measures are illustrated in Table 19.1. Clearly, the performance of the wavelet transform based method is better than that of the average based method according to Table 19.1. This argument can also be verified visually from Figure 19.2.

19.3 Mutual Information (MI), Fusion Factor (FF), and Fusion Symmetry (FS)

The measures introduced in the above section are mostly based on the quantitative calculation of the pixel deviation between the original image and the fused image. However, one goal of image fusion is to integrate complementary information from multiple sources so that the fused images are more suitable for the purpose of human visual perception and computer processing. Therefore, a measure should also estimate how much information is obtained from the individual input images. In this section mutual information (MI) is presented as a means of assessing image fusion performance.

It is well known that mutual information is a concept from information theory measuring the statistical dependence between two random variables or, in other words, the amount of information that one variable contains about the other. Let A and B be two random variables with marginal distributions $p_A(a)$ and $p_B(b)$ and joint probability distribution $p_{AB}(a, b)$. Mutual information measures the degree of dependence of the two random variables A and B. It is defined as follows [6]:

$$\text{MI}_{AB}(a, b) = \sum_{a,b} p_{AB}(a, b) \log \frac{p_{AB}(a, b)}{p_A(a) p_B(b)} \tag{19.5}$$

Considering the image intensity values a and b of a pair of corresponding pixels in two images to be samples generated from the random variables A and B, respectively, estimation of the joint and marginal distributions $p_{AB}(a, b)$, $p_A(a)$, and $p_B(b)$ can be obtained by normalisation of the joint and marginal histograms of both images [6].

The fused image should contain the important information from all of the input (source) images. Obviously the notion of 'important information' depends on the application and is difficult to define. Mutual information is the amount of information that one image contains about another. This inspires us to employ the mutual information as a measure of image fusion performance. Considering two input images A, B, and a fused image F, we can calculate the amount of information that F contains about A and B, according to (19.5):

$$\text{MI}_{FA}(f, a) = \sum_{f,a} p_{FA}(f, a) \log \frac{p_{FA}(f, a)}{p_F(f) p_A(a)} \tag{19.6}$$

$$\text{MI}_{FB}(f, b) = \sum_{f,b} p_{FB}(f, b) \log \frac{p_{FB}(f, b)}{p_F(f) p_B(b)} \tag{19.7}$$

Thus, an image fusion performance measure can be defined as

$$\text{MI}_F^{AB} = \text{MI}_{FA}(f, a) + \text{MI}_{FB}(f, b) \tag{19.8}$$

Equation (19.8) indicates that the proposed measure reflects a total amount of mutual information that the fused image F contains about A and B.

The authors in [5] also use the MI measurement with the name Fusion Factor (FF) and state that large FF indicates that more information has been transferred from the source images to the fused image. However, they point out that large FF still cannot indicate whether the source images are fused symmetrically. Therefore, they develop a concept called Fusion Symmetry (FS) given in the equation

$$\text{FS} = \text{abs}\left(\frac{\text{MI}_{FA}(f, a)}{\text{MI}_{FA}(f, a) + \text{MI}_{FB}(f, b)} - 0.5 \right) \tag{19.9}$$

to denote the symmetry of the fusion process in relation to two input images. The smaller the FS, the better the fusion process performs.

Based on their definition, the FF has to be given importance, when one of the two sensors is inferior. When both sensors are of high quality, then the FS parameter is also of importance and an algorithm with relatively smaller FS has to be chosen.

Figure 19.3 *Two AVIRIS source images to be fused.*

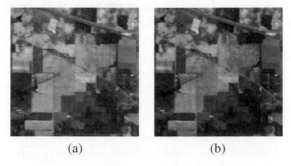

(a)   (b)

Figure 19.4 *The fused images using the averaging and the WTF fusion methods. (a) Average; (b) WTF fused.*

Table 19.2 *Performance evaluation results using MI, FF, and FS.*

Fusion method		Average image	WTF fused image
MI	S_A	2.3202	2.9627
	S_B	1.4969	1.2421
FF		3.8172	4.2049
FS		0.1078	0.2045

The above measures in their current form are developed to assess the fusion of two source images only. However, with proper modifications they could also assess the performance of a fusion technique applied to multiple source images.

19.3.1 Experiment

In this experiment, two source images are selected from the AVIRIS data set and they are displayed in Figure 19.3. The two image fusion algorithms to be analysed and compared are the same as those in Experiment 1. The fused images corresponding to the averaging method and the WTF method are displayed in Figure 19.4. The performance evaluation results using MI, FF, and FS measurements are presented in Table 19.2.

It can be seen from Table 19.2 that the FF of the WTF fused image is larger than that of the average image. As the FF indicates how much information is extracted from the source images, we can say that the WTF method performs better than the average method.

Moreover, as the average image is generated by taking equal information form the source images, its FS is larger than that of the WTF fused image, which is generated using a more complicated information extraction method. Thus, we can say that the performance evaluation results conform with the theory behind the proposed techniques, as we expected.

19.4 An edge information based objective measure

A measure for objectively assessing pixel level fusion performance derived in [7] is presented in this section. The goal in pixel level image fusion is to combine and preserve in a single output image all the 'important' visual information that is present in the input images. Thus, an objective fusion measure should extract all the perceptually important information that exists in the input images and measure the ability of the fusion process to transfer as accurately as possible this information into the output image. In [7] visual information is associated with the 'edge' information that is present in each pixel of an image. Note that this visual to edge information association is supported by human visual system studies and is extensively used in image analysis and compression systems. Furthermore, by evaluating the amount of edge information that is transferred from the individual input images to the fused image, a measure of fusion performance can be obtained. More specifically, consider two input images A and B and a resulting fused image F. A Sobel edge operator is applied to yield the edge strength $g(m, n)$ and orientation $a(m, n)$ information for each image location (pixel) (m, n), $1 \leqslant m \leqslant M$ and $1 \leqslant n \leqslant N$. Thus, for an input image A

$$g_A(m, n) = \sqrt{S_A^x(m, n)^2 + S_A^y(m, n)^2} \qquad (19.10)$$

$$a_A(m, n) = \tan^{-1} \frac{S_A^y(m, n)}{S_A^x(m, n)} \qquad (19.11)$$

where $S_A^x(m, n)$ and $S_A^y(m, n)$ are the responses of the Sobel masks centred at location (m, n).

The relative edge strength and orientation values $G^{AF}(m, n)$ and $A^{AF}(m, n)$ of an image A with respect to an image F at location (m, n) are formed as

$$G^{AF}(m, n) = \begin{cases} \frac{g_F(m,n)}{g_A(m,n)}, & g_A(m, n) > g_F(m, n), \\ \frac{g_A(m,n)}{g_F(m,n)}, & \text{otherwise} \end{cases} \qquad (19.12)$$

$$A^{AF}(m, n) = 1 - \frac{|a_A(m, n) - a_F(m, n)|}{\pi/2} \qquad (19.13)$$

These are used to derive the edge strength and orientation preservation values given below

$$Q_g^{AF}(m, n) = \frac{\Gamma_g}{1 + e^{\kappa_g(G^{AF}(m,n) - \sigma_g)}} \qquad (19.14)$$

$$Q_a^{AF}(m,n) = \frac{\Gamma_a}{1 + e^{\kappa_a(A^{AF}(m,n) - \sigma_a)}} \tag{19.15}$$

where $Q_g^{AF}(m,n)$ and $Q_a^{AF}(m,n)$ model the perceptual loss of information in F, in terms of how well the edge strength and orientation values of a pixel (m,n) in A are represented in the fused image. The constants $\Gamma_g, \kappa_g, \sigma_g$ and $\Gamma_a, \kappa_a, \sigma_a$, determine the exact shape of the sigmoid functions used to form the edge strength and orientation preservation values. Edge information preservation values are then defined as

$$Q^{AF}(m,n) = Q_g^{AF}(m,n) Q_a^{AF}(m,n) \tag{19.16}$$

with $0 \leqslant Q^{AF}(m,n) \leqslant 1$. A value of 0 corresponds to complete loss of edge information at location (m,n) while transferred from A into F. $Q^{AF}(m,n) = 1$ indicates edge information transferred from A to F without loss.

Having $Q^{AF}(m,n)$ and $Q^{BF}(m,n)$ for $M \times N$ size images, a normalised weighted performance metric $Q_P^{AB/F}$ of a given fusion process P that operates on images A and B and produces F is obtained as follows [7]:

$$Q_P^{AB/F} = \frac{\sum_{m=1}^{M} \sum_{n=1}^{N} (Q^{AF}(m,n) w^A(m,n) + Q^{BF}(m,n) w^B(m,n))}{\sum_{m=1}^{M} \sum_{n=1}^{N} (w^A(m,n) + w^B(m,n))} \tag{19.17}$$

Note that the edge preservation values $Q^{AF}(m,n)$ and $Q^{BF}(m,n)$ are weighted by $w^A(m,n)$ and $w^B(m,n)$, respectively. In general, edge preservation values which correspond to pixels with high edge strength should influence $Q_P^{AB/F}$ more than those of relatively low edge strength. Thus, $w^A(m,n) = [g_A(m,n)]^L$ and $w^B(m,n) = [g_B(m,n)]^L$, where L is a constant. Also note that $0 \leqslant Q_P^{AB/F} \leqslant 1$.

19.5 Fusion structures

In the case of multiple source images, these can be fused in various ways, but different fusion sequences will yield different fusion results. In this section we will discuss the effects of fusion structures on the outcomes of fusion schemes.

19.5.1 Fusion structures

Considering a general process of image fusion assume that we have N source images $S = \{S_i, i = 1, \ldots, N\}$ and M fusion structures $T = \{T_i, i = 1, \ldots, M\}$. When utilising some image fusion algorithm each fusion structure T_i will fuse the source images S into K_i resulting images $F_i = \{F_i^j, j = 1, \ldots, K_i\}$. Thus, the fused images can be expressed as follows:

$$F = \{T_i(S), i = 1, \ldots, M\} = \{F_i^j, i = 1, \ldots M, j = 1, \ldots, K_i\} \tag{19.18}$$

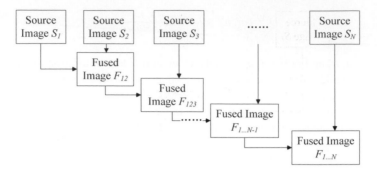

Figure 19.5 *Hierarchical image fusion structure.*

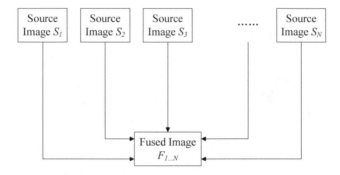

Figure 19.6 *Overall image fusion structure.*

As the image fusion structures vary from each other, the fused images arising via different fusion structures will vary from each other, too. Fusion structures can be generally divided into three classes, namely, (1) hierarchical fusion structures, (2) overall fusion structures, and (3) arbitrary fusion structures.

As shown in Figure 19.5, the hierarchical fusion structure aims at fusing source images in a predefined order and each time only two source images can be fused. This structure is suitable for fusion techniques that are deliberately designed to fuse only two source images.

For techniques that can fuse multiple images in a single fusion process the overall image fusion structure shown in Figure 19.6 is more suitable.

In most applications, the above mentioned fusion structures are generally used jointly to yield a so called arbitrary fusion structure, a possible generic diagram of which is illustrated in Figure 19.7.

19.5.2 Effects of fusion structures on image fusion performance

What effects such fusion structures will impose on the fusion result? Let us first consider a widely used multi-resolution analysis based image fusion technique which can fuse any

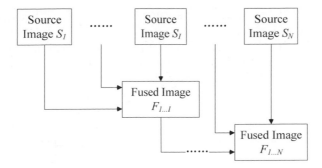

Figure 19.7 *Arbitrary fusion structure.*

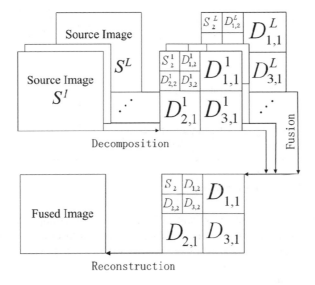

Figure 19.8 *Wavelet transform based image fusion.*

number of images in one fusion process, i.e. the wavelet transform based image fusion. The principle of the wavelet transform based image fusion (here a two-layer wavelet transform is used) can be illustrated in Figure 19.8 where S and D denote the so called approximate and detail images, respectively.

The detailed information about decomposition of the source image into approximate and detail sub-images, as well as the reconstruction from the corresponding fused sub-images to the final fused image can be found in [15], which uses the wavelet transform based image fusion in a recursive way for hyperspectral image fusion applications.

As far as image fusion is concerned, the key techniques of wavelet based methods are feature extraction, weight determination, and fusion.

Feature extraction is for determining the following weight which is based on the calculation of the local information in sub-images as follows

$$E^l_{j,k} = \mathrm{QI}(j,k,l) \tag{19.19}$$

where QI is the quantity of information on level j and orientation k in the lth approximate or detail image. The quantity of information can be defined for different applications according to special considerations. In [15], the standard energy is used as a measure for image information.

When the quantity of information $E^l_{j,k} = \mathrm{QI}(j,k,l)$ of the approximate image and detail images in the window is obtained, the weights of each sub-image can be calculated. The weight is defined as

$$W^l_{j,k} = f_l\big(E^1_{j,k}, \ldots, E^L_{j,k}\big), \quad l = 1, \ldots, L \tag{19.20}$$

where f_l is the weight determining function, which is usually a nonlinear function. However, for some applications a linear weight determining function can also be used obtained as follows:

$$W^l_{j,k} = E^l_{j,k} \Big/ \sum_{l=1}^{L} E^l_{j,k} \tag{19.21}$$

In a given window, the more information the area contains, the larger the value $E^l_{j,k}$ is and the larger the corresponding weight should be.

Image fusion is conducted on the approximate image and detail images. The fused images are defined as a weighted sum of the L approximate images and detail images, i.e.

$$S_J = \sum_{l=1}^{L} W^l_J S^l_J, \qquad D_{k,j} = \sum_{l=1}^{L} W^l_{k,j} D^l_{k,j} \tag{19.22}$$

where S_J and $D_{k,j}$ are approximate images and detail images, respectively.

If the lth image includes more information, its weight must be higher and this image must contribute more information to the final fused image.

In hierarchical fusion structures, when applying wavelet transform based image fusion, the main step that will affect the fusion performance is weight determination. When we first fuse the source images S^1 and S^2, the approximate and detail sub-images can be obtained as

$$\big(S^1_2, D^1_{1,2}, D^1_{2,2}, D^1_{3,2}, D^1_{1,1}, D^1_{2,1}, D^1_{3,1}\big) = \mathrm{DC}(S^1),$$
$$\big(S^2_2, D^2_{1,2}, D^2_{2,2}, D^2_{3,2}, D^2_{1,1}, D^2_{2,1}, D^2_{3,1}\big) = \mathrm{DC}(S^2) \tag{19.23}$$

where $\mathrm{DC}(\cdot)$ is the image decomposition process.

Now let us take the fusion of the detail sub-images $D^1_{1,2}$ and $D^2_{1,2}$ as an example. Their corresponding quantity of information can be obtained as $E^1_{1,2} = \mathrm{QI}(D^1_{1,2})$, $E^2_{1,2} =$

$\text{QI}(D^2_{1,2})$ and their weight is $W^1_{1,2} = f_1(E^1_{1,2} + E^2_{1,2})$, $W^2_{1,2} = f_2(E^1_{1,2} + E^2_{1,2})$. Therefore, the fused sub-image is $D_{1,2} = W^1_{1,2}D^1_{1,2} + W^2_{1,2}D^2_{1,2}$.

Hence, the intermediate fused image from S^1 and S^2 can be obtained from the fused approximate sub-image and detail sub-images as follows

$$F^{12} = \text{RC}(S_2, D_{1,2}, D_{2,2}, D_{3,2}, D_{1,1}, D_{2,1}, D_{3,1}) \tag{19.24}$$

where $\text{RC}(\cdot)$ is the image reconstruction process.

As hierarchical fusion progresses, another image S^3 might be fused with the intermediate image F^{12}. After decomposition, feature extraction, and weight determination of S^3 and F^{12} their weights (here we still select detail sub-image $D^3_{1,2}$ as an example) are

$$W^{F^{12}}_{1,2} = f_1\big(\text{QI}(D^{F^{12}}_{1,2}) + \text{QI}(D^3_{1,2})\big), \qquad W^3_{1,2} = f_2\big(\text{QI}(D^{F^{12}}_{1,2}) + \text{QI}(D^3_{1,2})\big) \tag{19.25}$$

and the corresponding fused detail sub-image is

$$D_{1,2} = W^{F^{12}}_{1,2}D^{F^{12}}_{1,2} + W^3_{1,2}D^3_{1,2} \tag{19.26}$$

On the other hand, in overall fusion structure the three source images S^1, S^2, and S^3 are fused in one single process in which the approximate sub-images and detail sub-images are the same as those in hierarchical fusion structure. But it can be seen from the process of overall structure that the corresponding fused sub-images are

$$D_{1,2} = \sum_{i=1}^{3} f_i\big(\text{QI}(D^1_{1,2}) + \text{QI}(D^2_{1,2}) + \text{QI}(D^3_{1,2})\big)D^i_{1,2} \tag{19.27}$$

As the inputs to feature extraction and weight determination function are different, the fused sub-images are different from those obtained from hierarchical structure and also the reconstructed final fused image is different. This is an example that justifies why different image fusion structures will yield different fusion performances even though the same fusion technique is applied.

As to arbitrary fusion structure, because it is a combination of hierarchical and overall fusion structure schemes, the different combinations of source images and intermediate images will yield different fusion results.

However, there are still some situations in which the hierarchical fusion structure and overall fusion structure can yield identical image fusion results. Here we use the linear weight determination function as a representative example; for a nonlinear function the analysis would be more complicated. Therefore, in hierarchical fusion structure

$$D_{12} = \frac{E^{F^{12}}_{1,2}\frac{(E^1_{1,2}D^1_{1,2} + E^2_{1,2}D^2_{1,2})}{E^1_{1,2} + E^2_{1,2}} + E^3_{1,2}D^3_{1,2}}{E^{F^{12}}_{1,2} + E^3_{1,2}} \tag{19.28}$$

Figure 19.9   *Four source images from hyperspectral data.*

and in overall fusion structure

$$D_{12} = \frac{E_{1,2}^1 D_{1,2}^1 + E_{1,2}^2 D_{1,2}^2 + E_{1,2}^3 D_{1,2}^3}{E_{1,2}^1 + E_{1,2}^2 + E_{1,2}^3} \tag{19.29}$$

Therefore, it can be found that if the feature extraction function satisfies $E_{1,2}^{F^{12}} = E_{1,2}^1 + E_{1,2}^2$, the fused detail sub-images from two different fusion structures by the same wavelet based image fusion technique are the same. Therefore, we can conclude that under specific conditions the different fusion structures might yield the same performance.

19.5.3 Experiments

In order to verify our analysis, experiments are conducted again on AVIRIS data.

Four images selected from the data set are fused by the wavelet transform based image fusion technique using the linear weight determination function. Two different feature extraction methods are applied. One feature is the standard image energy which has been used in [15]. The other is a deliberately designed feature extraction scheme, where the feature of approximate image and detail images of the source images and intermediate images is defined as the number of images from which the specific image is generated. For example, the feature of the source image is set to one, and for an intermediate image, say F^{12}, which is generated from two source images, its feature is set to two.

Figure 19.9 shows the four source images, and Figure 19.10 displays the two fused images by different fusion structures when using the standard image energy feature and their error image. The images in Figure 19.11 are the two fused images when using the latter feature extraction and their error image. In order to clearly present the differences between the two fused images, amplified error images are also displayed in Figures 19.10 and 19.11 which are generated by multiplying the original error image by 4.

It can be found from the above figures, that the two fused images of Figure 19.11 are exactly the same, as we deliberately designed the feature extraction function to meet the requirement that different fusion structures would yield the same results. Therefore, their error image is zero. However, the fused images of Figure 19.10 are different, as the feature extraction function is designed in such a fashion so that it does not satisfy the above requirement. Therefore, there is information in their error image.

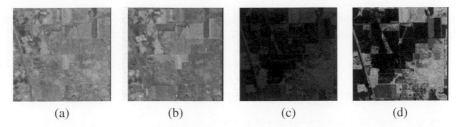

(a)   (b)   (c)   (d)

Figure 19.10 *Fused images when the standard image energy feature is used. (a) Fused image via hierarchical structure; (b) fused image via overall structure; (c) their difference (error) image; (d) amplified error image generated by multiplying the original error image by 4.*

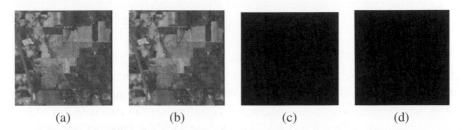

(a)   (b)   (c)   (d)

Figure 19.11 *Fused images when specially designed feature extraction is used. (a) Fused image via hierarchical structure; (b) fused image via overall structure; (c) their error image; (d) amplified error image generated by multiplying the original error image by 4.*

19.6 Fusion of multiple inputs

In this section we will discuss performance measures that can handle fusion techniques which fuse multiple input images into multiple output images. The evaluation schemes are based on the correlation degree and the information deviation degree between the source images and the fused images described below.

19.6.1 Nonlinear Correlation Coefficient (NCC)

For describing any type of correlation between two variables and not only linear correlation as the correlation coefficient does, the mutual information concept is widely used. Mutual information can be thought of as a generalised correlation, analogous to the linear correlation coefficient, but sensitive to any type of relationship and not just linear dependence [16].

Let us consider two discrete variables $X = \{x_i\}_{1 \leqslant i \leqslant N}$ and $Y = \{y_i\}_{1 \leqslant i \leqslant N}$. Their values are sorted in ascending order and placed into b bins with the first N/b values into the first bin, the second N/b values into the second bin, and so on. The value pairs $\{(x_i, y_i)\}_{1 \leqslant i \leqslant N}$ are placed into a $b \times b$ bin grid by finding the bins where the individual value pairs belong to. The revised joint entropy of the two variables X and Y is defined as

$$H(X, Y) = -\sum_{i=1}^{b}\sum_{j=1}^{b} \frac{n_{ij}}{N} \log_b \frac{n_{ij}}{N} \tag{19.30}$$

where n_{ij} is the number of sample pairs distributed in the ijth location of the bin grid. The Nonlinear Correlation Coefficient (NCC) is defined as

$$\text{NCC}(X, Y) = H(X) + H(Y) - H(X, Y) \tag{19.31}$$

where $H(X)$ is the revised entropy of the variable X, which is defined as

$$H(X) = -\sum_{i=1}^{b} \frac{n_i}{N} \log_b \frac{n_i}{N} \tag{19.32}$$

Notice that the number of samples distributed into each rank (bin) of X and Y is invariant, since

$$H(X) = -\sum_{i=1}^{b} \frac{N/b}{N} \log_b \frac{N/b}{N} = -b\frac{1}{b} \log_b \frac{1}{b} = 1$$

The total number of sample pairs is N and, therefore, the nonlinear correlation coefficient can be rewritten as

$$\text{NCC}(X, Y) = 2 + \sum_{i=1}^{b} \sum_{j=1}^{b} \frac{n_{ij}}{N} \log_b \frac{n_{ij}}{N} = 2 + \sum_{i=1}^{b^2} \frac{n_i}{N} \log_b \frac{n_i}{N} \tag{19.33}$$

where n_i is used to present the sample pairs distributed into the two-dimensional bin grid.

NCC is sensitive to nonlinear correlation between two variables. Its value lies within the closed interval $[0, 1]$, with 0 indicating the minimum correlation and 1 indicating the maximum. In the maximum correlation scenario sample sequences of the two variables are exactly the same, i.e. $x_i = y_i$, $i = 1, \ldots, N$. Therefore,

$$\text{NCC}(X, Y) = 2 + \sum_{i=1}^{b^2} p_i \log_b p_i = 2 + b\frac{N/b}{N} \log_b \frac{N/b}{N} = 1$$

Furthermore, under the minimum correlation situation, where the sample pairs are distributed equally into the $b \times b$ ranks, we have

$$\text{NCC}(X, Y) = 2 + \sum_{i=1}^{b^2} p_i \log_b p_i = 2 + b^2 \frac{N/b^2}{N} \log_b \frac{N/b^2}{N} = 0$$

19.6.2 Nonlinear Correlation Information Entropy (NCIE)

In a multivariate situation, the general relation between every two variables can be obtained according to the definition of NCC, and hence, the nonlinear correlation matrix of the concerned K variables can be defined as

$$R = \{\text{NCC}_{ij}\}_{1 \leqslant i \leqslant K, 1 \leqslant j \leqslant K} \tag{19.34}$$

where NCC_{ij} denotes the nonlinear correlation coefficient of the ith and jth variable. Since a variable is identical to itself, $\mathrm{NCC}_{ii} = 1, 1 \leqslant i \leqslant K$.

The diagonal elements of R, $\mathrm{NCC}_{ii} = r_{ii} = 1, 1 \leqslant i \leqslant K$, represent the autocorrelation of each variable. The rest of the elements of R, $1 \leqslant r_{ij} \leqslant 1, i \neq j, i \leqslant K, j \leqslant K$, denote the correlation of ith and jth variable. When the variables are uncorrelated to each other R is the unit matrix. When all the variables are identical, each element of R equals to 1. In the later scenario the correlation among the variables is the strongest possible. The general relation of the concerned K variables is implied in the nonlinear correlation matrix R. In order to quantitatively measure it, the nonlinear joint entropy H_R is defined as

$$H_R = -\sum_{i=1}^{K} \frac{\lambda_i^R}{K} \log_b \frac{\lambda_i^R}{K} \tag{19.35}$$

where $\lambda_i^R, i = 1, \ldots, K$, are the eigenvalues of the nonlinear correlation matrix. According to matrix eigenvalue theory, it can be deduced that $0 \leqslant \lambda_i^R \leqslant K, i = 1, \ldots, K$, and $\sum_{i=1}^{K} \lambda_i^R = K$.

The nonlinear correlation information entropy NCIE_R, used as a nonlinear correlation measure of the concerned variables, is defined as

$$\mathrm{NCIE}_R = 1 - H_R = 1 + \sum_{i=1}^{K} \frac{\lambda_i^R}{K} \log_b \frac{\lambda_i^R}{K} \tag{19.36}$$

NCIE has some excellent mathematical properties that further prove its suitability as a measure for the nonlinear type of correlation of multiple variables.

First, it lies within the interval $[0, 1]$, with 0 indicating the minimum nonlinear correlation among the K variables concerned and 1 indicating the maximum. If the variables are uncorrelated the nonlinear correlation matrix becomes the identity matrix, and therefore, $\lambda_i^R = 1, i = 1, \ldots, K$. As a result, the NCIE equals to zero. If the variables are identical, the nonlinear correlation coefficient of each two variables equals to 1. This leads to every element of the nonlinear correlation matrix being equal to one, and therefore, $\lambda_i^R = 0$, $i = 1, \ldots, K - 1$, and $\lambda_K^R = K$. In this case, the nonlinear correlation information entropy equals to 1.

Finally, it is sensitive to any type of relation of the K variables concerned, not merely linear relations. This characteristic will be testified by the following numerical simulations.

19.6.2.1 Numerical verification

Figure 19.12 shows the relations of three distributions, i.e. uniform distribution, normal distribution, exponential distribution, and the corresponding NCIE. As the three variables concerned are randomly distributed, their relations are weak, and therefore, NCIE is small. Figure 19.12 shows also the relations of three common functions, i.e. linear, circular, and square relations of three variables. Noise with different power is added to

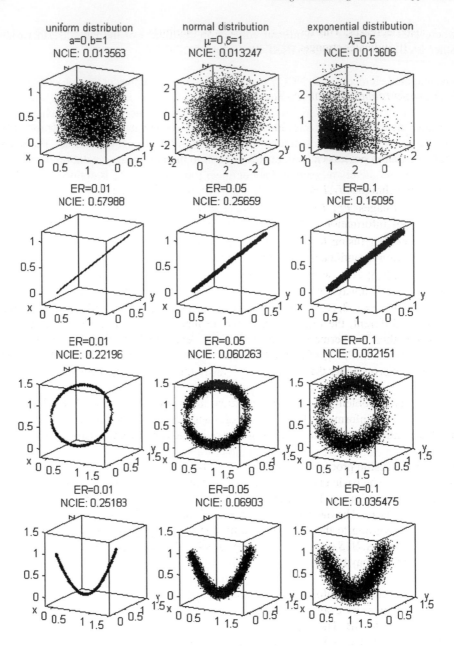

Figure 19.12 *NCIE of three random variables of three typical distributions, i.e. uniform distribution, normal distribution, and exponential distribution and three common relations, i.e. linear relation, circular relation, and square relation with different noise added to the variables. The number of values of each variable is N = 10 000 and the bin number b = 100.*

the functions in order to generate different correlation degrees. The NCIE of the corresponding relation can also be found in the figures. As the amplitude of the added noise increases, the correlation degree of the concerned three variables decreases, and their

NCIE also decreases. This result conforms to our definition of NCIE, which states that larger NCIE indicates stronger correlation.

19.6.3 Information Deviation Analysis (IDA)

The next measure we will introduce is based on the concept of information deviation between the source images and fused images. It can be applied to the general situation where the fusion process can handle multiple input images and multiple output images.

Let fusion techniques $T_i, i = 1, 2, \ldots, M$, fuse the source images S, $S = \{S_1, S_2, \ldots, S_N\}$, into result images F_i^j, $i = 1, 2, \ldots, M$; $j = 1, 2, \ldots, K_i$, respectively. In the ideal situation, all information in the source images can be transferred into the result image by a perfect fusion technique T_A, and $F_A^j = T_A(S)$, $j = 1, 2, \ldots, K_A$. If there is an information quantification scheme Q, then the information deviation $D(S, F_A) = |Q(S) - Q(F_A)| = 0$, where $F_A = \{F_A^j; \ j = 1, 2, \ldots, K_A\}$. Actually, in the fusion process, there is always, more or less, the loss of information in the image fusion. So $D(S, F_i) > 0, i = 1, 2, \ldots, M$. However, since the performances of different fusion techniques are different, the information deviation measurements $D(S, F_i)$ will also vary from each other. Furthermore, if the information deviation $D(S, F_i)$ of fusion technique T_i is larger than $D(S, F_j)$ of fusion technique T_j, then we can infer that the fusion technique T_j performs better than T_i, from the viewpoint of information deviation.

Now the main problem comes to the information quantification method which not only can quantify the information of source images and fused images from the same scheme, but also is independent of the number of source images and fused images.

In the image fusion process, the information is transferred from the source images to the fused images. The basic unit of the information transfer is each image itself. Let there be W pixels in each image. The pixels are assumed having L states, and the possibility of each state L_i is p_i. The mean number of pixels in each state $B = W/L$ is called the state-width. So according to the Shannon information theory, for one image S_A, the entropy can be defined as $H(S_A) = -\sum_{i=1}^{L} p_i \log p_i$, where the sum is over the L 'states' that the pixels of S_A may have.

In order to evaluate the information quantity of the N source images $S = \{S_1, S_2, \ldots, S_N\}$ in the same way as we evaluate a single image, we set state number equal to L, but expand the state-width to NB. Therefore, the information entropy of the N source images S can be defined as $H(S) = -\sum_{i=1}^{L} p_i \log p_i$. In the same way, the entropy of the fused images of an image fusion technique T_i, i.e. F_i^j, $j = 1, 2, \ldots, K_i$, can be defined with the state-width expanded to $K_i B$.

Now the method to quantify the information of source images and fused images has been defined. As far as an image fusion technique T_i is concerned, its information deviation in the fusion process can be defined as

$$D_i = \left| H(S) - H(F_i) \right| \tag{19.37}$$

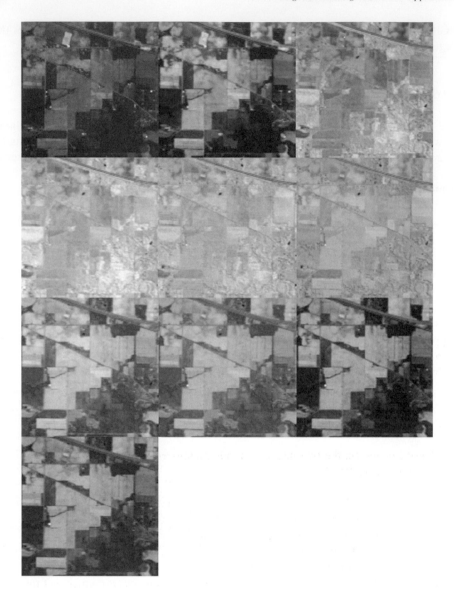

Figure 19.13   *Source images.*

As a result, the different image fusion techniques can be compared and evaluated according to (19.37).

19.6.4 Experiments on NCA and IDA

In order to verify the proposed NCA and IDA evaluation methods, an experiment is conducted in which two widely used multi resolution analysis based image fusion methods, the Wavelet Transform based Fusion (WTF) and the Pyramid Transform based fusion

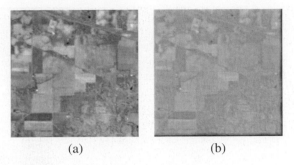

(a)   (b)

Figure 19.14   *Fused images using multi resolution analysis based fusion methods. (a) WTF; (b) PTF.*

Table 19.3   *Image fusion performance evaluation in the first experiment.*

Fusion method	NCA	IDA
WTF	0.08700	0.1770
PTF	0.08611	0.6547

(PTF), are evaluated. The detailed information about these fusion techniques can be found in [2,15].

Our experiments are conducted again on AVIRIS data. For computational simplicity, ten spectral bands are selected from the 220 bands as source images for the specific data set chosen, which are shown in Figure 19.13.

The fused images via the two fusion methods mentioned above, i.e. WTF and PTF, are displayed in Figure 19.14.

The performance evaluation and comparison of the fusion methods is conducted using the original information (ten source images) and fused information (two fused images). The amount of relation between source images and fused images is calculated, and the results are presented in Table 19.3. As stronger relationship indicates better performance of the concerned fusion method, we conclude that, from the information correlation perspective, the WTF performs better. Moreover, according to the IDA results presented in Table 19.3, we find that the deviation of WTF fused results is less than that of PTF fused results. Therefore, we can also say that, from the information deviation perspective, the WTF performs better.

The experiment results conform to the application conducted in [15] which focused on the classification of hyperspectral data. The results of classification accuracy in the particular application are presented in Table 19.4. From Table 19.4, it can be found that more accurate classification results are obtained by the WTF method, which also exhibits better results in the performance evaluation simulation.

The second experiment is conducted on another hyperspectral image set, which exhibits richer content and has larger dimension, i.e. 512 rows and 608 columns. The six source

Table 19.4 *Hyperspectral image classification accuracy (per cent).*

	Corn	Grass	Soybean	Forest	Average
WTF	97.40	99.60	90.60	99.00	96.65
PTF	95.60	97.40	86.40	98.40	94.45

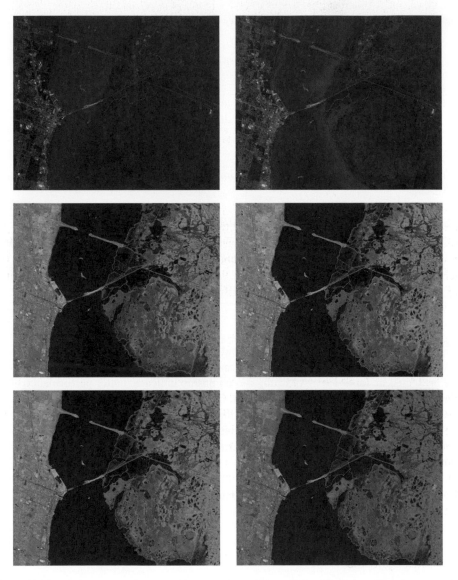

Figure 19.15 *Six hyperspectral images of the second experiment which exhibit larger dimension and richer content.*

(a)   (b)

Figure 19.16   *Fused images via (a) WTF and (b) PTF of the second experiment.*

Table 19.5   *Image fusion performance evaluation in the second experiment.*

Fusion method	NCA	IDA
WTF	0.1036	0.5019
PTF	0.0958	0.5645

images are presented in Figure 19.15 and the fused images via WTF and PTF are presented in Figure 19.16. The performance evaluation results are presented in Table 19.5. It can be concluded from the figures of Table 19.5 that the performance of WTF is better than that of PTF, as the NCA result of WTF is larger and the IDA is less than those of PTF. The conclusions agree with those of the first experiment.

19.6.5 Discussion

This chapter discusses various performance evaluation measures that have been proposed in the field of image fusion and also analyses the effects of fusion structures on the outcomes of fusion schemes. Indicative experiments on applying these measures to evaluate a couple of widely used image fusion techniques are also presented to demonstrate the usage of the measures, as well as to verify their correctness and effectiveness. It is important to stress out that there is not a single performance evaluation measure that can be classified as superior. Each measure highlights different features in an image and, therefore, the selection of a particular measure to evaluate an image fusion technique is based on the particular application.

Acknowledgements

This work is supported by the National Natural Science Foundation of China (No. 6060-4021).

References

[1] H. Li, B.S. Manjunath and S.K. Mitra, 'Multisensor image fusion using the wavelet transform', in *Proc. of the IEEE International Conference on Image Processing (ICIP)*, Vol. 1, 13–16 November 1994, pp. 51–55.

[2] T.A. Wilson, S.K. Rogers and M. Kabrisky, 'Perceptual-based image fusion for hyperspectral data', *IEEE Transactions on Geoscience and Remote Sensing*, Vol. 35, No. 4, 1997, pp. 1007–1017.

[3] J.-H. Park, K.-O. Kim and Y.-K. Yang, 'Image fusion using multiresolution analysis', in *Proc. of the International Geoscience and Remote Sensing Symposium*, Vol. 2, 2001, pp. 864–866.

[4] Z.L. Zhang, S.H. Sun and F.C. Zheng, 'Image fusion based on median filters and SOFM neural networks: A three-step scheme', *Signal Processing*, Vol. 81, No. 6, 2001, pp. 1325–1330.

[5] C. Ramesh and T. Ranjith, 'Fusion performance measures and a lifting wavelet transform based algorithm for image fusion', in *Proc. of the 5th International Conference on Information Fusion*, Vol. 1, 2002, pp. 317–320.

[6] G.H. Qu, D. Zhang and P. Yan, 'Information measure for performance of image fusion', *Electronics Letters*, Vol. 38, No. 7, 2002, pp. 313–315.

[7] C.S. Xydeas and V. Petrovic, 'Objective image fusion performance measure', *Electronics Letters*, Vol. 36, No. 4, 2000, pp. 308–309.

[8] Q. Wang, Y. Shen, Y. Zhang and J.Q. Zhang, 'A quantitative method for evaluating the performances of hyperspectral image fusion', *IEEE Transactions on Instrumentation and Measurement*, Vol. 52, No. 4, 2003, pp. 1041–1047.

[9] Q. Wang, Y. Shen, Y. Zhang and J.Q. Zhang, 'Fast quantitative correlation analysis and information deviation analysis for evaluating the performances of image fusion techniques', *IEEE Transactions on Instrumentation and Measurement*, Vol. 53, No. 5, 2004, pp. 1441–1447.

[10] Q. Wang and Y. Shen, 'Performances evaluation of image fusion techniques based on nonlinear correlation measurement', in *Proc. of IEEE Instrumentation and Measurement Technology Conference*, Vol. 1, 18–20 May 2004, pp. 472–475.

[11] Q. Wang, Y. Shen and J.Q. Zhang, 'A nonlinear correlation measure for multivariable data set', *Physica D: Nonlinear Phenomena*, Vol. 200, No. 3–4, 2005, pp. 287–295.

[12] L. Wald, T. Ranchin and M. Mangolini, 'Fusion of satellite images of different spatial resolutions: Assessing the quality of resulting images', *Photogrammetric Engineering and Remote Sensing*, Vol. 63, No. 6, 1997, pp. 691–699.

[13] M.E. Ulug and C.L. McCullough, 'A quantitative metric for comparison of night vision algorithms', in *Sensor Fusion: Architectures, Algorithms and Applications IV*, in *Proceedings of the SPIE*, Vol. 4051, 2001, pp. 80–88.

[14] L. Alparone, S. Baronti and A. Garzelli, 'Assessment of image fusion algorithms based on noncritically-decimated pyramids and wavelets', in *Proc. of the Geoscience and Remote Sensing Symposium (IGARSS)*, Vol. 2, 2001, pp. 852–854.

[15] Q. Wang, Y. Zhang and J.Q. Zhang, 'The reduction of hyperspectral data dimensionality and classification based on recursive subspace fusion', *Chinese Journal of Electronics*, Vol. 11, No. 1, 2002, pp. 12–15.

[16] M.S. Roulston, 'Significance testing of information theoretic functionals', *Physica D: Nonlinear Phenomena*, Vol. 110, 1997, pp. 62–66.

Subject index